NONLINEAR OPTICS

Nonlinear Optics

Proceedings of the Sixteenth Scottish
Universities Summer School in Physics, 1975

A NATO Advanced Study Institute

Edited by

PHILIP G. HARPER

and

BRIAN S. WHERRETT

Department of Physics
Heriot-Watt University
Edinburgh, Scotland

1977

ACADEMIC PRESS

LONDON · NEW YORK · SAN FRANCISCO

A Subsidiary of Harcourt Brace Jovanovich, Publishers

ACADEMIC PRESS INC. (LONDON) LTD.
24/28 Oval Road
London NW1

United States Edition published by
ACADEMIC PRESS INC.
111 Fifth Avenue
New York, New York 10003

Library of Congress Catalog Card Number: 77-71822
ISBN: 0-12-325950-9

FILM SET BY COMPOSITION HOUSE, SALISBURY, ENGLAND
PRINTED IN GREAT BRITAIN BY WHITSTABLE LITHO LTD., KENT

List of Participants

ORGANIZING COMMITTEE

SMITH, PROFESSOR S. D. S. F.R.S., *Heriot-Watt University, Edinburgh, Scotland.*
COLLES, DR. M. J. *Heriot-Watt University, Edinburgh, Scotland.*
TAYLOR, DR. W. *University of Edinburgh, Edinburgh, Scotland.*
HARPER, PROFESSOR P. G. *Heriot-Watt University, Edinburgh, Scotland (Joint Editor).*
WHERRETT, DR. B. S. *Heriot-Watt University, Edinburgh, Scotland (Joint Editor).*
DENNIS, DR. R. B. *Heriot-Watt University, Edinburgh, Scotland.*
FARAGO, PROFESSOR P. *University of Edinburgh, Edinburgh, Scotland.*
MILLIGAN, MR. R. *Heriot-Watt University, Edinburgh, Scotland.*
GRIEVE, MR. I. *Heriot-Watt University, Edinburgh, Scotland.*

LECTURERS

BREWER, DR. R. G. *IBM Research Laboratory, San Jose, California 95193, U.S.A.*
BYER, PROFESSOR R. L. *Department of Applied Physics, Stanford University, California U.S.A.*
DUCUING, PROFESSOR J. *Ecole Polytechnique, 91120, Palaiseau, France.*
HUGHES, DR. T. *Department of Physics, University of Essex, Wivenhoe Park, Colchester, Essex CO4 35Q, England.*
KAISER, PROFESSOR W. and A. LAUBERAU. *Physik-Department, Technische Universität, München, München, West Germany.*
MOORADIAN, DR. A. *Lincoln Laboratory, Massachusetts Institute of Technology, Massachusetts 02173, U.S.A.*
STOICHEFF, PROFESSOR B. P. *Department of Physics, University of Toronto, Canada.*
WALTHER, PROFESSOR, H. *Physik-Department, Universität München, München, West Germany.*
WOLFF, PROFESSOR P. A. *Department of Physics, Massachusetts Institute of Technology, Cambridge, Massachussets 02139, U.S.A.*

ALOBAIDI, DR. A. H. *Atomic Energy Commission, Iraq.*
ANDREOU, DR. D. *Culham Laboratory, Abingdon.*
ANGUS, DR. A. *Heriot-Watt University, Edinburgh, Scotland.*
ARUESEN, MR. A. *Fysikum, Uppsala, Sweden.*
AVERY, MR. R. K. *University College of Swansea, Wales.*
BECHTOLD, MR. P. S. *D-5 Köln 41, West Germany.*
BEIGANG, MR. R. *Universität Hanover, Germany.*
BERKDAR, MR. F. AL. *Heriot-Watt University, Edinburgh, Scotland.*
CAUDREY, DR. P. J. *UMIST, Department of Mathematics, Manchester, England.*
CHAKRABORTY, DR. A. K. *University College of Science & Technology, Calcutta, India.*
CHAMBON, MR. G. *Université Paris VI, France.*
COTTER, MR. D. *Department of Electronics, Southampton University, England.*
CROWDER, DR. J. G. *Brunel University, Middlesex, England.*
DUNNET, MR. W. *Heriot-Watt University, Edinburgh, Scotland.*
EILBECK, DR. C. *Heriot-Watt University, Department of Mathematics, Edinburgh, Scotland.*
FERGUSON, MR. A. I. *Imperial College, London, England.*
FEINLEIB, DR. J. *MIT, U.S.A.*
HANNA, DR. D. C. *Department of Electronics, University of Southampton, England.*
HAMZA, Mr. F. H. *Heriot-Watt University, Edinburgh, Scotland.*
HASSAN, DR. A. R. *Ain. Shams University, Egypt.*
HOSEA, Mr. T. J. *University of Edinburgh, Edinburgh, Scotland.*
HINZ, MR. W. *Erstes Physikalisches Institut, Universität Köln, Germany.*
IRONSIDE, MR. C. N. *Heriot-Watt University, Edinburgh, Scotland.*
JAMISON, MR. S. A. *Brown University, U.S.A.*
KOPAINSKY, MR. B. *Technische Universität, München, West Germany.*
KOPP, MR. A. *Université Paris VI, France.*
LAHMANN, DR. W. *Universität Hanover, Germany. Institut für Angewandte Physik.*
LECLERC, MR. A. *Université Paris VI, France. Laboratoire de Spectroscopie Hertzienne.*
LEUCHS, MR. G. *Universität Köln, Germany.*
LIGHTFOOT, DR. H. W. *Heriot-Watt University, Edinburgh, Scotland.*
LIPTON, MR. K. S. *Department of Nat. Philosophy, University of Strathclyde, Glasgow, Scotland.*
LIFTIN, MR. *Universität Hanover, Germany.*
LOCKWOOD, DR. D. J. *University of Edinburgh, Edinburgh, Scotland.*
MAK, MR. S. N. *Universtiy of Maryland, U.S.A.*
MARINERO, MR. E. *Heriot-Watt University, Edinburgh, Scotland.*
MILLER, DR. A. *Heriot-Watt University, Edinburgh, Scotland.*

MLYNEK, MR. J. *Fondation Suisse, Paris, France. Ecole Polytechnique-Laboratoire d'Optique Quantique.*
MOZOLOWSKI, MR. M. *Heriot-Watt University, Edinburgh, Scotland.*
MCGREGOR, MR. J. *Heriot-Watt University, Edinburgh, Scotland.*
MCKINNON, MR. A. *Heriot-Watt University, Edinburgh, Scotland.*
MCMILLAN, MISS J. L. *University of Edinburgh, Edinburgh, Scotland.*
NOWAKOWSKI, DR. A. *Heriot-Watt University, Edinburgh, Scotland.*
OUDAR, MR. J. L. *CNET/PEC, Paris, France.*
PARKS, DR. G. L. *North Georgia College, U.S.A.*
PIDGEON, DR. C. R. *Heriot-Watt University, Edinburgh, Scotland.*
PLACIDO, DR. F. *University of Edinburgh, Edinburgh, Scotland.*
QUADAR, MR. A. M. *Heriot-Watt University, Edinburgh, Scotland.*
ROBINSON, MR. P. *Heriot-Watt University, Edinburgh, Scotland.*
ROGERS, MR. B. D. *University of Bradford, Bradford, England.*
ROGERS, MR. S. *Heriot-Watt University, Edinburgh, Scotland.*
ROSS, MR. G. *Heriot-Watt University, Edinburgh, Scotland.*
RUSHWORTH, MR. A. J. *Clarendon Laboratory, Oxford, England.*
RUSTAGI, DR. K. *Universite des Sciences et Techniques des Solides, France.*
SCRAGG, MR. T. *Heriot-Watt University, Edinburgh, Scotland.*
SIMAAN, MR. H. D. *University of Essex, Essex, England.*
SIMMONDS, MR. L. *Heriot-Watt University, Edinburgh, Scotland.*
SMITH, MR. D. L. *Texas Tech. Univ., Texas, U.S.A.*
SMITH, MR. J. *Heriot-Watt University, Edinburgh, Scotland.*
SMITH, DR. R. C. *University of Southampton, Southampton, England.*
TURNER, MR. A. J. *University of Southampton, Southampton, England.*
TYLER, PROFESSOR I. L. *University of Missouri-Rolla, U.S.A.*
VALLAURI, DR. R. *Laboratorio di Elettronica Quantistica, Italy.*
VOGT, DR. H. *II. Physikalisches Institut der Universität Köln, Germany.*
WELLS, DR. J. S. *Nat. Bureau of Standards, U.S.A.*
WILLIAMS, DR. A. R. *Heriot-Watt University, Edinburgh, Scotland.*
WILSON, Dr. J. B. *Heriot-Watt University, Edinburgh, Scotland.*
YI YAN, MR. A. *Glasgow University, Glasgow, Scotland.*
YURATICH, MR. M. A. *University of Southampton, Southampton, England.*
ZYSS, MR. J. *CNET, Paris, France.*

ADDITIONAL PARTICIPANTS

DR. W. J. FIRTH, DR. D. L. WEAIRE, DR. P. M. MARSHALL, DR. G. E. PECKHAM, DR. D. L. WARD, MR. H. A. MACKENZIE, and DR. R. A. WOOD from the Department of Physics, Heriot-Watt University.

VISITOR

PROFESSOR R.A. SMITH. *Heriot-Watt University, Edinburgh, Scotland.*

Director's Preface

The Sixteenth Scottish Universities Summer School in Physics was held on the Riccarton Campus of Heriot-Watt University, from July 28th to 15th August 1975.

The subject of the School, Nonlinear Optics, is still of course very young (dating from the early 1960s) but has expanded so rapidly and successfully that our interpretation of the subject was extended to include a number of topics in Laser Spectroscopy. The notion of Nonlinear Optics deals with phenomena that occur in the high radiation intensities obtainable in laser beams and encompasses interaction with material systems varying from high temperature plasmas at very high density through conventional solid state material to very isolated atomic and molecular systems at low pressure. The prevailing common factor is of course that the interactions between light and matter can be expressed in terms of various orders of nonlinear susceptibilities.

Our lecturers were all well-known for recent contributions to current research on this fast moving subject. The demonstrations of their mastery, in a variety of styles, led to a very successful School attended by nearly 100 students varying from those just beginning research to those returning to research in retirement! At the end of the School we were all convinced that the topic of Nonlinear Optics continues to provide a useful unifying theme and I hope that this volume will appropriately confirm this feeling. The major organization of the School was capably undertaken by the Secretary, Dr. M. J. Colles; he reinforced this notion in some concluding remarks in which he showed that we did consider all non-linear susceptibilities up to seventh order.

It is a pleasure to acknowledge contributions made by many others: Dr. W. Taylor, as Treasurer; Dr. R. B. Dennis and Mr. I. Grieve, as Social Secretaries; Professor P. G. Harper and Dr. B. S. Wherrett, as editors, Miss Myrtle Lee and her staff in the University Residences; Mrs. J. McGregor and all the Secretaries in the Physics Department and many others whom space does not permit to mention individually.

The intense interest and time spent by the above contributed greatly to the professional, social and sporting success of the School.

S. D. Smith
Heriot-Watt University
Department of Physics

Editors' Note

Most of the articles in this volume have been prepared by the authors and follow closely the lectures presented at the Summer School. Exceptions are the synopsis of Professor Walther's lectures, and Professor Stoicheff's introductory lecture which serves here as a summary.

The boundaries separating the various topics are largely due to the specialization demanded by the rapid growth of Nonlinear Optics. It seemed appropriate in this volume to point out connections between apparently separate developments and wherever possible draw them together. The prefactory article by B. S. Wherrett attempts to provide a framework for such an integrated view.

We are grateful to Stella Sharpe for secretarial assistance in preparing the final version of the manuscripts.

P. G. Harper
B. S. Wherrett

Contents

From Linear Dispersion to Self-Induced Transparency: An Introduction to Nonlinear Optics

B. S. WHERRETT

Department of Physics, Heriot-Watt University, Edinburgh

A vast amount of new information has been amassed in the field of Quantum Electronics over the past fifteen years, since the development of lasers. It is the aim of this introduction to give an overall view of that section of quantum electronics which we have chosen to regard as Nonlinear Optics, for the purpose of the Summer School. The idea is to provide a framework in which may be placed the following contributions from the nine invited lecturers.

The underlying cause of all linear and nonlinear optical effects is of course the interaction of electromagnetic radiation with matter. Thus, each contributor dealt with one or more aspects of the *response* of matter to radiation. The effect on the radiation itself provides the detectable evidence of the interaction, in most instances, and in turn provides information on the reaction of the material system and on the physical characteristics of the material. Alternatively one could choose to look upon the effect on the field as being the important consequence of the interaction; for example as the provision of a new source of radiation (chapters 2, 4, 5 and 7). We have chosen to order the chapters of this proceedings by considering the "degree of alteration" of the material system appropriate in each. Thus in chapters 1 and 2 we are concerned primarily with phenomena in which the material plays (over a period of time) a passive role. Spontaneous processes in which the fractional excitation of the material is negligible are discussed in chapters 3 and 4. Stimulated absorption, emission and Raman scattering appear in chapters 2–7; for such processes the material is excited, ultimately to saturation. The populations of the material states can actually become reversed (on an appropriate time scale) for the coherent mechanisms described in chapter 8. Finally the role of the radiation is, in chapter 9, to change completely the state of the material.

We will discuss firstly a few ideas which are common throughout most of nonlinear optics.

Just as the radiation-matter interaction is the physical basis of nonlinear optics so is the interaction-Hamiltonian, which is discussed formally in chapter 4, the basis for the mathematical description of NLO phenomena. One uses this Hamiltonian to determine the redistribution of charge in thc material, on a *microscopic* scale, due to the radiation field. This redistribution corresponds, on a *macroscopic* scale, to a polarization. The complete radiation field must then be determined, in a self-consistent manner, by including this polarization in the electromagnetic wave equations. In practice the vast majority of NLO processes are described adequately by applying the "electric-dipole approximation" to the Hamiltonian. In order to use this approximation the interaction is considered firstly within a "polarizable unit" of the material system. This is a volume in which the electromagnetic field can be assumed to be uniform at any given time. For a gas this volume encloses a single atom or molecule, the dimensions of which are small compared to the radiation wavelength. In a solid the unit is a volume large by comparison to atomic dimensions but again small compared to the wavelength. The total polarization set up in the medium is then the summation over all polarizable units, each term in the sum containing a spatial factor appropriate to the average radiation field in that unit ($\mathbf{E}(t)$). As the field is uniform within a unit the radiation interaction looks formally like that of an electric-dipole in a constant field. For example in atomic systems it is appropriate to use the form:

$$H_{\text{interaction}} = -e\mathbf{r} \cdot \mathbf{E}(t).$$

The realm of nonlinear optics, as far as the radiation frequencies are concerned, is precisely that in which the electron-dipole approximation is relevant; that is the spectral region from the far-infrared to the ultraviolet. For shorter wavelengths $\mathbf{E}$ can no longer be assumed to be uniform over atomic dimensions whilst in the microwave and longer wavelength regions the "magnetic-dipole approximation" is more relevant.

It is because of the applicability of the electric-dipole approximation that one is able to frame much of nonlinear optics in terms of *susceptibilities*. That is, the macroscopic polarization set up in the material can be expanded in terms that are directly proportional (spatially) to some power of the field amplitude.

$$\mathbf{P}(\mathbf{r}, t) = \chi^{(1)}\mathbf{E}(\mathbf{r}, t) + \chi^{(2)}\mathbf{E}(\mathbf{r}, t)\mathbf{E}(\mathbf{r}, t) + \cdots.$$

The susceptibilities, $\chi^{(n)}$, are functions only of the radiation frequencies and of the material characteristics; they are independent of position when the electric-dipole approximation holds. The usefulness of such an expansion also relies on the dominance of the lower order terms. This dominance occurs in all but the high saturation cases (chapter 8). In the above formulation of H_{int} and of **P**, **E** and **P** are treated as classical parameters; it is essential to quantize the material states however in order to obtain expressions for the susceptibilities $\chi^{(n)}$. This "semi-classical" treatment of the interacting radiation-matter system is appropriate to nearly all of those processes which we have chosen to include in NLO; the exceptions are spontaneous emission and Raman scattering.

The common denominators in the description of most of the NLO phenomena with which our lecturers were concerned are thus: (i) the usefulness of the electric-dipole approximation and of the semi-classical treatment for the radiation-matter interaction, (ii) the consequent polarization expansion in which specific, low-order terms dominate and in which nonlinear susceptibilities are meaningful. We will now turn to discussion of the individual phenomena.

There are many independent parameters in nonlinear optics that can suitably be formulated in the dimension of time. In any particular experiment the relative values of these time scales determines the nature of the NLO process and, to a large extent, the mathematical technique that can most efficiently be used to describe the process. For this reason much of the remainder of this introduction will deal with time scales.

In order to run through the physical processes of NLO in some sort of logical manner it is convenient to begin with *passive* or "*frequency mixing*" processes, in which the material acts essentially as a catalyst. In all of NLO one approximate time restriction is that the period over which measurements are made be long compared to the radiation oscillation periods, ω^{-1}. Hence energy (or frequency) conservation is applicable.

For passive processes this conservation is amongst the frequency components of the radiation field. The lower order passive processes, all of which are described by nonlinear susceptibilities, are collated in Table I. In general a component of the polarization at frequency ω_r, ($P(\omega_r) \propto \exp -i\omega_r t$), set up by components of the radiation field at ω_a, ω_b etc., is expressed as:

$$P^{(n)}(\omega_r = \omega_a + \omega_b + \cdots) = \chi^{(n)}(\omega_r; \omega_a, \omega_b \cdots)E(\omega_a)E(\omega_b)\cdots,$$

with suitable directional dependences (see chapter 1).

In order to discuss the time inequalities appropriate in NLO it is instructive to consider a two-level electronic system (most NLO phenomena involve electronic processes). Supposing the distributions in the pure states to be given by ψ_0 and ψ_1 then in a perturbing field one can write the eigenfunctions generally as:

$$\psi(\mathbf{r}, t) = c_0(t)\psi_0(\mathbf{r}, t) + c_1(t)\psi_1(\mathbf{r}, t)$$

TABLE I. Passive nonlinear optical phenomena. ($\omega = 0$ indicates the presence of a uniform electric field.)

Frequencies of incident fields	Frequencies of fields generated by the polarization of the medium	Susceptibility	Process
ω_1	No polarization	$0\ (\varepsilon = 1)$	Vacuum propagation
ω_1	ω_1	$\chi^{(1)}(\omega_1; \omega_1)$	Linear dispersion
ω_1, ω_2	$\omega_3[\omega_3 = \omega_1 + \omega_2]$	$\chi^{(2)}(\omega_3; \omega_1, \omega_2)$	Sum mixing
ω_1	$\omega_3[\omega_3 = 2\omega_1]$	$\chi^{(2)}(\omega_3; \omega_1, \omega_1)$	Second harmonic generation
$\omega_1, 0$	ω_1	$\chi^{(2)}(\omega_1; \omega_1, 0)$	Electro-optic linear Kerr effect
ω_1	$\omega_2, \omega_3[\omega_1 = \omega_2 + \omega_3]$	$\chi^{(2)}(\omega_2; -\omega_3, \omega_1)$	Difference-frequency mixing
ω_1	$\omega_2[\omega_1 = 2\omega_2]$	$\chi^{(2)}(\omega_2; -\omega_2, \omega_1)$	Degenerate difference-frequency
ω_1	0	$\chi^{(2)}(0; -\omega_1, \omega_1)$	Inverse electro-optic effect
$\omega_1, \omega_2, \omega_3$	$\omega_4[\omega_4 = \omega_1 + \omega_2 + \omega_3]$	$\chi^{(3)}(\omega_4; \omega_1, \omega_1, \omega_1)$	Third harmonic generation
ω_1, ω_2	$\omega_3, \omega_4[\omega_1 + \omega_2 = \omega_3 + \omega_4]$	$\chi^{(3)}(\omega_3; -\omega_4, \omega_1, \omega_2)$ $\chi^{(3)}(\omega_4; -\omega_3, \omega_1, \omega_2)$	Four-wave difference-frequency mixing processes
ω_1	$\omega_2, \omega_3, \omega_4[\omega_1 = \omega_2 + \omega_3 + \omega_4]$	$\chi^{(3)}(\omega_2; -\omega_3, -\omega_4, \omega_1)$	
ω_1	ω_1	$\chi^{(3)}(\omega_1; \omega_1, -\omega_1, \omega_1)$	Intensity-dependent refractive index
$\omega_1, 0$	ω_1	$\chi^{(3)}(\omega_1; 0, 0, \omega_1)$	Quadratic Kerr effect

The major *microscopic* problem of NLO is to calculate the coefficients c_0, c_1.

One introduces a dephasing time T_2, corresponding to transitions out of the upper level ψ_1, and assumes the perturbing field to be weak enough that $c_0(t)$ is close to unity always. For times long compared to the dephasing time $|c_1(t)|^2$ is constant:

$$|c_1(t)|^2 = \frac{|ez_{10}E/\hbar|^2}{(\omega_{10} - \omega)^2 + T_2^{-2}}$$

ez_{10} is the transition dipole-moment, $\hbar\omega_{10}$ the energy level separation and E the component of the radiation field at frequency ω. The lifetime of electrons in the lower level, with respect to radiative excitation, is $T_2|c_1|^{-2}$ which we shall call τ_a.

One can now demand that for passive, linear dispersion the excitation lifetime is long compared to the observation time (so that the energy absorbed from the field is small compared to the energy, $\hbar\omega_{10}$ per particle, of the material) and that this in turn is long compared to T_2. Under these conditions $|c_1(t)|^2$ is automatically small, as required in the perturbation treatment. For nonlinear processes it is necessary that the above inequalities apply to all radiation frequencies present and to all pairs of material states. It is adequate to use standard perturbation theory to obtain the susceptibilities under two provisos. Firstly for frequencies in the vicinity of resonances one must introduce damping or dephasing times phenomenologically. Secondly one must never contravene Pauli exclusion, even between the intermediate levels involved in nonlinear processes. The latter proviso places restrictions on the summations appearing in susceptibilities. As the *density matrix* has elements $c_i^*c_j$, which are just those combinations of the c-coefficients which appear in populations and in polarization expressions, density matrix theory is a very useful method of describing NLO mechanisms. In this theory one automatically includes the initial state occupancies and the damping factors may be introduced in a relatively formal manner (chapter 1). Whichever way $\chi^{(n)}$ or $\mathbf{P}^{(n)}$ is obtained the appropriate nonlinear polarization must then be included in the wave-equations and these solved for the characteristics of the radiation field $\mathbf{E}$ in self-consistent manner. This is the *macroscopic* problem of NLO.

For the passive processes the susceptibilities are predominantly real. As resonances are approached susceptibilities become complex and under certain conditions may be considered as totally imaginary. It is the imaginary parts of $\chi^{(1)}$ and $\chi^{(3)}$ respectively which describe *linear absorption* and *Raman scattering*. In general energy can be exchanged between the radiation and the material (over a period of time $\gg \omega^{-1}$) only for those mechanisms described by non-linear susceptibilities with "paired frequencies" (see Table II). In

TABLE II. Active nonlinear optical phenomena

Susceptibility	Process
$\chi^{(1)}(\omega_1;\omega_1)$	Linear absorption ($\omega_1 \simeq \omega_{10}$)
$\chi^{(3)}(\omega_2;\omega_1,-\omega_1,\omega_2)$	Raman scattering ($\omega_2 \simeq \omega_1 \mp \omega_{10}$)
$\chi^{(3)}(\omega_1;\omega_1,-\omega_1,\omega_1)$	Two photon absorption ($2\omega_1 \simeq \omega_{10}$) or Saturable absorption ($\omega_1 \simeq \omega_{10}$)
$\chi^{(5)}(\omega_2;\omega_1,\omega_1,-\omega_1,-\omega_1,\omega_2)$	Hyper-Raman scattering ($\omega_2 \simeq 2\omega_1 \mp \omega_{10}$)

other situations $\chi^{(n)}$ may be resonant or imaginary but will lead only to enhancement of the relevant passive process.

If one thinks of a set of two level systems in which the ground states only are occupied initially, then under equilibrium conditions an upper level population is acquired such that the rate of excitation due to "stimulated" absorption from the ground state is balanced by de-excitation via (*a*) stimulated emission, (*b*) spontaneous emission and (*c*) non-radiative decay mechanisms. Setting a lifetime T_1 for relaxation by processes (*b*) and (*c*), with level populations N_0, N_1:

$$\frac{N_0 - N_1}{\tau_a} = \frac{N_1}{T_1}$$

For normal (Beer's law) absorption conditions τ_a must be large compared to T_1. As these times become comparable one moves into the region of *saturated absorption* where for example the effective absorption coefficient becomes intensity dependent. For a Lorentzian absorption line

$$\alpha = \left[\frac{2\pi\omega N(ez_{10})^2}{nc\hbar}\right] \frac{T_2^{-1}}{(\omega - \omega_{10})^2 + T_2^{-2} + \dfrac{2T_1}{T_2}\left(\dfrac{ez_{10}E}{\hbar}\right)^2}$$

The intense radiation (small τ_a) bleaches the sample by creating a large upper level population, in turn the rate of absorption must decrease, producing the "power broadened" lines indicated by the equation. This result is obtained by coupling the rate equation to the wave equation. Using the perturbation theory method one therefore appears to get (to lowest order) an effective susceptibility $\chi^{(3)}$ due to the saturation. This effective parameter differs from those discussed hitherto in that it contains T_1. Also it is not too difficult to pass into the strong saturation region, $T_1 > \tau_a$, in which case it is insufficient to use just the lower order correction.

Stimulated emission exceeds absorption, so providing nett *gain*, in a population inverted system (corresponding to a negative value of Im $\chi^{(1)}$); stimulated Stokes scattering in a normal system also provides gain (negative Im $\chi^{(3)}$). Hence both are ripe for *laser oscillation* using cavity feedback. Oscillation occurs providing the round trip gain exceeds losses due to both transmission through cavity ends and absorption processes. Under continuous wave conditions above oscillation threshold the gain will adjust itself so as to be equal to the nett loses. Any optical process which involves exponential gain, that is where the generation of radiation of some frequency ω depends on the prior existence (due to spontaneous noise say) of radiation at ω, will in principle allow for oscillation. Such is the parametric oscillator case (chapter 2). Conversely, for example, second harmonic production requires only the presence of the fundamental field initially, no oscillation is possible.

Everything we have said so far can apply to continuous operation, but a large number of NLO effects occur only as a consequence of *transients* or for *pulsed operation*. For example, in association with absorption there is always an anomalous dispersion (connected through the Kramers-Krönig relations). Hence associated with saturated absorption there is a nonlinear refractive index charge which is proportioned to the radiation intensity. The time-dependent refractive index that will therefore be created under pulsed operation leads to pulse reshaping effects (self-steepening and self-phase modulation). If the pulse rise-time is short compared to the instantaneous value of τ_a and to T_1 then the saturation is unable to follow the pulse. The above adiabatic following process will not occur but in this non-equilibrium case the leading edge of the pulse will experience excess absorption (because the material has been unable to bleach) and vice versa. Hence again one achieves pulse reshaping.

The non-equilibrium absorption is mirrored in the gain for pulsed emission (laser). If the radiation lifetime t_c in the cavity is long then for a small transient increase in the pumping rate one can obtain a build up of radiation. The lifetime τ_a is inversely proportional to the radiation intensity so it in turn decreases. This coupling of the electron population and the radiation produces oscillations in each; in the emission these are called relaxation oscillations or spiking, having period roughly $(t_c T_1)^{1/2}$ and being damped over a time of the order of T_1. The output of a Q-switched laser is essentially the initial spike of the relaxation oscillations. By decreasing the loss suddenly, rather than increasing a pumping rate (in this case sufficiently to take the gain well above threshold) then, if t_c is small by comparison to T_1 and to the pumping time, the radiation density in the cavity builds up and is emitted over a period of the order of t_c—the Q-switched pulse. A narrower pulse still is achieved by mode locking. This can be obtained if the gain profile of the laser is broad enough to allow several axial modes to oscillate simultaneously. By introducing, for example, a saturable absorber that bleaches only for the more intense fluctuations in the resulting cavity-radiation one achieves a series of pulses separated by the round-trip cavity period $2c/nL$ with pulse times limited by the recovery time of the absorber and by the dephasing time of the laser medium itself. Once the latter time scale is achieved coherent effects come into play causing decomposition of the pulse.

For times small compare to T_2 the perturbed state of the medium is a coherent mixture of ground and excited states. It is essential to include T_2 and T_1 in the c-coefficient equation rather than introducing T_1 in separate

rate-equations as has been done so far. As the material is grossly affected by the radiation (extreme saturation) it is no longer reasonable to discuss coherence effects in terms of susceptibilities. The polarization of the medium is no longer proportional (in time) to powers of the electric fields present. In the limit of all times short compared to T_2 the c-coefficients are given by

$$|c_1(t)|^2 = 1 - |c_0(t)|^2 = \left[\frac{\sin \mu t/2}{\mu/2}\right]^2 \left(\frac{ez_{10}E}{\hbar}\right)^2,$$

where $\mu = -\tfrac{1}{2}(\omega_{10} - \omega) \pm \tfrac{1}{2}[(\omega_{10} - \omega)^2 + (2ez_{10}E/\hbar)^2]^{1/2}$.

Hence the populations oscillate in time. Under near resonant or strong signal conditions, $ez_{10}E/\hbar \gg (\omega_{10} - \omega)$, oscillation occurs at the Rabi-flop frequency $ez_{10}E/\hbar$, actually producing total inversion after a time $t = \pi\hbar/ez_{10}E$. Note that in the rate equation treatments appropriate for times longer than T_2 one can never reverse the material populations of a two level system by radiation pumping. The equilibrium results are obtained from the above analysis by reintroducing T_2, T_1; the $|c_1(t)|^2$ oscillations decay to the steady state saturation value. One remarkable feature of the Rabi oscillations is that for a pulse of duration $2\hbar/ez_{10}E$ the material populations are unaffected overall and no absorption takes place. The resulting pulse propagation with no nett absorption is termed self-induced transparency. Several other coherent effects are discussed in chapters 7 and 8. For example, if an intense field is suddenly switched off then over a time T_2 the resulting coherent state emits directional, coherent radiation (free induction decay). Coherent anti-Stokes Raman scattering (chapter 7) is the Raman analogue of the induction decay. In this analogue stimulated Raman scattering creates the coherence then a weak probe pulse is applied within a period T_2. Coherent anti-Stokes scattering from the probe indicates the presence of material coherence.

If the incident pulse time t_p is short enough that two absorbing levels of separation ΔE fall within an energy band $\hbar t_p^{-1}$ then a coherent three-level system is created. Whilst the upper levels remain phased the subsequent fluoresence from the two levels will also be phased producing "*quantum beats*" of period $\hbar/\Delta E$. Again there is a Raman analogy, the two upper levels being excited by Stokes scattering and a probe pulse producing anti-Stokes emission from them (chapter 7). An associated experiment is the production of coherent Raman beats between the pump laser and Stokes emission (chapter 8).

In the large number of NLO experiments that are undertaken with gaseous samples *Doppler shifts* play a significant role. As well as broadening any absorption line there are various modifications, described in chapter 8, to coherent effects. In particular photon echos are observable as the free-induction-decay emission from different velocity groups becomes in phase. There have also been a whole class of experiments devised to penetrate the Doppler broadening, (chapters 3 and 6). They include Lamb-dip saturated spectroscopy, two-photon absorption, double resonances and level-crossing experiments and the coherent quantum-beat effects.

Finally we might note two effects which are associated with the *spatial* scale of experiments rather than temporal scales. The demand of energy conservation fixes the frequencies which may be generated in passive processes. By similar token, over a long distance one demands momentum (phase) conservation. Thus unless the wave vector of the freely propagating wave at some frequency is equal to the wave vector of the generated radiation (the wave vectors would be $\mathbf{k}(2\omega)$ and $2\mathbf{k}(\omega)$ respectively for second harmonic generation) the range over which efficient generation can occur is limited. The attainment of conditions such that the two waves travel together is known as phase-matching (chapter 2). The second spatial effect is a transverse one, self-focussing. The nonlinear refractive index changes associated with intense radiation produce an index which varies transversely across any incident beam leading in consequence to the focussing (or defocussing) of the radiation.

In conclusion, we have chosen to define Nonlinear Optics to cover those physical effects which, by and large, are describable by semiclassical theory in the electric-dipole approximation. The ideas of material-change, susceptibilities and time scales have been used to correlate different phenomena. It would appear in summarizing these ideas that, for a given material system with fixed relaxation and dephasing times (T_1, T_2), the time which best epitomizes the interaction between radiation and material is the lifetime for electron excitation (τ_a). τ_a is inversely proportional to the radiation intensity; as the latter is increased one passes through the various nonlinear optics régimes (Table III).

TABLE III. The time scales of nonlinear optics. (t represents, in some way, the observation (period of observation, time after a pulse etc.).)

Process	Time inequalities
Passive	$\tau_a > t > T_1, T_2$
Stimulated	$t > \tau_a > T_1, T_2$
Saturated (Laser)	$t > \tau_a \simeq T_1 > T_2$
Non-equilibrium	$\tau_a, T_1 > t > T_2$
Coherent	$T_1, T_2 > t > \tau_a$

Since the growth of nonlinear optics in the 1960s enormous use has been made of nonlinear optical effects, both in the development of coherent light sources and in the study of material systems. In this introduction we have restricted ourselves to an overall view of the NLO processes, no mention is made of their applications. In contrast the following nine chapters contain both the details of processes and discussion of many of their applications.

1. Microscopic and Macroscopic Nonlinear Optics

J. DUCUING

Ecole Polytechnique, 91120, Palaiseau, France

INTRODUCTION

The theory of electromagnetic processes rests upon a set of perfectly linear equations, the familiar Maxwell's equations, together with a set of "constitutive" relations connecting the moment of the charge and current distributions in the medium to the electric and magnetic field. These last relations which express the response of the medium to an electromagnetic excitation are *not* linear. However, in the description of traditional optical processes they are linearized. Thus conventional optics is a linear theory. The superposition principle can be applied: waves do not interact, do not change their frequency. Even so, nonlinear optical processes have been known to opticians for some time. In a sense, it is fair to say that all effects leading to light modulation like the Pockels and Kerr effects, and the Faraday effect, belong to this class. Even to some extent the photoelectric effect. Two-photon processes were considered explicitly by Maria Goeppert Mayer in 1931. Coulson in 1952 and later on Buckingham expanded the dipole moment of the molecules in powers of the field and considered some implication of this.

However a strong stimulation to study nonlinear optical effects came only in 1960 with the advent of the laser and the corresponding enormous increase of the brightness of the sources. Comparison between a high pressure mercury arc and a conventional Q-switched ruby laser shows the laser to be brighter by a factor 10^{12}. Perturbation theory indicates that the terms in the polarization expansion in powers of the field decrease as successive powers of E/E_{at} where E_{at} is a characteristic intra-atomic field. As the maximum electric field which can be obtained from a source is directly related to its brightness, it is natural that the higher order terms which are responsible for nonlinear effects would appear more readily with lasers. New frequencies will be generated and there will appear large energy transfer between waves. The theory of nonlinear optics has a dual aspect. On one hand it concerns itself with describing and analyzing the relation between the applied electromagnetic field and the moments of the induced charge and current distributions, i.e. determining the so-called nonlinear susceptibility. On the other hand it deals with the problems of determining the field radiated by these charge and current distributions, that is to say the problem of wave propagation in a nonlinear medium. Although they can be treated completely separately these two problems are from a practical point of view intimately related, as the susceptibility determines the strength of the interactions.

This chapter can be summarized as follows. Basically there are four sections. The first deals with the framework of nonlinear optics. It sets out the elementary notions of nonlinear polarization in its basic aspects, such as symmetry and tensor properties. The second section deals with the theory of nonlinear susceptibilities, illustrations of the physical significance are given in a number of instances. e.g. Two-photon processes or four-wave parametric processes. Section III sets out basic microscopic theories of these various susceptibilities. In particular attention is concentrated on molecular orbital calculations in solids and also the charge transfer model. The final section deals with a simple model to describe nonlinearities in a one-dimensional system of conjugated π-orbitals.

I. THE FRAMEWORK OF NONLINEAR OPTICS

A. WAVE PROPAGATION

We start with Maxwell's equations in matter (e.s. units).

$$\nabla \times \mathbf{E} = -\frac{1}{c}\frac{\partial \mathbf{H}}{\partial t} \tag{I.1a}$$

$$\nabla \times \mathbf{H} = \frac{1}{c}\frac{\partial \mathbf{E}}{\partial t} + \frac{4\pi}{c}\frac{\partial}{\partial t}(\mathbf{P} - \nabla \mathbf{Q} + \cdots) + 4\pi\nabla \times \mathbf{M} + \cdots \tag{I.1b}$$

$$\nabla \cdot \mathbf{B} = 0 \tag{I.1c}$$

$$\nabla \cdot \mathbf{D} = 0 \tag{I.1d}$$

In the multipole expansion we neglect $\nabla \mathbf{Q}$, $\nabla \times \mathbf{M}$ and all higher terms. Then with only $\mathbf{P}$ retained $\mathbf{D} = \mathbf{E} + 4\pi\mathbf{P}$. We separate $\mathbf{P}$ into a linear part: $\mathbf{P}_L = \chi^{(1)}\mathbf{E}$ where $\chi^{(1)}$ is the linear dielectric susceptibility, and a nonlinear one $\mathbf{P}_{\mathrm{NLS}}$. Then elimination of $\mathbf{H}$ between (I.1a) and (I.1b) yields the equation of propagation:

$$\nabla \times (\nabla \times \mathbf{E}) + \frac{1}{c^2}\frac{\partial^2}{\partial t^2}(\varepsilon\mathbf{E}) = -\frac{4\pi}{c^2}\frac{\partial^2}{\partial t^2}\mathbf{P}_{\mathrm{NLS}}. \tag{I.2}$$

The nonlinear polarization acts as a source for the electric field $\mathbf{E}$, to which it transfers energy at a rate φ

$$\varphi = -\left\langle \mathbf{E}\cdot\frac{\partial \mathbf{P}_{\mathrm{NLS}}}{\partial t}\right\rangle \tag{I.3}$$

The time average is taken over intervals long compared to the period of the field, but short compared to some characteristic time for the evolution of the field amplitude.

B. PHENOMENOLOGICAL DEFINITION OF THE SUSCEPTIBILITIES

(a) *Second-Order*

Let us first consider the lowest order nonlinearity. The simplest relation one could write between $P^{(2)}_{\mathrm{NLS}}$ and E is of the form

$$P^{(2)} = \chi^{(2)}E^2 \qquad \text{(we drop }_{\mathrm{NLS}}\text{ in the following)}$$

This however is not satisfactory as we must account for *dispersion* and *tensorial nature*, the former representing the inability of the material to respond instantly to changing field. With just two Fourier components at ω_1, ω_2

$$\mathbf{E}(t) = \mathscr{E}(\omega_1)e^{i\omega_1} + \mathscr{E}(\omega_2)e^{i\omega_2} + \text{c.c.}$$

the spectrum of $\mathbf{P}^{(2)}$ will include $\omega_1 + \omega_2$. We write

$$P^{(2)}_{\sigma}(\omega_1 + \omega_2) = \chi^{(2)}_{\sigma\lambda\mu}(\omega_1, \omega_2)E_{\lambda}(\omega_1)E_{\mu}(\omega_2) \tag{I.4}$$

This result can be obtained in a less empirical way by introducing[1] the second order response tensor $R^{(2)}_{\sigma\lambda\mu}(\tau_1, \tau_2)$:

$$P^{(2)}_{\sigma}(t) = \int_0^{\infty} d\tau_1 \int_0^{\infty} d\tau_2\, R^{(2)}_{\sigma\lambda\mu}(\tau_1, \tau_2)E_{\lambda}(t - \tau_1)E_{\mu}(t - \tau_2) \tag{I.5}$$

where $R^{(2)}_{\sigma\lambda\mu} = 0$ for $\tau_2, \tau_1 < 0$ (causality). For an arbitrary decomposition, the real field $\mathbf{E}(t)$ becomes

$$\mathbf{E}(t) = \int_{-\infty}^{\infty} d\omega\, \mathscr{E}(\omega)e^{i\omega t}, \quad \text{with} \quad \mathscr{E}(-\omega) = \mathscr{E}^*(\omega),$$

whence

$$P^{(2)}_{\sigma}(t) = \int_{-\infty}^{+\infty} d\omega_1 \int_{-\infty}^{+\infty} d\omega_2\, \chi^{(2)}_{\sigma\lambda\mu}(\omega_1, \omega_2)\mathscr{E}_{\lambda}(\omega_1)\mathscr{E}_{\mu}(\omega_2)e^{i(\omega_1+\omega_2)t}, \tag{I.6}$$

where

$$\chi^{(2)}_{\sigma\lambda\mu}(\omega_1, \omega_2) = \int_0^{\infty} d\tau_1 \int_0^{\infty} d\tau_2\, R^{(2)}_{\sigma\lambda\mu}(\tau_1, \tau_2)e^{-i\omega_1\tau_1 - i\omega_2\tau_2} \tag{I.7}$$

Note:

1. Since $\boldsymbol{R}$ is real $\chi^{(2)}_{\sigma\lambda\mu}(-\omega_1, -\omega_2) = (\chi^{(2)}_{\sigma\lambda\mu}(\omega_1, \omega_2))^*$
2. $\chi^{(2)}_{\sigma\lambda\mu}(\omega_1, \omega_2)$ is invariant under simultaneous permutations of λ, μ and ω_1, ω_2
3. Relation (I.4) is obtained by setting:

$$\mathscr{E}_{\lambda}(\omega) = E_{\lambda}\delta(\omega - \omega_1) \text{ etc.}\ldots$$

(b) *Higher-Orders*

The generalization to nth order is obvious:

$$P^{(n)}_{\sigma}(t) = \int_{-\infty}^{+\infty} d\omega_1 \cdots \int_{-\infty}^{+\infty} d\omega_n\, \chi^{(n)}_{\sigma\lambda\mu\cdots\nu}(\omega_1, \ldots, \omega_n) \tag{I.6$'$}$$

$$\mathscr{E}_{\lambda}(\omega_1), \ldots, \mathscr{E}_{\nu}(\omega_n)\exp i(\omega_1 + \cdots + \omega_n)t$$

or

$$P^{(n)}_{\sigma}(\omega_1 + \cdots + \omega_n) = K_n\chi^{(n)}_{\sigma\lambda\cdots\nu}(\omega_1, \ldots, \omega_n)\mathscr{E}_{\lambda}(\omega_1), \ldots, \mathscr{E}_{\nu}(\omega_n).$$

where K_n is the number of distinguishable permutations of frequencies $\omega_1, \ldots, \omega_n$.

C. FOURIER DECOMPOSITION OF THE WAVE EQUATION. PHASE MATCHING

Using the Fourier components of the field and polarization at ω_σ:

$$\nabla \times (\nabla \times \mathbf{E}(\mathbf{r}, \omega_\sigma)) - \frac{\omega_\sigma^2}{c^2}\,\varepsilon(\omega_\sigma)\mathbf{E}(\mathbf{r}, \omega_\sigma) = \frac{4\pi}{c^2}\,\omega_\sigma^2\,\mathbf{P}_{\mathrm{NLS}}(\mathbf{r}, \omega_\sigma) \tag{I.8}$$

The homogeneous wave equation defines modes of the radiation field. For instance, for plane waves at ω_σ

$$\mathbf{E}(\mathbf{r}, \omega_\sigma) = E_\sigma \hat{\mathbf{e}}_\sigma \exp(-i\mathbf{k}_\sigma \cdot \mathbf{r})$$

where $\hat{\mathbf{e}}_\sigma$, $\mathbf{k}_\sigma$ must satisfy:

$$\mathbf{k}_\sigma \times (\mathbf{k}_\sigma \times \hat{\mathbf{e}}_\sigma) + \frac{\omega_\sigma^2}{c^2} \varepsilon(\omega_\sigma)\hat{\mathbf{e}}_\sigma = 0. \tag{I.9}$$

The source polarization will radiate energy into these modes with a rate (see Equation I.3)

$$\varphi = -\left\langle \mathbf{E} \cdot \frac{\partial \mathbf{P}}{\partial t} \right\rangle = i\omega_\sigma(\mathbf{E} \cdot \mathbf{P}^* - \mathbf{E}^* \cdot \mathbf{P}) = 2\omega_\sigma \operatorname{Im}(\mathbf{P} \cdot \mathbf{E}^*); \tag{I.10}$$

the sign of φ will be determined by the relative phase of **P** and **E**. Consider now the case where the Fourier components of the fields creating $\mathbf{P}_{\mathrm{NLS}}$ can be represented by plane waves. Then

$$P_{\sigma,\mathrm{NLS}}(\omega_\sigma) = \chi^{(n)}_{\sigma,\lambda\mu\cdots\nu} E_\lambda(\omega_1)E_\mu(\omega_2), \ldots, E_\mu(\omega_n)\exp - i(\mathbf{k}_1 + \mathbf{k}_2 + \cdots + \mathbf{k}_n)\cdot\mathbf{r} \tag{I.11}$$

$$\mathbf{P} \cdot \mathbf{E}^* \propto \exp i(\mathbf{k}_\sigma - \mathbf{k}_1 - \mathbf{k}_2 - \cdots - \mathbf{k}_n) \cdot \mathbf{r} = \exp i\Delta\mathbf{k} \cdot \mathbf{r} \tag{I.12}$$

If $\Delta\mathbf{k} \neq 0$, the relative phase will vary in space and there will be no net energy transfer in an infinite medium (see chapter 2).

For the situation of plane quasi-monochromatic waves of finite transverse extent, interacting in a nonlinear medium, one can under generally satisfied conditions derive an approximate wave equation. The electric field associated with a wave at ω can be written as the product of a plane monochromatic wave solution of the wave equation by a complex *amplitude* function depending on space and time (envelope)

$$\mathbf{E}(\omega, \mathbf{r}, t) = \hat{\mathbf{e}}A(\mathbf{r}, t)\exp[-i(\mathbf{k} \cdot \mathbf{r} - \omega t)] + \text{c.c.}$$

where

$$\mathbf{k} \times (\mathbf{k} \times \hat{\mathbf{e}}) + \left(\frac{\omega}{c}\right)^2 \varepsilon(\omega)\hat{\mathbf{e}} = 0$$

If the relative variation of A within an optical period and over a wavelength is small, it is possible to neglect its second partial derivatives. One then obtains the approximate first-order equation[2]:

$$\frac{\partial A}{\partial z} - \tan\alpha \frac{\partial A}{\partial x} + \frac{1}{v_g}\frac{\partial A}{\partial t} = -i\frac{4\pi^2}{n\lambda}\frac{\hat{\mathbf{e}} \cdot \mathbf{P}_{\mathrm{NLS}}}{\cos^2\alpha} \exp(i\mathbf{k} \cdot \mathbf{r}), \tag{I.13}$$

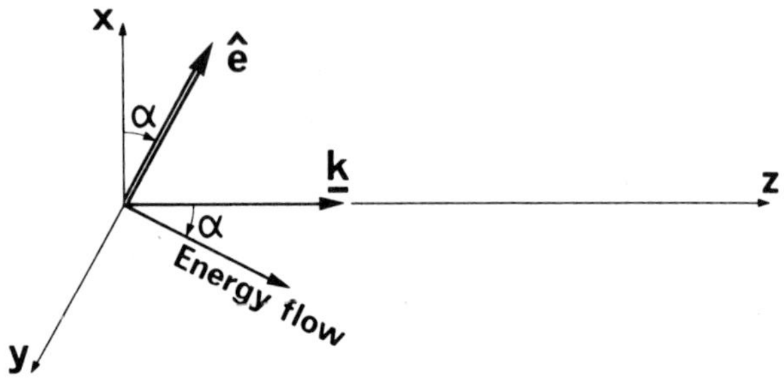

FIG. 1.

where the coordinate system is indicated in Fig. 1. the wave vector **k** being along z; n, v_g are respectively the refractive index and the group velocity corresponding to frequency ω. The angle α between the energy flow and the normal to the phase plane is zero in an optically isotropic medium. When the expression (I.11) for $\mathbf{P}_{\mathrm{NLS}}$ is taken into account one sees clearly the influence of phase matching on the growth of A. For the steady state case when infinite transverse extension is assumed (plane homogeneous wave) Equation (I.13) reduces to the familiar form

$$\frac{\partial A}{\partial z} = -i\,\frac{4\pi^2}{n\lambda}\,\frac{\hat{\mathbf{e}}\cdot\mathbf{P}_{\mathrm{NLS}}}{\cos^2\alpha}\exp(i\mathbf{k}\cdot\mathbf{r})$$

derived by Armstrong *et. al.*[3]

D. A GLANCE AT NONLINEAR PROCESSES

The notation used to write the susceptibilities clearly indicates the nature of the physical process. Thus, $\chi^{(2)}(\omega_1, \omega_2)$ relates to the process in which two fields $\mathbf{E}(\omega_1)$, $\mathbf{E}(\omega_2)$ create a source $\mathbf{P}_{\mathrm{NLS}}\,(\omega_1 + \omega_2)$ at the sum frequency (algebraic frequencies).

In third-order some cases of special interest arise. Consider

$$P_\sigma(\omega_\sigma) = \chi^{(3)}_{\sigma\lambda\mu\nu}(\omega_1, \omega_2, \omega_3)E_\lambda(\omega_1)E_\mu(\omega_2)E_\nu(\omega_3),$$

and take

$$\omega_3 = -\omega_2,\ E_\nu(\omega_3) = E_\nu^*(\omega_2),\ \omega_\sigma = \omega_1$$

We obtain

$$P_\sigma(\omega_1) = \chi^{(3)}_{\sigma\lambda\mu\nu}(\omega_1, \omega_2, -\omega_2)E_\lambda(\omega_1)E_\mu(\omega_2)E_\nu^*(\omega_2)$$

This describes the modification of the dielectric susceptibility at ω_1 by a field at ω_2. One has:

$$\Delta\chi_{\sigma\lambda}(\omega_1) = \chi^{(3)}_{\sigma\lambda\mu\nu} E_\mu(\omega_2) E^*_\nu(\omega_2)$$

In particular the real part of $\chi^{(3)}_{\sigma\lambda\mu\nu}$ describes the modification of the refractive index due to an optical field, the so called optical Kerr effect. The imaginary part describes induced absorption (Im $\chi^{(3)} > 0$) or gain. To this class belong stimulated scattering for which gain is induced at the Stokes frequency, and two-photon absorption (or emission for an inverted medium).

Symmetry of the Medium

We consider briefly, the influence of the spatial symmetry of the nonlinear medium on the susceptibility tensors. The relation

$$\mathbf{P}^{(n)}(\omega_\sigma) = \boldsymbol{\chi}^{(n)}(\omega_1, \omega_2, \ldots, \omega_n)\mathbf{E}(\omega_1), \ldots, \mathbf{E}(\omega_n)$$

implies that $\boldsymbol{\chi}^{(n)}$ is a polar tensor of $(n+1)$th order since $\mathbf{P}, \ldots, \mathbf{E}$ transform as polar tensors of first order under linear orthogonal transformations of the coordinate system. If we call T such a transformation represented by the orthogonal matrix $T_{\kappa\lambda}$, we have:

$$E'_\kappa = T_{\kappa\lambda} E_\lambda, \quad P^{(n)\prime}_\kappa = T_{\kappa\lambda} P^{(n)}_\lambda$$

(′ in the new coordinate system), substitution yields:

$$\chi^{(n)\prime}_{\sigma'\lambda'\mu'\cdots\nu'} = T_{\sigma'\sigma} T_{\lambda'\lambda}, \ldots, T_{\nu'\nu} \chi^{(n)}_{\sigma\lambda\mu\cdots\nu}$$

which shows that $\boldsymbol{\chi}^{(n)}$ transforms like a $(n+1)$th order tensor.

Now $\boldsymbol{\chi}^{(n)}$ must be invariant under the symmetry operations which transform the medium into itself. With each symmetry operation is associated a coordinate transformation. For such a transformation

$$\boldsymbol{\chi}^{(n)\prime} \equiv \boldsymbol{\chi}^{(n)}.$$

This yields a number of relations between the components of $\boldsymbol{\chi}^{(n)}$ from which one can extract the non-zero independent components. This analytical method has been used by Butcher[1] to determine the structure of the second- and third-order tensors for all crystal classes.

Let us illustrate this procedure by a straightforward but important example. Assume that the medium is invariant under inversion. To this transformation corresponds $T_{\lambda'\lambda} = (-1)\delta_{\lambda'\lambda}$ and

$$\chi^{(n)\prime}_{\sigma\lambda\mu\cdots\nu} = (-1)^{n+1} \chi^{(n)}_{\sigma\lambda\mu\cdots\nu}$$

but now we must have

$$\chi^{(n)'}_{\sigma\lambda\mu\cdots\nu} \equiv \chi^{(n)}_{\sigma\lambda\mu\cdots\nu}$$

so for *n even* χ vanishes. An important practical consequence is that for media with inversion symmetry $\chi^{(2)} \equiv 0$.

Thus in a medium with inversion symmetry $\chi^{(3)}$ represents the lowest-order nonlinearity in the electric-dipole approximation. In an isotropic medium $\chi^{(3)}$ has only 21 non-zero elements; one can show that these are:

$$\begin{aligned} &xxxx = yyyy = zzzz \\ &xxyy = yyzz = zzxx = zzyy = yyxx = xxzz \\ &xyxy = yzyz = zxzx = zyzy = yxyx = xzxz \\ &xyyx = yzzy = zxzx = zyyz = yxxy = xzzx \end{aligned}$$

with

$$xxxx = xxyy + xyxy + xyyx,$$

where *ijkl* stands for $\chi^{(3)}_{ijkl}$. There are thus only 3 independent elements.

Let us use these symmetry properties to show that a circularly polarized beam generates *no* third harmonic in an isotropic medium[4].

$$E = \hat{\mathbf{x}}E_x e^{\omega} + \hat{\mathbf{y}}E_y e^{\omega} + \text{c.c.}$$

$$P_x = \{\chi^{(3)}_{xxxx}E_x^3 + (\chi^{(3)}_{xxyy} + \chi^{(3)}_{xyxy} + \chi^{(3)}_{xyyx})E_x E_y^2\}e^{i3\omega t}.$$

For a circularly polarized beam $E_x = E$, $E_y = iE$. As

$$\chi^{(3)}_{xxxx} = \chi^{(3)}_{xxyy} + \chi^{(3)}_{xyxy} + \chi^{(3)}_{xyyx}$$

we see that $P_x = 0$ and similarly $P_y = 0$. This can readily be interpreted in terms of conservation of the angular momentum associated with the photons[4].

II. The Theory of Susceptibilities

The generalized susceptibility tensors $\chi^{(n)}$ have now been defined phenomenologically. In this section we shall try to relate them to the microscopic structure of the medium. The first problem is to determine the effective field acting on the polarizable charges in the medium. The second to determine the dipole moment induced in the charge distribution by this effective field. After we consider these two questions, we illustrate these formal developments by discussing some experimental implications of resonance in the third-order susceptibility.

A. THE "EFFECTIVE" OR "LOCAL" FIELD PROBLEM

The microscopic field acting on charges inside matter differs from the macroscopic field appearing in Maxwell's equations. It is in fact the sum of the external fields produced by the sources outside the medium and which would exist *in vacuo* and of the field produced by the currents and charges induced in the medium. At the turn of the century, Lorentz discussed this matter[5]. He showed, in particular, that for atoms or ions situated at the sites of a cubic lattice, the effective field E_{eff} is related to the macroscopic field E and to the polarization P by:

$$E_{\text{eff}} = E + \frac{4\pi}{3} P \tag{II.1}$$

which immediately leads to:

$$E_{\text{eff}} = \frac{\varepsilon + 2}{3} E \tag{II.2}$$

This relation is found to apply well to ionic crystals. It can be used to describe the polarizability of their strongly localized electrons.

Darwin[6] considered the case of the strongly delocalized nearly-free electrons of a metal. He showed that in this case, the rapid spatial fluctuations of the microscopic field were averaged out over the extended electron orbitals and that the effective field acting on the electron was given by:

$$E_{\text{eff}} = E \tag{II.3}$$

In intermediate situations, with which we will be dealing in the following, the problem becomes extremely complicated. Adler[7] has given a formal solution extended to the nonlinear case by Pine[8]. The corresponding expressions are unfortunately difficult to use and we shall not consider them here. We shall instead follow an empirical approach. We assume that the relation between effective and macroscopic fields and the polarization can be written as :

$$E_{\text{eff}} = E + \mathscr{L} P \tag{II.4}$$

where for the sake of simplicity we take $\mathscr{L}$ as a scalar. Then:

$$E_{\text{eff}} = E + \mathscr{L} \chi_m^{(1)} E_{\text{eff}}$$

where $\chi_m^{(1)}$ refers to the "microscopic" susceptibility (response of the polarizable units to the effective field). From this it follows that:

$$E_{\text{eff}} = \frac{1}{1 - \mathscr{L}\chi_m^{(1)}} E = fE \tag{II.5}$$

and:

$$\chi^{(1)} = f\chi_m^{(1)} \tag{II.6}$$

which expresses the effective field correction to the first-order susceptibility.

In the nonlinear situation, things become a little more complicated. Consider the case of second harmonic generation. An applied macroscopic field $E(\omega)$ creates in the medium a nonlinear polarization:

$$P_{\mathrm{NL}}(2\omega) = \chi_m^{(2)}(\omega, \omega) f^2(\omega) E^2(\omega)$$

This however is *not* the *source* polarization appearing in Equation (I.2). Consider however the total polarization

$$P_T(2\omega) = \chi_m^{(1)}(2\omega) E_{\mathrm{eff}}(2\omega) + P_{\mathrm{NL}}(2\omega) \tag{II.7}$$

where:

$$E_{\mathrm{eff}}(2\omega) = E(2\omega) + \mathscr{L} P_T(2\omega) \tag{II.8}$$

solving for P_T,

$$\begin{aligned} P_T(2\omega) &= f(2\omega)\chi_m^{(1)}(2\omega)E(2\omega) + f(2\omega)P_{\mathrm{NL}} \\ &= \chi^{(1)}(2\omega)E(2\omega) + f(2\omega)P_{\mathrm{NL}} \end{aligned} \tag{II.9}$$

In Section II we defined the source polarization as:

$$P_{\mathrm{NLS}}(2\omega) = P_T(2\omega) - \chi^{(1)}(2\omega)E(2\omega)$$

This immediately leads to:

$$\chi^{(2)}(2\omega) = f(2\omega) f^2(\omega) \chi_m^{(2)}(2\omega) \tag{II.10}$$

This relation[3] can be generalized in a straightforward fashion. Bedeaux and Bloembergen[9] have recently given a more rigorous formulation of these considerations.

B. MICROSCOPIC SUSCEPTIBILITIES

We turn now to the problem of calculating the microscopic susceptibilities χ_m, i.e. to determine the response of the polarizable units to an electromagnetic excitation. In this we use a *classical* description of the field which is completely appropriate for most of the nonlinear effects discussed in this course:

$$\mathbf{E}(\mathbf{r}, t) = \sum_{\omega} \mathbf{E}(\mathbf{r}, \omega) \exp i\omega t + \text{c.c.}$$

The thermodynamic response of the system perturbed by the electric field is conveniently expressed through use of the quantum statistical density operator ρ[10,11]. The quantum statistical average of a dynamic variable X is then given as

$$\langle X \rangle = \mathrm{Trace}(\rho X)$$

The operator ρ satisfies the equation of motion:

$$i\hbar \frac{d\rho}{dt} = [H, \rho] - i\hbar\Gamma\rho \tag{II.11}$$

Here:

$$H = H_0 - \mathbf{E}\cdot\mathbf{P} = H_0 + H' \tag{II.12}$$

where H_0 is the Hamiltonian for the isolated system, $\mathbf{P}$ the electric-dipole operator and $-\mathbf{E}\cdot\mathbf{P}$ describes the interaction of the field with the polarizable medium, within the electric-dipole approximation[12,13]. The term $-i\hbar\Gamma\rho$ which appears in Equation (II.11) provides a semiphenomenological way of including dissipation processes. For further discussion, see Refs [10,11].

Although the trace (ρX) is independent of representation, in practice it is usual to work with the eigenstates of H_0, namely energy levels E_r and matrix elements $H'_{rs}(t)$. A solution for ρ then appears in successive powers of H'_{rs}, $\rho = \rho^{(0)} + \rho^{(1)} + \rho^{(2)} + \cdots$, and $\rho^{(0)}$ becomes a diagonal matrix whose elements $\rho_{rr}^{(0)}$ are simply the equilibrium statistical-mechanical weights of the levels E_r. In general the perturbation H'_{rs} causes non-diagonal elements to appear, representing the development of phase correlation among the states.

The matrices $\rho^{(1)}$, $\rho^{(2)}$ etc., are determined by the equation of motion for ρ, namely,

$$i\hbar \frac{d}{dt}\rho^{(1)} + [\rho^{(1)}, H_0] - i\hbar\Gamma\rho^{(1)} = [H', \rho^{(0)}] \tag{II.13}$$

$$i\hbar \frac{d}{dt}\rho^{(2)} + [\rho^{(2)}, H_0] - i\hbar\Gamma\rho^{(2)} = [H', \rho^{(1)}] \tag{II.14}$$

and so on.

H' is now resolved into its temporal Fourier components representing the frequencies of the exciting radiation,

$$H'_{rs} = \sum_{\lambda} H'_{rs}(\omega_\lambda)\exp(+i\omega_\lambda t)$$

Similar expressions are assumed for $\rho^{(1)}$, $\rho^{(2)}$ etc. Substitution then leads to the "forced motion" solutions,

$$D_{rs}(\omega_\lambda)\rho_{rs}^{(1)}(\omega_\lambda) = H'_{rs}(\omega_\lambda)(\rho_{ss}^{(0)} - \rho_{rr}^{(0)}) \tag{II.15}$$

where

$$D_{rs}(\Omega) \equiv E_s - E_r - \hbar\Omega + i\hbar\Gamma_{rs} = \hbar(\omega_{sr} - \Omega + i\Gamma_{rs}) \qquad \text{(II.16)}$$

A phenomenological damping Γ_{rs} is assumed for each transition $r \to s$. In the same way:

$$D_{tu}(\omega_\lambda + \omega_\mu)\rho_{tu}^{(2)}(\omega_\lambda + \omega_\mu) = [H'(\omega_\mu), \rho^{(1)}(\omega_\lambda)]_{tu} + [H'(\omega_\lambda), \rho^{(1)}(\omega_\mu)]_{tu}$$

and:

$$\rho_{tu}^{(2)}(\omega_\lambda + \omega_\mu) = D_{tu}^{-1}(\omega_\lambda + \omega_\mu)\Bigg(\sum_r H'_{tr}(\omega_\mu)H'_{ru}(\omega_\lambda)\frac{(\rho_{uu}^{(0)} - \rho_{rr}^{(0)})}{D_{ru}(\omega_\lambda)} - H'_{tr}(\omega_\lambda)H'_{ru}(\omega_\mu)\frac{(\rho_{rr}^{(0)} - \rho_{tt}^{(0)})}{D_{tr}(\omega_\lambda)} + \text{terms } \lambda \rightleftarrows \mu\Bigg)$$

or equivalently:

$$\rho_{tu}^{(2)}(\omega_\lambda + \omega_\mu) = D_{tu}^{-1}(\omega_\lambda + \omega_\mu)\Bigg(\sum_r H'_{tr}(\omega_\mu)H'_{ru}(\omega_\lambda) \times \left(\frac{(\rho_{uu}^{(0)} - \rho_{rr}^{(0)})}{D_{ru}(\omega_\lambda)} + \frac{(\rho_{tt}^{(0)} - \rho_{rr}^{(0)})}{D_{tr}(\omega_\mu)}\right) + \text{terms } \lambda \rightleftarrows \mu\Bigg)$$

similarly:

$$D_{ji}(\omega_\lambda + \omega_\mu + \omega_\nu)\rho_{ji}^{(3)}(\omega_\lambda + \omega_\mu + \omega_\nu) = [H'(\omega_\nu), \rho^{(2)}(\omega_\mu + \omega_\lambda)]_{ji} + 2 \text{ terms } (\nu \to \mu \to \lambda)$$

This can be written concisely:

$$\rho_{ji}^{(3)}(\omega_\lambda + \omega_\mu + \omega_\nu) = D_{ji}^{-1}(\omega_\lambda + \omega_\mu + \omega_\nu) \times \mathscr{P}_{\lambda,\mu,\nu}\sum_{t,r}\Bigg\{\frac{H'_{jt}(\omega_\nu)H'_{tr}(\omega_\mu)H'_{ri}(\omega_\lambda)}{D_{ti}(\omega_\lambda + \omega_\mu)}\left(\frac{\rho_{ii}^{(0)} - \rho_{rr}^{(0)}}{D_{ri}(\omega_\lambda)} - \frac{\rho_{tt}^{(0)} - \rho_{rr}^{(0)}}{D_{tr}(\omega_\mu)}\right) + \frac{H'_{jr}(\omega_\lambda)H'_{rt}(\omega_\mu)H'_{ti}(\omega_\nu)}{D_{jt}(\omega_\lambda + \omega_\mu)}\left(\frac{\rho_{rr}^{(0)} - \rho_{tt}^{(0)}}{D_{rt}(\omega_\mu)} + \frac{\rho_{rr}^{(0)} - \rho_{jj}^{(0)}}{D_{jr}(\omega_\lambda)}\right)\Bigg\}$$

where $\mathscr{P}_{\lambda,\mu,\nu}$ designates the sum over *all distinguishable* permutations of $\omega_\lambda, \omega_\mu, \omega_\nu$.

Having determined ρ, we can now calculate the nth order polarization $P^{(n)}$ through the relation $P^{(n)} = N\langle P\rangle^{(n)} = N\,\mathrm{Tr}(P\rho^{(n)})$.

Linear term

$$\langle P_\sigma^{(1)}\rangle = N\,\mathrm{Tr}\langle P_\sigma \rho^{(1)}\rangle$$

$$P_\sigma^{(1)}(\omega) = N\sum_{i,j} P_{ij,\sigma}\rho_{ji}^{(1)}(\omega) = N\sum_{ij}\frac{P_{ij,\sigma}H'_{ji}(\rho_{ii}^{(0)}-\rho_{jj}^{(0)})}{D_{ji}(\omega)}$$

$$P_\sigma^{(1)}(\omega) = -E_\lambda(\omega)N\sum_{ij}\rho_{ii}^{(0)}\left(\frac{P_{ij,\sigma}P_{ji,\lambda}}{D_{ji}(\omega)} - \frac{P_{ij,\lambda}P_{ji,\sigma}}{D_{ij}(\omega)}\right)$$

which yields the well known expression of the linear susceptibility:

$$\chi_{\alpha\beta}^{(1)}(\omega) = N\sum_{ij}\rho_{ii}^{(0)}\frac{1}{\hbar}\left\{\frac{P_{ij,\alpha}P_{ji,\beta}}{\omega_{ji}+\omega-i\Gamma_{ij}} + \frac{P_{ij,\beta}P_{ji,\alpha}}{\omega_{ji}-\omega+i\Gamma_{ij}}\right\} \qquad \text{(II.17)}$$

Second-order

Using the expression for $\rho^{(2)}(\omega_\lambda+\omega_\mu)$

$$P_\sigma^{(2)}(\omega_\sigma=\omega_\lambda+\omega_\mu) = N\sum_{ij}P_{ij,\sigma}\rho_{ij}^{(2)}(\omega_\lambda+\omega_\mu) = E_\lambda E_\mu N\mathscr{P}_{\lambda\mu}\sum_{i,j,k}\rho_{ii}^{(0)}$$

$$\times\left\{\frac{P_{ij,\sigma}P_{jk,\mu}P_{ki,\lambda}}{D_{ji}(\omega_\lambda+\omega_\mu)D_{ki}(\omega_\lambda)} - \frac{P_{ij,\lambda}P_{jk,\sigma}P_{ki,\mu}}{D_{jk}(\omega_\lambda+\omega_\mu)D_{ik}(\omega_\lambda)}\right.$$

$$\left. - \frac{P_{ij,\lambda}P_{jk,\sigma}P_{ki,\mu}}{D_{jk}(\omega_\lambda+\omega_\mu)D_{ji}(\omega_\mu)} - \frac{P_{ij,\mu}P_{jk,\lambda}P_{ki,\sigma}}{D_{ij}(\omega_\lambda+\omega_\mu)D_{ik}(\omega_\lambda)}\right\} \qquad \text{(II.18)}$$

In a transparency range, away from resonances:

$$\frac{1}{D_{jk}(\omega_\lambda+\omega_\mu)D_{ik}(\omega_\lambda)} + \frac{1}{D_{jk}(\omega_v+\omega_\mu)D_{ji}(\omega_\mu)} = \frac{1}{D_{ik}(\omega_\lambda)D_{ji}(\omega_\mu)}$$

and one obtains:

$$\chi_{\sigma\lambda\mu}^{(2)}(\omega_\lambda,\omega_\mu) = \frac{N\mathbb{P}_{\lambda\mu}}{2^{\mathbf{k}}}\sum_{i,j,k}\frac{\rho_{ii}^{(0)}}{\hbar^2}$$

$$\times\left\{\frac{P_{ij,\sigma}P_{jk,\lambda}P_{ki,\mu}}{(\omega_{ji}+\omega_\lambda+\omega_\mu)(\omega_{ki}+\omega_\mu)} + \frac{P_{ij,\lambda}P_{jk,\sigma}P_{ki,\mu}}{(\omega_{ki}-\omega_\lambda)(\omega_{ji}+\omega_\mu)}\right.$$

$$\left. + \frac{P_{ij,\mu}P_{jk,\lambda}P_{ki,\sigma}}{(\omega_{ji}-\omega_\lambda-\omega_\mu)(\omega_{ki}-\omega_\lambda)}\right\} \qquad \text{(II.19)}$$

where $\mathbb{P}_{\lambda\mu}$ designates the sum over *all* permutations of ω_λ, ω_μ.

This can be made still more compact and generalized to the nth order, by introducing[1] the permutations over the set of frequencies $\omega_\lambda, \omega_\mu, \ldots, \omega_\nu, \omega_\sigma$, where $-\omega_\sigma = \omega_\lambda + \omega_\mu + \cdots + \omega_\nu$

$$\chi^{(n)}_{\sigma\lambda\mu\cdots\nu}(\omega_\lambda, \omega_\mu, \ldots, \omega_\nu) = \frac{N}{n!}\,\mathbb{P}_{\sigma\lambda\mu\cdots\nu} \sum_{ijk\cdots l} \frac{\rho^{(0)}_{ii}}{\hbar^n}$$

$$\times \frac{P_{ij,\sigma} P_{jk,\lambda} P_{kr,\mu} \cdots P_{li,\nu}}{(\omega_{ji} + \omega_\lambda + \omega_\mu + \omega_\nu)(\omega_{ki} + \omega_\mu + \cdots + \omega_\nu)\cdots(\omega_{li} + \omega_\nu)} \tag{II.20}$$

where $\mathbb{P}_{\sigma\lambda\mu\cdots\nu} = \sum \Pi_{\sigma\lambda\mu\cdots\nu}$, $\Pi_{\sigma\lambda\mu\cdots\nu}$ being any permutation of $\sigma\lambda\mu\nu$ and of the corresponding frequencies. This expression leads immediately to the important relation[3]:

$$\Pi_{\sigma\lambda\mu\cdots\nu}\chi^{(n)}_{\sigma\lambda\cdots\nu}(\omega_\lambda, \ldots, \omega_\nu) = \chi^{(n)}_{\sigma\lambda\cdots\nu}(\omega_\lambda, \ldots, \omega_\nu)$$

This relates the components of tensors describing different physical effects. Consider for instance in second order the case: $\omega_\lambda = \omega_1$, $\omega_\mu = -\omega_2$. We have:

$$\begin{aligned}\chi^{(2)}_{\sigma\lambda\mu}(\omega_1, -\omega_2) &= \chi^{(2)}_{\lambda\sigma\mu}(-(\omega_1 - \omega_2), -\omega_2)\\ &= [\chi^{(2)}_{\lambda\sigma\mu}((\omega_1 - \omega_2), \omega_2)]^*\end{aligned}$$

The tensor component on the left side describes difference-frequency generation whereas that on the right-hand side describes sum-frequency generation. For instance ω_1 and ω_2 can be two optical frequencies and $\omega_1 - \omega_2$ a radio or microwave frequency. Then $\chi^{(2)}_{\lambda\sigma\mu}(\omega_1 - \omega_2, \omega_2)$ describes the Pockels effect.

In the case where dispersion can be neglected we can ignore the frequencies altogether. Then $\chi^{(2)}_{\lambda\sigma\mu}$ is invariant under any permutation of the indexes, this is known as Kleinman's symmetry condition[14].

We finally give the expression for the third-order polarization when damping must be kept explicitly:

$$P^{(3)}_\sigma(\omega_\sigma = \omega_\lambda + \omega_\mu + \omega_\nu) = -E_\lambda E_\mu E_\nu N \mathscr{P}_{\lambda\mu\nu} \sum_{i,j,k,l}$$

$$\times \left\{ \frac{P_{ij,\sigma} P_{jk,\nu} P_{kl,\mu} P_{li,\lambda}}{D_{ji}(\omega_\sigma) D_{ki}(\omega_\lambda + \omega_\mu)} \left(\frac{\rho^{(0)}_{ii} - \rho^{(0)}_{ll}}{D_{li}(\omega_\lambda)} + \frac{\rho^{(0)}_{kk} - \rho^{(0)}_{ll}}{D_{kl}(\omega_\mu)} \right) \right.$$

$$\left. + \frac{P_{ij,\sigma} P_{jk,\lambda} P_{kl,\mu} P_{li,\nu}}{D_{ji}(\omega_\sigma) D_{jl}(\omega_\lambda + \omega_\mu)} \left(\frac{\rho^{(0)}_{kk} - \rho^{(0)}_{ll}}{D_{kl}(\omega_\mu)} + \frac{\rho^{(0)}_{kk} - \rho^{(0)}_{jj}}{D_{jk}(\omega_\lambda)} \right) \right\} \tag{II.21}$$

From this expression, the third-order susceptibility $\chi^{(3)}_{\sigma\lambda\mu\nu}(\omega_\lambda, \omega_\mu, \omega_\nu)$ can be easily deduced. Note that (II.18) and (II.21) are different from the expressions which would result by introducing damping *a posteriori* in (II.20). In particular they exhibit resonances between unpopulated levels.

Resonant interactions

We turn now to the important special case where there exists a "two-photon" resonance. In this case the sum (or difference) of two frequencies of the exciting field is approximately equal to a transition frequency of the medium, say $\omega_\lambda + \omega_\mu \simeq \omega_{ba}$. We restrict our consideration to the third-order case. In the expression for $P^{(3)}_\sigma$ the resonant terms involving $D_{ab}(\omega_\lambda + \omega_\mu) = \hbar(\omega_{ba} - \omega_\lambda - \omega_\mu + i\Gamma_{ba})$ will dominate. Retaining only these and assuming distinct frequencies one obtains:

$$P^{(3)}_\sigma(\omega_\sigma) = E_\lambda E_\mu E_\nu N \frac{(\rho^{(0)}_{aa} - \rho^{(0)}_{bb})}{D_{ab}(\omega_\lambda + \omega_\mu)} \left\{ \sum_{jl} \left(\frac{P_{bj,\nu} P_{ja,\sigma}}{D_{aj}(\omega_\sigma)} - \frac{P_{bj,\sigma} P_{ja,\nu}}{D_{jb}(\omega_\sigma)} \right) \right.$$
$$\left. \times \left(\frac{P_{al,\mu} P_{lb,\lambda}}{D_{al}(\omega_\mu)} + \frac{P_{al,\lambda} P_{lb,\mu}}{D_{al}(\omega_\lambda)} \right) \right\} \qquad \text{(II.25)}$$

In this we have assumed that there is no other (intermediate) resonance and used $D_{lb}(\omega_\mu) \approx -D_{al}(\omega_\lambda)$. It should also be noted that the resonant term will dominate $P^{(3)}_\sigma(\omega_\sigma)$ only if a had b are connected through some intermediate state j. If the system has inversion symmetry this implies that a and b have the same parity.

Let us restrict further the choice of frequencies and consider the case of only two incident beams with well-defined polarization. In the spectrum of the exciting field we have components $\pm\omega_1$, $\pm\omega_2$ and in the spectrum of the polarization we find $\omega_\sigma = \omega_1 + \omega_2 - \omega_2 = \omega_1$. Taking $\sigma - \lambda, \mu = \nu$ we have $E_\mu = E^*_\nu$ and:

$$P^{(3)}_\lambda(\omega_1) = E_\lambda |E_\mu|^2 N \frac{(\rho^{(0)}_{aa} - \rho^{(0)}_{bb})}{\hbar^3(\omega_{ba} - \omega_1 - \omega_2 + i\Gamma_{ba})} \left| \sum_j \frac{P_{bj,\lambda} P_{ja,\mu}}{\omega_{ja} - \omega_2} + \frac{P_{bj,\mu} P_{ja,\lambda}}{\omega_{ja} - \omega_1} \right|^2 \qquad \text{(II.26)}$$

which describes two-photon absorption ($\rho_{aa} > \rho_{bb}$) *or* emission ($\rho_{bb} > \rho_{aa}$) between states a and b.

The case of stimulated Raman scattering is very similar. For instance the induced amplification of ω_1 produced by the scattering of ω_2 by the material excitation $b \to a$, $\omega_1 = \omega_2 - \omega_{ab}$, is obtained by making $\omega_2 \to -\omega_2$

in the preceding calculation:

$$P_\lambda^{(3)}(\omega_1) = E_\lambda |E_\mu|^2 N \frac{(\rho_{aa}^{(0)} - \rho_{bb}^{(0)})}{\hbar^3(\omega_2 - \omega_1 - \omega_{ab} + i\Gamma_{ab})} \left| \sum_j \frac{P_{bj,\lambda} P_{ja,\mu}}{\omega_{ja} + \omega_2} + \frac{P_{bj,\mu} P_{ja,\lambda}}{\omega_{ja} - \omega_1} \right|^2 \tag{II.27}$$

These two processes can be represented by the following graphs:

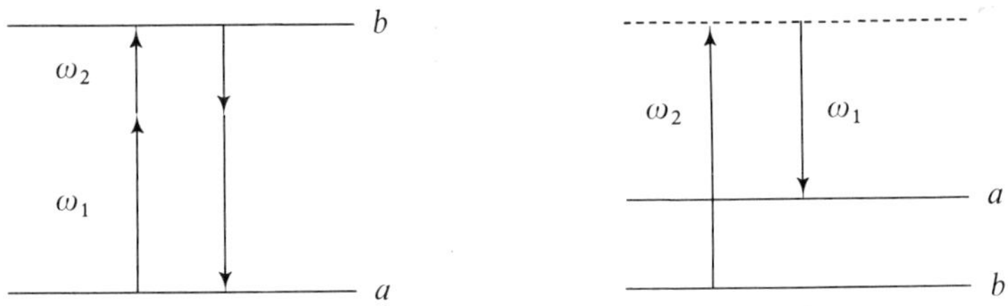

(a) Two-photon absorption or emission (b) Raman Stokes emission

FIG. 2.

It can be useful to rewrite Equation (II.27) in terms of the spontaneous Raman cross section per unit solid angle $(d\sigma_{\mu\nu}/d\Omega)$

$$P_\lambda^{(3)}(\omega_1) = E_\lambda |E_\mu|^2 \frac{(\rho_{aa}^{(0)} - \rho_{bb}^{(0)})}{\hbar(\omega_2 - \omega_1 - \omega_{ab} + i\Gamma_{ab})} \frac{c^4}{\omega_2 \omega_1^3} \frac{d\sigma_{\mu\nu}}{d\Omega}$$

In dilute media where the relative linewidths Γ_{ba}/ω_{ba} are small, resonances can strongly increase the susceptibility. Enhancement of third harmonic generation by a resonance at the second harmonic frequency was considered early by Manykin and Afanas'ev[15]. Harris and coworkers[16] have used near coincidences of the fundamental and harmonic frequencies with atomic transitions to obtain efficient third harmonic generation in metal vapours. They have extended this technique using frequency mixing to generate coherent light in the vacuum U. V. beyond 1000 Å. Resonant enhancement of the susceptibility allows the use of gases and vapours which offer the important advantage of an extended transparency range both in the U. V. and in the I. R. A particularly attractive scheme is that of four-wave interactions in which resonance is achieved at the sum or difference of two frequencies ω_λ, ω_μ. The third frequency ω_ν can then be varied providing a tunable output at $\omega_\lambda + \omega_\mu + \omega_\nu = \omega_\sigma$. Sorokin *et al.*[17] used this method to build a coherent I. R. source in alkali metal vapours. Figure 3 illustrates the atomic scheme for their

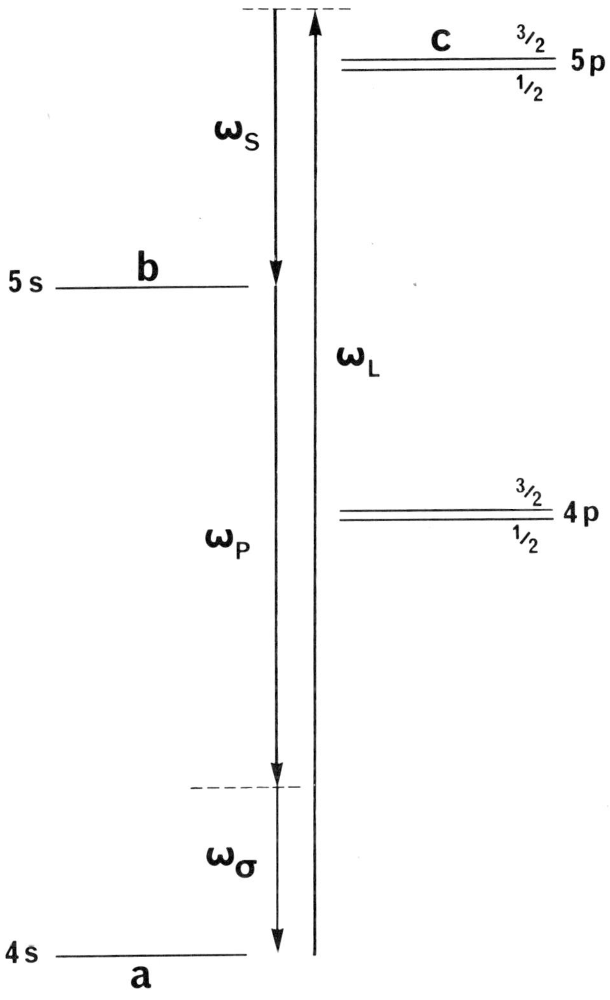

FIG. 3. The atomic scheme in potassium vapour for the four-wave mixing experiment of Sorokin *et al.* (Ref. 17).

experiment in potassium vapour. Two nitrogen-pumped dye lasers provide two tunable intense beams at ω_L, ω_P. The first at ω_L falls close to the strong 4*s*-5*p* resonance and generates ω_S through stimulated Raman Scattering to the 5*s* level. The mixing of these frequencies through third-order nonlinearities produces a tunable output at $\omega_\sigma = \omega_L - \omega_S - \omega_P$. Note that in this case $\chi^{(3)}(\omega_L, -\omega_S, \omega_P)$ is increased through the exact resonance $\omega_L - \omega_S = \hbar^{-1}(E_{5s} - E_{4s})$ but also by having ω_L very close to the $4s \rightarrow 5p$ transition.

However to obtain efficient generation at ω_σ one must satisfy the phase matching condition $\mathbf{k}_\sigma = \mathbf{k}_L - \mathbf{k}_S - \mathbf{k}_P$. In this collinear beam experiment, because of the strong dispersion of the vapour in the vicinity of the $4s \rightarrow 5p$ transition, this can be achieved by fine tuning of ω_L (note that ω_σ is determined only by ω_P). Hodgson *et al.*[19] have used a similar arrangement to produce tunable coherent radiation in the vacuum U. V. With two incident beams at ω_1 and ω_2 irradiating strontium vapour, these authors observed enhancement of the $2\omega_1 + \omega_2$ signal through tuning of $2\omega_1$ through two-photon resonances and also further enhancement corresponding to the coincidence of $\hbar(2\omega_1 + \omega_2)$ with the energy of autoionizing states.

III. Susceptibility Calculations

Formal expressions for the nonlinear-susceptibility tensors have been given in the previous sections. Their complicated structure which involves summations over infinite sets of intermediate states prevents any attempt to extract accurate numerical values from them. Yet the susceptibilities determine the magnitude of the energy transfer between waves, and a quantitative understanding of the underlying polarisation mechanisms is essential to nonlinear optics.

The first attempt to a quantitative evaluation of $\chi^{(2)}$ seems to be found in an empirical rule proposed by Miller[19]. This relates in a simple fashion $\chi^{(2)}$ and $\chi^{(1)}$:

$$\chi^{(2)}_{ijk}(\omega_1, \omega_2) = \delta_{ijk}(\omega_1, \omega_2)\chi^{(1)}_{ii}(\omega_1 + \omega_2)\chi^{(1)}_{jj}(\omega_1)\chi^{(1)}_{kk}(\omega_2)$$

in which δ is a constant, at least for materials of a given symmetry. In fact a survey of experimental data shows that although δ is not a constant its variations are considerably smaller than that of $\chi^{(2)}$. Miller's rule clearly shows the influence of the electronic polarizability: the more polarizable the medium, the stronger the optical nonlinearity. However it does not incorporate in a satisfactory way the symmetry of the medium. An obvious case is that of a centrosymmetric medium for which $\chi^{(2)}$ vanishes identically. In spite of this Miller's rule has been of very valuable help in the search for materials with large nonlinearities.

In this chapter we shall consider, as an illustration, the calculation of the low-frequency limit of the electronic susceptibility tensor $\chi^{(2)}$ (0, 0) in the frequency range above the lattice resonances but below the electronic transitions. This quantity determines the strength of the lowest order interactions such as second harmonic generation, difference-frequency generation and parametric amplification. It is thus of obvious practical significance.

The simplest class of crystals possessing a non-zero second-order susceptibility $\chi^{(2)}$ is the $\bar{4}3m$ class which includes the tetrahedrally coordinated binary semiconductors and insulators of the III–V and II–VI families. We shall restrict our considerations to these compounds whose structure (zinc blende) is recalled on Fig. 4. In the frequency range of interest the dominant contribution to the linear susceptibility is known[20] to come by far from the valence electrons: the core electrons more tightly bound are considerably less polarizable. The same situation occurs for $\chi^{(2)}$ even more strongly as an additional factor comes into play. The second-order susceptibility is zero for a free atom which possesses inversion symmetry. In a solid this symmetry can be removed because of the interactions with the neighbouring atoms. These interactions will essentially affect the valence electrons and the mechanism of the optical nonlinearities must be sought in the dynamics of these electrons in interaction with the incident radiation. To describe them one might think of using the band picture, but this leads to computational difficulties. Moreover in the low frequency range considered here there is no contribution from real transitions and no particular domain of the Brillouin zone is singled out. Thus the detailed band structure has little influence on the value of $\chi^{(2)}$. Furthermore, it can be shown[21] that under generally satisfied conditions $\chi^{(2)}$ is determined exclusively by the knowledge of the ground state electronic density. It should also be remembered that for a fully saturated compound the total (antisymmetrized) wave function is

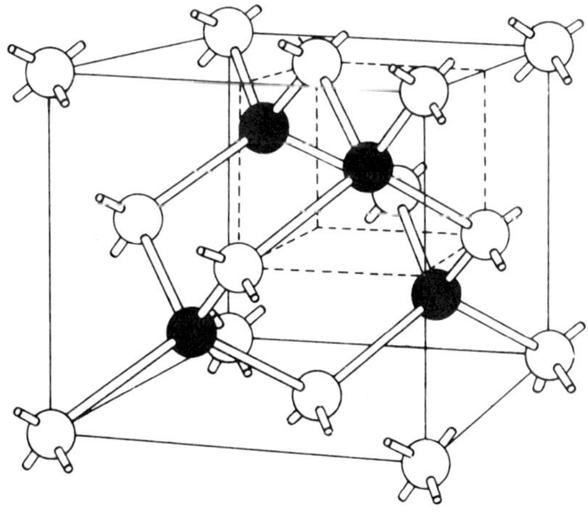

● = III ○ = V

FIG. 4. Zinc blende structure.

not affected by the choice of one-electron wave functions as localized orbitals or suitable linear combinations of these[22]. Thus it seems justified to decompose the crystal into bonds considered as independent polarizable units.

In this model the valence electrons are considered as a system of separated electron pairs accommodated in effectively localized bonds between adjacent atoms. There are four equivalent tetrahedral bonds A · B per unit cell (Fig. 5). If one assumes that the bond possesses axial symmetry and

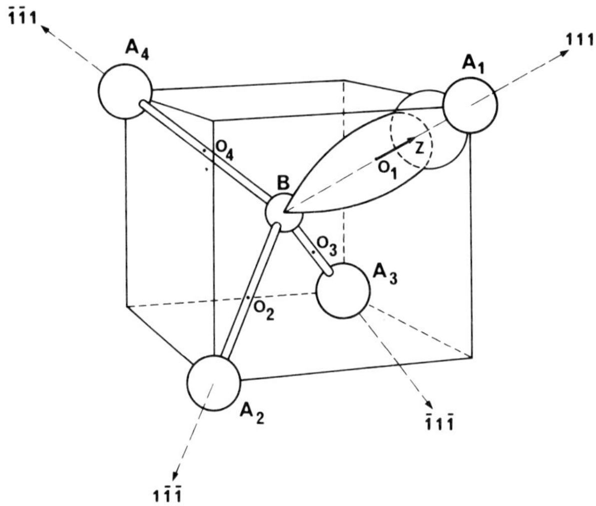

FIG. 5. Four equivalent bonds in a unit cell of a III–V compound. A is the non-metal, B the metal atom.

defines the linear bond-polarizabilities $\alpha_{\|}$ and $\alpha_{\perp}$ for fields along the bond axis or normal to it, and similarly the nonlinear bond-polarizabilities $\beta_{\|} = \beta_{\zeta\zeta\zeta}$ and $\beta_{\perp} = \beta_{\xi\xi\zeta}$ then simple geometrical considerations show that:

$$\chi^{(1)} = \tfrac{4}{3}N(\alpha_{\|} + 2\alpha_{\perp}) \qquad \text{(III.1a)}$$

$$\chi_{14}^{(2)} = \frac{4}{3\sqrt{3}}N(\beta_{\|} - 3\beta_{\perp}) \qquad \text{(III.1b)}$$

where N is the number of unit cells per unit volume $(4/a^3)$.

There is a two-fold problem in calculating these bond susceptibilities. One must first find the "effective" or local field acting on the bond. Then one must find the response of the electrons to this effective field (The so-called microscopic susceptibility of section II, p. 19). The many published attempts to solve this problem can be divided into two categories *ab initio* and model calculations. In the following, we describe some examples.

A. ABINITIO CALCULATIONS

Jha and Bloembergen[23] and Flytzanis and Ducuing[21,24] have used LCAOMO wave functions to calculate the bond polarizabilities. In this method each bond A·B is represented in its ground state by two uncorrelated electrons with opposite spins ascribed to a bonding orbital $\psi^{(0)} = N^{-1/2}(\psi_A + \lambda\psi_B)$, where N is a normalization constant ψ_A and ψ_B are tetrahedral sp^3 orbitals centred respectively on atoms A and B, λ is a parameter which expresses the polarity of the bond. The two-electron wave function of the ith bond is:

$$\psi_i(1, 2) = \psi^{(0)}(1)\psi^{(0)}(2)\frac{\alpha(1)\beta(2) - \beta(1)\alpha(2)}{\sqrt{2}}$$

where α and β are spin wave functions. In the following we outline the work of Flytzanis and Ducuing[21]. These authors first assume that the effective field is a constant over the polarizable unit and that for the cubic crystal under investigation it can be represented by

$$E_{\text{eff}} = fE,$$

where f is a scalar. Note that for these semiconductor electrons which are neither confined to atomic sites nor completely free to wander around the whole unit cell, one expects for f a value intermediate between the Lorentz value $f = \varepsilon + 2/3$ which applies to ionic crystals and $f = 1$ which applies to metals. With these assumptions the relations between the susceptibilities and the *microscopic* bond-polarizabilities (response to the effective field) α^m, β^m are:

$$\chi^{(1)} = \tfrac{4}{3}Nf(\alpha_{\parallel}^m + 2\alpha_{\perp}^m) \qquad \text{(III.2a)}$$

$$\chi_{14}^{(2)} = \frac{4}{3\sqrt{3}}Nf^3(\beta_{\parallel}^m - 3\beta_{\perp}^m) \qquad \text{(III.2b)}$$

The assumptions made here are very crude. Unfortunately an *a priori* calculation (or even estimate) of the local field factor f is not tractable. One way to partially circumvent this difficulty is to determine f by requiring that the calculated value of $\chi^{(1)}$ is equal to the experimental one.

Using the low frequency limits of the expressions given in the preceding sections the microscopic bond-polarizabilities can be written:

$$\alpha_{\parallel}^{m} = 4e^2 \sum_i{}' \frac{\langle 0|z|i\rangle\langle i|z|0\rangle}{E_i - E_0} \tag{III.3}$$

$$\beta_{\parallel}^{m} = 6e^3 \left\{ \sum_{i,j}{}'' \frac{\langle 0|z|i\rangle\langle i|z|j\rangle\langle j|z|0\rangle}{(E_i - E_0)(E_j - E_0)} - \langle 0|z|0\rangle \sum_i{}' \frac{|\langle 0|z|i\rangle|^2}{(E_i - E_0)^2} \right\} \tag{III.4}$$

where z is the coôrdinate along the bond. And similar expressions for the perpendicular components $\alpha_{\perp}^{m}$ and $\beta_{\perp}^{m}$. These expressions can be estimated by using the so-called *Unsöld approximation*[25]. In this, all the energy denominators are approximated by a common value E_u which can be determined for instance, by using the Thomas-Kuhn sum rule.

$$\sum_i \frac{2m}{\hbar^2} (E_i - E_0) |\langle 0|z|i\rangle|^2 = 1$$

Then:

$$\alpha_{\parallel} = 8 \frac{(\langle 0|\tilde{z}^2|0\rangle)^2}{a_0} \tag{III.5}$$

$$\beta_{\parallel} = 24 \frac{\langle 0|\tilde{z}^3|0\rangle}{ea_0^2} (\langle 0|\tilde{z}^2|0\rangle)^2 \tag{III.6}$$

where: $\tilde{z} = z - \langle 0|z|0\rangle$. Equations (III.5) and (III.6) express α and β in terms of the moments of the ground state charge density[24,26].

This is only a rough approximation. It can be refined by noting that if one can determine an operator F such that:

$$\langle i|F|0\rangle = \frac{\langle i|z|0\rangle}{E_i - E_0}$$

then the polarizabilities can be expressed simply as:

$$\alpha_{\parallel}^{m} = 4e^2 \langle 0|\tilde{z}\tilde{F}|0\rangle \tag{III.7}$$

$$\beta_{\parallel}^{m} = 6e^3 \langle 0|\tilde{z}\tilde{F}\tilde{z}|0\rangle \tag{III.8}$$

It can be shown that if spin-orbit coupling is neglected $\tilde{F}$ can be determined from the knowledge of the ground state charge density[27,28]. This is the approach followed in Ref. [21] for the III–V compounds, using the LCAO-MO determined by Coulson *et al.*[29], and a variational procedure due to Schwarz[30]. For further details on the calculations the reader is referred to the original paper[21] and to the thesis of Flytzanis[31]. The results, shown on Table I, are in good agreement with experimental values. There are two interesting by-products of these calculations. The first one is that for all the compounds studied $\beta_{\parallel}^{m} \gg \beta_{\perp}^{m}$. This is physically reasonable if one considers that a field applied along the bond will redistribute the electronic charge density *along the bond*. In such a rearrangement the parallel component of the polarizability will undergo a larger modification ($\Delta\alpha_{\parallel} = \beta_{\parallel} E$) than the transverse component ($\Delta\alpha_{\perp} = \beta_{\perp} E$).

TABLE I. Molecular orbital calculations of $\chi^{(2)}$ in III–IV semiconductors (from Refs 24 and 31). Experimental values from: (a) R. K. Chang, J. Ducuing and N. Bloembergen. *Phys. Rev. Lett.* **15**, 415 (1965); (b) C. K. N. Patel, *Phys. Rev. Lett.* **16**, 613, (1966); (c) N. Bloembergen, R. K. Chang, J. Ducuing and P. Lallemand. *In* "Physics of Semiconductors". Dunod, Paris (1964); (d) D. F. Nelson and E. M. Turner. *J. Appl. Phys.* **39**, 3337 (1968)

		Theoretical		Experimental
		$10^7 \times \chi_{14}^{(2)}$	$\frac{10^9}{(4\pi)^3} \times \delta$	$10^7 \times \chi_{14}^{(2)}$
Compound	f	e.s.u.	e.s.u.	e.s.u.
InSb	1.95	65	2.0	30 ± 15(a)
InAs	1.41	41	3.3	20 ± 8(b)
InP	1.20	28	3.8	
GaSb	1.93	16	0.7	(18)(c)
GaAs	1.50	19	1.8	18 ± 8(b)
GaP	1.24	14	2.8	5(d)
AlSb	1.30	7	1	
BP		0	0	

The second by-product resides with a strong correlation found between Miller's parameter δ and the dipole moment p of the bond (Fig. 6). This dipole moment can be considered as a measure of the asymmetry of the total charge distribution. An interesting case is that of BP. In this compound the electronic charge distribution is symmetrical but its centre is slightly shifted away from that of the ionic charges. As a result although $p \neq 0$, the electronic $\chi^{(2)}$ vanishes and $\delta \approx 0$. These results cast light on the charge symmetry considerations which are missing from Miller's initial formulation.

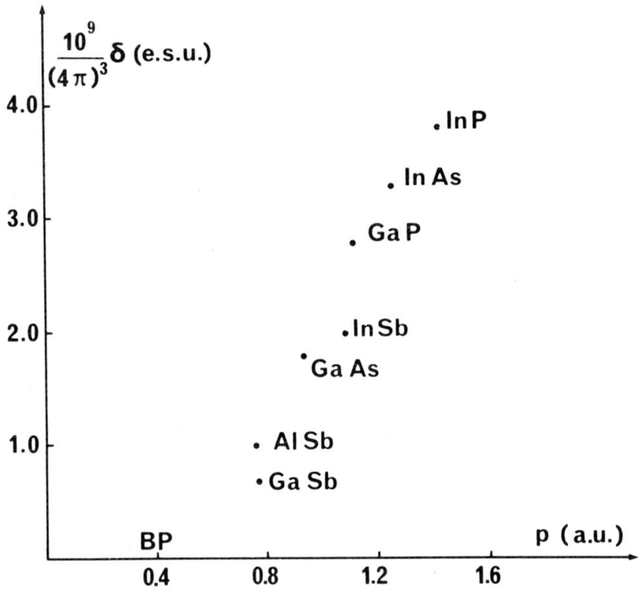

FIG. 6. Variation of Miller's δ with the electronic bond dipole moment for III–V compounds (from Refs 24 and 31).

B. MODEL CALCULATIONS

In this low frequency limit, if the influence of an electric field on the dielectric constant is known, $\chi^{(2)}$ can be simply determined through the first derivative of $\chi^{(1)}$. The model calculations which have appeared in the literature in the last few years rely on simple approximate expressions for $\varepsilon_1(0)$. The form of such an expression can be obtained on very general grounds by considering the following sum-rules:

$$\varepsilon_1(0) = 1 + \frac{2}{\pi}\int_0^\infty \frac{\varepsilon_2(\omega)}{\omega}\,d\omega \qquad \text{(Kramers-Krönig)}$$

$$\frac{\pi}{2}\Omega_p^2 = \int_0^\infty \omega\varepsilon_2(\omega)d\omega \qquad \text{(Thomas-Kuhn)}$$

Where Ω_p is the electronic (total) plasma frequency. The contribution of the valence electron transitions dominate the first sum[20]. If one keeps only this in the second sum rule, introducing some appropriate cut-off frequency ω_c then one obtains

$$\frac{\pi}{2}\omega_p^2 = \int_0^{\omega_c} \omega\varepsilon_2(\omega)d\omega \qquad \text{(III.9)}$$

where now ω_p is the valence electron plasma frequency $\omega_p = 4\pi n_v e^2/m$. Assuming now that ε_2 peaks strongly around some average frequency $E_g/\hbar$ (see Fig. 7) one obtains:

$$\varepsilon_1(0) = 1 + \left(\frac{\hbar\omega_p}{E_g}\right)^2 \qquad \text{(III.10)}$$

This expression which has also been deduced by Penn[32] from the study of a spherical nearly-free electron gas with gap E_g, has been extensively used by Phillips and Van Vechten[33,34] in their dielectric theory of semiconductors. These authors propose that in expression (III.10) the average energy E_g be determined in the following way. If the compound is homopolar (AA),

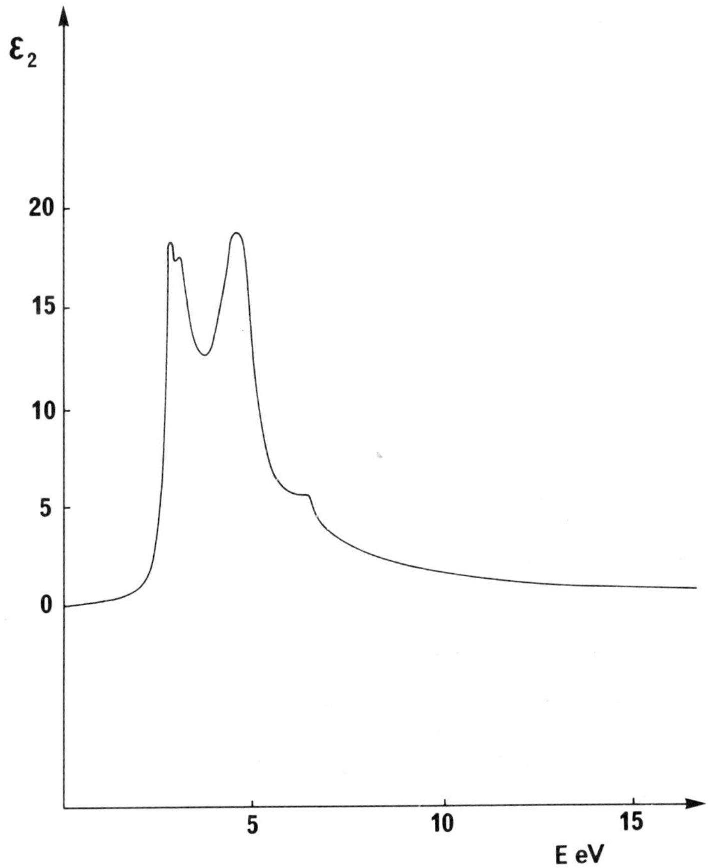

FIG. 7. A sketch of the variation of the imaginary part of the dielectric constant in GaAs.

E_g determined experimentally by relation (III.10) is found to be a simple function of bond length, $E_g = E_h(d) = Ad^{-2.5}$. For a heteropolar compound

$$E_g^2 = E_h^2 + C^2$$

where

$$C(A, B) = 4b\left(\frac{\zeta_A}{r_A} e^2 - \frac{\zeta_B}{r_B} e^2\right)e^{-k_s d} \tag{III.11}$$

where b is a dimensionless constant ≈ 1.5, $4\zeta_i$ is the number of electrons on atom i, the r_i's are the *covalent* radii, k_s the Thomas-Fermi screening constant corresponding to the valence electrons. This expression gives a good fit of the known values of $\varepsilon_1(0)$ for tetrahedrally coordinated compounds. It obviously describes the reduction of the susceptibility $(\varepsilon_1(0) - 1)4\pi$ with increasing polarity $\zeta_A - \zeta_B$. It is the basis for Phillips dielectric scale of electronegativity. Phillips defines the quantity

$$f_i = \frac{C^2}{E_g^2} \tag{III.12}$$

which expresses the ionic character of the bond. From expression (III.9) and (III.10) and Equation (III.1a) we deduce:

$$\alpha_{\parallel} = \frac{3}{16\pi N} \frac{1}{1 + 2K} \frac{(\hbar\omega_p)^2}{E_h^2 + C^2} \tag{III.13}$$

where $K = \alpha_{\perp}/\alpha_{\parallel}$ characterizes the bond anisotropy.

An applied electric field induces an additional asymmetric potential over the bond and thus modifies the value of E_g, hence of $\alpha_{\parallel}$. If this variation can be determined, then β and $\chi^{(2)}$ will be found.

C. THE CHARGE TRANSFER MODEL[35]

When an electric field E is applied a redistribution of valence charges occurs. The charges associated with atoms A and B become $\zeta_A - \Delta\zeta$ and $\zeta_B + \Delta\zeta$. This in turn modifies the heteropolar component C to the value $C + \Delta C$.

This is now readily related to the applied field E. Indeed the induced dipole (from B to A) is:

$$\alpha_{\parallel} E = \Delta\zeta e\, d_e \tag{III.14}$$

where d_e corresponds to the effective distance of motion of the charge $d_e = d - r_{iA} - r_{iB}$, where $r_{iA,B}$ are the *ionic* radii (core) Equation (III.14) determines $\Delta\zeta$ and ΔC as functions of E. Now from (III.13)

$$\alpha_{\|}(E) = \frac{3}{16\pi N} \frac{1}{1 + 2K} \frac{(\hbar\omega_p)^2}{E_h^2 + (C + \Delta C)^2} = \alpha_{\|} + \beta_{\|} E + \cdots$$

From which:

$$\chi_{14}^{(2)} = \frac{2}{(1 + 2K)^2} \left(\frac{d}{d_e}\right) \frac{\Gamma e d}{\sqrt{3}} \frac{C}{E_g^2} (\chi^{(1)})^2 \qquad \text{(III.15)}$$

where

$$\Gamma = \frac{16bd}{\sqrt{3}} e^{-k_s d}\left(\frac{1}{r_A} + \frac{1}{r_B}\right)$$

All quantities in Equation (III.15) are determined. So there is no adjustable parameter in this charge transfer theory. The results obtained by Tang and Flytzanis are shown on Table II .Their agreement with experimental values

TABLE II. Calculated charge transfer contribution to the nonlinear optical susceptibility $\chi_{14}^{(2)}$ for various semiconductors of Zinc-blende structure (from Ref. 35). Experimental values from: (a) R. K. Chang, J. Ducuing and N. Bloembergen. *Phys. Rev. Lett.* **15**, 415 (1965); (b) J. J. Wynne and N. Bloembergen. *Phys. Rev.* **188**, 1211 (1969); (c) C. K. N. Patel. *Phys. Rev. Lett.* **16**, 613 (1969); (d) R. A. Soref and H. W. Moos, *J. Appl. Phys.* **35**, 2152 (1964)

		$\chi_{14}^{(2)}$ (10^{-7} e.s.u.)	
Compound	Γ	Calc.	Expt.
InSb	4.03	40	33[a]
InAs	4.03	24	20[b]
GaSb	4.04	36	30[b]
GaAs	4.00	10	9[b]
GaP	4.03	6.2	5.2[b]
CdTe	4.00	7.1	8.0[c]
ZnTe	4.01	7.3	7.3[d]
ZnSe	4.01	3.0	2.2[d]
ZnS	4.04	1.7	1.7[d]

is strikingly good in view of the simplicity of the model. In this calculation one obtains an approximate expression for the Miller parameter

$$\delta = \frac{\chi^{(2)}}{[\chi^{(1)}]^3}$$

$$\delta = D(Z - R)\frac{d^3}{ea_0} + O(Z^2, \Delta^2) \qquad \text{(III.16)}$$

where D is a nearly constant factor, a_0 the Bohr radius and

$$Z = \frac{Z_A - Z_B}{Z_A + Z_B}, \qquad R = \frac{r_B - r_A}{r_B + r_A}$$

Expression (III.16) shows that if the term R could be ignored, δ would be proportional to the dipole moment of the bond[36]. But the spatial distribution of the charge which determines R interferes with this. This for symmetrical charge distribution might decrease appreciably the value of δ (see the comments made earlier concerning the *ab initio* calculation).

An interesting question is that of the sign of δ. According to Gatos and Lavine's sign conventions[37] the 111 bond axis points from the metal (Ga) to the non-metal (As) atom. As long as only s and p electrons are present the metal atom will have fewer electrons. An electric field oriented positively along the bond axis will bring back some electrons to the metal atom reducing the polarity of the bond, decreasing the gap and increasing the dielectric constant. Thus $\chi_{14}^{(2)}$ will be positive. This picture can change completely in presence of the d electrons which will appear with heavier elements. These d electrons will be localised around the atom from which they originate, if this is a metal atom they will tend to re-equilibriate the charge distribution and might even *reverse its polarity*. This appears to be the situation for ZnO and CuCl[35,38,39].

Further comments

1. C. L. Tang[38] in a very clearly written article uses simple molecular orbital theory to show that the average energy E_g can be expressed by the quadratic relation

$$E_g^2 = E_h^2 + C^2$$

where E_h is an homopolar "gap" and C is simply related to the parameter λ of the molecular orbital expression for $\chi_{14}^{(2)}$.

2. Choy *et al.*[40] have performed a similar calculation (charge redistribution) using the more recent bond-orbital model of Harrison instead of Phillips' dielectric theory of semiconductors.

3. The present discussion is limited in scope and obviously far from exhaustive. Among the numerous contributions which have been left out, the important work of Levine[41] must be singled out. This author has used Phillips' bond charge model to make extensive calculations of nonlinear susceptibilities of many compounds.

IV. Optical Nonlinearities in One-Dimensional Crystals

To conclude this chapter we consider a somewhat different situation, that of an infinite chain of conjugated π-orbitals[42]. Such chains which exhibit extended electron delocalization are encountered in "one-dimensional" solids: the TCNQ salts and transition metal complexes[43] and also conjugated polymers such as the polydiacetylenes[44]. We attempt here to describe their nonlinear properties with a very simple model, that of the one-dimensional Mathieu crystal[45]. In this model electrons are assumed to move in a periodic pseudopotential:

$$V(x) = V_0 \cos \pi \frac{x}{d}$$

which couples free-electron states with wave vectors k and $(\pi/d) - k$. Restricting ourselves to a two-band model[46] we find for these valence and conduction bands, the energies

$$E_v(k) = E_0(k) - \frac{V_0}{2} f(k)$$

$$E_c(k) = E_0\left(k - \frac{\pi}{d}\right) + \frac{V_0}{2} f(k)$$

and wave functions in Bloch form:

$$\psi_{c,v}(k; x) = e^{ikx} u_{c,v}(k; x)$$

$$u_v(k; x) = \frac{1}{\sqrt{(1 + f^2)2d}} (1 - f e^{-i(\pi x/d)})$$

$$u_c(k; x) = \frac{1}{\sqrt{(1 + f^2)2d}} (f + e^{-i(\pi x/d)})$$

where $E_0(k) = (\hbar^2/2m)k^2$

$$f(k) = \frac{a}{\left(1 - \dfrac{2kd}{\pi}\right) + \sqrt{\left(1 - \dfrac{2kd}{\pi}\right)^2 + a^2}}$$

and the delocalization parameter $a = [V_0/(\hbar^2\pi^2/2md^2)] = V_0/4w_F$, where w_F is the Fermi energy. The interband transition energy $E_{cv} = E_c(k) - E_v(k)$ has its maximum value $\simeq \hbar^2\pi^2/2md^2$ at $k = 0$ and decreases to the finite value V_0 at the edge of the Brillouin zone $k = \pi/2d$. At this edge the joint density of states per unit length

$$\rho(E_{cv}) = \frac{1}{2\pi}\left(\frac{dE_{cv}}{dk}\right)^{-1} \propto \frac{1}{\sqrt{E_{cv}^2 - V_0^2}} \tag{IV.1}$$

exhibits a singularity, characteristic of one-dimensional systems.

In the case of bands, the straightforward application of the formulae of Section II leads to difficulties connected with the application of perturbation theory to a continuum of states. One can circumvent this by using the approach of Genking and Mednis[47]. The following expressions are obtained for the low frequency limit of the susceptibilities per unit length $\kappa^{(1)}$ and $\kappa^{(3)}$ (for one chain):

$$\kappa^{(1)} = \frac{2e^2}{\pi}\int_{-\pi/2d}^{+\pi/2d} dk\, \frac{\Omega_{vc}\Omega_{cv}}{E_{cv}} \tag{IV.2}$$

$$\kappa^{(3)} = \frac{4e^4}{\pi}\int_{-\pi/2d}^{+\pi/2d} dk\left(\frac{1}{E_{cv}}\left|\frac{d}{dk}\left(\frac{\Omega_{cv}}{E_{cv}}\right)\right|^2 - \frac{\Omega_{cv}^4}{E_{cv}^3}\right) \tag{IV.3}$$

where

$$\Omega_{rs} = \int_0^{2d} u_r^* \frac{\partial}{\partial k} u_s\, dx$$

and the result $\Omega_{cc} = \Omega_{vv} = 0$ has been taken into account.

For $a < 1$, this gives, for the linear susceptibility

$$\kappa^{(1)} = \frac{4}{3\pi^2}\frac{e^2 d}{aV_0} \tag{IV.4}$$

The two terms under the integral sign in Equation (IV.3) correspond respectively to the intra and inter-band contributions to $\kappa^{(3)}$. They are of opposite sign, as already noted by Van Vechten *et al.*[48]. Again for $a < 1$, a straightforward integration gives:

$$\kappa^{(3)} = \frac{2^{14}}{45\pi^4}\frac{e^4 d^3}{w_F^3}\left(\frac{w_F}{V_0}\right)^6 \tag{IV.5}$$

or equivalently

$$\kappa^{(3)} = \frac{2^5}{45}\pi^2\left(\frac{a_0}{d}\right)^3 \frac{e^{10}}{V_0^6} \tag{IV.6}$$

The *positive* sign of $\kappa^{(3)}$ is due to the stronger intraband contribution. This contrasts with the result obtained for a localized two-level system.

The salient feature in these results is the strong enhancement of $\kappa^{(3)}$ with respect to a system of localized electrons. This enhancement goes as $(w_F/V_0)^6$ and can reach an extremely large value for systems having a small bond alternance. This dependence of $\kappa^{(3)}$ on V_0^{-6} can be traced directly to the variation of the integrand in Equation (IV.3), which is strongly concentrated in a range $\simeq a(\pi/2d)$ around the limits of the Brillouin zone. In this range the matrix element $\Omega_{cv} = (1/1 + f^2)(df/dk)$ has a value $(d/\pi)(4w_F/V_0)$ which is enhanced by delocalization. Van Vechten *et al.*[48] have studied the third-order susceptibility $\chi^{(3)}$ of a three-dimensional semiconductor. They use a three part model band structure[49] and find that one of the contributions coming from the Penn "threshold" E_G[50] is proportional to E_G^{-6}. A similar result is obtained in a direct study[51] of the three-dimensional Penn's model. This is not surprising as Penn's model postulates expressions for the energy and wave functions which are identical to those of the present one-dimensional model. Two points should be noted, however. First, whereas in Penn's model E_G has the meaning of an average interband energy, V_0 is the lowest interband energy and coincides with the optical gap. In germanium $E_G \simeq 4.5$ eV whereas the optical gap E_0 lies around 0.6 eV. This is of practical importance. Second, the representation of the E_2 structure by a Penn threshold is only approximate and it is not obvious that the features which in Penn's model give rise to the E_G^{-6} dependence are all reproduced in three-dimensional semiconductors.

The solid state polymerization of diacetylenes R—C≡C—C≡C—R leads to a highly conjugated structure exhibiting infinite linear chains[44]. Upon thermal annealing or γ irradiation each monomer molecule joins with two neighbouring molecules (Fig. 8). The electron redistribution leads to an infinite conjugated chain similar to that of polyene. Polymer single crystals of large size and good optical quality can be obtained. The third-order susceptibilities of compounds of this family have been measured in our laboratory[52] using third harmonic generation from a Raman-shifted dye laser at 2.62 μ. In the case of TCDU where $R = (CH_2)_6OCONHC_6H_5$ we studied the variation of $\chi^{(3)}$ under polymerization. The colourless monomer single crystals exhibit a normal nonlinearity, only slightly higher than predicted by Miller's rule with $\chi^{(3)}$ components $\simeq 1 \times 10^{-13}$ e.s.u. In strong contrast the yellow polymer crystals obtained upon γ-ray irradiation showed

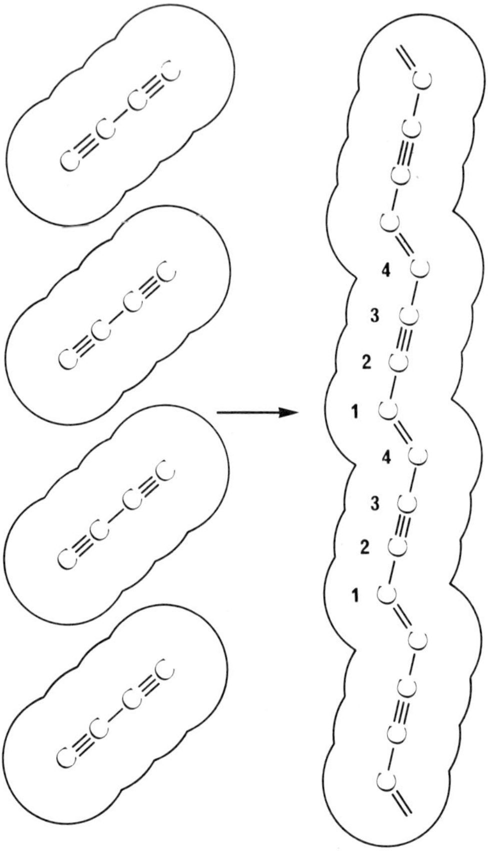

FIG. 8. Polymerization of diacetylenes (after Baughman and Yee).

a strong highly anisotropic nonlinearity with the longitudinal component along the chain $\chi^{(3)}_{xxxx} = 3.7 \times 10^{-11}$ e.s.u. dominating all other components by more than two orders of magnitude. Poly-PTS which has a slightly lower gap than poly-TCDU shows an even larger nonlinearity with a low frequency value of the component along the chain $\chi^{(3)}_{xxxx} = 1.6 \times 10^{-10}$ e.s.u.

Neglecting local field corrections, the expression for $\chi^{(3)}$ deduced from Equation (IV.6) is:

$$\chi^{(3)} = \frac{2^5}{45}\frac{\pi^2}{\sigma}\left(\frac{a_0}{d}\right)^3 \frac{e^{10}}{E_g^6} \qquad \text{(IV.7)}$$

where σ is the cross-sectional area per chain and V_0 has been identified to the energy gap E_g. Using the values $d = 1.35$ Å, $\sigma = 118$ Å^2, $E_g = 17\,900$ cm^{-1}

for poly-TCDU and $d = 1.35$ Å, $\sigma = 95$ Å^2, $E_g = 15\,800$ cm^{-1} for poly-PTS in Equation (IV.7) one obtains:

$$\chi^{(3)}_{zzzz}(\text{TCDU}) = 1.2 \times 10^{-11} \text{ e.s.u.}$$

$$\chi^{(3)}_{zzzz}(\text{PTS}) = 3.0 \times 10^{-11} \text{ e.s.u.}$$

This gives a good order of magnitude agreement. Obviously one should not expect more in view of the simplicity of the model which neglects the supermodulation of bond lengths which takes place in polydiacetylenes. Whereas in polyenes a simple alternance occurs, the situation is more complicated here. Owing to the presence of an additional π-bond between atoms 2 and 3 the corresponding interatomic distance is shortened with respect to that between atoms 4 and 1 (Fig. 8). Additional Fourier components should be introduced in the pseudopotential to take this into account. A treatment of this problem by the Hückel method can be found in Ref. [51]. Also the local field correction has been omitted: although difficult to estimate it should not be very large for highly delocalized electrons.

The high values of $\chi^{(3)}$, comparable to those observed in semiconductors, obtained in these polydiacetylenes are specially remarkable in view of the small relative number of π-electrons: roughly 1 out of 40 valence electrons takes part in the conjugated chain. They produce a comparatively small infrared anisotropy. However they completely dominate the nonlinear properties as could be expected from Equation (IV.4) and (IV.5) which predict a much stronger dependence on delocalization for $\kappa^{(3)}$ than for $\kappa^{(1)}$. The small density of π-electrons is due to the bulky side groups

$$R = (CH_2)_4OCONHC_6H_5.$$

Proper choice of these groups could reduce the chain cross-section: this with a lowering of the optical gap could lead to very large values of $\chi^{(3)}$. From a practical point of view the moderate values of the refractive index are of importance. It should be noted that for nonlinear interactions the important quantity is not $\chi^{(3)}$ but the ratio $\chi^{(3)}/n^2$. This factor of merit is larger for poly-PTS than for Ge. Furthermore the transmission ranges are very different for the two compounds. Whereas Ge has an optical gap of 0.6 eV, poly-PTS presents a very steep absorption edge (characteristic of one-dimensional compounds) at 2 eV. A last favourable point can be mentioned. In the picosecond regime, the damage threshold measured at 1.06 μ, is in the range of 50 GW/cm^2, about 20 times the corresponding value for Ge.

Further systematic studies of semiconducting one-dimensional compounds should uncover very large nonlinearities. The large variety of systems which are available or can be designed offers attractive prospects. The considerations developed here could be extended to the case when resonances take place. Large multiphoton cross-sections are expected. The sharp edges and peaks observed in the linear properties of one-dimensional compounds[55] should also appear in the dispersion of $\chi^{(3)}$. In order to underline the essential physical ideas we have limited ourselves to simple centrosymmetric systems. It would be interesting to study $\chi^{(2)}$ in one-dimensional non-centrosymmetric chains such as asymmetrically substituted carotenoides along the lines followed by Levine[53] and Oudar and Chemla[54] for two-dimensional systems. This raises the problem of the influences of asymmetry on electron delocalization.

References

1. P. N. Butcher. "Nonlinear Optical Phenomena", Ohio State University, Engineering Publications, Columbus, Ohio (1965).
2. J. Ducuing and C. Flytzanis. *In* "Optical Properties of Solids" (F. Abeles, Ed.), North Holland, Amsterdam (1972).
3. J. A. Armstrong, N. Bloembergen, J. Ducuing and P. S. Persham. *Phys. Rev.* **127**, 1918 (1962).
4. P. P. Bey and H. Rabin. *Phys. Rev.* **162**, 794 (1967).
5. H. A. Lorentz. "The Theory of Electrons", Teubner, Leipzig (1916).
6. C. G. Darwin. *Proc. Roy. Soc.* (*London*) **146A**, 17 (1934).
7. S. Adler. *Phys. Rev.* **126**, 413 (1962).
8. A. S. Pine. *Phys. Rev.* **139**, **A901** (1966).
9. D. Bedeaux and N. Bloembergen. *Physica* **69**, 57 (1973).
10. A. Abragam. "The Principles of Nuclear Magnetism", Oxford (1961).
11. C. P. Slichter. "Principles of Magnetic Resonance", Harper and Row (1963).
12. M. Goeppert Mayer. *Ann. Phys.* (*Leipzig*) **9**, 273 (1931).
13. T. Fivtak. *Can. J. Phys.* **41**, 12 (1963).
14. D. A. Kleinman. *Phys. Rev.* **126**, 1761 (1962).
15. E. A. Manykin and A. M. Afanas'ev. *Sov. Phys. JETP.* **21**, 619 (1965).
16. S. E. Harris and R. B. Miles. *Appl. Phys. Lett.* **19**, 385 (1971). J. F. Young, G. C. Borklund, A. H. Kung, R. B. Miles and S. E. Harris. *Phys. Rev. Lett.* **27**, 1551 (1971). R. B. Miles and S. E. Harris. *IEEE J. Quantum Electron.* **9**, 470 (1973).
17. P. P. Sorokin, J. J. Wynne and J. R. Lankard. *Appl. Phys. Lett.* **22**, 342 (1973).
18. R. T. Hodgson, P. P. Sorokin and J. J. Wynne. *Phys. Rev. Lett.* **32**, 343 (1974).
19. R. C. Miller. *Appl. Phys. Lett.* **5**, 17 (1964).
20. See for instance H. R.Philipp and H. Ehrenreich. *Phys. Rev.* **129**, 1550 (1963).
21. C. Flytzanis and J. Ducuing. *Phys. Rev.* **178**, 1218 (1969).
22. See for instance G. H. Wannier. "Elements of Solid State Theory", Cambridge University Press (1959).
23. S. Jha and N. Bloembergen. *Phys. Rev.* **171**, 891 (1968).
24. C. Flytzanis and J. Ducuing. *Phys. Lett.* **26A**, 315 (1968).
25. A. Unsöld. *Ann. Phys.* (*Leipzig*) **43**, 388 (1927).

26. F. N. H. Robinson. *Bell System Tech. J.* **46**, 913 (1967).
27. A. Dalgarno and J. Lewis. *Proc. Roy. Soc. (London)* **A233**, 70 (1956).
28. J. Pople and P. Schofield. *Phil. Mag.* **2**, 591 (1957).
29. C. A. Coulson, L. Redei and D. Stocker. *Proc. Roy. Soc. (London)* **A270**, 357 (1962).
30. C. Schwartz. *Ann. Phys. (N.Y.)* **6**, 156 (1959).
31. C. Flytzanis. Thesis, Université de Paris (Orsay) 1970.
32. D. R. Penn. *Phys. Rev.* **128**, 2093 (1962).
33. J. C. Phillips. *Rev. Mod. Phys.* **42**, 317 (1970) and references quoted therein. *See also* J. C. Phillips. "Bonds and Bands in Semiconductors", Academic Press, New York and London (1973).
34. J. A. Van Vechten. *Phys. Rev.* **182**, 891 (1969); J. A. Van Vechten. *Phys. Rev.* **187**, 1007 (1969).
35. C. L. Tang and C. Flytzanis. *Phys. Rev.* **B4**, 2520 (1971).
36. J. Ducuing and C. Flytzanis. *Comptes rendus Ac. Sci. (Paris)* **266B**, 810 (1968).
37. H. C. Gatos and H. C. Levine. *J. Electrochem. Soc.* **107**, 427 (1960).
38. C. L. Tang. *IEEE J. Quantum Electron.* **QE9**, 755 (1973).
39. R. C. Miller *et al. Solid State Comm.* **9**, 1463 (1971).
40. M. Choy, S. Ciraci and R. L. Byer. *IEEE J. Quantum Electron.* **QE11**, 40 (1975).
41. B. F. Levine. *Phys. Rev. Lett.* **22**, 787 (1969). *See also* B. F. Levine. *Phys. Rev.* **B7**, 2591 (1973) and references quoted therein.
42. *See also* J. Ducuing. *In* "Nonlinear Spectroscopy, Proceedings of the International School of Physics 'Enrico Fermi', Course LXIV" (N. Bloembergen, Ed.), Academic Press, New York and London (1975).
43. I. F. Shchegolev. *Phys. Stat. Sol.* (a) **12**, 9 (1972).
44. G. Wegner. *Z. Naturforsch.* **24b**, 924 (1969); G. Wegner. *Makromol. Chem.* **134**, 219 (1970). R. H. Baughman. *J. Appl. Phys.* **43**, 4362 (1972).
45. J. C. Slater. *Phys. Rev.* **81**, 385 (1951).
46. L. Pincherle. "Electronic Energy Bonds in Solids", MacDonald, London (1971).
47. V. N. Genkin and P. M. Mednis. *Soviet Physics JETP* **27**, 609 (1968).
48. J. A. Van Vechten, M. Cardona, D. E. Aspenes and R. M. Martin. "Proceedings of the Tenth International Conference on Semiconductors", Cambridge, Mass. (1970).
49. M. Cardona, K. L. Shaklee and F. H. Pollak. *Phys. Rev.* **154**, 696 (1967).
50. D. R. Penn. *Phys. Rev.* **128**, 2093 (1962).
51. G. Agrawal and C. Flytzanis. *Chem. Phys. Lett.* **44**, 366 (1976).
52. C. Sauteret, J.-P. Hermann, R. Frey, F. Pradère, J. Ducuing, R. H. Baughman and R. R. Chance. *Phys. Rev. Lett.* **36**, 956 (1976).
53. B. F. Levine. *J. Chem. Phys.* **63**, 115 (1975); B. F. Levine. *Chem. Phys. Lett.* To be published.
54. J. L. Oudar and H. Le Person. *Opt. Comm.* **15**, 258 (1975).
55. D. Bloor, D. J. Ando, F. H. Preston and G. C. Stevens. *Chem. Phys. Lett.* **24**, 407 (1974).

2. Parametric Oscillators and Nonlinear Materials†

R. L. BYER

Department of Applied Physics, Stanford University, California, U.S.A.

I. INTRODUCTION

It is now possible with only a handful of nonlinear crystals to generate tunable coherent radiation over a very broad spectral range extending from 2200 Å in the ultraviolet to beyond 25 μm in the infrared. In this chapter the author will discuss nonlinear phenomena and devices that allow the production of coherent radiation by second harmonic generation, mixing, and parametric oscillation. Material requirements for device applications are also considered and the properties of important nonlinear materials summarized. The field has grown rapidly and a brief historical review is in order at the outset to provide perspective.

A. HISTORICAL REVIEW

In 1961, shortly after the demonstration of the laser, Franken *et al.*[1] generated the second harmonic of a Ruby laser in crystal quartz. The success of this

† Support is acknowledged of NSF-RANN, ARO, ERDA and the Sloan Foundation during the work.

experiment relied directly on the enormous increase of power spectral brightness provided by a laser source compared to incoherent sources. Power densities greater than 10^9 W/cm^2 became available; these correspond to an electric field strength of 10^6 V cm^{-1}. This field strength is comparable to atomic field strengths and, therefore, it was not too surprising that materials responded in a nonlinear manner to the applied fields.

The early work in nonlinear optics concentrated on second harmonic generation. Harmonic generation in the optical region is similar to the more familiar harmonic generation at radio frequencies, with one important exception. In the radio frequency range the wavelength is usually much larger than the harmonic generator, so that the interaction is localized in a volume much smaller than the dimensions of a wavelength. In the optical region the situation is usually reversed and the nonlinear medium extends over many wavelengths. This leads to the consideration of propagation effects since the electromagnetic wave interacts over an extended distance with the generated nonlinear polarization. The situation is similar to a propagating wave interacting with a phased linear dipole array. If this interaction is to be efficient, the phase of the propagating wave and the generated polarization must be proper. In nonlinear optics this is referred to as phasematching. For second harmonic generation, phasematching implies that the phase velocity of the fundamental and second harmonic waves are equal in the nonlinear material. Since optical materials are dispersive, it is not possible to achieve equal phase velocities in isotropic materials. Shortly after Franken *et al.*'s first relatively inefficient nonphasematched second harmonic generation experiment, Kleinman[2], Giordmaine[3], and Maker *et al.*[4], and later Akhmanov *et al.*[5] showed that phase velocity matching could be achieved in birefringent crystals by using the crystal birefringence to offset the dispersion.

Along with the important concept of phasematching, other effects leading to efficient second harmonic generation were studied. These included focussing[6,7], double refraction[8–10], and operation of second harmonic generators with an external resonator[11,12] and within a laser cavity[13,14].

An important extension of nonlinear interactions occurred in 1965 when Wang and Racette[15] observed significant gain in a three-frequency mixing experiment. The possibility of optical parametric gain had been previously considered theoretically by Kingston[16], Kroll[17], Akhmanov and Khokhlov[18], and Armstrong *et al*[19]. It remained for Giordmaine and Miller[20] in 1965 to achieve adequate parametric gain in $LiNbO_3$ to overcome losses and reach threshold for coherent oscillation. This early work led to considerable activity in the study of parametric oscillators as tunable coherent light sources.

Simultaneously with the activity in nonlinear devices, the theory of nonlinear interactions received increased attention. It was recognized quite early that progress in the field depended critically upon the availability of quality nonlinear materials. Initially, the number of phasematchable nonlinear crystals with accurately measured nonlinear coefficients was limited to a handful of previously known piezoelectric, ferroelectric, or electro-optic materials. An important step in the problem of searching for new nonlinear materials was made when Miller[21] recognized that the nonlinear susceptibility was related to the third power of the linear susceptibility by a factor now known as Miller's delta. Whereas nonlinear coefficients of materials span over four orders of magnitude, Miller's delta is constant to within 50%. To the crystal grower and nonlinear materials scientist, this simple rule allows the prediction of nonlinear coefficients based on known crystal indices of refraction and symmetry, without having to carry out the expensive and time-consuming tasks of crystal growth, accurate measurement of the birefringence to predict phasematching, orientation, and finally second harmonic generation.

The early progress in nonlinear optics has been the subject of a number of monographs (Akhmanov and Khokhlov[22], Bloembergen[23], Butcher[24], Franken and Ward[25]) and review articles (Ovander[26], Bonch-Bruevich and Khodovoi[27], Minck *et al.*[28], Pershan[28a], Akhmanov *et al.*[29], Terhune and Maker[30], Akhmanov and Khokhlov[31], and Kielich[32]). In addition, nonlinear materials have been reviewed by Suvorov and Sonin[33], Rez[34], and Hulme[34a], and a compilation of nonlinear materials is provided by Singh[35]. Two books have appeared, one a brief introduction by Baldwin[36], and the second a clearly written text by Zernike and Midwinter[37]. Finally, a text covering all aspects of nonlinear optics has recently appeared[38].

B. NONLINEAR DEVICES

The primary application of nonlinear materials is the generation of new frequencies not available with existing laser sources. The variety of applications for nonlinear optical devices is so large that the author will touch only the highlights here.

Second harmonic generation (SHG) received early attention primarily because of early theoretical understanding and its use for measuring and testing the nonlinear properties of crystals. Efficient SHG has been demonstrated using a number of materials and laser sources. In 1968 Geusic *et al.*[39] obtained efficient doubling of a continuous wave (cw) Nd:YAG laser using the crystal $Ba_2NaNb_5O_{15}$. That same year, Dowley[40] reported efficient SHG of an argon ion laser operating at 0.5145 μm in ADP. Later Hagen *et al.*[41] reported 70% doubling efficiency of a high energy Nd:glass laser in KDP (potassium dihydrogen phosphate), and Chesler *et al.*[42] reported efficient SHG of a *Q*-switch Nd:YAG laser using $LiIO_3$. An efficiently doubled *Q*-switched Nd:YAG laser is now available as a commercial laser source[43]. In addition, $LiIO_3$ has been used to efficiently double a Ruby laser[44]. Recently the 10.6 μm CO_2 laser has been doubled in Tellurium with 5% efficiency[45] and in a ternary semiconductor $CdGeAs_2$ with 15% efficiency[46].

Three-frequency nonlinear interactions include sum generation, difference-frequency generation or mixing, and parametric generation and oscillation. An interesting application of sum generation is infrared up-conversion and image up-conversion. For example, Smith and Mahr[47] report achieving a detector noise equivalent power of 10^{-14} W at 3.5 μm by up-converting to 0.447 μm in $LiNbO_3$ using an argon ion laser pump source. This detection method is being used for infrared astronomy. Numerous workers have efficiently up-converted 10.6 μm to the visible range[48–52] for detection by a photomultiplier. An extension of single beam up-conversion is image up-conversion[37,53,54]. Resolution to 300 lines has been achieved, but at a cost in up-conversion efficiency.

Combining two frequencies to generate the difference frequency by mixing was first demonstrated by Wang and Racette[15]. Zernike and Berman[55] used this approach to generate tunable far infrared radiation. Recently a number of workers have utilized mixing in proustite[56,57], CdSe[58], $ZnGeP_2$[52,52a], $AgGaS_2$[59], and recently $AgGaSe_2$[60,61] to generate tunable coherent infrared output from near infrared or visible sources.

Perhaps the most unique aspect of nonlinear interactions is the generation of coherent continuously-tunable laser-like radiation by parametric oscillation in a nonlinear crystal. Parametric oscillators were well known in the microwave region[62,63] prior to their demonstration in the optical range. To date, parametric oscillators have been tuned across the visible and near infrared in KDP[29,64,65] and ADP[66] when pumped at the second harmonic and fourth harmonic of the 1.06 μm Nd:YAG laser, and they have been tuned over the infrared range from 0.6 μm to 3.7 μm in $LiNbO_3$[67–72]. The above parametric oscillators were pumped by *Q*-switched, high peak power, laser sources. Parametric oscillators have also been operated in a cw manner in $Ba_2NaNb_5O_{15}$[73–76] and in $LiNbO_3$[77,78]. However, the low gains inherent in cw pumping have held back research in this area.

In 1969 Harris[79] reviewed the theory and devices aspects of parametric oscillators. Up to that time oscillation had been achieved in only three materials: KDP, $LiNbO_3$, and $Ba_2NaNb_5O_{15}$. Since 1969 parametric oscillation has been extended to four new materials: ADP[80], $LiIO_3$[81,82], proustite (Ag_3AsS_3)[227], and CdSe[83]. The new materials have extended the available tuning range. However, the development of oscillator devices still has remained materials-limited. At this time $LiNbO_3$ is the only nonlinear crystal used in a commercially available parametric oscillator.† Smith[84] and recently Byer[85] have discussed parametric oscillators in review papers and Byer[86] has reviewed their application to infrared spectroscopy.

Nonlinear interactions allow the extension of coherent radiation by second harmonic generation, sum generation, and difference-frequency mixing over a wavelength range from 2200 Å to beyond 1 mm in the far infrared. In addition, tunable coherent radiation can be efficiently generated from a fixed frequency pump laser source by parametric oscillation. The very wide spectral range and efficiency of nonlinear interactions assures that they will become increasingly important as coherent sources.

II. Nonlinear Phenomena

A. Introduction

When a medium is subjected to an electric field the electrons in the medium are polarized. For weak electric fields the polarization is linearly proportional to the applied field

$$\mathbf{P} = \varepsilon_0 \chi^{(1)} \mathbf{E}$$

where $\chi^{(1)}$ is the linear optical susceptibility and ε_0 is the permittivity of free space with the value 8.85×10^{-12} F/m in mks units. The linear susceptibility is related to the medium's index of refraction n by $\chi^{(1)} = n^2 - 1$.

In a crystalline medium the linear susceptibility is a tensor that obeys the symmetry properties of the crystal. Thus for isotropic media there is only one value of the index, and for uniaxial crystals two values, n_o the ordinary and n_e the extraordinary indices of refraction, and for biaxial crystals three values n_α, n_β, and n_γ.

A linear polarizability is an approximation to the complete constitutive relation which can be written as an expansion in powers of the applied field, as

$$\mathbf{P} = \varepsilon_0[\boldsymbol{\chi}^{(1)} + \boldsymbol{\chi}^{(2)}\mathbf{E} + \boldsymbol{\chi}^{(3)}\mathbf{E}^2 + \cdots]\mathbf{E}$$

where $\boldsymbol{\chi}^{(2)}$ is the second-order nonlinear susceptibility and $\boldsymbol{\chi}^{(3)}$ is the third-order nonlinear susceptibility. A number of interesting optical phenomena arise from the second- and third-order susceptibilities. For example, $\boldsymbol{\chi}^{(2)}$ gives

† Chromatix Inc., Mountain View, California.

rise to second harmonic generation[1], dc rectification[87], the linear electro-optic effect or Pockels effect[25], parametric oscillation[20], and three-frequency processes such as mixing[15] and sum generation. The third-order susceptibility is responsible for third harmonic generation[88], the quadratic electro-optic effect or Kerr effect[28], two-photon absorption[89], and Raman[90], Brillouin[91], and Rayleigh[92] scattering.

We are primarily interested in effects that arise from $\chi^{(2)}$. For a review of the nonlinear susceptibility $\chi^{(2)}$ and the resulting interactions in a nonlinear medium see Wemple and DiDomenico[93] and Ducuing and Flytzanis[93a].

To see how $\chi^{(2)}$ gives rise to second harmonic generation and other nonlinear effects, consider an applied field

$$\mathbf{E} = \mathbf{E}_1 \cos(k_1 x - \omega t) + \mathbf{E}_2 \cos(k_2 x - \omega t)$$

incident on the nonlinear medium. The nonlinear polarization is proportional to $\chi^{(2)}\mathbf{E}^2$, giving

$$\tfrac{1}{2}\chi^{(2)}\mathbf{E}_1^2[1 + \cos(2k_1 x - 2\omega_1 t)]$$

plus a similar term for frequency ω_2. This term describes both dc rectification and second harmonic generation. In addition, there are sum and mixing terms of the form

$$\chi^{(2)}\mathbf{E}_1\mathbf{E}_2[\cos\{(k_1 - k_2)x - (\omega_1 - \omega_2)t\} + \cos\{(k_1 + k_2)x - (\omega_1 + \omega_2)t\}]$$

present in the expansion. These terms describe difference-frequency and sum-frequency generation. All of the above processes take place simultaneously in the nonlinear medium. The question that naturally occurs is how one process is singled out to proceed efficiently relative to the competing processes. In nonlinear interactions phasematching selects the process of interest to the exclusion of the other possible processes. Thus, if the crystal birefringence is adjusted (by temperature or angle of propagation) such that second harmonic generation is the phasematched process, then it proceeds with relatively high efficiency compared to the remaining processes involving sum- and difference-frequency generation.

B. CRYSTAL SYMMETRY

Like the linear susceptibility, the second-order nonlinear susceptibility must display the symmetry properties of the crystal medium. An immediate consequence of this fact is that in centrosymmetric media the second order nonlinear coefficients must vanish. Thus nonlinear optical effects are restricted to acentric materials. This is the same symmetry requirement as for the piezoelectric $\tilde{\mathbf{d}}$ tensors[94] and therefore the non-zero components of the second-order susceptibility can be found by reference to the listed $\tilde{\mathbf{d}}$ tensors. However, the nonlinear coefficient tensors have been listed in a number of references[24,35,37,95].

The tensor property of $\boldsymbol{\chi}^{(2)}$ can be displayed by writing the nonlinear polarization in the form

$$P_i(t) = \varepsilon_0 \sum_{ijk} \chi_{ijk}^{(2)} E_j(t) E_k(t) \tag{1}$$

where $\boldsymbol{\chi}^{(2)}$ is the nonlinear susceptibility tensor coupling the instantaneous fields.

It is customary to define the nonlinear susceptibility tensor in terms of the Fourier components of the fields. We define the field component at frequency ω by the Fourier relation

$$\mathbf{U}(t) = \tfrac{1}{2}[\mathbf{U}(\omega) e^{i(\mathbf{k}\cdot\mathbf{r} - \omega t)} + cc].$$

Substituting for the fields and polarizations in Equation (1) we find

$$P_i(-\omega_3) = \varepsilon_0 \sum_{ijk} \frac{\chi_{ijk}^{(2)}}{2} (-\omega_3, \omega_2, \omega_1) E_j(\omega_2) E_k(\omega_1) \times \exp i[(\mathbf{k}_2 + \mathbf{k}_1 - \mathbf{k}_3)\cdot\mathbf{r}] \tag{2}$$

where we have assumed $\omega_3 = \omega_2 + \omega_1$.

In addition to crystal symmetry restrictions $\chi_{ijk}^{(2)}$ satisfies two additional symmetry relations. The first is an intrinsic symmetry relation which can be derived for a lossless medium from general energy considerations[23,96]. This relation states that $\chi_{ijk}^{(2)}(-\omega_3, \omega_2, \omega_1)$ is invariant under any permutation of the three pairs of indices $(-\omega_3, i); (\omega_2, j); (\omega_1, k)$ as was first shown by Armstrong *et al.*[19]. The second symmetry relation is based on a conjecture by Kleinman[2] that in a lossless medium the permutation of the frequencies is irrelevant and therefore $\chi_{ijk}^{(2)}$ is symmetric under any permutation of its indices.

It is customary to write the driving polarization for SHG in terms of a nonlinear tensor **d** defined by

$$P_i(\omega_3) = \varepsilon_0 \sum_{ijk} d_{ijk}(\omega_3, \omega_1, \omega_1) E_j(\omega_1) E_k(\omega_1) \exp i[(2\mathbf{k}_1 - \mathbf{k}_3)\cdot\mathbf{r}]$$

where $\omega_3 = 2\omega_1$. Comparison of Equation (2) with the above definition of d_{ijk} show that

$$\chi_{ijk}^{(2)}(-\omega_3, \omega_1, \omega_1) = 2 d_{ijk}(-\omega_3, \omega_1, \omega_1). \tag{3}$$

In addition, d_{ijk} can be written in reduced notation $d_{i(jk)} = d_{im}$ where m runs from 1–6 with the correspondence

				32	31	21
(jk)	11	22	33	23	13	12
m	1	2	3	4	5	6.

The 3 × 6 d_{im} matrix operates on a column vector $(EE)_m$ given by

$$(EE)_1 = E_x^2, \qquad (EE)_2 = E_y^2, \qquad (EE)_3 = E_z^2$$

$$(EE)_4 = 2E_yE_z, \qquad (EE)_5 = 2E_xE_z, \qquad (EE)_6 = 2E_xE_y.$$

As an example, the nonlinear **d** tensor for the $\bar{4}2m$ point group to which KDP and the chalcopyrite semiconductor crystals belong has the components

$$P_x = \varepsilon_0 2d_{14}E_yE_z; \qquad P_y = \varepsilon_0 2d_{14}E_xE_z; \qquad P_z = \varepsilon_0 2d_{36}E_xE_y.$$

However, Kleinman's symmetry conjecture states that $d_{14} = d_{123}$ equals $d_{36} = d_{312}$ since any permutation of indices is allowed. This is experimentally verified. The definition of the nonlinear susceptibility has been discussed in detail by Boyd and Kleinman[97] and by Bechmann and Kurtz[95].

For the 21 out of 32 crystal classes that lack a center of inversion the contracted form of the SHG tensors is given in Table I. The nonlinear crystal axes are chosen according to the IRE convention[99].

C. COUPLED WAVE EQUATIONS

We are now in a position to derive a coupled set of wave equations for the electromagnetic fields generated by the nonlinear driving polarization.

Our starting point is of course Maxwell's equations. In SI units

$$\nabla \times \mathbf{E} = -\frac{\partial \mathbf{B}}{\partial t} \tag{4a}$$

$$\nabla \times \mathbf{H} = \frac{\partial \mathbf{D}}{\partial t} + \mathbf{J} \tag{4b}$$

with the constitutive relations

$$\begin{aligned} \mathbf{D} &= \varepsilon_0 \mathbf{E} + \mathbf{P} \\ \mathbf{J} &= \sigma \mathbf{E} \\ \mathbf{B} &= \mu_0 \mathbf{H} \end{aligned} \tag{5}$$

TABLE I. Second-order susceptibility tensors

TRICLINIC SYSTEM

Class 1—C_1

$$\begin{pmatrix} d_{11} & d_{12} & d_{13} & d_{14} & d_{15} & d_{16} \\ d_{21} & d_{22} & d_{23} & d_{24} & d_{25} & d_{26} \\ d_{31} & d_{32} & d_{33} & d_{34} & d_{35} & d_{36} \end{pmatrix}$$

MONOCLINIC SYSTEM

Class m—C_{1h}

$$\begin{pmatrix} d_{11} & d_{12} & d_{13} & 0 & 0 & d_{16} \\ d_{21} & d_{22} & d_{23} & 0 & 0 & d_{26} \\ 0 & 0 & 0 & d_{34} & d_{35} & 0 \end{pmatrix} m \perp Z$$

Class m—C_{1h}

$$\begin{pmatrix} d_{11} & d_{12} & d_{13} & 0 & d_{15} & 0 \\ 0 & 0 & 0 & d_{24} & 0 & d_{26} \\ d_{31} & d_{32} & d_{33} & 0 & d_{35} & 0 \end{pmatrix} \begin{matrix} m \perp Y \\ \text{(IRE-convention)} \end{matrix}$$

Class 2—C_2

$$\begin{pmatrix} 0 & 0 & 0 & d_{14} & d_{15} & 0 \\ 0 & 0 & 0 & d_{24} & d_{25} & 0 \\ d_{31} & d_{32} & d_{33} & 0 & 0 & d_{36} \end{pmatrix} 2 \parallel Z$$

Class 2—C_2

$$\begin{pmatrix} 0 & 0 & 0 & d_{14} & 0 & d_{16} \\ d_{21} & d_{22} & d_{23} & 0 & d_{25} & 0 \\ 0 & 0 & 0 & d_{34} & 0 & d_{36} \end{pmatrix} \begin{matrix} 2 \parallel Y \\ \text{(IRE-convention)} \end{matrix}$$

ORTHORHOMBIC SYSTEM

Class mm2—C_{2v}

$$\begin{pmatrix} 0 & 0 & 0 & 0 & d_{15} & 0 \\ 0 & 0 & 0 & d_{24} & 0 & 0 \\ d_{31} & d_{32} & d_{33} & 0 & 0 & 0 \end{pmatrix}$$

Class 222—D_2

$$\begin{pmatrix} 0 & 0 & 0 & d_{14} & 0 & 0 \\ 0 & 0 & 0 & 0 & d_{25} & 0 \\ 0 & 0 & 0 & 0 & 0 & d_{36} \end{pmatrix}$$

TETRAGONAL SYSTEM

Class 4—C_4

$$\begin{pmatrix} 0 & 0 & 0 & d_{14} & d_{15} & 0 \\ 0 & 0 & 0 & d_{15} & -d_{14} & 0 \\ d_{31} & d_{31} & d_{33} & 0 & 0 & 0 \end{pmatrix}$$

(Continued)

TABLE I. (*Continued*)

Class $\bar{4}$—S_4

$$\begin{pmatrix} 0 & 0 & 0 & d_{14} & d_{15} & 0 \\ 0 & 0 & 0 & -d_{15} & d_{14} & 0 \\ d_{31} & -d_{31} & 0 & 0 & 0 & d_{36} \end{pmatrix}$$

Class 4mm—C_{4v}

$$\begin{pmatrix} 0 & 0 & 0 & 0 & d_{15} & 0 \\ 0 & 0 & 0 & d_{15} & 0 & 0 \\ 0 & 0 & 0 & 0 & 0 & d_{36} \end{pmatrix}$$

Class $\bar{4}2m$—D_{2d}

$$\begin{pmatrix} 0 & 0 & 0 & d_{14} & 0 & 0 \\ 0 & 0 & 0 & 0 & d_{14} & 0 \\ 0 & 0 & 0 & 0 & 0 & d_{36} \end{pmatrix}$$

Class 422—D_4

$$\begin{pmatrix} 0 & 0 & 0 & d_{14} & 0 & 0 \\ 0 & 0 & 0 & 0 & -d_{14} & 0 \\ 0 & 0 & 0 & 0 & 0 & 0 \end{pmatrix}$$

TRIGONAL SYSTEM

Class 3—C_3

$$\begin{pmatrix} d_{11} & -d_{11} & 0 & d_{14} & d_{15} & -d_{22} \\ -d_{22} & d_{22} & 0 & d_{15} & -d_{14} & -d_{11} \\ d_{31} & d_{31} & d_{33} & 0 & 0 & 0 \end{pmatrix}$$

Class 3m—C_{3v}

$$\begin{pmatrix} 0 & 0 & 0 & 0 & d_{15} & -d_{22} \\ -d_{22} & d_{22} & 0 & d_{15} & 0 & 0 \\ d_{31} & d_{31} & d_{33} & 0 & 0 & 0 \end{pmatrix} \quad m \perp X \text{ (IRE-convention)}$$

Class 3m—C_{3v}

$$\begin{pmatrix} d_{11} & -d_{11} & 0 & 0 & d_{15} & 0 \\ 0 & 0 & 0 & d_{15} & 0 & -d_{11} \\ d_{31} & d_{31} & d_{33} & 0 & 0 & 0 \end{pmatrix} \quad m \perp Y$$

Class 32—D_3

$$\begin{pmatrix} d_{11} & -d_{11} & 0 & d_{14} & 0 & 0 \\ 0 & 0 & 0 & 0 & -d_{14} & -d_{11} \\ 0 & 0 & 0 & 0 & 0 & 0 \end{pmatrix}$$

HEXAGONAL SYSTEM

Class $\bar{6}$—C_{3h}

$$\begin{pmatrix} d_{11} & -d_{11} & 0 & 0 & 0 & -d_{22} \\ -d_{22} & d_{22} & 0 & 0 & 0 & -d_{11} \\ 0 & 0 & 0 & 0 & 0 & 0 \end{pmatrix}$$

TABLE I. (*Continued*)

Class 6—C_6

$$\begin{pmatrix} 0 & 0 & 0 & d_{14} & d_{15} & 0 \\ 0 & 0 & 0 & d_{15} & -d_{14} & 0 \\ d_{31} & d_{31} & d_{33} & 0 & 0 & 0 \end{pmatrix}$$

Same as Class 4—C_4

Class $\bar{6}m2$—D_{3h}

$$\begin{pmatrix} 0 & 0 & 0 & 0 & 0 & -d_{22} \\ -d_{22} & d_{22} & 0 & 0 & 0 & 0 \\ 0 & 0 & 0 & 0 & 0 & 0 \end{pmatrix} \quad m \perp X \text{ (IRE-convention)}$$

Class $\bar{6}m2$—D_{3h}

$$\begin{pmatrix} d_{11} & -d_{11} & 0 & 0 & 0 & 0 \\ 0 & 0 & 0 & 0 & 0 & -d_{11} \\ 0 & 0 & 0 & 0 & 0 & 0 \end{pmatrix} \quad m \perp Y$$

Class 6mm—C_{6v}

$$\begin{pmatrix} 0 & 0 & 0 & 0 & d_{15} & 0 \\ 0 & 0 & 0 & d_{15} & 0 & 0 \\ d_{31} & d_{31} & d_{33} & 0 & 0 & 0 \end{pmatrix}$$

Same as Class 4mm—C_{4v}

Class 622—D_6

$$\begin{pmatrix} 0 & 0 & 0 & d_{14} & 0 & 0 \\ 0 & 0 & 0 & 0 & -d_{14} & 0 \\ 0 & 0 & 0 & 0 & 0 & 0 \end{pmatrix}$$

Same as Class 422—D_4

CUBIC SYSTEM

Class 23—T

$$\begin{pmatrix} 0 & 0 & 0 & d_{14} & 0 & 0 \\ 0 & 0 & 0 & 0 & d_{14} & 0 \\ 0 & 0 & 0 & 0 & 0 & d_{14} \end{pmatrix}$$

Class $\bar{4}3m$—T_4

$$\begin{pmatrix} 0 & 0 & 0 & d_{14} & 0 & 0 \\ 0 & 0 & 0 & 0 & d_{14} & 0 \\ 0 & 0 & 0 & 0 & 0 & d_{14} \end{pmatrix}$$

Class 432—O

All elements vanish

Taking the curl of Equation (4a) and assuming a nonmagnetic medium we have

$$\nabla \times (\nabla \times \mathbf{E}) = -\frac{\partial}{\partial t}(\nabla \times \mathbf{B})$$

$$= -\mu_0 \frac{\partial}{\partial t}(\nabla \times \mathbf{H})$$

$$= -\mu_0 \frac{\partial}{\partial t}\left(\frac{\partial \mathbf{D}}{\partial t} + \mathbf{J}\right)$$

Rewriting the left-hand side and noting that $\nabla \cdot \mathbf{E} = 0$ we find

$$\nabla \times (\nabla \times \mathbf{E}) = \nabla(\nabla \cdot \mathbf{E}) - \nabla^2 \mathbf{E} = -\mu_0 \frac{\partial}{\partial t}\left(\frac{\partial \mathbf{D}}{\partial t} + \mathbf{J}\right)$$

or

$$\nabla^2 \mathbf{E} - \mu_0 \sigma \frac{\partial \mathbf{E}}{\partial t} - \mu_0 \varepsilon_0 \frac{\partial^2 \mathbf{E}}{\partial t^2} = \mu_0 \frac{\partial^2 \mathbf{P}}{\partial t^2} \tag{6}$$

This wave equation describes the electric field in the medium, generated by the driving polarization **P**. The equation is general and applies to all processes that can be described in terms of a polarization, including laser action, acousto-optic interaction, Brillouin and Raman scattering, and, of course, nonlinear optical interactions.

The fields in Equation (6) are the instantaneous fields. As described above, it is customary to define the nonlinear susceptibility not in terms of the instantaneous fields but their Fourier components defined by

$$\mathbf{E}(\mathbf{r}, t) = \tfrac{1}{2}[\mathbf{E}(\mathbf{r}, \omega)\exp i(\mathbf{k} \cdot \mathbf{r} - \omega t) + cc]$$

and (7)

$$\mathbf{P}(\mathbf{r}, t) = \tfrac{1}{2}[\mathbf{P}(\mathbf{r}, \omega)\exp i(\mathbf{k} \cdot \mathbf{r} - \omega t) + cc]$$

We now assume propagation in the z-direction (not related to any crystal axis) to reduce Equation (6) to a one-dimensional equation. Substituting Equation (7) for the electric field and polarization into Equation (6) we find upon differentiation that

$$\frac{1}{2}\frac{\partial^2 E}{\partial z^2} + ik\frac{\partial E}{\partial z} - \frac{\mu_0 \sigma}{2}\frac{\partial E}{\partial t} + \frac{i\omega\mu_0\sigma}{2}E - \frac{\mu_0 \varepsilon_0}{2}\frac{\partial^2 E}{\partial t^2} + i\omega\mu_0\varepsilon_0\frac{\partial E}{\partial t}$$

$$= \frac{\mu_0}{2}\frac{\partial^2 P}{\partial t^2} - i\omega\mu_0\frac{\partial P}{\partial t} - \frac{\omega^2 \mu_0}{2}P$$

where we have already cancelled two terms. In the slowly-varying-amplitude approximation

$$\omega^2 P \gg \omega \frac{\partial P}{\partial t} \gg \frac{\partial^2 P}{\partial t^2}$$

and

$$k \frac{\partial E}{\partial z} \gg \frac{\partial^2 E}{\partial z^2}$$

$$\omega E \gg \frac{\partial E}{\partial t}$$

With these approximations the wave equation reduces to

$$ik \frac{\partial E}{\partial z} + \frac{i\omega\mu_0 \sigma}{2} E + i\omega\mu_0 \varepsilon_0 \frac{\partial E}{\partial t} = -\frac{\omega^2 \mu_0}{2} P$$

Dividing by ik and noting that $k = n\omega/c$ and $\mu_0 \varepsilon_0 = 1/c^2$ we find that

$$\frac{\partial E}{\partial z} + \alpha E + \frac{1}{c}\frac{\partial E}{\partial t} = \frac{i\mu_0 c\omega}{2n} P \tag{8}$$

where $\alpha = \mu_0 \sigma c/2$ is the electric-field loss coefficient. This equation relates the envelopes of the electric field and driving polarization. We are now in a position to derive the coupled wave equations driven by the nonlinear polarization. The procedure is to find the effective driving polarization for each frequency in the interaction and to substitute into Equation (8) to derive the field equations. In the next section we apply this procedure successively to second harmonic generation, sum generation, mixing and parametric generation.

III. NONLINEAR INTERACTIONS

A. SECOND HARMONIC GENERATION

In writing the nonlinear polarization in Equation (1) we included the frequency and spatial properties of the field. For second harmonic generation with $\omega_3 = 2\omega$ and $\omega_1 = \omega_2 = \omega$ the polarization reduces to

$$P(2\omega) = \varepsilon_0 d_{\text{eff}} E(\omega)E(\omega)\exp[i(2\mathbf{k}_\omega - \mathbf{k}_{2\omega}) \cdot \mathbf{r}]$$

and (9)

$$P(\omega) = 2\varepsilon_0 d_{\text{eff}} E(2\omega)E^*(\omega)\exp[i(\mathbf{k}_{2\omega} - 2\mathbf{k}_\omega) \cdot \mathbf{r}]$$

where d_{eff} is the effective nonlinear coefficient for the interaction dependent on crystal symmetric and propagation direction in the medium. We will delay discussion of d_{eff} until later.

Substituting the driving polarizations $P(2\omega)$ and $P(\omega)$ into Equation (8) and assuming steady state we find

$$\frac{dE(\omega)}{dz} + \alpha_{\omega} E(\omega) = i\kappa E(2\omega)E^*(\omega)e^{i\Delta kz} \tag{11a}$$

$$\frac{dE(2\omega)}{dz} + \alpha_{2\omega} E(2\omega) = i\kappa E(\omega)E(\omega)e^{-i\Delta kz} \tag{11b}$$

where

$$\Delta \mathbf{k} = \mathbf{k}_{2\omega} - 2\mathbf{k}_{\omega}$$

and

$$\kappa = \frac{\omega d_{\text{eff}}}{nc}$$

When $\Delta k = 0$ (phase velocity matching) $n_{\omega} = n_{2\omega}$ since $|\mathbf{k}| = 2\pi n/\lambda$, where n is the index of refraction of the crystal. Equations (11a) and (11b) have been solved exactly[19]. We will proceed by solving the equations in a low conversion efficiency limit and then consider the high conversion efficiency case. We will simplify the analysis by assuming a lossless interaction.

(a) *Low Conversion-Efficiency Limit*

In the low conversion limit, the fundamental wave is constant with distance. Therefore, we set $dE(\omega)/dz = 0$ and integrate Equation (11b)

$$\int_0^{E(z=l)} dE(2\omega) = \int_{-l/2}^{l/2} i\kappa E^2(\omega)\exp(-i\Delta kz)dz$$

which gives

$$E(2\omega)|_{z=l} = i\kappa E^2(\omega)\frac{[\exp(i\Delta kl/2) - \exp(i\Delta kl/2)]}{i\Delta k}$$

$$= \kappa E^2(\omega)l\,\frac{\sin(\Delta kl/2)}{(\Delta kl/2)}$$

If we denote $(\sin x)/x$ by sinc x, note that the intensity is given by

$$I = \frac{nc\varepsilon_0}{2}|\mathbf{E}|^2$$

then in the low conversion limit, the conversion efficiency is

$$\frac{I(2\omega)}{I(\omega)} = \Gamma^2 l^2 \operatorname{sinc}^2\left(\frac{\Delta k l}{2}\right) \tag{12}$$

where

$$\Gamma^2 l^2 = \kappa^2 |E(\omega)|^2 l^2 = \frac{2\omega^2 |d_{\text{eff}}|^2 l^2 I(\omega)}{n^3 c^3 \varepsilon_0} \tag{13}$$

This example shows that phasematching enters into the nonlinear conversion process through the phase synchronism factor $\operatorname{sinc}^2(\Delta kl/2)$ which is unity at $\Delta kl = 0$. Also, the second harmonic conversion efficiency is proportional to $|d|^2$ and l^2, as expected, and varies as the fundamental intensity. The above result holds in the plane-wave focussing limit where $I = P/A$ and the area $A = \pi w_0^2/2$ with w_0 the Gaussian beam electric field radius.

(b) *High Conversion-Efficiency Limit*

Again we start with Equations (11a) and (11b) assuming a lossless interaction. In addition, we assume exact phasematching so that $\Delta kz = 0$. If we multiply Equation (11a) by $E^*(\omega)$ and Equation (11b) by $E^*(2\omega)$ then

$$E^*(\omega)\frac{dE(\omega)}{dz} = i\kappa E^*(\omega)E^*(\omega)E(2\omega) \tag{14a}$$

$$E^*(2\omega)\frac{dE(2\omega)}{dz} = i\kappa E(\omega)E(\omega)E^*(2\omega) \tag{14b}$$

Taking the complex conjugate of Equation (14b) and adding it to Equation (14a) we find

$$\frac{d}{dz}(|E(\omega)|^2 + |E(2\omega)|^2) = 0$$

or

$$|E(\omega)|^2 + |E(2\omega)|^2 = \text{const} = |E_{\text{inc}}|^2 \tag{15}$$

Returning now to Equation (11b) and letting

$$E(2\omega) = \mathscr{E}_{2\omega} e^{i\varphi_{2\omega}}$$
$$E(\omega) = \mathscr{E}_{\omega} e^{i\varphi_{\omega}}$$

where $\mathscr{E}_{2\omega}$ and $\mathscr{E}_{\omega}$ are real, we note that Equation (11b) becomes

$$\frac{d\mathscr{E}_{2\omega}}{dz} = \kappa \mathscr{E}_{\omega}\mathscr{E}_{\omega}[ie^{i2\varphi_{\omega} - i\varphi_{2\omega}}]$$

For growth of the second harmonic field we let $-i = e^{2i\varphi_\omega - i\varphi_{2\omega}}$ so that the coupled equations describing the real field amplitudes become

$$\frac{d\mathscr{E}_\omega}{dz} = -\kappa \mathscr{E}_\omega \mathscr{E}_{2\omega} \tag{16a}$$

$$\frac{d\mathscr{E}_{2\omega}}{dz} = \kappa \mathscr{E}_\omega \mathscr{E}_\omega \tag{16b}$$

Substituting from Equation (15) into Equation (16b) and carrying out the integration we find

$$\begin{aligned}
\frac{d\mathscr{E}_{2\omega}}{dz} &= \kappa \mathscr{E}_\omega \mathscr{E}_\omega = \kappa(\mathscr{E}_{\text{inc}}^2 - \mathscr{E}_{2\omega}^2) \\
\int_0^{\mathscr{E}_{2\omega}(l)} \frac{d\mathscr{E}'_{2\omega}}{\mathscr{E}_{\text{inc}}^2 - \mathscr{E}_{2\omega}^2} &= \int_0^l \kappa\, dz \\
\frac{1}{\mathscr{E}_{\text{inc}}} \tanh^{-1}\left[\frac{\mathscr{E}_{2\omega}(l)}{\mathscr{E}_{\text{inc}}}\right] &= \kappa l \\
\mathscr{E}_{2\omega} &= \mathscr{E}_{\text{inc}} \tanh(\mathscr{E}_{\text{inc}} \kappa l) \\
I(2\omega) &= I_{\text{inc}} \tanh^2(\Gamma l)
\end{aligned} \tag{17}$$

By conservation of energy (Equation (15)) and the trigonometric relation $\tanh^2 + \operatorname{sech}^2 = 1$ we find for the fundamental field that $I(\omega) = I_{\text{inc}} \operatorname{sech}^2 \Gamma l$ as expected. In theory SHG should proceed toward 100% conversion efficiency for plane-wave interactions. In practice Gaussian intensity laser beams are used instead of plane waves and peak power conversion efficiencies are limited to near 60%. Finally, note that at small incident fields such that $\Gamma l < 1$, $\tanh \Gamma l \approx \Gamma l$ and Equation (17) reduces to the previously derived low conversion efficiency result given by Equation (12).

B. PHASEMATCHING AND EFFECTIVE NONLINEAR COEFFICIENT

This is a good place in the discussion to answer some previously apparent questions concerning both phasematching and the effective nonlinear coefficient. We will discuss these aspects of nonlinear interactions as they apply to SHG, but the results are easily extended to the three-frequency process to be discussed later.

Our derivation of the SHG conversion efficiency (Equation (12)) showed that the efficiency is proportional to d_{eff}^2 and to l^2. Thus we should like to carry out the process in a nonlinear crystal to maximise these factors. We will first give our attention to the interaction length which is governed by the phase synchronism factor sinc($\Delta kl/2$). When

$$\Delta kl = \pi$$

the interaction is said to have occurred over one coherence length. Using the definition of $\Delta k = k_{2\omega} - 2k_{\omega}$ for SHG and solving for the coherence length we find

$$l_{\text{coh}} = \frac{\lambda_{\omega}}{4(n_{2\omega} - n_{\omega})} \tag{18}$$

For typical non-birefringent crystals (GaAs for example), l_{coh} varies between 10 μm to 100 μm. Since the SHG efficiency varies as the interaction length squared, non-phasematched $\Delta kl \neq 0$ interactions are between 10^{-6}–10^{-4} times as efficient as phasematched interactions over a 1 cm length crystal.

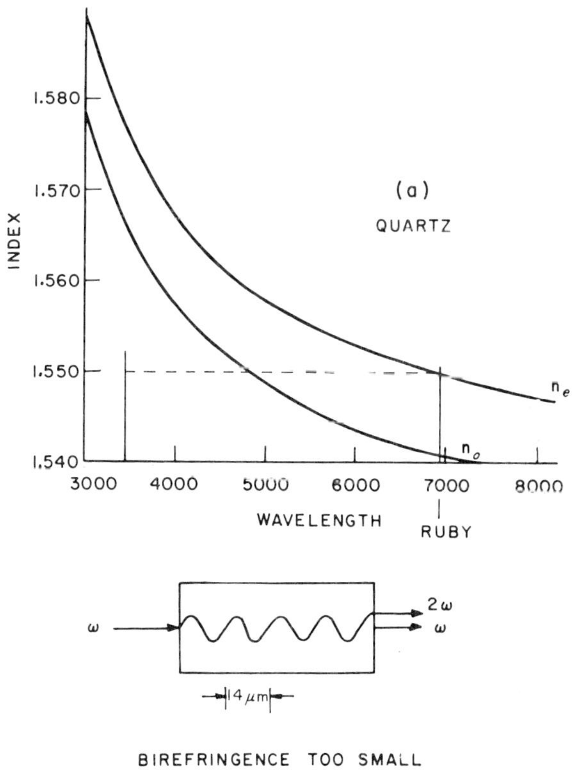

FIG. 1. (a) Phasematching for SHG of ruby laser in crystal quartz which lacks adequate birefringence.

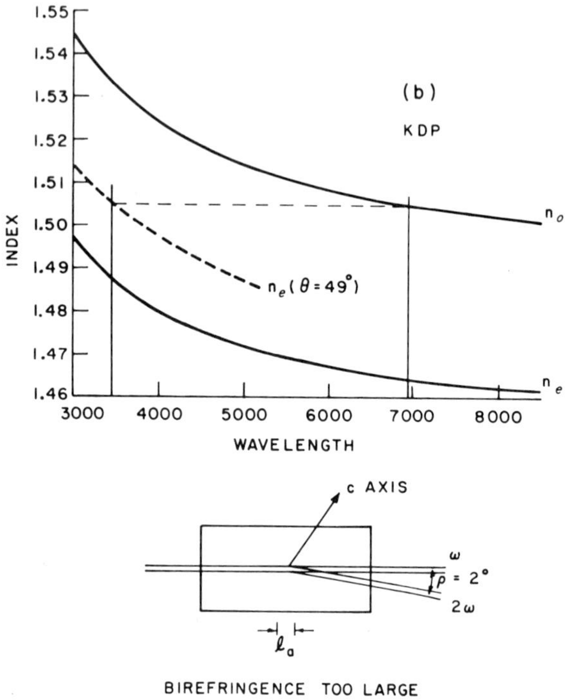

FIG. 1. (b) Phasematching for SHG of ruby laser in KDP which has more than adequate birefringence.

Thus achieving phasematching is of utmost importance for efficient SHG.

Since most materials are positively dispersive over the frequency range of interest, typically $n_{2\omega} > n_\omega$ unless special steps are taken. The special step most often taken is to use a birefringent crystal to offset the effect of dispersion[2–5]. However, other methods such as inverting the crystal about its symmetry axis every coherence length or using waveguide dispersion may also be used to achieve phasematching.

Figures 1(a), (b), and (c) illustrate the phasematching process in quartz which lacks sufficient birefringence, KDP(KH_2PO_4) which has more than adequate birefringence and $LiNbO_3$ which by an adjustment in crystal temperature has exactly the right amount of birefringence to offset dispersion.

In crystal quartz the generated SHG has the proper phase for $l_{coh} = 14\ \mu m$ and then is out of phase and converts back to the fundamental frequency. In KDP a propagation angle θ to the optic axis can be chosen such that $n_{2\omega}^e(\theta) = n_\omega^o$ and phasematching is achieved. The angle is $\theta = 49^\circ$ for SHG of 6943 Å

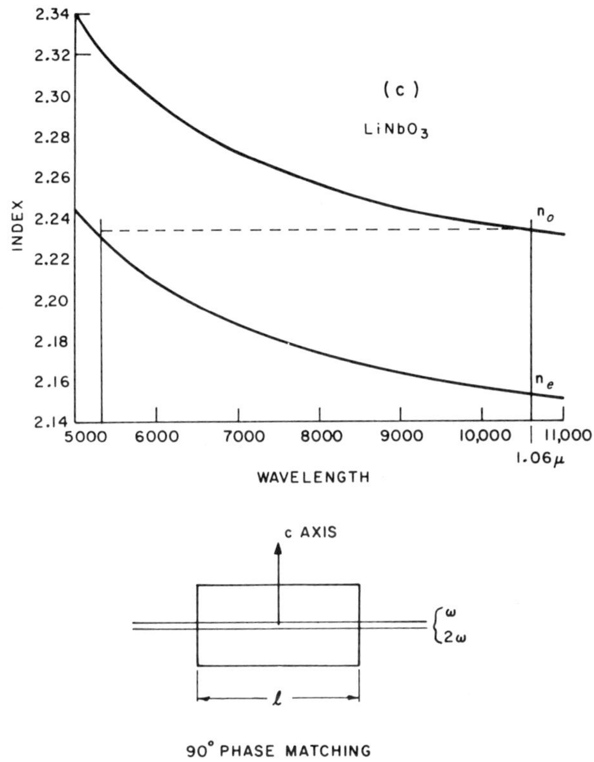

FIG. 1. (c) Phasematching for SHG of Nd:YAG laser in $LiNbO_3$ which through temperature adjustment has exactly the right amount of birefringence to offset dispersion.

ruby light. Note that the fundamental wave is polarized ordinary and the generated SH wave is extraordinary to take advantage of the birefringence.

For wave propagation in birefringent media, an incident wave of arbitrary polarization is resolved into two polarization components upon entering the media. The polarizations are orthogonal and in uniaxial crystals are resolved into ordinary and extraordinary components. The ordinary wave propagates with an index of refraction n^o which is independent of propagation direction. The extraordinary wave (polarized in the plane containing the optic axis and the propagation direction) has an index of refraction which is direction-dependent given by

$$\frac{1}{n^e(\theta)} = \left(\frac{\sin^2\theta}{n_e^2} + \frac{\cos^2\theta}{n_o^2}\right)^{1/2} \tag{19}$$

The extraordinary index may be either greater than or less than the ordinary index (positive or negative birefringence) and equals the ordinary index at $\theta = 0$ (propagation along the optic axis). In biaxial crystals the situation is more complex since the wave may propagate as n_α, n_β, or n_γ. However, Hobden[100] has given a complete description of the phasematching process including a careful definition of the three indices of refraction.

Unless the crystal birefringence is exactly correct such as the case with $LiNbO_3$ shown in Fig. 1(c), the propagation direction in the crystal must be chosen at a phasematching angle θ such that $n^e_{2\omega}(\theta) = n^o_\omega$ for a negative birefringent crystal or $n^o_{2\omega} = n^e_\omega(\theta)$ for a positive birefringent crystal. This is referred to as type I phasematching and for $\theta \neq 90^\circ$ as critical phasematching for reasons to be discussed. The case of $LiNbO_3$ where $\theta = 90^\circ$ achieves phasematching is called non-critical phasematching or 90° phasematching and offers some advantages in acceptance angle and effective nonlinear coefficient over the critical phasematching case.

There is also the possibility of achieving phasematching by averaging the birefringence. This is called type II phasematching and for negative birefringent crystals obeys the relation $n^e_{2\omega}(\theta) = \frac{1}{2}[n^e_\omega(\theta) + n^o_\omega]$. For positive birefringent crystal the ordinary and extraordinary waves are reversed. Note that type I phasematching uses the full birefringence to offset dispersion while type II uses approximately one-half.

The second parameter of importance in choosing propagation direction and the type of phasematching to be used in the nonlinear interaction is the effective nonlinear coefficient. As indicated by Equation (1) and Table I the nonlinear coefficient is a tensor and thus must be evaluated for a particular crystal and propagation direction within the crystal, first to see that it remains non-zero and second for its magnitude.

For a wave propagating at an angle θ to the optic axis and φ to the crystal x axis in a crystal with small birefringence such that $\rho \ll \theta$ (see Equation (23)), the electric field components along the x, y, and z axes for the ordinary and extraordinary polarization components are given by[37]

$$E^o_j = |\mathbf{E}^o| \begin{vmatrix} \sin\varphi \\ -\cos\varphi \\ 0 \end{vmatrix} = |\mathbf{E}^o|\, a_j \tag{20a}$$

$$E^e_j = |\mathbf{E}^e| \begin{vmatrix} -\cos\varphi\cos\theta \\ -\sin\varphi\cos\theta \\ \sin\theta \end{vmatrix} = |\mathbf{E}^e|\, b_j \tag{20b}$$

Similarly, the components of the generated nonlinear polarization are given by

$$P^o_{\text{eff}} = a_i P_i \tag{21a}$$

$$P^e_{\text{eff}} = b_i P_i \tag{21b}$$

Thus for type I phasematching in negative uniaxial crystals the effective nonlinear polarization is

$$\text{Type I(neg)} \qquad |\mathbf{P}^e| = (2 - \delta_{jk}) b_i d_{ijk} a_j a_k |\mathbf{E}^o|^2 \tag{22a}$$

For other cases we have

$$\text{Type I(pos)} \qquad |\mathbf{P}^o| = (2 - \delta_{jk}) a_i d_{ijk} b_j b_k |\mathbf{E}^e|^2 \tag{22b}$$

$$\text{Type II(neg)} \qquad |\mathbf{P}^e| = (2 - \delta_{jk}) b_i d_{ijk} a_j b_k |\mathbf{E}^o||\mathbf{E}^e| \tag{22c}$$

$$\text{Type II(pos)} \qquad |\mathbf{P}^o| = (2 - \delta_{jk}) a_i d_{ijk} a_j b_k |\mathbf{E}^o||\mathbf{E}^e| \tag{22d}$$

We can now derive the expressions for d_{eff} for the 13 uniaxial crystal classes that lack inversion symmetry. For each class we find four expressions for the effective nonlinear coefficient in the most general case assuming lack of Kleinman's symmetry. However, if we assume that Kleinman's symmetry holds then the expressions reduced to type I and type II for each crystal class. Table II lists the effective nonlinear coefficients assuming Kleinman symmetry.

As an example consider the point group $\bar{4}2m$ to which KDP and its isomorphs belong (negative birefringent crystals) and also the chalocopyrite

TABLE II. Equations for d_{eff} for cases where Kleinman's symmetry is valid

Crystal class	Two e rays and one o ray	Two o rays and one e ray
6 and 4	0	$d_{15} \sin\theta$
622 and 422	0	0
$6mm$ and $4mm$	0	$d_{15} \sin\theta$
$\bar{6}m2$	$d_{22} \cos^2\theta \cos 3\varphi$	$-d_{22} \cos\theta \sin 3\varphi$
$3m$	$d_{22} \cos^2\theta \cos 3\varphi$	$d_{15} \sin\theta - d_{22} \cos\theta \sin 3\varphi$
$\bar{6}$	$\cos^2\theta(d_{11} \sin 3\varphi + d_{22} \cos 3\varphi)$	$\cos\theta(d_{11} \cos 3\varphi - d_{22} \sin 3\varphi)$
3	$\cos^2\theta(d_{11} \sin 3\varphi + d_{22} \cos 3\varphi)$	$d_{15} \sin\theta + \cos\theta(d_{11} \cos 3\varphi - d_{22} \sin 3\varphi)$
32	$d_{11} \cos^2\theta \sin 3\varphi$	$d_{11} \cos\theta \cos 3\varphi$
$\bar{4}$	$\sin 2\theta(d_{14} \cos 2\varphi - d_{15} \sin 2\varphi)$	$-\sin\theta(d_{14} \sin 2\varphi + d_{15} \cos 2\varphi)$
$\bar{4}2m$	$d_{14}(\sin 2\theta \cos 2\varphi)$	$-d_{14} \sin\theta \sin 2\varphi$

ternary semiconductor compounds ($CdGeAs_2$ for example with positive birefringence). For type I phasematching in KDP ($n_o > n_e$) two waves are polarized ordinary and the generated polarization is extraordinary. Thus the effective nonlinear coefficient is

$$d_{\text{eff}} = -d_{14} \sin \theta \sin 2\varphi$$

Obviously to maximize d_{eff}, φ should be 45° to the x or y axes. Since θ is restricted by the phasematching condition it cannot be maximized. However, 90° phasematching, when possible, such as in doubling the 5145 Å argon-ion laser line at a crystal temperature of −13°C does give the maximum value of d_{eff}. For type II phasematching where two extraordinary waves and one ordinary wave are used the effective nonlinear coefficient is

$$d_{\text{eff}} = d_{14} \sin 2\theta \cos 2\varphi$$

Now, it is desirable to set $\varphi = 0$ and have θ near 45°. The same KDP crystal cut that maximizes type I d_{eff} gives a zero d_{eff} for type II phasematching.

A second example of some importance is $LiNbO_3$ which belongs to crystal class $3m$. For type I phasematching (the only type of phasematching with adequate birefringence in $LiNbO_3$)

$$d_{\text{eff}} = d_{15} \sin \theta - d_{22} \cos \theta \sin 3\varphi$$

Now we must in addition to maximizing d_{22} through the correct choice of $\varphi(\varphi = 30°$ or propagation in the $y - z$ plane in trigonal $LiNbO_3$) choose θ correctly so that d_{15} and d_{22} add and not subtract for $\theta \neq 90°$. Recent measurements have shown that $|d_{15}| = -|d_{22}|$ so that θ must lie in the negative $y - z$ quadrants to maximize d_{eff}.

Two further aspects of phasematching should be considered. They are double refraction and acceptance angle in the nonlinear interaction.

As the extraordinary wave propagates in a crystal its power flow direction differs by the double refraction angle ρ from its phase velocity direction. The effect referred to as Poynting vector walk-off leads to a walk-off of the beam energy at an angle ρ. The double refraction angle is given by

$$\rho \approx \tan \rho = \frac{n_\omega^o}{2}\left[\frac{1}{(n_{2\omega}^e)^2} - \frac{1}{(n_{2\omega}^o)^2}\right]\sin 2\theta \tag{23}$$

For typical birefringent crystals at $\theta \sim 45°$, $\rho \sim 1$–2°. Thus for focussed laser beams in the crystal the power flow leads to a separation of the extraordinary and ordinary waves after a distance

$$l_a = \frac{\sqrt{\pi}\, w}{\rho} \tag{24}$$

called the aperture distance. Here w is the focussed laser beam electric field radius. For example, for KDP $\rho \sim 1^\circ$ and at $w = 50\ \mu$m we find $l_a = 0.5$ cm. This may be less than the full crystal length and thus lead to a substantial reduction in SHG efficiency. Note that at $\theta = 90^\circ$, $\rho = 0$ and there is no Poynting vector walk-off and thus an infinite aperture length. Therefore, 90° phasematching is desirable when possible.

The acceptance angle of the nonlinear crystal can be calculated by expanding the argument of the phase synchronism factor sinc($\Delta kl/2$) in a Taylor series. For SHG

$$\Delta k = \Delta k(\theta = \theta_{\text{match}}) + \left.\frac{\partial \Delta k}{\partial \theta}\right|_{\theta_m} \delta\theta$$

where

$$\Delta k = k_{2\omega} - 2k_\omega = \frac{4\pi}{\lambda_\omega}\,[n^e_{2\omega}(\theta) - n^o_\omega]$$

for negative birefringent type I phasematching. Since at $\theta = \theta_m$, $\Delta k = 0$ we find

$$\Delta k = \frac{4\pi}{\lambda_\omega} \left.\frac{\partial n^e_{2\omega}(\theta)}{\partial \theta}\right|_{\theta_m} \delta\theta \tag{25}$$

where $n^e_{2\omega}(\theta)$ is given by Equation (19). For type II phasematching

$$\Delta k = \frac{4\pi}{\lambda_\omega}\frac{\partial}{\partial\theta}\left.\{n^e_{2\omega}(\theta) - \tfrac{1}{2}[n^e_\omega(\theta) + n^o_\omega]\}\right|_{\theta_m} \delta\theta$$

Carrying out the differentiation implied in Equation (25) we find

$$\Delta k = \frac{2\pi}{\lambda_\omega}\,[n^e_{2\omega}(\theta)]^3\,\frac{[(n^e_{2\omega})^2 - (n^o_{2\omega})^2]}{(n^e_{2\omega})^2 (n^o_{2\omega})^2}\,\sin 2\theta\delta\theta$$

Evaluating at $\theta = \theta_m$ we have $n^e_{2\omega}(\theta) = n^0_\omega$ so that

$$\Delta k = \frac{2\pi}{\lambda_\omega}\,(n^o_\omega)^3 \left[\frac{1}{(n^e_{2\omega})^2} - \frac{1}{(n^o_{2\omega})^2}\right] \sin 2\theta_m \delta\theta \tag{26}$$

For type II phasematching we find a similar result but reduced by a factor of two. For crystals with small birefringence this expression simplifies to

$$\Delta k = \frac{2\pi}{\lambda_\omega}\,(n^o_{2\omega} - n^e_{2\omega}) \sin 2\theta_m \delta\theta \tag{27}$$

Finally, to find the acceptance angle we note that the sinc x function reaches its first zero for $x = \pi$. Thus $\Delta kl/2 = \pi$ is approximately the half-width. Using this criteria the half angular acceptance width inside the nonlinear crystal is

$$\delta\theta = \frac{\lambda_\omega}{l(n_{2\omega}^o - n_{2\omega}^e)\sin 2\theta_m} \tag{28}$$

For a 1 cm crystal of KDP at $\lambda_\omega = 1\ \mu$m, $\Delta n = 0.04$, $\sin 2\theta_m \approx 1$ we find $\delta\theta = 2.5$ mr. The external acceptance angle is $\delta\theta_{\text{ext}} = n\delta\theta = 3.75$ mr.

For 90° phasematching $\sin 2\theta$ approaches zero so that the expansion must be carried to the next order term. In this case

$$\Delta k = \frac{2\pi}{\lambda_\omega} \frac{\partial^2 n_{2\omega}^e(\theta)}{\partial \theta^2}\bigg|_{\theta_m} (\delta\theta)^2$$

Since $\theta_m \sim 90°$ we can approximate the expansion by noting that $\sin 2\theta_m \delta\theta \sim 2(\delta\theta)^2$ so that

$$\Delta k = \frac{4\pi}{\lambda_\omega}(n_{2\omega}^o - n_{2\omega}^e)(\delta\theta)^2$$

from Equation (27). Setting $\Delta kl/2 = \pi$ and solving for $\delta\theta$ we find

$$\delta\theta = \sqrt{\frac{\lambda_\omega}{2l(n_{2\omega}^o - n_{2\omega}^e)}} \tag{29}$$

As an example for a 1 cm $LiNbO_3$ crystal with $\Delta n = 0.08$ and $\lambda_\omega = 1\ \mu$m we find $\delta\theta = 25$ mr or an order of magnitude greater than the acceptance angle for critically phasematched KDP. Thus non-critical phasematching offers the additional advantage of increased acceptance angle compared to critical phasematching.

C. FOCUSSING

Our analysis to this point has assumed interacting plane waves of infinite extent. In fact, the beams are usually focussed into the nonlinear crystal to maximize the conversion efficiency. In addition, the waves are usually generated by laser sources and thus have a Gaussian amplitude profile with electric field beam radius w. Gaussian beam propagation theory and laser resonators have been treated by a number of authors and will not be considered here. However, for the present discussion we will define the important beam parameters and then consider focussing for the SHG case. Historically, the problem of focussing was first solved for SHG[6–10], and later extended to three-frequency interactions[97]. We will treat focussing for the case of SHG in successively more detail starting with Gaussian beam interactions in the near field without double refraction, with double refraction, and then treat the general case.

The following discussion considers the near-field focussing limit in detail since the near-field results apply to most experimental situations. The near-field results are then extended to the general focussing case by introducing Boyd and Kleinman's focussing parameter[97]. Finally, practical aspects of focussing are discussed.

The fundamental mode or Gaussian mode electric field is described by

$$E(x, y, z) = E_0 \frac{w_0}{w(z)} \exp[-i(kz - \varphi)] \exp\left\{-r^2\left[\frac{1}{w^2(z)} + \frac{ik}{2R}\right]\right\} \tag{30}$$

where

$$\varphi = \tan^{-1}\left(\frac{z}{z_R}\right)$$

is the phase factor for a TEM_{00} mode and

$$2z_R = b = w_0^2 k$$

describes the confocal parameter b as twice the Rayleigh range, z_R. Here R is the beam radius of curvature and $k = 2\pi n/\lambda$ is the propagation factor in the medium.

In general the beam radius and wavefront curvature as a function of the distance z along the beam propagation direction are given by

$$w^2(z) = w_0^2\left[1 + \left(\frac{z}{z_R}\right)^2\right]$$

and

$$R(z) = z\left[1 + \left(\frac{z_R}{z}\right)^2\right]$$

where w_0 is the electric field radius at the beam waist at $z = 0$. The far-field $(z \gg z_R)$ beam diffraction angle for the fundamental mode is

$$\theta = \frac{\lambda}{\pi w_0}$$

For $z > z_R$ diffraction is important and leads to beam spreading with distance. In this far field limit the electric field amplitude shows an amplitude-dependence on z in addition to its radial-dependence. The z-dependent electric field leads to complications in solving the coupled nonlinear equations such that only an integral solution is possible. Solutions of the coupled equations including beam diffraction have been considered by Boyd and Kleinman[97] and are discussed later.

In the near-field limit $z \ll z_R$ the expression for the TEM_{00} Gaussian beam reduces to

$$E(x, y) = E_0 \exp\left(-\frac{r^2}{w_0^2}\right) \tag{31}$$

where $r^2 = x^2 + y^2$. In this limit $w(z) \to w_0$, $\varphi \to 0$, and $R(z) \to \infty$ so that the fundamental mode is a plane wave with a Gaussian amplitude profile.

The power in the beam is

$$\begin{aligned} P &= \frac{nc\varepsilon_0}{2} \int |E(x, y)|^2 \, dx \, dy \\ &= \frac{nc\varepsilon_0}{2} \int_0^{2\pi} \int_0^{\infty} |E(r)|^2 r \, dr \, d\varphi \\ &= \frac{nc\varepsilon_0}{2} |E_0|^2 \left(\frac{\pi w_0^2}{2}\right) \\ &= I_0 \left(\frac{\pi w_0^2}{2}\right) \end{aligned} \tag{32}$$

where I_0 is the peak laser intensity and $\pi w_0^2/2$ is the effective area of the Gaussian beam.

To find the power generated by SHG in the near-field limit we consider the low conversion-efficiency limit and integrate Equation (11b) as before. Simplifying the notation

$$E_2(r) = \kappa E_1^2(r) \operatorname{sinc}\left(\frac{\Delta k l}{2}\right)$$

We now introduce the radial field variation by letting $E_1(r) = E_1 \exp(-r^2/w_1^2)$, where w_1 is the electric field radius of the fundamental field. Thus

$$E_2(r) = \kappa E_1^2 \, l \exp\left(-\frac{2r^2}{w_1^2}\right) \tag{33}$$

where we have assumed $\Delta k l/2 = 0$ and set $\operatorname{sinc}(\Delta k l/2) = 1$. Note that the electric field radius of the generated SH field is $w_1/\sqrt{2}$. Thus the fundamental and SH fields have equal confocal parameters.

The power generated at the SH is

$$P_2 = \frac{nc\varepsilon_0}{2} \int_0^{2\pi} \int_0^{\infty} |E_2(r)|^2 r \, dr \, d\varphi$$

Substituting from Equation (33) and carrying out the integration we find

$$P_2 = \frac{nc\varepsilon_0}{2} E_1^4 \kappa^2 l^2 \left(\frac{\pi w_1^2}{4} \right)$$

For a Gaussian mode the power is given by Equation (32) so that

$$P_2 = P_1 |E_1|^2 \kappa^2 l^2 / 2$$

or

$$\frac{P_2}{P_1} = P_1 \frac{2\kappa^2 l^2}{nc\varepsilon_0} \frac{1}{\pi w_1^2}$$

In terms of intensity with $I_1 = P_1/(\pi w_1^2/2)$ and $I_2 = P_2/(\pi w_2^2/2)$ with $w_2 = (1/\sqrt{2})w_1$ we find

$$\frac{I_2}{I_1} = I_1 \frac{2k^2 l^2}{nc\varepsilon_0} = \Gamma^2 l^2$$

which agrees with the previous conversion efficiency derived for the plane-wave case given by Equation (12).

It should be mentioned here that a similar analysis can be carried out for a pulsed field assuming a Gaussian beam in time. The result is analogous with the spatial case; the SH pulse width is less than the fundamental pulse width by $1/\sqrt{2}$. The space-time analogy extends considerably further than this simple example illustrates.

We now consider focussing in the near-field but in the presence of double refraction. We again assume perfect phasematching in the low conversion limit and write

$$E_2(x, y, l) = i\kappa \int_0^l E_1^2(x, y) dz \tag{34}$$

To account for Poynting vector walk-off at an angle ρ we let the walk-off occur in the x-z plane so that observed field coordinates must be related to the generating field by

$$\begin{aligned} y &= y' \\ x &= x' + \rho(l - z') \end{aligned} \tag{35}$$

Thus Equation (34) becomes

$$E_2(x, y, l) = i\kappa \int_0^l E_1^2[x - \rho(l - z'), y]dz'$$

The fundamental driving field has the form

$$E_1(x'y') = E_1 \exp\left(-\frac{y'^2}{w_1^2}\right)\exp\left(-\frac{x'^2}{w_1^2}\right)$$

Substituting we have

$$E_2(x, y, l) = i\kappa E_1^2 \exp\left(-\frac{2y}{w_1^2}\right) \int_0^l \exp\left(-\frac{2}{w_1^2}[x - \rho(l - z')]^2\right)dz' \quad (36)$$

Following Boyd, *et al.*[101] we introduce a set of normalized coordinates

$$u = \sqrt{2}\,\frac{x - \rho l}{w_1}$$

$$\tau = \sqrt{2}\,\frac{\rho z'}{w_1} \quad (37)$$

$$t = \sqrt{2}\,\frac{\rho l}{w_1}$$

and define the integral

$$F(u, t) = \frac{1}{t}\int_0^t \exp - [u + \tau]^2\, d\tau \quad (38)$$

In terms of these parameters Equation (36) becomes

$$E_2(x, y, l) = i\kappa E_1^2 \exp\left(-\frac{2y^2}{w_1^2}\right) lF(u, t) \quad (39)$$

The integral $F(u, t)$ describes the skewed second harmonic field resulting from double refraction. In the absence of double refraction ($\rho = 0$), $F(u, t) = \exp(-2x^2/w_1^2)$ and the generated field reduces to the previous case. For non-zero ρ, $F(u, t)$ leads to generated SH power skewed in the direction of ρ. Figure 2 illustrates the generated SH power as a function of u with t as a parameter.

A useful parameter for describing the effects of double refraction is the aperture length

$$l_a = \frac{\sqrt{\pi}\, w_1}{\rho} \quad (24)$$

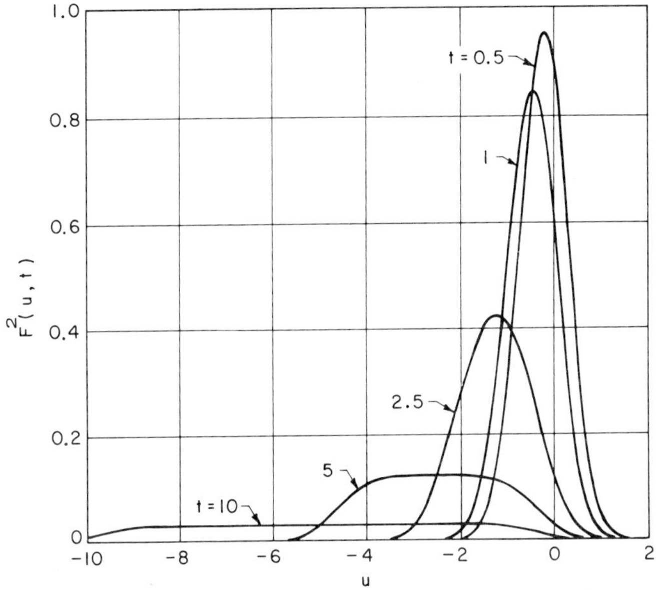

FIG. 2. The function $F^2(u, t, q)$ defined by (38) as a function of u for $q = 0$ and several values of t. (After Boyd *et al.* Ref. 101).

which was introduced previously to describe the double refraction-limited interaction length. The parameter t is normalized to l_a by

$$t = \sqrt{2\pi}\,\frac{l}{l_a}$$

and thus is a measure of the effects of beam walk-off.

To calculate the total SH power we substitute Equation (39) into the expression for power

$$P_2 = \frac{nc\varepsilon_0}{2}\int_{-\infty}^{\infty}\int_{-\infty}^{\infty} |E_2(x, y, l)|^2\,dx\,dy$$

Carrying out the y integral using $\int_0^\infty e^{-a^2x^2}\,dx = \sqrt{\pi}/2a$ we find

$$P_2 = \kappa^2\,\frac{nc\varepsilon_0}{2}\,E_1^4\,\frac{\sqrt{\pi}\,w_1}{2}\,l^2\int_{-\infty}^{\infty} F^2(u, t)dx$$

$$\frac{P_2}{P_1} = \frac{\Gamma^2 l^2}{2}\sqrt{\frac{2}{\pi}}\int_{-\infty}^{\infty} F^2(u, t)du \qquad (40)$$

$$= \frac{\Gamma^2 l^2}{2}\,G(t)$$

we have let

$$G(t) = \sqrt{\frac{2}{\pi}} \int_{-\infty}^{\infty} F^2(u, t) du. \tag{41}$$

Figure 3 plots $G(t)$ as a function of t. In the absence of walk-off $F(u, 0) = \exp(-2x^2/w_1^2) = \exp(-u^2)$ and $G(t) = 1$ so that the conversion efficiency again reduces to that found for the plane-wave case. For $\rho \neq 0$, $G(t)$ leads to a reduction in the second harmonic efficiency. $G(t)$ is approximately 0.5 for $t = 4$. Thus in the presence of walk-off the beam should be focussed such that

$$w_1 \gtrsim \frac{1}{2\sqrt{2}} \rho l \tag{42}$$

to avoid excessive loss of efficiency.

We are now ready to discuss the extension of focussing to the far-field case. To include the effects of beam divergence in focussing we introduce the Boyd

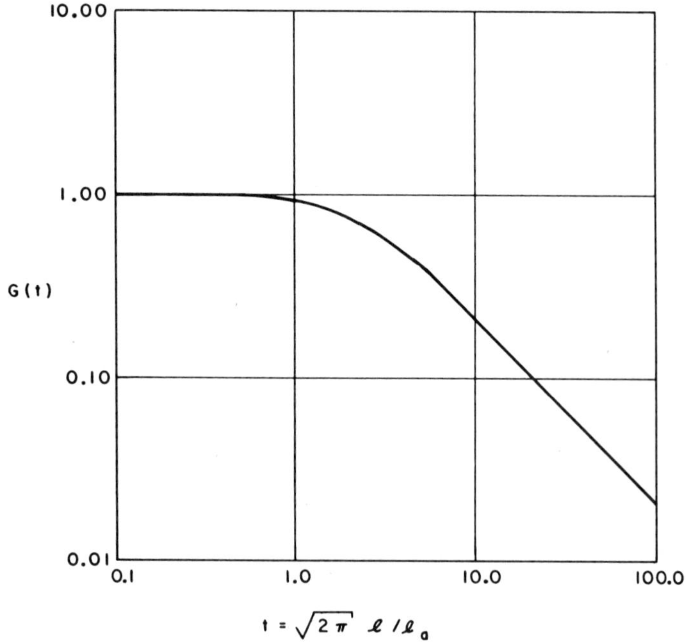

FIG. 3. The function of $G(t)$ defined by Equation (41).

and Kleinman focusing factor $h(B, \xi)$. The second harmonic generation efficiency can be written in terms of $h(B, \xi)$ as

$$\frac{P_2}{P_1} = \left(\frac{2\omega^2 d_{\rm eff}^2}{\pi n_\omega^2 n_{2\omega} \varepsilon_0 c^3}\right) P_1 l k_\omega h(B, \xi) \tag{43}$$

where B is the double refraction parameter

$$B = \tfrac{1}{2}\rho(lk_\omega)^{1/2} \tag{44}$$

and ξ is the focussing parameter

$$\xi = \frac{l}{b} \tag{45}$$

and b is the confocal distance defined earlier.

We now proceed to discuss the gain reduction factor and its limiting forms. For second harmonic generation Equation (43) shows that the total second harmonic efficiency is reduced by $h(B, \xi)$. In parametric processes we are usually interested in the parametric gain coupled into the fundamental mode. For this case $h(B, \xi)$ is reduced further to $\bar{h}(B, \xi)$. Figure 4 shows the values of $h(B, \xi)$ and $\bar{h}(B, \xi)$ as a function of the double refraction parameter B which can be written in terms of the aperture length as

$$B = \frac{\sqrt{\pi}}{2}\left(\frac{l}{l_a}\right)\xi^{-1/2} \tag{46}$$

Figure 4 shows that for small B, or equivalently for small walk-off angles

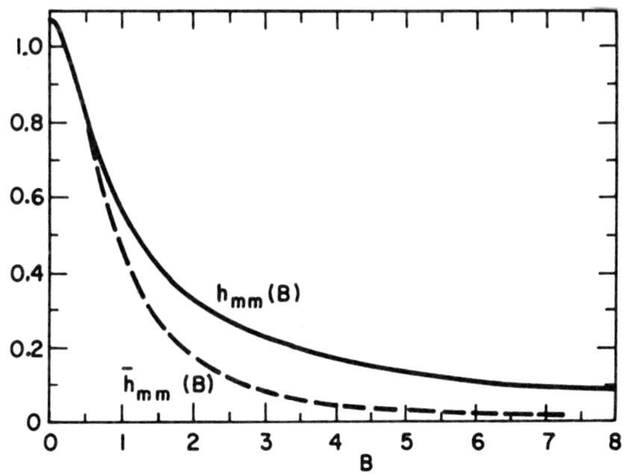

FIG. 4. Conversion efficiency reduction factor for total SHG given by $h_{mm}(B)$ and gain reduction factor for parametric generation, $\bar{h}_{mm}(B)$. The dashed curve is closely approximated by Equation (48). (After Boyd and Kleinman, Ref. 97).

for crystal lengths less than the aperture length, $h(B, \xi) \to \bar{h}(B, \xi)$. For second harmonic generation this corresponds to the total second harmonic output becoming less multimode due to Poynting vector walk-off and more fundamental-mode.

Fortunately, $\bar{h}(B, \xi)$ can usually be approximated by one of its limiting forms. In the near-field approximation with negligible double refraction

$$\bar{h}(B, \xi) \to \xi \left(\xi < 0.4, \xi < \frac{1}{6B^2} \right) \tag{47}$$

Substituting this limit into Equation (43) reduces it to previous near-field result given by Equation (12). In the limit of negligible double refraction but for confocal focussing where $\xi = 1$ the gain reduction factor becomes

$$\bar{h}(0, 1) \approx 1$$

which corresponds to the previous near-field result first derived by Boyd and Ashkin[102]. In fact, the maximum value of $\bar{h}(B, \xi) = \bar{h}_{mm}(0, 2.84) = 1.068$ occurs at $\xi = 2.84$ instead of $\xi = 1$. However, practical considerations usually limit $\xi < 1$ so that the confocal approximation is an excellent one.

In the tight-focussing limit and neglecting double refraction the gain reduction factor becomes

$$\bar{h}(0, \xi) \to \frac{1.187\pi^2}{\xi} \qquad \left(80 < \xi < \frac{\pi^2}{16B} \right)$$

which shows that the gain reduction factor $\bar{h}(0, \xi)$ is proportional to b/l in this limit so that the gain reduces to b^2.

When double refraction is important the gain reduction factor can be represented by the empirical approximation

$$\bar{h}_{mm}(B) \approx \frac{\bar{h}_{mm}(0)}{\left[1 + \left(\frac{4B^2}{\pi} \right) \bar{h}_{mm}(0) \right]} \tag{48}$$

to better than 10% accuracy over the entire range of B. Here $\bar{h}_{mm}(0) = 1.068$ is the maximum value of $\bar{h}(B, \xi)$ at optimum focussing where $\xi = 2.84$. The condition

$$\left(\frac{4B^2}{\pi} \right) \bar{h}_{mm}(0) \approx 1$$

designates the point at which the gain begins to limit due to double refraction effects. We use this condition to define an effective useful crystal length

$$l_{\text{eff}}(\rho) = \frac{\lambda_0}{2n_\omega \rho^2 \bar{h}_{mm}(0)} \approx \frac{\lambda_0}{2n_\omega \rho^2} \tag{49}$$

For a crystal longer than this length the spot size must be increased to keep $l_a \geq l$ so that the SHG efficiency per input power remains constant.

In terms of l_{eff} the gain reduction factor becomes

$$\bar{h}_{mm}(B) \approx \frac{\bar{h}_{mm}(0)}{\left[1 + \frac{l}{l_{\text{eff}}}\right]}$$

In the limit of strong double refraction such that $l_{\text{eff}} \ll l$ or $(4B^2/\pi)\bar{h}_{mm}(0) \gg 1$ the gain reduction factor is

$$\bar{h}_m(B, \xi) \to \frac{\pi}{4B^2} \qquad \left(\frac{B^2}{4} > \xi > \frac{2}{B^2}\right) \tag{50}$$

Referring back to the SHG expression given by Equation (43), we can see that the strong, double-refraction limit has the effect of replacing the crystal length l by l_{eff}.

In addition, the large, double-refraction limit maintains $\bar{h}_m(B, \xi)$ at a nearly constant value over a wide focussing range. Thus to minimize crystal damage problems ξ can be chosen such that $\xi \gtrsim 2/B^2$ which is approximately the condition given by Equation (49). This condition corresponds to a focal spot size w_1 derived earlier

$$w_1 \lesssim \frac{1}{2\sqrt{2}} \rho l \tag{42}$$

and an effective area

$$\frac{\pi w_1^2}{2} \lesssim \frac{\pi}{16} \rho^2 l^2 \tag{51}$$

For larger spot sizes the gain is reduced and for smaller spot sizes the gain is practically constant but the intensity increases. Focussing for minimum intensity has been discussed by Harris[79].

Figure 5 shows the gain reduction factor as a function of the focussing parameter ξ for various values of the double refraction parameter B. The

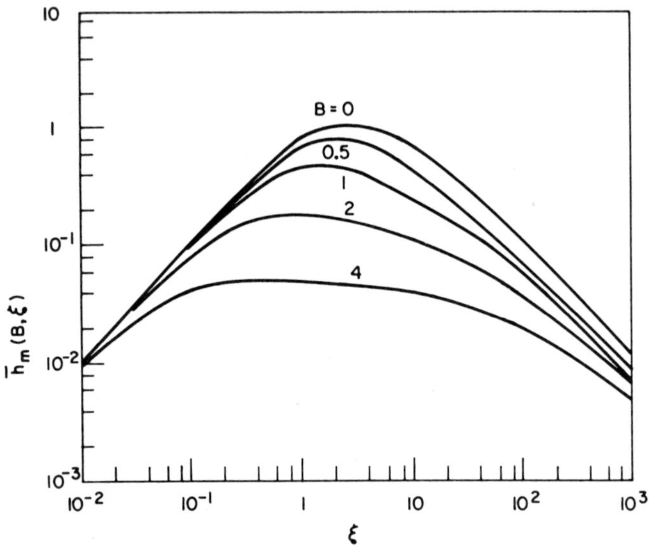

FIG. 5. Gain reduction factor $\bar{h}_{mm}(B, \xi)$ versus ξ. (From Boyd and Kleinman, Ref. 97).

figure shows the significant reduction in gain for non-zero values of B. In addition, the near field limit $\bar{h}(0, \xi) \rightarrow \xi$ is evident for small ξ.

The gain reduction factor and the effective crystal length are seriously reduced for non-collinear phasematching. Referring to Equation (13) we can write a crystal figure of merit for SHG in the limits of collinear and non-collinear phasematching as

$$M_{\rho=0} = \frac{d^2}{n_\omega^2 n_{2\omega}} \tag{52}$$

and

$$M_{\rho\neq 0} = \frac{d^2}{n_\omega^2 n_{2\omega}} \cdot \frac{l_{\text{eff}}}{l} \tag{53}$$

For large double refraction the conversion efficiency is reduced by $\pi/4B^2 \approx l/l_{\text{eff}}$. For $LiNbO_3$ at room temperature with phasematching achieved by angle tuning to $\theta = 43^\circ$ for a 2.12 μm fundamental wavelength, $\rho = 0.037$ radians and $B = 4.7\, l^{1/2}$. The SHG conversion efficiency is reduced by 28 times for a 1 cm crystal length and 140 times for a 5 cm crystal length compared to the 90° phasematched case. For this double refraction angle the maximum effective crystal length is only $l_{\text{eff}} \approx 0.36$ mm. For a 1 cm crystal the focussing parameter ξ can vary between $5.5 > \xi > 0.09$ without affecting the gain. This corresponds to a confocal parameter variation between 11 cm and 0.2 cm. For experimental ease and reduction of the intensity at the crystal surface the larger value of b would be utilized. For a 1 mm crystal length the gain remains the same but the confocal parameter varies between 1.1 mm $< b <$ 1.8 mm and thus the intensity is considerably increased. The pumping intensity decreases with crystal length as l^2 in agreement with Equation (51).

If we introduce an effective length due to focussing.

$$l_f = \frac{\pi w_1^2 k}{2} = \frac{2}{\pi} b \tag{54}$$

and use the aperture length $l_a = \sqrt{\pi} w_1/\rho$ then the SHG conversion efficiency can be summarized by a series of limiting forms

$$\frac{P(2\omega)}{P(\omega)} = \frac{KP(\omega)}{w_2} \cdot \begin{cases} l^2 & \text{for} \quad (l_a, l_f \gg 1) \\ l l_a & \text{for} \quad (l_f \gg l \gg l_a) \\ l_f l_a & \text{for} \quad (l \gg l_f \gg l_a) \\ 4 l_f^2 & \text{for} \quad (l \gg l_a \gg l_f) \\ 4.75\, l_f^2 & \text{for} \quad (l_a \gg l \gg l_f) \end{cases}$$

The first limit corresponds to the plane-wave focussing case where $h(\xi, B) \approx l/b$. This case holds until $l/b \approx 1$, which is called the confocal focussing limit. The second harmonic conversion efficiency is optimum at $l/b = 2.84$ where $h(2.84, 0) = 1.06$[97].

If $\rho \neq 0$, then the aperture length may limit the interaction length to the second and third cases. In practice, even a small walk-off angle may reduce the second harmonic efficiency by 30 times for a given crystal length and focussing. It is therefore very desirable to adjust the crystal indices of refraction to achieve 90° phasematching if possible. In general, the last three of the above limits are not encountered experimentally. This discussion on focussing is intended to serve as a general introduction. A more detailed presentation is given by Boyd and Kleinman[97].

D. THREE-FREQUENCY PROCESSES

There are three processes of interest involving three-frequency interactions. They are sum generation or up-conversion, difference-frequency generation or mixing, and parametric generation. The processes are shown schematically in Fig. 6.

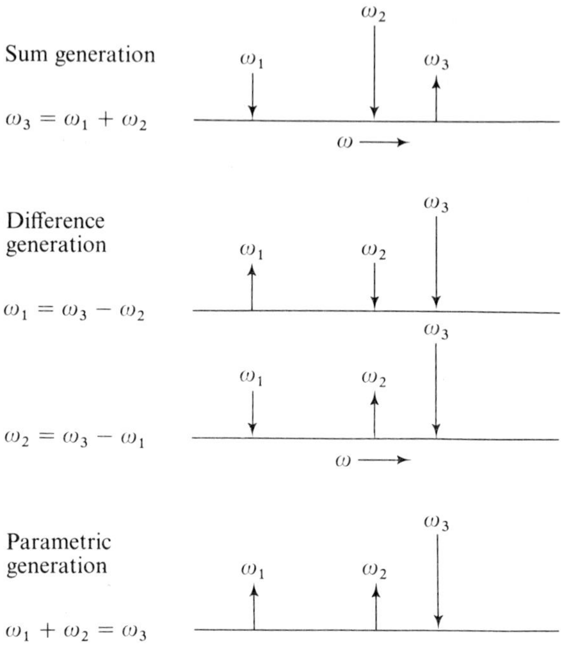

FIG. 6. Schematic diagrams of the possible three-frequency interactions.

Our starting point is again the travelling-wave equation in the slowly-varying-amplitude approximation. As an example, we assume the driving polarization appropriate for 90° phasematching in $LiNbO_3$ so that

$$\begin{aligned} P(\omega_1) &= \varepsilon_0 2d_{31} E(\omega_3)E^*(\omega_2) \\ P(\omega_2) &= \varepsilon_0 2d_{31} E(\omega_3)E^*(\omega_1) \\ P(\omega_3) &= \varepsilon_0 2d_{31} E(\omega_1)E(\omega_2) \end{aligned} \tag{55}$$

Substituting the polarizations into Equation (8) and simplifying the notation gives three coupled equations

$$\frac{dE_1}{dz} + \alpha_1 E_1 = i\kappa_1 E_3 E_2^* e^{i\Delta kz} \tag{56a}$$

$$\frac{dE_2}{dz} + \alpha_2 E_2 = i\kappa_2 E_3 E_1^* e^{i\Delta kz} \tag{56b}$$

$$\frac{dE_3}{dz} + \alpha_3 E_3 = i\kappa_3 E_1 E_2 e^{-i\Delta kz} \tag{56c}$$

where

$$\kappa_i = \frac{\omega_i d}{n_i c}$$
$$\alpha_i = \tfrac{1}{2}\mu_0 \sigma_i c \tag{57}$$

and

$$\omega_3 = \omega_2 + \omega_1 \tag{58a}$$
$$k_3 = k_2 + k_1 + \Delta k \tag{58b}$$

In the following discussion we assume perfect phasematching and set $\Delta k = 0$. If we choose to keep Δk non-zero, the solutions would generate the phase synchronism factor $\mathrm{sinc}(\Delta kl/2)$. We also assume plane-wave interactions and leave the extension of focusing from SHG to three-frequency cases as an exercise. We consider in turn sum, difference, and parametric generation.

(a) *Sum Generation (Up-Conversion)*

In this interaction we assume an input at ω_1 sums with an intense pump field at ω_2 to generate a wave at ω_3. We further assume that the interaction proceeds without depletion of the pump field at ω_2. Thus $dE_2/dz = 0$ and Equations (56) reduce to a pair of coupled equations

$$\frac{dE_1}{dz} = i\kappa_1 E_3 E_2^*$$
$$\frac{dE_3}{dz} = i\kappa_3 E_1 E_2 \tag{59}$$

where we have assumed $\Delta k = 0$ and $\alpha = 0$.

To solve these equations we look for a solution of the form $e^{\Gamma' z}$. Substituting and setting the determinant of the coefficients equal to zero we find

$$\Gamma' E_1 - (i\kappa_1 E_2^*)E_3 = 0$$
$$(-i\kappa_3 E_2)E_1 + \Gamma' E_3 = 0$$

which results in propagation constants

$$\Gamma'_\pm = \pm i[\kappa_1\kappa_3 |E_2|^2]^{1/2}$$

The solutions are therefore of the form

$$E_1 = E_1^+ e^{\Gamma' z} + E_1^- e^{-\Gamma' z}$$
$$E_3 = E_3^+ e^{\Gamma' z} + E_3^- e^{-\Gamma' z}$$

where the coefficients $E_1^\pm$ and $E_3^\pm$ are determined by the boundary conditions. If we assume that in addition to the strong pump field at ω_2 we have input only at ω_1 then

$$E_3(z = 0) = 0$$
$$E_1(z = 0) = E_1(0)$$

The solutions for E_1 and E_3 with these boundary conditions become

$$\begin{aligned} E_1(z) &= E_1(0)\cos \Gamma z \\ E_3(z) &= C \sin \Gamma z \end{aligned} \tag{60}$$

where $\Gamma' = i\Gamma$. To solve for the constant we substitute Equation (60) for $E_3(z)$ into Equation (59) and find

$$C = iE_1(0)\sqrt{\frac{\kappa_3}{\kappa_1}} = iE_1(0)\sqrt{\frac{\omega_3 n_1}{\omega_1 n_3}}$$

The intensity ratios are thus

$$\begin{aligned} \frac{I_1(z)}{I_1(0)} &= \cos^2 \Gamma z \\ \frac{I_3(z)}{I_1(0)} &= \frac{\omega_3}{\omega_1} \sin^2 \Gamma z \end{aligned} \tag{61}$$

The solution shows that the input field at ω_1 is converted with 100% photon efficiency to the field at ω_3. Furthermore, the conversion takes place over an interaction length defined by

$$\Gamma l = \frac{\pi}{2}$$

At longer lengths the field at ω_3 begins to reconvert back to the field at ω_1. Under the present assumptions the pump field at ω_2 is not depleted in the interaction. Note also, that the fields oscillate as $\sin \Gamma z$ and $\cos \Gamma z$ and thus exhibit no gain.

In the low conversion-efficiency limit $\sin \Gamma z \sim \Gamma z$ so that the sum generation conversion efficiency reduces to

$$\begin{aligned} \frac{I_3(z)}{I_1(0)} &= \frac{\omega_3}{\omega_1} \Gamma^2 z^2 \\ &\simeq \frac{2\omega_3^2 d^2 I_2 z^2}{n_1 n_2 n_3 c^3 \varepsilon_0} \operatorname{sinc}^2\left(\frac{\Delta k z}{2}\right) \end{aligned} \tag{62}$$

where we have included the phase synchronism factor.

In the low conversion-efficiency limit the sum-generation efficiency equals the SHG efficiency as one would expect from physical arguments. However, in the high conversion-efficiency limit the SHG process approaches 100% efficiency as a $\tanh^2 x$ function while sum generation oscillates as a $\sin^2 x$ function.

(b) *Difference Generation (Mixing)*

For difference-frequency generation we assume a strong pump field at ω_3 mixes with an input field at ω_1 to generate an output at ω_2. Assuming no pump depletion, $dE_3/dz = 0$ so that the coupled equations reduce to

$$\frac{dE_1}{dz} = i\kappa_1 E_3 E_2$$
$$\frac{dE_2^*}{dz} = -i\kappa_2 E_3^* E_1 \tag{63}$$

where we have taken the complex conjugate of Equation (56b). Proceeding as before, we assume a solution of the form $e^{\Gamma z}$ and find

$$\Gamma = \pm[\kappa_1\kappa_2|E_3|^2]^{1/2}$$

For boundary conditions at the input of the nonlinear crystal $E_1(z = 0) = E(0)$ and $E_2(z = 0) = 0$ we find solutions

$$E_1(z) = E_1(0)\cosh \Gamma z$$
$$E_2^*(z) = C \sinh \Gamma z$$

where

$$C = -i\sqrt{\frac{\kappa_2}{\kappa_1}}\, E_1(0)$$

The intensity conversion efficiencies are thus

$$\frac{I_1(z)}{I_1(0)} = \cosh^2 \Gamma z$$
$$\frac{I_2(z)}{I_1(0)} = \frac{\omega_2}{\omega_1} \sinh^2 \Gamma z \tag{64}$$

We now find that both the input field and the generated field grow at the expense of the pump field. In the limit of small conversion efficiency $\sinh x \sim x$ so that

$$\frac{I_2(z)}{I_1(0)} = \frac{\omega_2}{\omega_1} \Gamma^2 z^2 \operatorname{sinc}^2\left(\frac{\Delta k z}{2}\right)$$

$$= \frac{2\omega_2^2 |d|^2 z^2 I_3}{n_1 n_2 n_3 c^3 \varepsilon_0} \operatorname{sinc}^2\left(\frac{\Delta k z}{2}\right) \tag{65}$$

which again agrees with the SHG efficiency. In this case, however, both waves exhibit gain and thus suggest the possibility of building an amplifier or oscillator.

(c) *Parametric Generation*

For parametric generation we assume a strong pump field at the highest frequency ω_3 and look for generation of both the field at ω_2 (signal field) and at ω_1 (idler field). We begin by investigating the coupled equations in the limit of no loss

$$\frac{dE_1}{dz} = i\kappa_1 E_3 E_2^* e^{i\Delta k z}$$

$$\frac{dE_2}{dz} = i\kappa_2 E_3 E_1^* e^{i\Delta k z}$$

$$\frac{dE_3^*}{dz} = -i\kappa_3 E_1^* E_2^* e^{+i\Delta k z}$$

Multiplying successively by

$$\left(\frac{nc\varepsilon_0}{2}\right)\frac{E_1^*}{\omega_1}, \quad \left(\frac{nc\varepsilon_0}{2}\right)\frac{E_2^*}{\omega_2}, \quad \text{and} \quad \left(\frac{nc\varepsilon_0}{2}\right)\frac{E_3}{\omega_3}$$

we note that the right-hand sides of each equation are equal so that

$$\frac{1}{\omega_1}\frac{dI_1}{dz} = \frac{1}{\omega_2}\frac{dI_2}{dz} = -\frac{1}{\omega_2}\frac{dI_3}{dz} \tag{66}$$

which is a form of photon conservation or Manley-Rowe relation. Simply stated, parametric generation splits one pump photon into two photons which satisfy conservation of energy at every point in the nonlinear crystal. (This prompted one chemist to call a parametric generator a "photon cutter".) It should also be noted that the signal and idler fields (ω_2 and ω_1) are symmetric under interchange of indices.

We now proceed to solve Equations (56a) and (56b) including loss and momentum mismatch. We assume, however, no pump depletion so that $dE_3/dz = 0$. Again, we look for an exponentially varying solution. A little forethought shows that the expected solutions are of the form

$$E_1^*(z) = E_1^* e^{(\Gamma' - i\Delta k/2)z}$$
$$E_2(z) = E_2 e^{(\Gamma' + i\Delta k/2)z}$$

where the complex conjugate of Equation (56a) is used. Substituting into Equations (56a) and (56b) results in the characteristic equations

$$E_2\left(\Gamma' + \alpha_2 + \frac{i\Delta k}{2}\right) + E_1^*\left(-\frac{i\omega_2\, dE_3}{n_2 c}\right) = 0$$

$$E_2\left(\frac{i\omega_2\, dE_3^*}{n_2 c}\right) + E_1^*\left(\Gamma' + \alpha_1 - \frac{i\Delta k}{2}\right) = 0$$

Solving for Γ' gives

$$\Gamma'_\pm = \frac{-(\alpha_1 + \alpha_2)}{2} \pm \frac{1}{2}\left[(\alpha_1 - \alpha_2)^2 + \frac{i\Delta k}{2}(\alpha_1 - \alpha_2) - 4\left(\frac{\Delta k}{2}\right)^2 + \frac{4\omega_1\omega_2|d|^2|E_3|^2}{n_1 n_2 c^2}\right]^{1/2}$$

We now define a parametric gain coefficient as

$$\Gamma^2 = \frac{\omega_1\omega_2|d|^2|E_3|^2}{n_1 n_2 c^2} = \frac{2\omega_1\omega_2|d|^2 I_3}{n_1 n_2 n_3 \varepsilon_0 c^3} \qquad (67)$$

which is identical to the SHG conversion efficiency factor given by Equation (13) when $\omega_1 = \omega_2 = \omega$ which is referred to as the degeneracy condition. Of course, for parametric generation, $\omega_1 + \omega_2 = \omega_3$ is the only restriction the generated frequencies must obey.

The general solution to the coupled equations is therefore

$$E_1^*(z) = (E_{1+}^* e^{gz} + E_{1-}^* e^{-gz})e^{-\alpha z}e^{-i\Delta kz/2} \qquad (68a)$$

$$E_2(z) = (E_{2+} e^{gz} + E_{2-} e^{-gz})e^{-\alpha z}e^{+i\Delta kz/2} \qquad (68b)$$

where we have set $\alpha_1 = \alpha_2 = \alpha$ and define a total gain coefficient

$$g = \left[\Gamma^2 - \left(\frac{\Delta k}{2}\right)^2\right]^{1/2} \qquad (69)$$

For parametric amplification we assume input fields at ω_1 and ω_2 of $E_1^*(z = 0) = E_1^*(0)$ and $E_2(z = 0) = E_2(0)$. These input fields give the boundary conditions that determine the $E_{1\pm}$ and $E_{2\pm}$ coefficients. Carrying out the algebra results in the solutions for the field amplitude at $z = l$

$$E_1(l)e^{\alpha l} = E_1(0)e^{i\Delta kl/2}\left[\cosh gl - \frac{i\Delta k}{2g}\sinh gl\right] + i\frac{\kappa_1 E_3}{g}E_2^*(0)e^{i\Delta kl/2}[\sinh gl] \tag{70a}$$

and

$$E_2(l)e^{\alpha l} = E_2(0)e^{i\Delta kl/2}\left[\cosh gl - \frac{i\Delta k}{2g}\sinh gl\right] + i\frac{\kappa_2 E_3}{g}E_1^*(0)e^{i\Delta kl/2}[\sinh gl], \tag{70b}$$

where $\kappa_i = \omega_i d/n_i c$ and $\Gamma^2 = \kappa_1\kappa_2|E_3|^2$ as before. These expressions are symmetric under the interchange of subscripts $1 \leftrightarrow 2$ as expected.

When a single input field at frequency ω_2 is incident on the parametric generator it encounters a single pass power gain

$$G_2(l) = \frac{|E_2(l)|^2}{|E_2(0)|^2} - 1 = \Gamma^2 l^2 \frac{\sinh^2 gl}{(gl)^2} \tag{71}$$

In this case the gain is independent of the phase of the input field $E_2(0)$ relative to the pump field E_3 since the generated idler field assumes the proper phase to maximize the gain. However, if two fields $E_1(0)$ and $E_2(0)$ are simultaneously incident on the amplifier the gain depends on their phases relative to the pump field.

Superfluorescence parametric emission is an extension of the parametric amplifier to the case of extremely high gain such that the input noise fields are amplified to a output intensity on the order of that of the pump field. In the high-gain limit Equation (71) becomes

$$G_2(l) = \left[1 + \left(\frac{\Delta k}{2g}\right)^2\right]\sinh^2 gl$$

or

$$G_2(l) = \tfrac{1}{4}\exp 2\Gamma l \tag{72}$$

when $(\Delta k/2) < g$.

The noise input per mode for a parametric amplifier is equivalent to one photon in either the signal or idler channel, or half a photon in each. It is interesting to note that $\frac{1}{2}\hbar\omega$ input into a laser amplifier also leads to the correct amplified spontaneous emission output. For efficient superfluorescent operation the input noise field must be amplified by approximately 10^{16} so that $\Gamma l \sim 20$. In practice, very high gains have been achieved[65,86] such that superfluorescence operation is possible[103,104].

Parametric amplifiers, like all linear amplifiers, have inherent noise. Since the parametric frequencies can be in the visible range and are on the order of 10^{-10} W per watt of pump power, the noise emission is intense enough to be easily visible as parametric fluorescence[105,106]. A striking colour photograph of parametric fluorescence generated in $LiNbO_3$[107], is reproduced in a review article by Giordmaine[108]. Parametric noise emission can be considered generated by the amplified zero-point fluctuations of the electromagnetic field[109,110]. It can be shown that the fluorescence noise power is given simply by P_{noise} = (energy per photon) × gain × bandwidth or approximately by $P_{noise} = \hbar\omega \times \Gamma^2 l^2 \times (c/2\Delta n l)$ where Δn is the crystal birefringence and l the crystal length. For $LiNbO_3$ the emitted noise power per watt is approximately 10^{-10} W, or one order of magnitude larger than the spontaneous Raman scattering power. The noise power can be used to measure the gain, bandwidth, and tuning characteristics of a parameter oscillator prior to ever achieving threshold[105,106]. Parametric fluorescence also provides a very accurate method of measuring the nonlinear coefficient of a crystal since only a power ratio need be measured and not an absolute power as for SHG measurements[109]. However, one of the most interesting applications of parametric generation is the achievement of coherent, tunable laser like output from an optical parametric oscillator.

In the low-gain limit where $\Gamma^2 l^2 < (\Delta k l/2)^2$ the gain of a parametric amplifier is

$$G_2(l) = \Gamma^2 l^2 \operatorname{sinc}^2\left\{\left[\left(\frac{\Delta k}{2}\right)^2 - \Gamma^2\right]^{1/2} l\right\} \tag{73}$$

Equations (71) and (73) show that as the gain varies from $\Gamma^2 l^2 > (\Delta k/2)^2$ to $\Gamma^2 l^2 < (\Delta k/2)^2$ the $(\sinh x)/x$ function continuously goes to $(\sin x)/x$. Note that for small gain the argument of the sinc function in Equation (73) reduces to the expected $\Delta k l/2$.

For parametric amplification we define a gain bandwidth by

$$\left|\left(\frac{\Delta k}{2}\right)^2 - \Gamma^2\right|^{1/2} l = \pi \tag{74}$$

For low gain such that $\Gamma^2 l^2 \ll \pi$ the bandwidth expression reduces to

$$\frac{\Delta k l}{2} = \pi \tag{75}$$

The ratio of the bandwidth in the low-gain to high-gain limit is

$$\frac{\text{BW(high gain)}}{\text{BW(low gain)}} = \left(1 + \frac{\Gamma^2 l^2}{\pi^2}\right)^{1/2}$$

Therefore, for superfluorescence gains such that $\Gamma^2 l^2 \sim 400$ the gain bandwidth increases by a factor of six over the low gain bandwidth.

This completes our discussion of nonlinear interactions. With this background we can proceed to consider in detail parametric oscillators and devices.

IV. Parametric Oscillators

A. Introduction

A parametric oscillator is schematically represented by a nonlinear crystal within an optical cavity. The nonlinear crystal when pumped provides gain at the two frequencies ω_2 and ω_1 (the signal and idler fields). When the gain exceeds the loss the device reaches threshold and oscillates. At threshold the output power increases dramatically, similar to the behaviour of a laser. The generated output is coherent and collinear with the pump laser beam. Once above threshold, the parametric oscillator efficiently converts the pump radiation to continuously tunable signal and idler frequencies.

There are a number of configurations for an optical parametric oscillator (OPO). The first distinction is between cw and pulsed operation. Due to the much higher gains, we consider only pulsed operation where the pump is typically generated by a Q-switched laser source. Parametric oscillators have also operated internal to the pump laser cavity, but the external configuration is more common. Finally, there are different parametric oscillator cavity configurations. The two most important are the doubly resonant oscillator (DRO), where both the signal and idler waves are resonated by the cavity mirrors, and the singly resonant oscillator (SRO), where only one wave is resonant. The operation characteristics of parametric oscillators with these cavity configurations have been discussed in detail in the review articles by Harris[79], Smith[84], and Byer[85].

The important differences between DRO and SRO operation include threshold, frequency stability, and pump laser frequency requirements. The threshold of a DRO occurs when the gain equals the product of the signal and idler losses or

$$\Gamma^2 l^2 \operatorname{sinc}^2\left(\frac{\Delta k l}{2}\right) = a_1 a_2 \tag{76}$$

where $\Gamma^2 l^2$ given by Equation (67) is the single pass parametric gain and a_2, a_1 are the single pass power losses at the signal and idler. The SRO threshold is greater than that of the DRO. If the idler wave at ω_1 is not resonated, then the SRO threshold becomes

$$\Gamma^2 l^2 \operatorname{sinc}^2\left(\frac{\Delta k l}{2}\right) = 2a_2 \tag{77}$$

which is $2/a_1$ times greater than the DRO threshold. As an example consider a 5 cm long 90° phasematched $LiNbO_3$ crystal pumped by the second harmonic

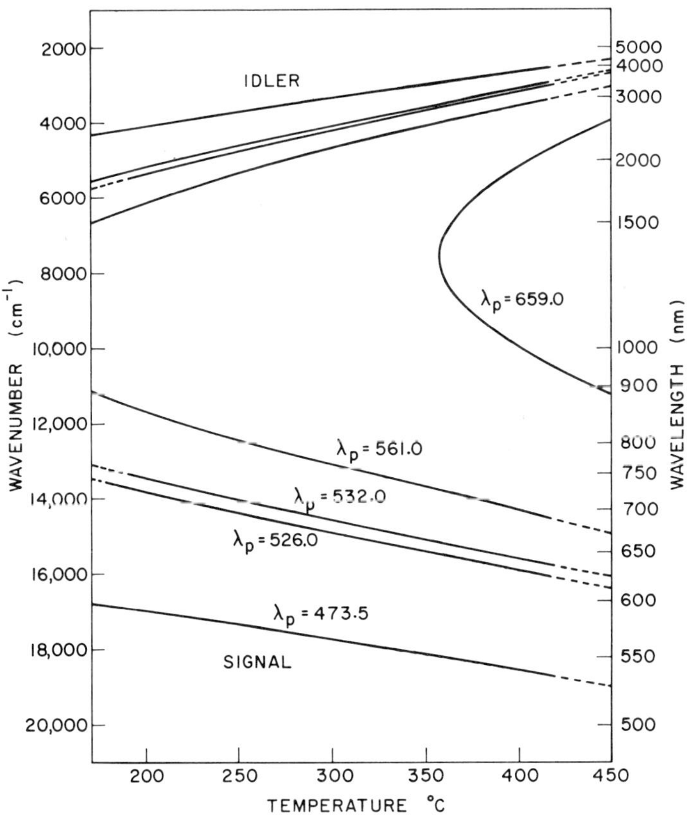

FIG. 7. $LiNbO_3$ temperature tuning curves for doubled Nd:YAG laser pump wavelengths.

of a Nd:YAG laser at 0.532 μm. The DRO pump power threshold for 2% losses at ω_1 and ω_2 is only 38 mW. The SRO threshold is increased to 3.8 W. For cw pump lasers the DRO is required in order to achieve threshold. However, for pulsed parametric oscillator operation where kilowatts of pump power is available, the higher SRO threshold is not a disadvantage and singly resonant offers significant advantages over doubly resonant operation.

In general, a parametric oscillator is tuned by altering the crystal birefringence. This is usually done by crystal rotation, thus changing the extraordinary index of refraction, or by controlling the crystal temperature and thus the birefringence. As an example Fig. 7 shows the temperature tuning curves of a $LiNbO_3$ parametric oscillator pumped by various doubled wavelengths of a Q-switched Nd:YAG laser. The tuning range extends from 0.55 to 3.7 μm. Tuning curves for a number of nonlinear crystals have been given in the review article by Byer[85].

The important parameters of parametric oscillator operation include threshold, rise time, conversion efficiency, tuning and bandwidth. We discuss these factors here with particular emphasis on recent developments.

B. THRESHOLD

In a typical parametric oscillator the pump makes a single pass through the nonlinear crystal. The generated idler and signal waves $E_1(l)$ and $E_2(l)$ grow according to Equations (70a) and (70b) during the single pass in the pump wave direction. Following reflection and the backward trip in the cavity, the waves are again travelling with the pump and are amplified. The gain is thus single-pass and the corresponding loss is the round-trip electric field loss α,

For the small-gain case and $\Delta k = 0$, the threshold condition is

$$E_1(l = 0)e^{\alpha_1 l} = E_1(0)\cosh \Gamma l + i\frac{\kappa_1}{\Gamma} E_3 E_2^*(0)\sinh \Gamma l, \tag{78a}$$

and

$$E_2(l = 0)e^{\alpha_2 l} = E_2(0)\cosh \Gamma l + i\frac{\kappa_2}{\Gamma} E_3 E_1^*(0)\sinh \Gamma l. \tag{78b}$$

Letting $e^{\alpha_1 l} \sim 1 + \alpha_1 l$ and $e^{\alpha_2 l} \sim 1 + \alpha_2 l$ and taking the complex conjugate of Equation (78), the solution of the two simultaneous equations is

$$\cosh \Gamma l = 1 + \frac{\alpha_1 \alpha_2 l^2}{2 + \alpha_1 l + \alpha_2 l}. \tag{79}$$

For low loss at both waves, or the doubly resonant oscillator case (DRO), Equation (79) reduces to

$$\Gamma^2 l^2 \sim \alpha_1 \alpha_2 l^2 = a_1 a_2, \tag{80}$$

where for small losses the round-trip electric field loss αl equals the single-pass power loss a. Inclusion of the phase synchronism factor into Equation (80) leads to the DRO threshold condition given by Equation (76).

If only one wave is resonant, the singly resonant oscillator (SRO) condition, Equation (79) with $\alpha_1 l \gg \alpha_2 l$ reduces to

$$\Gamma^2 l^2 \sim 2\alpha_2 l = 2a_2 \tag{81}$$

which again leads to the SRO threshold condition given by Equation (77).

A comparison of Equations (80) and (81) shows that the singly resonant oscillator has a threshold that is $2/a_1$ times that of the doubly resonant oscillator. For a single pass power loss of 1% the SRO threshold is thus 200 times the DRO threshold. This of course assumes that both waves of the DRO are resonant simultaneously. In practice, this requires a single frequency pump source and careful cavity control to assure resonance at both parametric waves. Lack of cavity length control or pump frequency control leads to instabilities in the threshold condition resulting in amplitude fluctuations of the oscillator. For these reasons, the lower threshold advantage of the DRO is not utilized except where necessary such as for cw operation.

The possibility of cw operation was first pointed out by Boyd and Ashkin[102]. It was not until 1968 that Smith *et al.*[73] using a 5 mm crystal of $Ba_2NaNb_5O_{15}$ pumped at 0.532 μm achieved cw operation at degeneracy. The threshold for that experiment was 45 mW. Later, Smith demonstrated a cw pumped oscillator with only 3 mW threshold.

Using an idea proposed by Harris[112], that the full power of a multiaxial-mode laser may be used to pump a parametric oscillator if the DRO axial-mode interval equals that of the pump, Byer *et al.*[77] constructed a visible cw DRO using a 1.7 cm $LiNbO_3$ crystal. This oscillator tuned from 0.68–0.71 μ and 1.9–2.1 μ for a 0.5145 μ argon ion laser pump. It achieved a threshold of less than 500 mW in a long cavity that included an internal lens to achieve proper focussing.

The low gains and resultant difficulty of constructing cw oscillators has held back work in this area. However, recently Laurence and Tittel[75] and Weller and Andrews[76] have extended work on cw argon-ion laser pumped $Ba_2NaNb_5O_{15}$ parametric oscillators. Weller and Andrews achieved 60 mW of output at a total efficiency of 15% in a tuning range from 0.93 μ to 1.15 μ.

The resultant net gain of the round trip SRO has been considered by Harris[79]. The analysis is a straightforward extension of Equations (78a) and

(78b) to include the net gain after an arbitrary phase shift suffered upon reflection from the end mirror of the SRO cavity. The importance of phase on the gain of an amplifier and DRO has also been considered by Smith[113] and Bjorkholm *et al.*[114].

At $\Delta k = 0$ the net round-trip SRO gain for a zero phase shift is $\Gamma^2(2l)^2$. Thus the threshold is reduced by a factor of four relative to the single-pass gain case. For a phase shift upon reflection of $\Delta(\varphi_p - \varphi_s - \varphi_i) = \pi/2$ the gain of the first pass is exactly cancelled by the return pass so that the net gain is zero. However, as shown by Harris[79] the gain is not zero at $\Delta k \neq 0$ and has a maximum of $0.4\Gamma^2(2l)^2$ at $|\Delta k|(2l) \lesssim 2\pi$. Therefore, the round trip SRO reduces threshold compared to the single pass SRO by 1.6 to 4 times. Here we have assumed that all three frequencies are perfectly reflected—a situation that may be difficult to achieve in practice.

In addition to the DRO and SRO with a single-pass pump wave generating single-pass parametric gain, there exist obvious extensions involving a round-trip pump wave. The advantage of reflecting the pump wave back through the nonlinear crystal is a reduction in threshold and possibly a higher conversion efficiency.

The round-trip DRO has been discussed by Bjorkholm *et al.*[114]. The threshold is

$$\Gamma^2 l^2 \operatorname{sinc}^2\left(\frac{\Delta kl}{2}\right) = \frac{a_1 a_2}{1 + R_3^2 + 2R_3 \cos(\Delta\varphi + \Delta kl)}, \tag{82}$$

where $\Delta\varphi = \varphi_3 - \varphi_2 - \varphi_1$ and R_3 is the electric field reflectivity of the second cavity mirror at the pump frequency. For $R_3 = 0$ this reduces to Equation (76). The threshold depends on $\Delta\varphi$ and Δkl, however, for $R_3^2 > 0.12$ the double pass DRO always has a lower threshold. For the ideal case of $\Delta\varphi = 0$ the threshold is reduced by a factor of four.

There are two variations of the round-trip SRO; the first where only the pump wave is reflected by the second cavity mirror which is transparent to the non-resonant waves and the second where all three waves are reflected by the second cavity mirror.

In the first case Bjorkholm *et al.*[114] show that the threshold is

$$\Gamma^2 l^2 \operatorname{sinc}^2\left(\frac{\Delta kl}{2}\right) = \frac{2a_1}{1 + R_3^2}. \tag{83}$$

For this case the threshold is independent of the phase and for $R_3 \sim 1$ is a factor of two lower than the single pass SRO. The only disadvantage of this device is the requirement for mirror coatings that reflect both the pump and the resonant wave.

The SRO where all three waves are reflected from the second cavity mirror has been discussed previously. For this case the threshold is dependent on the relative phase of the reflected waves. In the low gain limit the reduction in threshold is the same as that for the round-trip DRO given by Equation (82) if the reflectance is equal for all three waves at the second mirror. (Here $\cos(\Delta\varphi + \Delta kl) = \pm 1$ for proper and improper phase upon reflection.) As the gain is increased toward the high gain limit, the net double pass SRO gain approaches $(\cosh^2 \Gamma l \pm \sinh^2 \Gamma l)^2 - 1$. Thus the net gain varies between zero and $e^{4\Gamma l}$ depending on phase.

C. RISE-TIME

Thus far we have discussed the threshold of a parametric oscillator under steady state conditions. In fact, most optical parametric oscillators are presently constructed using a Q-switched pump source with pump pulse lengths that typically vary from 10 nsec to 1 μsec. When the pump is incident on the nonlinear crystal the initial signal and idler fields are amplified. However, the oscillator is not "on" until the fields are amplified from the initial noise level to a magnitude of the order of the pump field.

The number of round-trips in the optical cavity necessary to amplify fully the signal and idler waves, multiplied by the cavity round trip time, leads to the rise time necessary to achieve threshold. The excess gain which is proportional to the incident pump power, the pump pulse length, and the cavity length are important parameters in the threshold of a pulsed parametric oscillator.

The rise-time of a DRO for a step input pump pulse has been discussed by Byer[85] and Kreuzer[115]. Recently, Pearson *et al.*[116] extended the analysis to include an arbitrary input pulse shape for both the DRO and SRO cases.

Before discussing these results, I give an approximate solution to the rise time of a DRO and SRO based on the model discussed above. In particular, the rise-time can be expressed as an additional equivalent loss to be exceeded to reach threshold.[117]

For equal signal and idler losses, Equation (78a) reduces to

$$P_1(t) = P_1(t = 0)e^{2\Gamma(t)l - 2a}, \tag{84}$$

where $P_1(t = 0)$ is the initial noise power present at the time the pump pulse is incident on the crystal. The power P_1 grows for the duration of the pump pulse τ_p at a rate proportional to the excess gain. Letting the number of single-pass trips during the pump pulse duration be $n = \tau_p/\tau_{cav} = \tau_p c/L_c$ where L_c is the equivalent length of the cavity plus crystal, we have

$$\ln\left(\frac{P_1(t)}{P_1(0)}\right) = \frac{c}{L_c}\int_0^{\tau_p} (2\Gamma(t')l - 2a)dt'$$

$$\sqrt{\Gamma^2 l^2} = a + \frac{1}{2n}\ln\left(\frac{P_1(t)}{P_1(0)}\right), \tag{85}$$

where the second term is an equivalent loss due to the finite rise-time of the DRO. As an example, if we let $P_1(t)/P_1(0) = 10^{12}$, $\tau_p = 200$ nsec, and $L_c = 6$ cm, the equivalent loss is 1.4%. However, if $\tau_p = 20$ nsec the loss increases to 14%.

The exact value of $P_1(t)/P_1(0)$ is not very critical since the ratio appears in a logarithm. However, we have let $P_1(0)$ equal the noise power present in the cavity for a single mode. As discussed previously, the noise is an equivalent single photon so that $P_1(0) \approx \hbar\omega_1/\tau_{cav}$. For definiteness we let $P_1(t) = a_1 P_3$ where P_3 is the peak pump power.

Applying the above procedure to the SRO we have

$$P_1(t) = P_1(0)e^{\Gamma^2 l^2 - 2a_1}, \tag{86}$$

so that

$$\Gamma^2 l^2 = 2a_1 + \frac{1}{n}\ln\left(\frac{P_1(t)}{P_1(0)}\right). \tag{87}$$

For the SRO with the above parameters the equivalent loss is 2.8% and 28% for $\tau_p = 200$ nsec and $\tau_p = 20$ nsec.

A more direct way to determine the rise time of a parametric oscillator is to integrate Equation (56) with time dependence included. For the DRO we assume that the spatial variation of E_1 and E_2 are negligible so that the coupled equations reduce to

$$\begin{aligned} \frac{1}{c_1}\frac{\partial E_1}{\partial t} + \alpha_1 E_1 &= i\kappa_1 E_3 E_2^* \exp(i\Delta k z) \\ \frac{1}{c_2}\frac{\partial E_2}{\partial t} + \alpha_2 E_2 &= i\kappa_2 E_3 E_1^* \exp(i\Delta k z), \end{aligned} \tag{88}$$

where $c_1 = c/n_1$ and $c_2 = c/n_2$ are the phase velocities. Assuming $\Delta k = 0$, $\alpha_1 = \alpha_2 = a/L_c$ where L_c is the effective cavity length, and a step pump source turned on at $t = 0$ the calculated rise-time is

$$\tau_R = \frac{(L_c/c)\ln(P_1(t)/P_1(0))}{2a[\sqrt{N} - 1]}, \tag{89}$$

where $N = \Gamma^2 l^2/a^2$ is the number of times above threshold. When $\tau_R = \tau_p$ this solution agrees with Equation (85). At degeneracy Equation (88) integrates directly for pump pulses of the form $E_p(t) = E_p f(t)$. For example, a Gaussian pump pulse $E_p(t) = E_p e^{-t^2/\tau^2}$ where $\tau^2 = \tau_p^2/2 \ln 2$ and τ_p is the full width at half-maximum of the pump intensity, modifies Equation (89) by

$$\sqrt{N} \to \sqrt{N}\frac{1}{\tau_R}\int_0^{\tau_R} f(t)dt \sim 0.88\sqrt{N}, \tag{90}$$

for $\tau_R < \tau_p$.

Pearson *et al.*[116] have considered the rise-time of parametric oscillators in detail. Figure 8a shows the calculated ratio of rise-time to pump pulse length τ_R/τ_p versus cavity length for various values of N_0 where $N_0 = N(100\,\alpha')$ with N the number of times above threshold and α' the percentage single pass power loss of the resonant wave. For 1 % loss N_0 is the number of times above the cw threshold. For the DRO Pearson *et al.*[116] plot a minimum number of times above cw threshold the oscillator must be pumped in order to achieve a pulsed threshold. This result is shown in Fig. 8b. The solid lines are the exact result corresponding to the use of integral Equation (90). The dashed results assume a steady state square pump pulse solution given by Equation (85). For most practical cases the steady state and exact results are in good argeement. Figures 8a and 8b illustrate the importance of keeping the cavity length to a minimum in pulsed parametric oscillator operation.

Operation of pulsed parametric oscillators leads to consideration of a minimum threshold energy density rather than power density. The peak gain must remain on long enough to allow the fields to build up to the threshold value. For the DRO the minimum threshold energy density follows from Equation (85)

$$W_{\min}^{\text{DRO}} > \frac{I_{pk}}{\sqrt{N} - 1}\left(\frac{\tau_{\text{cav}}}{2a}\right)\ln\left(\frac{P_1(t)}{P_1(0)}\right). \tag{91}$$

Where the DRO is pumped at peak intensities many times above threshold such that $\sqrt{N} > 1$ the minimum energy necessary to reach threshold approaches a limiting value proportional to $\sqrt{I_{pk}}$ and independent of the loss given by

$$W_{\min}^{\text{DRO}} \sim \frac{I_{pk}}{2\Gamma l}\frac{\tau_{\text{cav}}}{2}\ln\left(\frac{P_1(t)}{P_1(0)}\right). \tag{92}$$

For the SRO the minimum pump energy density is

$$W_{\min}^{\text{SRO}} = \frac{I_{pk}}{(N-1)}\frac{\tau_{\text{cav}}}{2a}\ln\left(\frac{P_1(t)}{P_1(0)}\right), \tag{93}$$

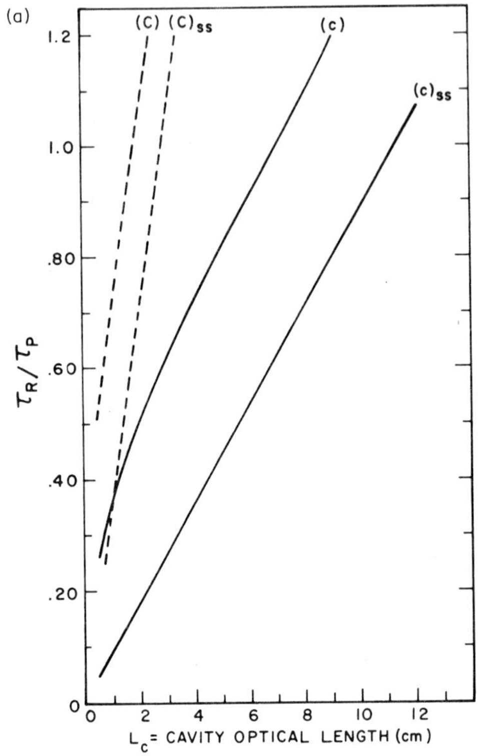

FIG. 8. (a) DRO rise time τ_R per pump pulse width τ_p versus cavity length. Case c is exact and c_{ss} steady state analysis at $\tau_p = 200$ nsec, and case C is exact and C_{ss} is steady state analysis at $\tau_p = 50$ nsec. Here $\alpha' = 1\%$ (after Pearson *et al.*[116]).

which in the limit $N > 1$ becomes

$$W_{\min}^{\text{SRO}} \sim \frac{I_{pk}}{\Gamma^2 l^2} \tau_{\text{cav}} \ln\left(\frac{P_1(t)}{P_1(0)}\right) \tag{94}$$

It is interesting to note that the minimum threshold energy ratio for the DRO to SRO in the limit of $N > 1$ reduces to

$$\frac{W_{\min}^{\text{DRO}}}{W_{\min}^{\text{SRO}}} = \frac{N^{\text{SRO}} - 1}{\sqrt{N^{\text{DRO}}} - 1} \approx \frac{\Gamma l}{2}, \tag{95}$$

which as expected shows the low threshold advantage of the DRO at low gains. However, for high gain operation the SRO requires the minimum energy to achieve oscillation from consideration of oscillation build-up time. For a 5 cm $LiNbO_3$ crystal pumped at 20 MW/cm^2, a typical operating

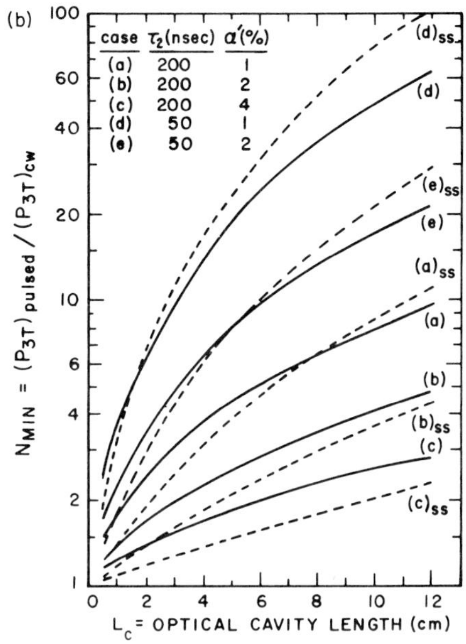

FIG. 8. (b) Minimum number of times the cw DRO threshold to reach threshold for the pulse DRO versus cavity length. Dashed lines are steady state analysis (after Pearson *et al.*[116])

intensity, $\Gamma^2 l^2 = 25$ or $\Gamma l/2 = 2.5$. Thus we may expect that a partially reflecting DRO using $LiNbO_3$ at this gain will reach threshold and oscillate as a SRO. This may be important in determining the frequency and power stability of the oscillator in view of the expected frequency fluctuations for a low gain DRO.

Recent experiments with a 1.06 μm pumped SRO $LiNbO_3$ parametric oscillator have confirmed the expected operation as an SRO in spite of feedback at both the non-resonant and resonant waves[118]. The frequency instability expected for the cw SRO with feedback as calculated by Falk[119] was not observed for the high gain SRO oscillator operation. This important result will be discussed further in the section on tuning and bandwidth.

D. CONVERSION EFFICIENCY

Parametric oscillators are potentially efficient devices for converting fixed frequency pump radiation to tunable signal and idler output. In this section we examine the theoretical conversion efficiency for various parametric oscillator configurations. We proceed by considering a steady state solution for interacting plane waves. In addition, we assume that the resonant fields vary slowly in passing through the nonlinear medium. This assumption allows solutions to be obtained that are valid even for high conversion efficiency without having to obtain exact solutions for coupled nonlinear equations.

Following the above discussion, modifications to the plane-wave results for multimode Gaussian amplitude beams are discussed. In addition, non-steady state results and exact solutions of the coupled equations are compared to the steady state and slowly-varying resonant-wave approximation solutions.

The conversion efficiency for a DRO was first treated by Siegman[120]. His assumption that the spatial variation of the resonant waves is nearly constant is the basis for the following conversion efficiency calculations. For the DRO we assume that dE_2/dz and dE_1/dz are negligible so that Equation (56) for the remaining forward and backward travelling pump waves becomes

$$\frac{d\mathscr{E}_3^+}{dz} = i\kappa_3 \mathscr{E}_2(0)\mathscr{E}_1(0)e^{-i(\Delta kz + \Delta\varphi_+)}, \tag{96a}$$

$$\frac{d\mathscr{E}_3^-}{dz} = -i\kappa_3 \mathscr{E}_2(0)\mathscr{E}_1(0)e^{i(\Delta kz - \Delta\varphi_-)}, \tag{96b}$$

where $E_i = \mathscr{E}_i e^{i\varphi_i}$ and $+$ and $-$ refer to the forward and backward directions.

These equations can be integrated under the present assumption that $\mathscr{E}_2(0)$ and $\mathscr{E}_1(0)$ are independent of z. Integrating Equation (96a) gives

$$\mathscr{E}_3^+(l) = \mathscr{E}_p(0) + i\kappa_3 \mathscr{E}_2(0)\mathscr{E}_1(0)l \,\mathrm{sinc}\left(\frac{\Delta kl}{2}\right)e^{-i(\Delta kl/2 + \Delta\varphi_+)}, \tag{97}$$

where $\mathscr{E}_3(0)$ is the incident pump field at $z = 0$. A similar expression follows integration of Equation (96b) for $\mathscr{E}_3^-$ except that $\mathscr{E}_3^-(l) = 0$ is the assumed boundary condition. For maximum energy transfer from the pump to the signal and idler $\Delta\varphi_+ + \Delta kl/2 = -\pi/2$.

We now use an extension of Equation (66) to relate the signal and idler powers by

$$\frac{\mathscr{E}_2^2(0)}{\mathscr{E}_1^2(0)} = \frac{\omega_2 n_1 a_1}{\omega_1 n_2 a_2}, \tag{98}$$

where as before a is the single-pass power loss. In addition, conservation of intensity leads to

$$n_3[\mathscr{E}_3^2(0) - \mathscr{E}_3^{+2}(l) - \mathscr{E}_3^{-2}(0)] = 2a_2 n_2 \mathscr{E}_2^2(0) + 2a_1 n_1 \mathscr{E}_1^2(0). \tag{99}$$

Solving Equations (97), (98), and (99) for the signal and idler output powers gives

$$\frac{\omega_3}{\omega_2}\frac{P_2}{P_3(0)} = \frac{\omega_3}{\omega_1}\frac{P_1}{P_3(0)} = 2\frac{1}{N}[\sqrt{N} - 1], \tag{100}$$

where N is the number of times above threshold, defined by $\Gamma^2 l^2 = a_1 a_2$, where Γ^2 is given by Equation (67).

The conversion efficiency of the DRO is therefore

$$\eta = \frac{P_1 + P_2}{P_3(0)} = \frac{2}{N}[\sqrt{N} - 1], \tag{101}$$

which reaches a maximum of 50% at four times above threshold. The generated power is divided between the two waves in the ratio ω_2/ω_1 since equal signal and idler photons are generated.

The actual output power transmitted through the cavity mirrors with transmittance $T_1^2 = 1 - R_1^2$ is

$$\frac{\omega_3}{\omega_1}\frac{P_1}{P_3(\mathrm{TH})} = 2\left(\frac{T_1^2}{a_1}\right)[\sqrt{N} - 1], \tag{102}$$

where $P_3(\mathrm{TH})$ is the threshold pump power and $a_1 = T_1^2 + a_D$ is the sum of the mirror transmittance and dissipative losses.

The transmitted pump power P_3' limits at the threshold power $P_3(\mathrm{TH})$ which forms the basis for the optical limiter first proposed by Siegman[120]. Finally, the reflected pump power is given by

$$\frac{P_3(\mathrm{R})}{P_3(\mathrm{TH})} = [\sqrt{N} - 1]^2. \tag{103}$$

The reflected pump power is not truly "reflected" but is generated by sum generation of the back-travelling signal and idler waves. The back-generated pump acts to reduce the signal and idler powers and simultaneously feed power back into the pumping laser which induces pump laser frequency and power instabilities.

Bjorkholm[121] analyzed the DRO without the back-generated pump wave and showed that its efficiency increases from 50% to 100% at $N = 4$. One method to insure no backward pump generation is to construct a ring cavity oscillator. This oscillator is referred to as a ring resonator oscillator (RRO) which implies that it is doubly resonant. Recently Fischer[122] has treated the RRO in detail.

The conversion efficiency calculation for the RRO proceeds as for the DRO except there is only forward interaction. Equating the signal and idler powers coupled out of the cavity to the power lost by the pump and using Equation (98) gives

$$\frac{\omega_3}{\omega_2}\frac{P_2}{P_3(0)} = \frac{\omega_3}{\omega_1}\frac{P_1}{P_3(0)} = \frac{4}{N}[\sqrt{N} - 1], \tag{104}$$

where the threshold remains the same as for the DRO. Maximizing Equation (104) with respect to N shows that the conversion efficiency $(P_2 + P_1)/P_3(0) = 100\%$ at $N = 4$. Thus in addition to eliminating feedback into the pump laser, the conversion efficiency is increased from 50% to 100% for the RRO.

The transmitted pump power for the RRO is

$$\frac{P'_3}{P_3(0)} = 1 - \frac{4}{N}[\sqrt{N} - 1]. \tag{105}$$

Figure 9 shows the conversion efficiency and transmitted pump power for the DRO and the RRO. The results are valid in the limit of low gain and interacting plane waves. Modifications of these results for Gaussian beams and also for high gains are discussed later.

There exists one further form of the DRO in which the pump is reflected from the second cavity mirror back through the nonlinear crystal. The threshold for the double-pass DRO is reduced compared to that for the DRO

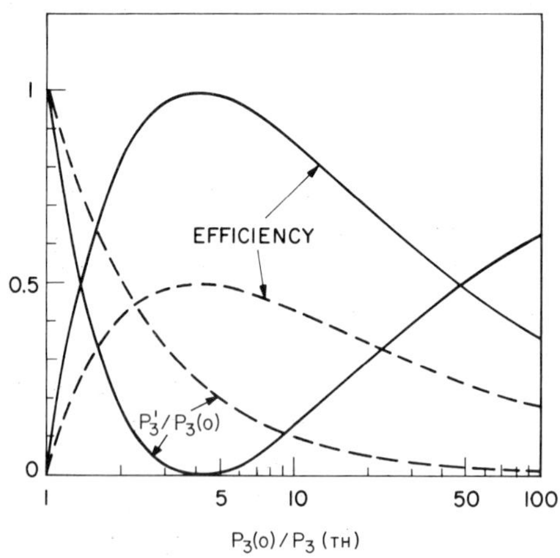

FIG. 9. Transmitted pump power, $P'_3/P_3(0)$, and efficiency of a DRO and ring resonator oscillator versus number of times above threshold in the plane wave limit. The dashed curves are for the DRO and the solid curves are for the ring resonator oscillator.

by between two and four times for $R_3 \sim 1$. In addition, the conversion efficiency is increased as shown by Bjorkholm *et al.*[114] to

$$\eta = \frac{1 + R_3^2 + 2R_3 \cos(\Delta\varphi + \Delta kl)}{1 + R_3 \cos(\Delta\varphi + \Delta kl)} \frac{2}{N} [\sqrt{N} - 1], \tag{106}$$

which for $R_3 = 1$ and the proper phase reaches 100% at $N = 4$. The double-pass DRO thus has 100% conversion efficiency similar to the ring resonator but in addition a decreased threshold. However, it has a disadvantage that the threshold conversion efficiency and optimum Δk depend on the phase change $\Delta\varphi$ induced upon reflection.

All of the DRO configurations suffer from the disadvantage of requiring simultaneous resonance at the signal and idler waves within a single cavity. Due to crystal dispersion, the axial-mode interval at the signal and idler frequencies are not the same so that the particular signal and idler cavity-modes that exactly sum to the pump frequency may occur off gain-centre. Furthermore, slight shifts in crystal index or cavity length causes the modes that align to shift about rapidly. Typically a few modes within a "cluster" are aligned or nearly aligned and the next "cluster" of modes is spaced a number of axial-mode intervals away. Giordmaine and Miller[67,68] first discussed the cluster effect which has since been observed by Bjorkholm[123]. The frequency instability due to double resonance is the main reason for sacrificing the low threshold of the DRO for operation as a single resonant oscillator whenever possible.

The SRO was first demonstrated by Bjorkholm[69] and later analyzed by Kreuzer[124]. It has since become the most common parametric oscillator configuration for a number of reasons. They include: lack of the cluster effect, use of the full multiaxial mode power of the pump laser, ease of mirror and cavity design, good conversion efficiency, and frequency and output power stability. These advantages more than offset the disadvantage of increased threshold relative to the DRO.

For the SRO in which the signal is resonant we assume $\mathscr{E}_2(z) = \mathscr{E}_{20}$ and have for the pump and idler waves that

$$\frac{d\mathscr{E}_3^+(z)}{dz} = i\kappa_3 \mathscr{E}_3(0)\mathscr{E}_1^+(z)e^{-i(\Delta kz + \Delta\varphi_+)}, \tag{107a}$$

and

$$\frac{d\mathscr{E}_1^+(z)}{dz} = i\kappa_1 \mathscr{E}_2(0)\mathscr{E}_3^+(z)e^{i(\Delta kz + \Delta\varphi_+)}. \tag{107b}$$

Solving these equations with the boundary condition $\mathscr{E}_1(0) = 0$ gives

$$|\mathscr{E}_1^+(z)|^2 = \mathscr{E}_3^2(0)\left(\frac{\omega_1 n_3}{\omega_3 n_1}\right)\frac{\sin^2 \beta z}{1 + \left(\frac{\Delta k}{2\alpha}\right)^2}, \tag{108a}$$

and

$$|\mathscr{E}_3^+(z)| = \mathscr{E}_3^2(0)\left[\cos^2 \beta z + \left(\frac{\Delta k}{2\beta}\right)^2 \sin^2 \beta z\right], \tag{108b}$$

where

$$\beta^2 = \alpha^2 + \left(\frac{\Delta k}{2}\right)^2, \tag{109}$$

and

$$\alpha^2 = \kappa_1 \kappa_3 |\mathscr{E}_2(0)|^2. \tag{110}$$

Using Equation (66) to relate the signal power to the idler power, Equation (108) can be written as

$$|\mathscr{E}_2(l)|^2 - |\mathscr{E}_2(0)|^2 = \kappa_1\kappa_2|\mathscr{E}_3(0)|^2|\mathscr{E}_2(0)|^2 l^2 \operatorname{sinc}^2 \beta l.$$

Finally, setting the signal gain $|\mathscr{E}_2(l)|^2/|\mathscr{E}_2(0)|^2 - 1$ equal to the round-trip loss $2a_2$ the signal power as a function of the incident pump power is given by the implicit relation

$$\kappa_1\kappa_2|\mathscr{E}_3(0)|^2 l^2 \operatorname{sinc}^2 \beta l = 2a_2, \tag{111}$$

or

$$\operatorname{sinc}^2 \beta l = \frac{1}{N},$$

where N is the number of times above threshold. Note that for $\beta = 0$ this reduces to the previously derived SRO threshold condition $\Gamma^2 l^2 = 2a_2$.

At line centre the SRO reaches 100% efficiency for $P_{30} = (\pi/2)^2 P_3(\mathrm{TH})$. As the pumping level increases beyond $(\pi/2)^2$ the pump wave begins to grow at the expense of the signal. Figure 10 shows the transmitted pump power for the SRO as a function of the number of times above threshold in the low-gain plane-wave approximation.

The SRO can also be extended to the double-pass configuration. For the case where all three waves are reflected the threshold is reduced by 2.0 to 4 times in the low-gain limit depending upon the phase change introduced by the reflection. However, if only the pump and resonant wave are reflected the

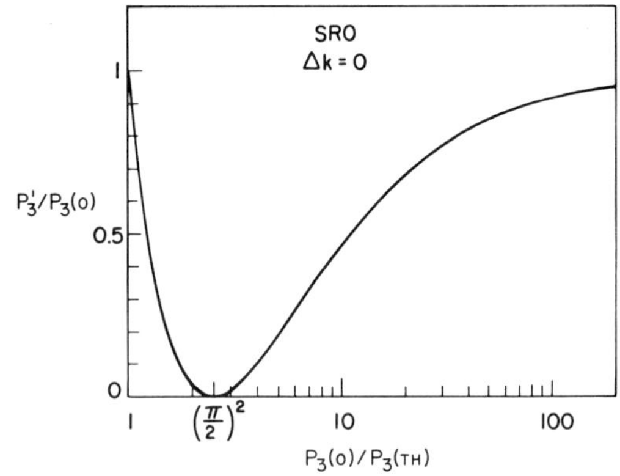

FIG. 10. Transmitted pump power, $P'_3/P_3(0)$, versus the number of times above threshold for a singly resonant oscillator in the plane-wave limit.

threshold is reduced by up to two times without phase dependence. The threshold is independent of phase since the back generated non-resonant wave builds up from noise in proper phase for maximum gain. Of more interest is the high conversion efficiency over a wider range of pump powers. Bjorkholm *et al.*[114] analyzed the SRO with pump reflectivity and showed that the resonant power is determined by

$$\operatorname{sinc}^2(\beta l) = \frac{2}{(1 + R_3^2 \cos \beta l)} \frac{1}{N}, \tag{112}$$

where R_3^2 is the pump intensity reflectivity of the second cavity mirror.

The conversion efficiency to the nonresonant wave is proportional to $1 - P_3^+(l)/P_3(0)$, and to the resonant wave, to the total pump depletion. Of particular interest is the resonant wave efficiency of nearly 100% over the wide range $2.5 \leq N \leq 8.5$. This could lead to improved conversion efficiency for non-plane wave pump sources used in actual experiments.

Early comparisons of experimental results to the calculated SRO efficiency and pump depletion did not show close agreement. Bjorkholm[125] and later Fischer[122] extended the present plane-wave analysis to multimode beams with Gaussian intensity profile. Bjorkholm then demonstrated that many previous observations agreed very well with predictions based on the Gaussian intensity model.

The model considers a multimode plane-wave pump beam with a Gaussian intensity profile.

$$I(r) = I_0 \exp\left(-\frac{2r^2}{w_0^2}\right), \tag{113}$$

where w_0 is the beam radius, incident on the parametric oscillator. Each part of the beam is taken to act independently so that parts of the incident pump above threshold take part in the parametric conversion and suffer depletion while the remainder of the beam is not affected. The analysis consists of integrating over the Gaussian profile treating the above threshold and below threshold regions independently.

For the DRO the transmitted intensity I'_3 is given by

$$\begin{aligned} I'_3 &= I_3(\mathrm{TH}); \qquad (I_3(0) > I_3(\mathrm{TH})) \\ I'_3 &= I_3(0); \qquad (I_3(0) < I_3(\mathrm{TH})). \end{aligned} \tag{114}$$

The transmitted power through the DRO is therefore

$$\begin{aligned} P'_3 &= \int_0^\infty I'_3(r) 2\pi r \, dr \\ &= 2\pi \int_0^R I_3(\mathrm{TH}) r \, dr + 2\pi \int_R^\infty I_3(0) e^{-2r^2/w_3^2} r \, dr, \end{aligned} \tag{115}$$

where R is the beam radius at which $I_3(0) = I_3(\mathrm{TH})$. Carrying out the integration

$$P'_3 = 2\pi I_3(\mathrm{TH}) R^2 + I_3(0)\left(\frac{\pi w_3^2}{2}\right) e^{-2R^2/w_3^2}. \tag{116}$$

The total power in a Gaussian beam is given by $P = (\pi w_0^2/2) I_0$ so that the threshold power is $P_3(\mathrm{TH}) = (\pi w_3^2/2) I_3(\mathrm{TH})$. Using these definitions Equation (116) reduces to

$$\frac{P'_3}{P_3(\mathrm{TH})} = 1 + \ln(N). \tag{117}$$

The efficiency

$$\eta = \frac{P_1 + P_2}{P_3(0)} = \frac{4}{N}[\sqrt{N} - 1 - \ln\sqrt{N}], \tag{118}$$

as a function of the number of times above threshold N, is less than the plane-wave efficiency by $-(4/N)\ln N$. Compared to the plane-wave case, the DRO pumped with a Gaussian intensity does not show power limiting and reaches a maximum efficiency of only 41% at $N = 12.5$.

For the ring resonator the transmitted pump power is found to be

$$\frac{P'_3}{P_3(0)} = 1 - \frac{4}{N}[\sqrt{N} - 1] + \frac{4}{N}[\ln N - (\sqrt{N} - 1)], \tag{119}$$

and the efficiency becomes

$$\eta = 1 - \frac{P'_3}{P_3(0)}, \tag{120}$$

which is only 82% at $N = 4$ instead of 100% as in the plane-wave analysis.

Figure 11(a) shows the transmitted pump power and efficiency for the DRO and Fig. 11(b) shows the transmitted pump power for the ring resonator

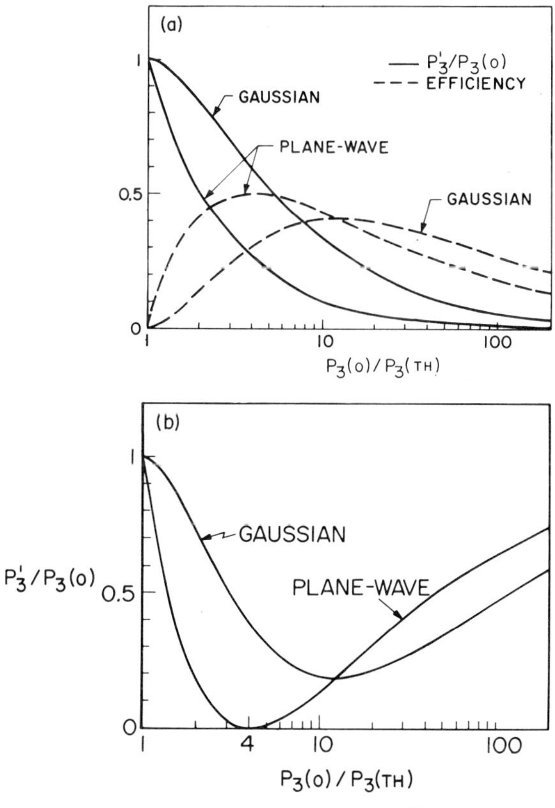

FIG. 11. (a) Transmitted pump power $P'_3/P_3(0)$ and efficiency versus number of times above threshold for a DRO. (b) Transmitted pump power $P'_3/P_3(0)$ versus number of times above threshold for a ring resonator oscillator for a Gaussian intensity wave (after Bjorkholm[125]).

with a Gaussian pump intensity. The efficiency for the plane-wave analysis is shown for comparison.

The SRO can be treated in a similar manner. In this case, however, the solution remains in an integral form since the resonant signal power is not solved for explicitly. For further discussion see Bjorkholm[125]. Figure 12 shows a comparison of the transmitted pump power for the plane-wave and Gaussian SRO as a function of N. For a multimode Gaussian intensity profile beam the maximum efficiency only reaches 71 % at 6.5 times above threshold.

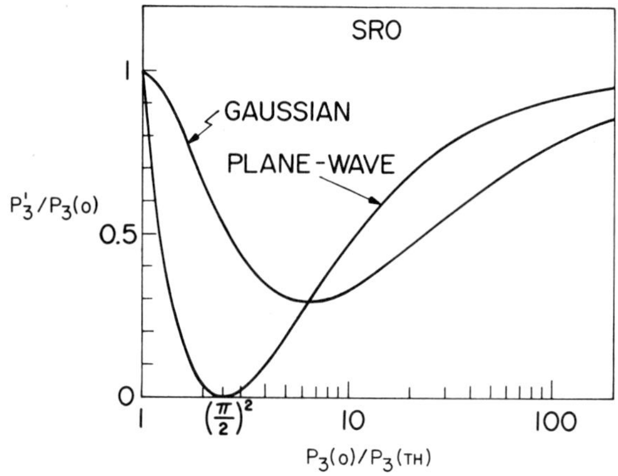

FIG. 12. Transmitted pump power $P'_3/P_3(0)$ versus number of times above threshold for a SRO for a Gaussian intensity wave (after Bjorkholm[125]).

The conversion efficiencies calculated for the various OPO configurations apply to operation in the low-gain, essentially steady state condition. In practice, the steady state condition is achieved only after a finite build-up time. Thus the energy conversion efficiency is less than the power conversion efficiency by approximately the fraction of the pumping time that the oscillator is above threshold. Again, operation of an OPO with a short cavity length to reduce build-up time leads to improvements in the conversion efficiency.

Up to this point we have considered OPO conversion efficiency with the approximation that the resonant wave is slowly varying within the cavity. The exact solution for a three-wave parametric interaction has been given by Armstrong *et al.*[19]. Bey and Tang[126] have extended Armstrong's results to the singly resonant parametric oscillator and to an OPO with intracavity up-converter and doubler.

If we let $E_i(z, \omega) = \mathscr{E}_i e^{i\varphi_i}$, the real and imaginary parts of Equation (56) become

$$\frac{d\mathscr{E}_1}{dz} = -\kappa_1 \mathscr{E}_2 \mathscr{E}_3 \sin\theta$$

$$\frac{d\mathscr{E}_2}{dz} = -\kappa_2 \mathscr{E}_3 \mathscr{E}_1 \sin\theta \tag{121}$$

$$\frac{d\mathscr{E}_3}{dz} = \kappa_3 \mathscr{E}_1 \mathscr{E}_2 \sin\theta$$

$$\frac{d\theta}{dz} = \Delta k + \left(\frac{\kappa_3 \mathscr{E}_1 \mathscr{E}_2}{\mathscr{E}_3} - \frac{\kappa_2 \mathscr{E}_3 \mathscr{E}_1}{\mathscr{E}_2} - \frac{\kappa_1 \mathscr{E}_2 \mathscr{E}_3}{\mathscr{E}_1}\right)\cos\theta, \tag{122}$$

where

$$\theta = \Delta k z + \varphi_3(z) - \varphi_2(z) - \varphi_1(z). \tag{123}$$

The total power flow is

$$I = \frac{c\varepsilon_0}{2}\left[n_1 \mathscr{E}_1^2 + n_2 \mathscr{E}_2^2 + n_3 \mathscr{E}_3^2\right]. \tag{124}$$

If we define new variables

$$u_1 = \left(\frac{n_1 \varepsilon_0 c}{2\omega_1 I}\right)^{1/2} \mathscr{E}_1$$

$$u_2 = \left(\frac{n_2 \varepsilon_0 c}{2\omega_2 I}\right)^{1/2} \mathscr{E}_2 \tag{125}$$

$$u_3 = \left(\frac{n_3 \varepsilon_0 c}{2\omega_3 I}\right)^{1/2} \mathscr{E}_3$$

where u^2 is the photon flux and

$$\zeta = \frac{d}{c}\left(n_1 \mathscr{E}_1^2 + n_2 \mathscr{E}_2^2 + n_3 \mathscr{E}_3^2\right)^{1/2} \left(\frac{\omega_1 \omega_2 \omega_3}{n_1 n_2 n_3}\right)^{1/2} z,$$

is a normalized length then Equations (121) and (122) reduce to

$$\frac{du_1}{d\zeta} = -u_2 u_3 \sin\theta$$

$$\frac{du_2}{d\zeta} = -u_3 u_1 \sin\theta \qquad (126)$$

$$\frac{du_3}{d\zeta} = u_1 u_2 \sin\theta$$

$$\frac{d\theta}{d\zeta} = \Delta s + \frac{d}{c^2}\cot\theta \frac{d}{d\zeta}(\ln u_1 u_2 u_3), \qquad (127)$$

where $\Delta s = \Delta kz/\zeta$.

The product $[u_3(0)\zeta]^2 = \Gamma^2 z^2$ where $u_3^2(0)$ is the input photon flux density. Thus ζ is a normalized nonlinear interaction length.

The solution of Equation (126) for various input and phasematching conditions is given by Armstrong *et al.*[19] Bey and Tang[126] solved the equations for the particular case of the singly resonant parametric oscillator. The results of their analysis gives important parameters for SRO operation.

Figure 13 shows the signal photon flux normalized to the input pump photon flux versus number of times above threshold. The figure shows that the circulating signal power may be many times the incident pump power

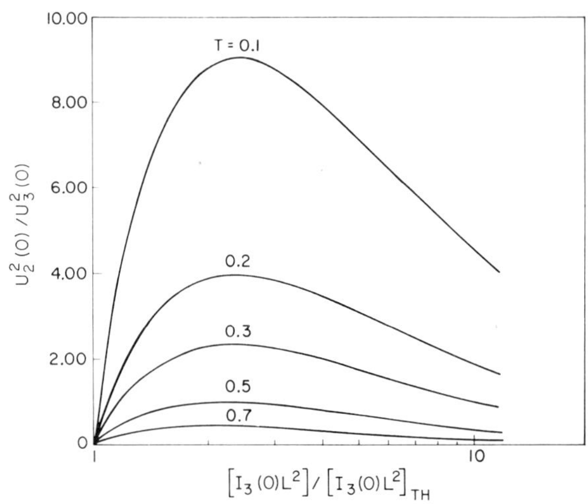

FIG. 13. Ratio of the intracavity forward-travelling signal-photon flux to the pump-photon flux versus number of times above threshold. Here $T = 1 - R_1R_2$ where R_1 and R_2 are the power reflectivities of the input and output mirrors (after Bey and Tang[126]).

especially for low loss cavities. This is important in parametric oscillators operating near the crystal or mirror burn density limit. The circulating signal power density may be reduced by resonating the lower frequency wave since

$$\frac{I_2}{I_3} = \left(\frac{\omega_2 u_2^2}{\omega_3 u_3^2}\right)\frac{A_2}{A_3},$$

where ω_2/ω_3 is the frequency ratio and A_2/A_3 is the area ratio of the signal to pump waves. Bey and Tang[126] extended their results to include a calculation of SRO threshold versus mirror transmission and quantum efficiency versus mirror transmission. They also discussed the conditions under which a second mode of the SRO reaches threshold and begins to oscillate.

Experimental measurements of OPO conversion efficiency have been made for both cw and pulsed operation and for DRO, SRO and ring resonators. Bjorkholm[69] and Kreuzer[127] investigated DRO conversion efficiencies. Using a single mode Ruby pump Kreuzer obtained 36% conversion efficiency for a DRO and 6% conversion for a SRO. Belyaev *et al.*[128] reported similar conversion efficiencies for both a DRO and SRO. Using a Ruby laser pump and a non-collinear SRO cavity arrangement, Falk and Murray[70] obtained 70% power conversion and 50% energy conversion in a $LiNbO_3$ oscillator. The energy conversion was limited by the finite build-up time of the oscillator.

Early conversion efficiency results were not in very good agreement with the conversion efficiency expected from plane-wave theory. As previously discussed, Bjorkholm[125] investigated the spatial dependence of pump depletion for multimode Gaussian amplitude waves and showed that much better agreement was obtained if the non-uniform pumping intensity was taken into account.

Wallace[72,129] described a $LiNbO_3$ SRO pumped by a Q-switched Nd:YAG laser. The oscillator operated at typically 50% conversion efficiency with threshold powers near 600 watts. Average output powers up to 70 mW have been obtained.

Using $LiNbO_3$ internal to a Q-switched Nd:YAG laser enabled Ammann *et al.*[71] to obtain an 8% conversion efficiency at 2.13 μm with up to 17 mW of average power. Recently these results were improved to over 1.2 W of average power at 2.1 μm and 60 mW at 3.1 μm.

Efficient travelling-wave oscillators have been operated in ADP and recently α-HIO_3. Yarborough and Massey[130] reported up to 25% power conversion in an ADP oscillator pumped with the fourth harmonic of Nd:YAG laser. The output pulses were typically 2 nsec in duration at 100 kW peak power. Kovrigin and Nikles[131] obtained 57% energy conversion at 1 to 1.10 μm in α-HIO_3 pumped with a doubled Nd:glass laser. The pulse length was 30 nsec at 20 MW/cm^2 peak power. At higher pump power densities the α-HIO_3 crystal damaged.

Recently Byer *et al*[119] measured a 40% energy conversion efficiency in an angle-tuned $LiNbO_3$ SRO pumped by 1.06 μm radiation from a Nd:YAG oscillator-amplifier. The parametric oscillator which uses a special [01.4] growth direction 5 cm long $LiNbO_3$ crystal[132] operated at up to 40 mJ input and 16 mJ output energies at 10 pps. Figure 14 shows the measured conversion efficiency versus the number of times above threshold.

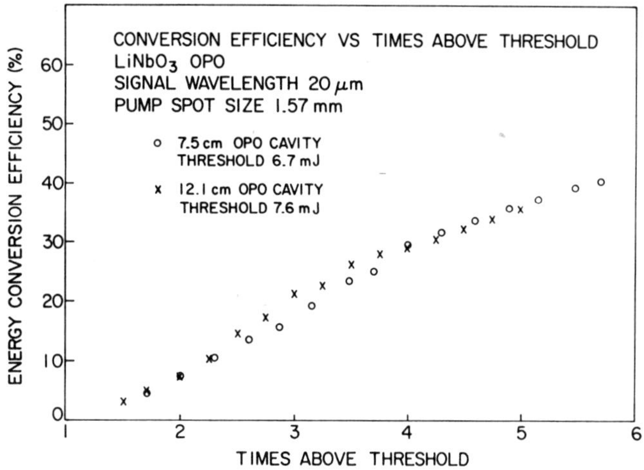

FIG. 14. Conversion efficiency versus number of times above threshold for a 1.06 μm pumped $LiNbO_3$ SRO.

The results agree well with the predictions of a Gaussian amplitude. SRO shown in Fig. 11(b), if account is taken of the rise-time on peak power reduction.

In general parametric oscillators have operated at near 50% conversion efficiency on a pulse-to-pulse basis. The peak power is limited by burn density of materials and the average power by thermal heating induced phase-mismatch. The attainable conversion efficiency especially in the near-infrared spectral region makes parametric oscillators a practical source of tunable radiation.

E. TUNING AND BANDWIDTH

The spectral properties of a parametric oscillator include tuning range and method of tuning, gain bandwidth and the detailed spectral structure within the gain bandwidth determined by the optical cavity.

Tuning is achieved by varying the crystal birefringence while simultaneously satisfying the frequency and phase matching conditions

$$\omega_3 = \omega_2 + \omega_1 \tag{58a}$$

$$\mathbf{k}_3 = \mathbf{k}_2 + \mathbf{k}_1 + \Delta\mathbf{k} \tag{58b}$$

In general, temperature tuning of the index or angle tuning by changing $n^e(\theta)$ give a wide tuning range, while pressure tuning through the photoelastic effect or electric field tuning through the electro-optic effect result in small tuning ranges.

Expanding Equation (58b) in a tuning variable $\mathfrak{m}$ and the signal frequency ω_2 gives the rate of tuning as a function of the tuning variable

$$\frac{d\omega_2}{d\mathfrak{m}} = \frac{1}{\beta_{12}}\left[\frac{\partial k_3}{\partial \mathfrak{m}} - \frac{\partial k_2}{\partial \mathfrak{m}} - \frac{\partial k_1}{\partial \mathfrak{m}}\right]. \tag{128}$$

Here $\beta_{12} = (1/u_1 - 1/u_2)$ where u_1 and u_2 are the group velocities defined by $u_n = (\partial k_n/\partial \omega_n)^{-1}$.

As shown previously (c.f. Equation (75)) the gain bandwidth of a parametric oscillator is determined by

$$\frac{\Delta kl}{2} = \pi \tag{75}$$

It is convenient at this point to expand Δk in Equation (75) as a function of frequency and write an expression for the bandwidth in the low gain limit. To second-order

$$k_n = k_{n0} + u_n^{-1}\delta\omega_n + \tfrac{1}{2}g_n(\delta\omega_n)^2,$$

where

$$u_n = \left(\frac{\partial k_n}{\partial \omega_n}\right)^{-1}, \tag{129}$$

is the group velocity and

$$g_n = \frac{\partial^2 k_n}{\partial \omega_n^2}, \tag{130}$$

is the group velocity dispersion. Since $\Delta k = |\mathbf{k}_3 - \mathbf{k}_2 - \mathbf{k}_1|$ and $k_{30} - k_{20} - k_{10} = 0$ at phasematching, we have

$$\begin{aligned}\Delta k &= \left(\frac{1}{u_1} - \frac{1}{u_2}\right)\delta\omega_2 + \tfrac{1}{2}(g_1 + g_2)(\delta\omega_2)^2 \\ &= \beta_{12}\delta\omega_2 + \tfrac{1}{2}\gamma_{12}(\delta\omega_2)^2 \end{aligned} \tag{131}$$

where $\beta_{12} = (1/u_1 - 1/u_2)$ and $\gamma_{12} = g_1 + g_2$. Here we have used $\delta\omega_2 = -\delta\omega_1$. Except for operation close to degeneracy ($\omega_2 = \omega_1$) or near a turning point in the dispersion curves $|4\pi\gamma_{12}/l\beta_{12}^2| < 1$ so that the bandwidth is

$$\delta\omega_2 = \frac{1}{2}\left|\left(\frac{\beta_{12}}{\gamma_{12}}\right)\left\{\sqrt{1 + \frac{4\pi\gamma_{12}}{l\beta_{12}^2}} - \sqrt{1 - \frac{4\pi\gamma_{12}}{l\beta_{12}^2}}\right\}\right| \approx \frac{2\pi}{l\beta_{12}}, \tag{132}$$

which is the usual expression for the gain bandwidth[79]. For the special case when $\beta_{12} \approx 0$ (group velocity matching) and $|4\pi\gamma_{12}/l\beta_{12}^2| > 1$ the bandwidth is

$$\delta\omega_2 = \left(\frac{\beta_{12}}{\gamma_{12}}\right)\sqrt{1 + \left|\frac{4\pi\gamma_{12}}{l\beta_{12}^2}\right|} \approx \sqrt{\left|\frac{4\pi}{l\gamma_{12}}\right|} \tag{133}$$

where γ_{12} is the group velocity dispersion. The dispersion constant β_{12} for $LiNbO_3$ has been calculated for various pump wavelengths [see Harris[79], Fig. 12]. The calculation of γ_{12} shows that the second-order term is not significant until tuning is within 200 cm^{-1} of degeneracy.

Comparison of Equations (128) and (132) show that β_{12} appears in both the tuning rate and bandwidth expressions. If β_{12} is small the tuning rate and bandwidth are both relatively large. In general a small tuning rate implies a small bandwidth. Expanding β_{12} in terms of the crystal index of refraction

$$\beta_{12} = \frac{1}{c}\left[n_2 - n_1 + \lambda_1\frac{\partial n_1}{\partial \lambda_1} - \lambda_2\frac{\partial n_2}{\partial \lambda_2}\right] \approx \frac{\Delta n}{c}, \tag{134}$$

and substituting into the bandwidth expression Equation (132) we find

$$|\delta\nu(\mathrm{cm}^{-1})| \approx \frac{1}{\Delta n l(\mathrm{cm})}, \tag{135}$$

where l is the crystal length and Δn the birefringence. Thus crystals with small birefringence have larger bandwidths and tuning rates than crystals with large birefringence. Equation (35) predicts a bandwidth of 12 cm^{-1} for a 1 cm crystal of $LiNbO_3$. The actual bandwidth is 4 cm^{-1}. Therefore the dispersion term can only be neglected to within this accuracy.

Figure 15 shows the measured pump, signal and idler bandwidths for a 0.659 μm pumped $LiNbO_3$ parametric oscillator. The measured oscillator bandwidth of 1.55 – 1.65 cm^{-1} for a 4 cm crystal agrees very well with the calculated half-power bandwidth of 1.60 cm^{-1}. This data was taken for oscillator operation approximately 10 times above threshold. For operation nearer threshold the resonant wave bandwidth decreases to a few tenths of a wavenumber.

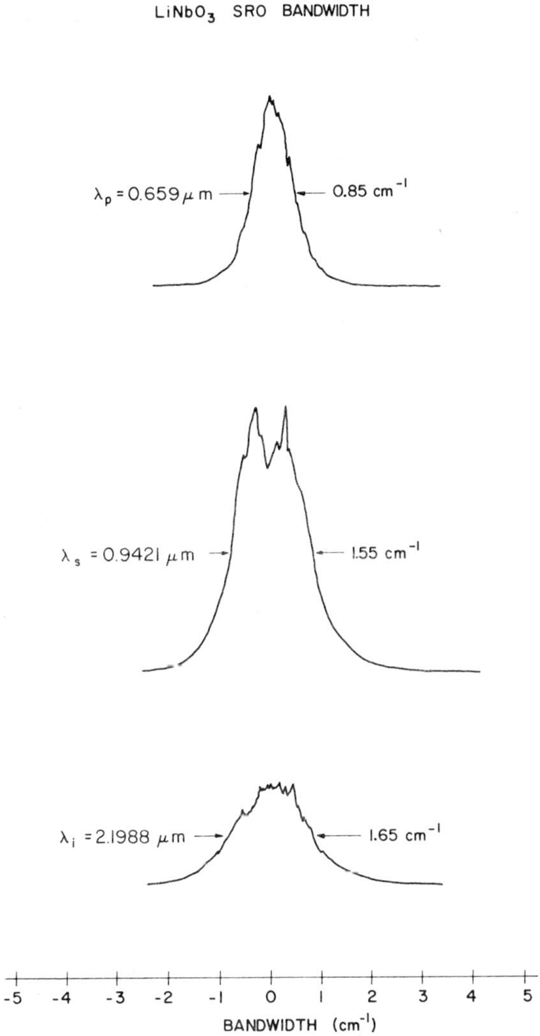

FIG. 15. Measured bandwidth for a 0.659 μm pumped $LiNbO_3$ SRO. The calculated linewidth for the 5 cm crystal is 1.60 cm^{-1}.

It is interesting to take a closer look at the rate of tuning. As an example consider electric field tuning through the electro-optic effect. The phase retardation $\Delta\varphi = \Delta kl$ is given by

$$\frac{\pi \Delta V}{V_\pi} \frac{l}{d} = \Delta kl, \tag{136}$$

where ΔV is the change in the applied field, V_π is the half-wave voltage, and l/d is the crystal aspect ratio. Substituting into Equation (128) and forming the ratio $d\omega_2/|\delta\omega_2|$, which is the number of bandwidths tuned, we find

$$n = \frac{d\omega_2}{|\delta\omega_2|} = \frac{1}{2}\frac{\Delta V}{V_\pi}\left(\frac{l}{d}\right) \quad \text{(transverse field)}, \tag{137}$$

$$= \frac{1}{2}\frac{\Delta V}{V_\pi} \quad \text{(longitudinal field)}, \tag{138}$$

for a transverse and longitudinal applied electric field. For $LiNbO_3$ $V_\pi = \lambda/(n_e^2 r_{33} - n_o^3 r_{13}) \simeq 2820$ V for the transverse field. Thus for a 10 to 1 aspect ratio crystal $n = 5\,\Delta V/V_\pi$ or approximately 5 cm^{-1} of tuning per 2820 V for a 5 cm crystal. This tuning method may have practical importance for rapid fine tuning of an oscillator about the gain line centre.

Electric field control of phasematching was first demonstrated in KDP by Adams and Barrett[133]. Kreuzer[134a] applied the technique to a Ruby-pumped OPO and observed a 6.2 Å/kV/cm tuning rate at 2.0 μm which is in good agreement with the predicted rate. Krivoshchekov *et al.*[134b] extended electro-optic tuning to a KDP OPO operating near 1 μm wavelength. To achieve a significant tuning range the KDP crystal was cooled to its Curie temperature near 123 K.

Tuning curves for parametric oscillators usually are determined prior to their operation. The straightforward approach is to solve the phasematching equations (Equations (58a) and (58b)) for signal and idler frequencies at a given pump frequency as a function of the tuning variable. To carry out the calculation the indices of refraction must be given by an analytical expression in the form of a Sellmeier equation.

For accurate calculation of tuning curves the indices of refraction must be known over the entire tuning range to an accuracy of 0.1 – 0.01 %.

Tuning curves can be measured experimentally without constructing an oscillator by observing parametric fluorescence. This method for obtaining tuning curves was first applied to $LiNbO_3$[105,109] and then extended to include ADP[106], $LiIO_3$[135] and Proustite[136]. In general the method is more useful in the visible spectral region where photomultiplier detectors or the eye can be used in conjunction with a spectrometer to measure the wavelength.

The spectral properties of an OPO are determined by the gain bandwidth, pump laser spectrum, and OPO cavity characteristics. The OPO has a finite acceptance bandwidth for the pump field that depends on the operating mode of the oscillator. For a DRO only a single longitudinal mode of the pump field is effective in driving the oscillator[112]. This limitation is due to the cavity resonance conditions that have to be met simultaneously at the signal and idler fields. If the length of the OPO cavity equals that of the pumping laser, then the full multimode power of the pump laser pumps the oscillator[112]. This condition was satisfied in the visible cw oscillator demonstrated by Byer *et al.*[77]. If the pump field is mode-locked then the signal and idler round-trip travel-time in the oscillator must equal the time between pump mode-locked pulses. This case has been treated by Harris[79].

For the SRO the maximum pump bandwidth effective in driving the oscillator depends upon the resonated oscillator field. For a resonant signal field, the signal cavity fixes the signal frequency but the free idler wave frequency varies to compensate for pump frequency changes. In addition, the full multimode power of the pump effectively couples to a single frequency through corresponding idler modes to satisfy the condition $\omega_3 = \omega_2 + \omega_1$ (see Harris[79], Fig. 13). Holding the signal frequency fixed and letting the idler frequency vary with pump wavelength leads to a pump acceptance bandwidth given by

$$\Delta\omega_3 = \frac{2\pi}{\beta_{23} l}, \tag{139}$$

where $\beta_{23} = (\partial k_2/\partial\omega_2 - \partial k_3/\partial\omega_3)$; for a resonant idler wave $\beta_{23} \to \beta_{13}$. The dispersion constant β_{23} in general does not equal β_{13} so that the pump bandwidth depends on the resonated wave. For example, a 3.2 cm $LiNbO_3$ crystal has $\Delta\omega_3 = 0.98\ cm^{-1}$ (signal resonant) and $\Delta\omega_3 = 5.8\ cm^{-1}$ (idler resonant) for a 0.473 μm pump wavelength. The doubled 0.946 μm Nd:YAG laser line has a 4 cm^{-1} bandwidth so that only one-fourth of the pump radiation effectively pumps the oscillator if the visible wave is resonant, but all of it pumps the oscillator if the infrared idler is resonant.

A useful variation of the broad band pumping of parametric oscillators is pumping with a tunable source such as a dye laser. This method of tuning proposed by Herbst and Byer[86] and demonstrated in $LiNbO_3$ by Wallace[80] has the advantages of rapid tuning over an extended wavelength region not available to the pumping source. In addition, pump tuning remains phase-matched without crystal rotation or temperature tuning since the signal and idler frequencies are not restricted. This allows the possibility of remaining at a 90° phasematch condition without adjusting the crystal temperature or angle. Figure 16 shows the theoretical and experimental tuning curve for a rhodamine-6G dye laser pumped $LiNbO_3$ parametric oscillator. In the experiment performed by Wallace the overall conversion efficiency was approximately 20% from the laser input through the optically pumped dye laser to the output of the parametric converter.

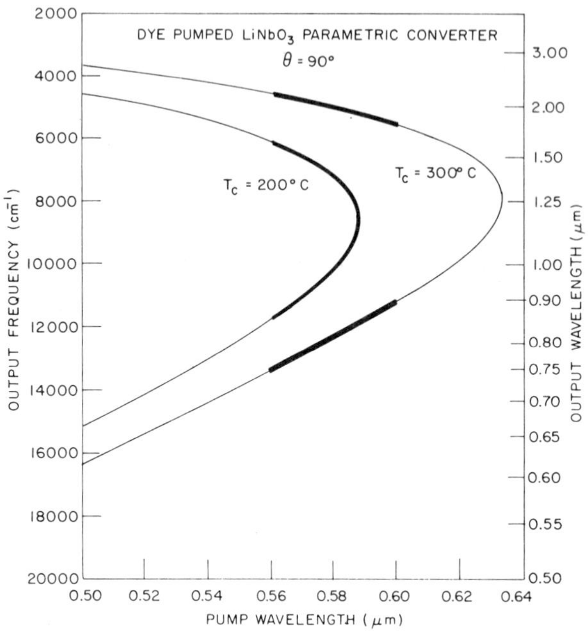

FIG. 16. Tuning curves for a 90° phasematched $LiNbO_3$ parametric oscillator pumped with a tunable source. Heavy curve indicates pump range of a rhodamine-6G dye laser.

Detailed spectral properties of a parametric oscillator depend on the longitudinal mode structure of the pump and on the $c/2L$ frequency spacing of the parametric oscillator cavity modes at the signal and idler. For the DRO there is a $c/2L$ frequency spacing for both oscillator waves and for the pump. Only when $\omega_3 = \omega_2 + \omega_1$ is satisfied does the oscillator see full gain. This usually occurs in "clusters" of modes spaced a few mode intervals apart depending on crystal dispersion[67]. The DRO spectral properties were experimentally investigated by Bjorkholm[123].

For the SRO with only one wave resonant, the non-resonant frequency is free to adjust so that $\omega_3 = \omega_1 + \omega_2$ is satisfied. Thus the SRO shows improved spectral stability compared to the DRO. This was experimentally verified by Bjorkholm[123]. Kreuzer[124] first narrowed the SRO spectrum to a single longitudinal mode. He used a tilted etalon in the resonant cavity of a Ruby laser pumped $LiNbO_3$ SRO to reduce the linewidth from 5 cm^{-1} to a single frequency.

Unlike other laser sources which show power loss as the bandwidth is narrowed, the OPO is expected to remain at full efficiency even for a single-mode operation. In this respect it behaves as a homogeneously saturated source.

Wallace[80] has demonstrated a long-term frequency stability of better than 0.005 cm^{-1} by use of a temperature-scanned infrared etalon internal to the oscillator cavity. In this experiment the OPO remained at 80% of its wideband power as expected showing a slight decrease in output power due to the etalon insertion loss.

The tilted etalon is useful for mode control in oscillators with relatively wide beam cross-sections. This led Pinard and Young[137] to study an alternate interferometric mode control technique based on an interferometer proposed by Damaschini[138a] which is a dual of the better known Smith[138b] interferometer. Pinard and Young[137] obtained a bandwidth of 0.001 cm^{-1} in the 2.5 μm spectral region tunable over a 0.015 cm^{-1} region before a second cavity mode appeared. Figure 17(a) shows a schematic of the parametric

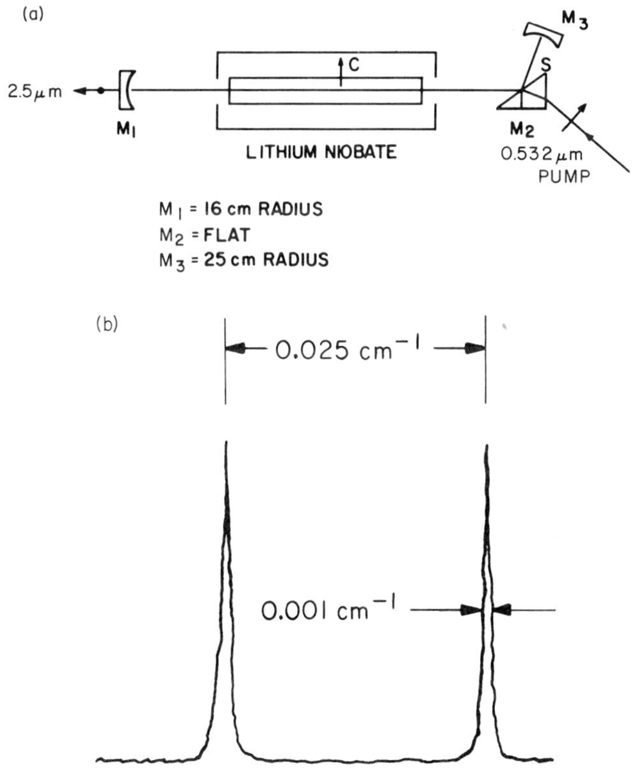

FIG. 17. (a) Schematic of an interferometrically stabilized oscillator. (b) Scanning Fabry-Perot analysis of the oscillator output at 2.5 μm (after Pinnard and Young[137]).

oscillator and the interferometer, and Fig. 17(b) shows the measured output spectrum. Operation of this oscillator was aided by the high gain properties of the $LiNbO_3$ oscillator which allowed oscillation off the crystal surface prior to the alignment of the interferometer.

One difficulty in obtaining single frequency output from an OPO is the requirement to control both the etalon centre frequency and the cavity mirror position. It is possible to tune the single frequency output over a narrow band ($\sim 0.02\ cm^{-1}$) by simultaneously scanning the interferometer and cavity mirror. For a wider tuning range the gain bandwidth of the oscillator must be reset prior to scanning the interferometer.

For operation with an etalon it is desirable to have an OPO with a narrow gain bandwidth to avoid multiple etalon transmission bands within the gain linewidth. Recently Byer *et al.*[118], used a thin tilted etalon, a two-element birefringent filter and a grating for primary line-narrowing elements within a $LiNbO_3$ parametric oscillator.

Figure 18(a) shows the $LiNbO_3$ SRO cavity design which utilizes an input dielectric beam splitter "in and around the corner" configuration. This cavity allows the insertion of line narrowing elements into the resonant signal beam but out of the 1.06 μm pump beam. The beamsplitter transmits up to 90 % of the incident "*p*" polarized pump beam and reflects 99 % of the "*s*" polarized signal beam over a 1.4 – 2.1 μm wavelength range. The output sapphire mirror is between 50 % and 70 % transmitting over the same signal wavelength range. Figure 18(b) illustrates the cavity mirror transmittance which enables SRO operation over the very wide 1.4 μm to 4.2 μm tuning range. When operated with a thin etalon or birefringent filter the grating is replaced by a broadband high reflector mirror which also reflects at the signal wavelength.

Figure 19 shows the tuning range versus crystal angle for the 1.06 μm pumped $LiNbO_3$ SRO. Also shown is the gain bandwidth. A single layer SiO_2 antireflection coating centred at 1.7 μm is sufficient to reduce the crystal reflection losses to approximately 1 % over the entire signal wavelength tuning range. The angle-tuning required to cover the tuning range is small enough that a single $LiNbO_3$ crystal is used.

Figure 20 shows the $LiNbO_3$ SRO linewidth with a 1 mm tilted etalon inserted in the cavity. The resonant signal linewidth under each etalon transmission mode is less than the resolution of the spectrometer while the non-resonant idler wave spectrum reproduces the Nd:YAG pump linewidth as expected. For operation on a single etalon mode a primary line-narrowing element is needed to reduce the operating linewidth of the oscillator to near 1 cm^{-1} from the 10–130 cm^{-1} gain bandwidth.

A thin tilted etalon with a free spectral range of 50 cm^{-1} was investigated.

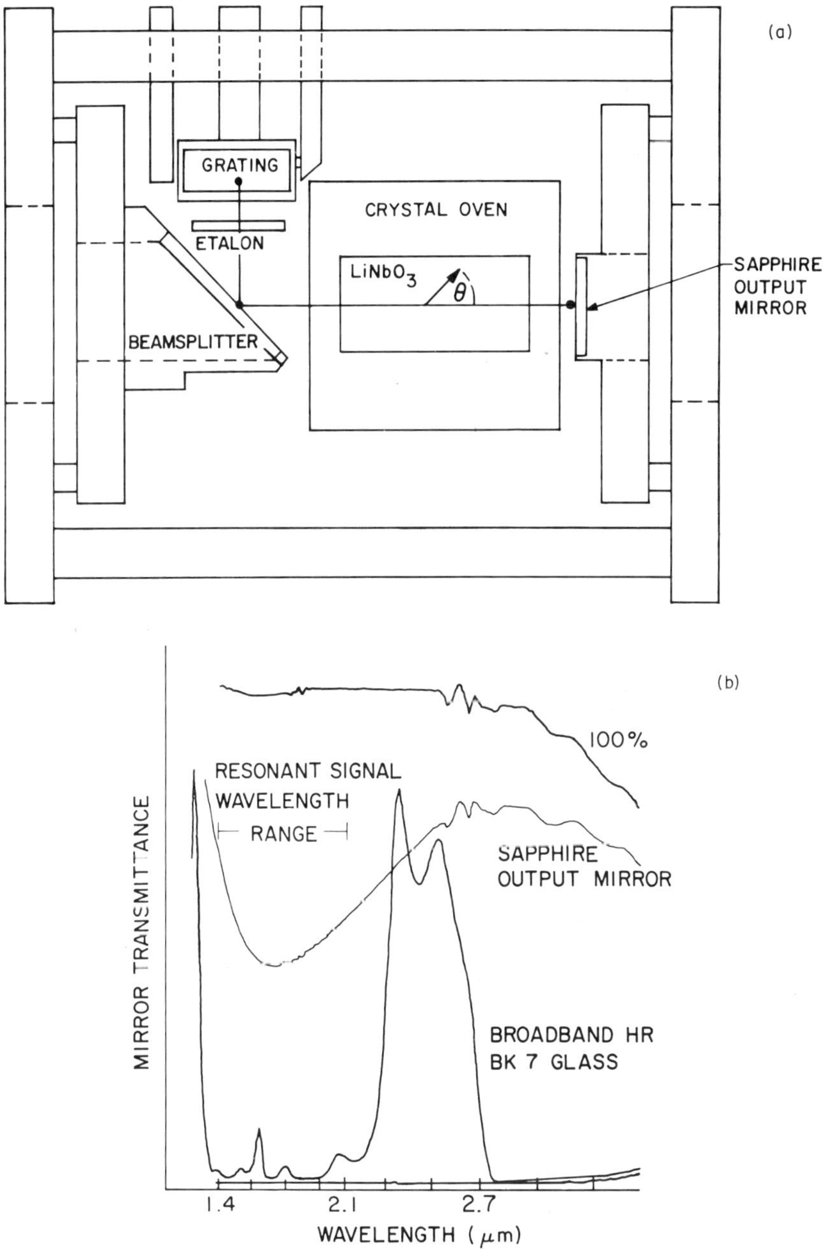

FIG. 18. (a) Cavity design for the 1.06 μm angle tuned $LiNbO_3$ SRO. (b) Broadband mirror coating transmittance at the signal wavelength 1.4 μm–2.1 μm.

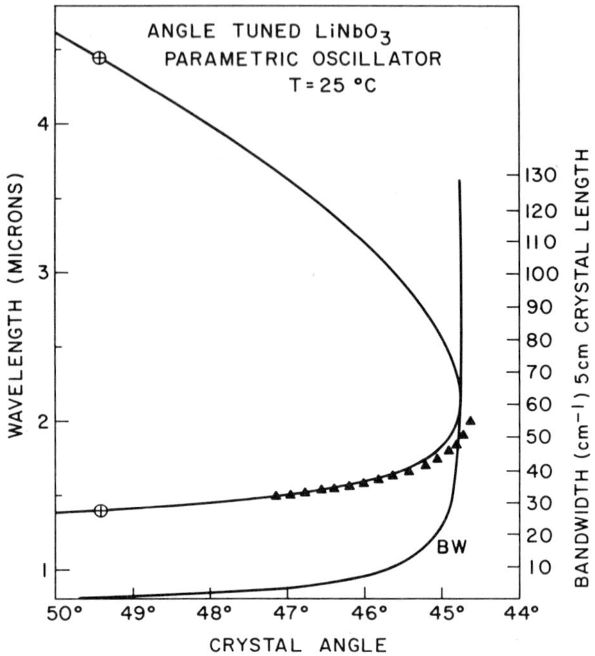

FIG. 19. Tuning curve and bandwidth for the 1.06 μm pumped angle-tuned $LiNbO_3$ SRO.

It operated as expected to reduce the oscillator linewidth to near 1 cm^{-1}. The insertion loss of the etalon is[139a]

$$l = \frac{R}{(1 - R)^2}\left(\frac{4\alpha t}{nw}\right)^2 \tag{140}$$

where R is the reflectivity, t the etalon thickness, α the tilt angle, n the index of refraction and w the oscillator spot size. As an example a 1 mm thick solid etalon with $R = 60\%$ at $w = 1.57$ mm has an insertion loss of 1.4 % at $\alpha = 2°$. Large spot sizes are therefore useful in reducing tilted-etalon insertion loss.

The second primary line-narrowing element tried was a two-element $LiNbO_3$ birefringent filter[139b]. The tuning rate of the filter when operated at Brewster's angle and rotated through an angle A about an axis through the Brewster angle is

$$\delta = \frac{2\pi T}{\lambda \sin \theta_B}[n_o - n_e(\gamma)] \tag{141}$$

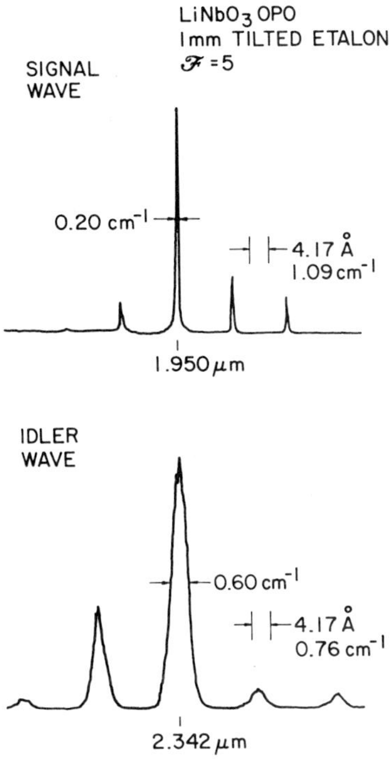

FIG. 20. $LiNbO_3$ SRO output spectrum at the resonant signal and non-resonant idler wave with a 1 mm, finesse of 5, tilted etalon.

where

$$\gamma = \tan^{-1}\left[\frac{n}{\cos A} - E\right]$$

with T the plate thickness, θ_B the Brewster angle and E the angle from the plate to the crystal optic axis. The two-element $LiNbO_3$ birefringent filter used plates of thickness $T_1 = 1.5$ mm and $T_2 = 3.0$ mm. It operated at a free spectral range of 48 cm^{-1}, which could be tuned by a 6° rotation of the plates. The tuning rate is remarkably linear, near 8 cm^{-1}/deg. As with the tilted etalon, rotation was accomplished by mounting the filter on a galvanometer which was electrically scanned over a $\pm 4°$ angular range with a resolution of 8 arc seconds.

Figure 21 shows the $LiNbO_3$ SRO bandwidth with the birefringent filter (dashed curve) and with an additional 1 mm tilted etalon. The strong modulation on the birefringent filter spectrum is due to resonant feedback from the cavity output sapphire mirror. Note that the intensity scales are arbitrary

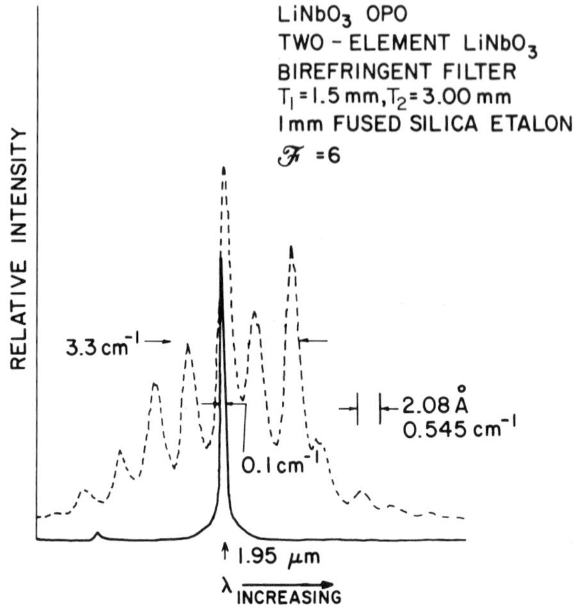

FIG. 21. The output spectrum of the $LiNbO_3$ SRO operating with a two-element $LiNbO_3$ birefringent filter (dashed) and with an additional etalon (solid).

since the oscillator did not lose significant power with the insertion of either or both line-narrowing elements. Again, the output linewidth of the oscillator with the etalon is less than the resolution of the 1 meter grating spectrometer.

The third primary line-narrowing element utilized was a grating operating in the Littrow mode. The resolution of the grating is

$$\frac{\lambda}{\Delta\lambda} = \frac{m}{2d\cos\varphi}\pi w_0$$

which reduces to an equivalent linewidth in cm^{-1} of

$$\Delta\nu = \frac{1}{\pi w_0 \tan\varphi} \tag{142}$$

For the $600l/mm$ 1.8 μm blaze grating used, $\varphi \sim 20°$ and $\Delta\nu = 5.6\ cm^{-1}$ for $w_0 = 0.157$ cm. Again a larger spot size is advantageous for improving resolution. The 5.6 cm^{-1} resolution is broader than the expected oscillator linewidth due to multipass line-narrowing by a factor of three. Figure 22 shows the measured output spectrum with the grating only (dashed) and with the additional tilted etalon operating at two angles to illustrate tuning. The

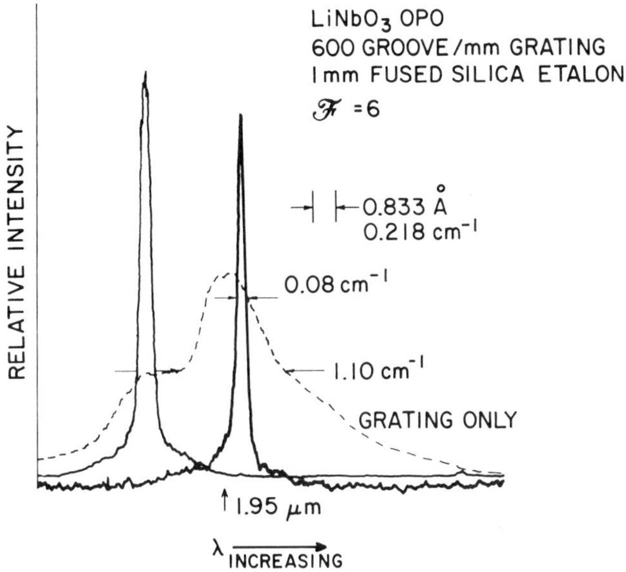

FIG. 22. The output spectrum of the $LiNbO_3$ with a grating (dashed) and a grating plus a tilted etalon (solid). The etalon was set at two angles to illustrate tuning.

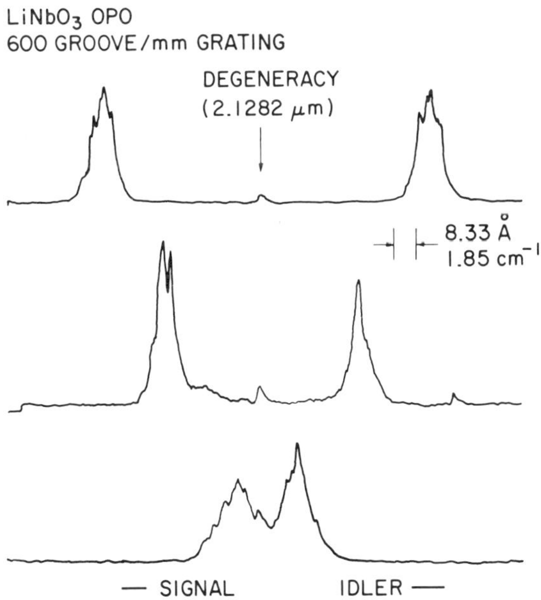

FIG. 23. Tuning of the $LiNbO_3$ SRO near degeneracy using the $600l$/mm grating.

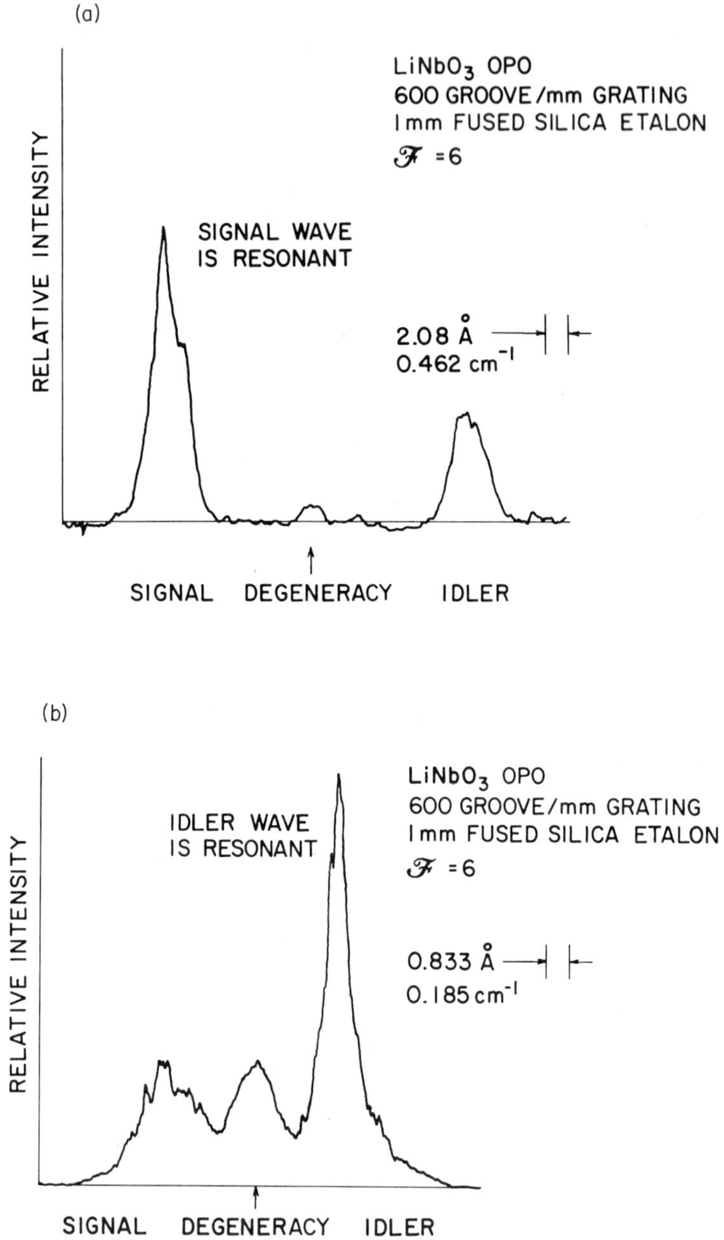

FIG. 24. (a) Grating plus tilted etalon operation with the signal resonant near the degeneracy. (b) Operation with the idler resonant.

linewidth of the etalon-narrowed output is less than 0.08 cm^{-1}. Somewhat surprisingly, the replica grating operated at over 100 MW/cm^2 incident intensity without damage. Due to the large TEM_{00} mode spot size of the parametric oscillator a beam expanding telescope is not needed prior to the grating. The grating angle was set by a stepper motor driven sin-drive stage with a resetability of ± 0.1 Å.

The operation of the parametric oscillator with a grating offers significant advantages in controlling the output spectrum. One of the advantages is accurate control of wavelength especially near the degeneracy region. Figure 23 illustrates the $LiNbO_3$ SRO tuning near degeneracy with the grating.

Finally, to illustrate the frequency control available with the grating plus etalon, Figs 24(a) and (b) show tuning at degeneracy with the signal and then the idler wave resonant. The figures show that the oscillator operates stably at degeneracy in a singly resonant mode in agreement with the predictions made by the rise-time calculations. For these measurements the oscillator operated at 10 pps with a 20 nsec pulse width 1.06 μm pump beam. Details of this oscillator, which forms a primary tuning element in a computer controlled coherent spectrometer, are discussed by Byer *et al.*[118].

V. Nonlinear Materials

A. Material Requirements

Nonlinear crystals must satisfy four criteria if they are to be useful for nonlinear optical applications. These criteria are adequate nonlinearity and optical transparency, proper birefringence for phasematching, and sufficient resistance to optical damage by intense optical irradiation. These properties are briefly discussed in the next four sections. They are then illustrated by descriptions of useful nonlinear crystals.

(a) *Nonlinearity*

In the early days of nonlinear optics adequate laser power was not always available to take full advantage of the potential conversion efficiency of a nonlinear crystal. Under those circumstances, the highest crystal nonlinearity was a very important factor. Since the late 1960s the situation has changed chiefly due to the availability of well controlled, high peak power laser sources at many wavelengths across the ultraviolet, visible, and infrared spectral regions. Under present circumstances, the crystal nonlinearity becomes only one factor in determining the crystal's potential for nonlinear applications. For example, if more than adequate laser power is available, then the nonlinear conversion efficiency is determined by the maximum incident intensity the crystal can withstand without damaging. In addition, the nonlinear conversion efficiency is then more usefully expressed at a given intensity rather than power level. Finally, in comparing nonlinear crystals with more than five orders of magnitude range in their figures of merit as illustrated by Fig. 25 one finds that most crystals have very nearly

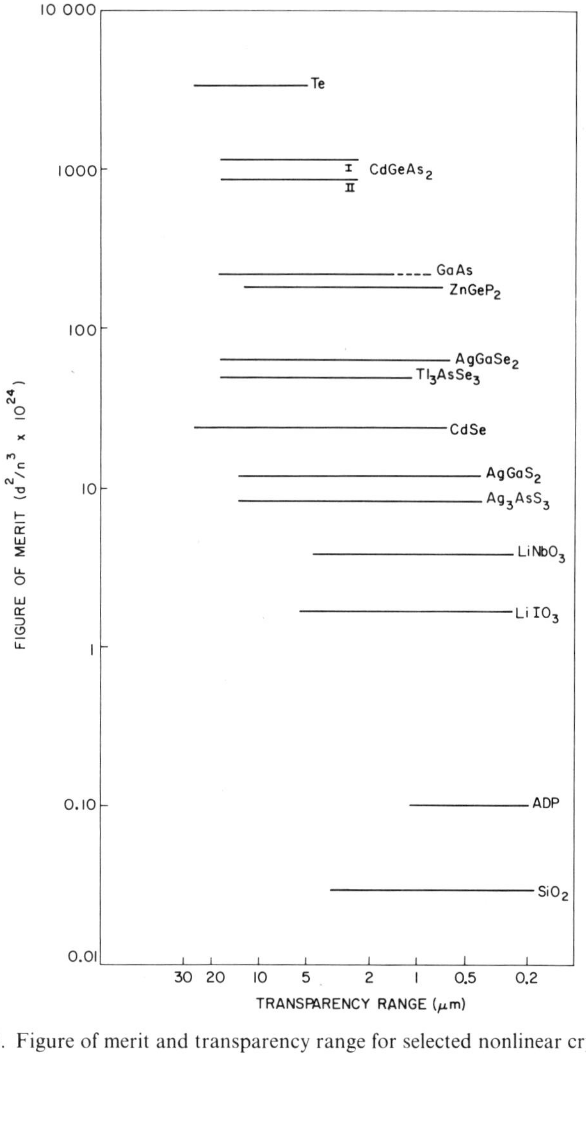

FIG. 25. Figure of merit and transparency range for selected nonlinear crystals.

the same conversion efficiency at a fixed intensity. This remarkable fact is illustrated by Fig. 26 which shows the parametric gain for a number of nonlinear materials. Under the present circumstances, the magnitude of a crystal's nonlinear coefficient is not of paramount importance, but must be taken into account along with the other material parameters.

The measurement of crystal nonlinear coefficients has continued at a rapid pace so that there are extended tables available listing nonlinear coefficients[35,85,95]. The measurement of absolute nonlinear coefficients has been carried out by a number of methods including second harmonic generation[140,141], mixing[58], and parametric fluorescence[107]. The latter method has the advantage of being a relative power measurement and not requiring the absolute measurement of power as for SHG. Relative nonlinear coefficient measurements have also been made by three methods: the powder technique[142], the Maker fringe method[4,143], and the wedge technique[144–146]. The latter method is particularly useful in that it gives, in addition to the relative magnitude of the nonlinear coefficient, the relative

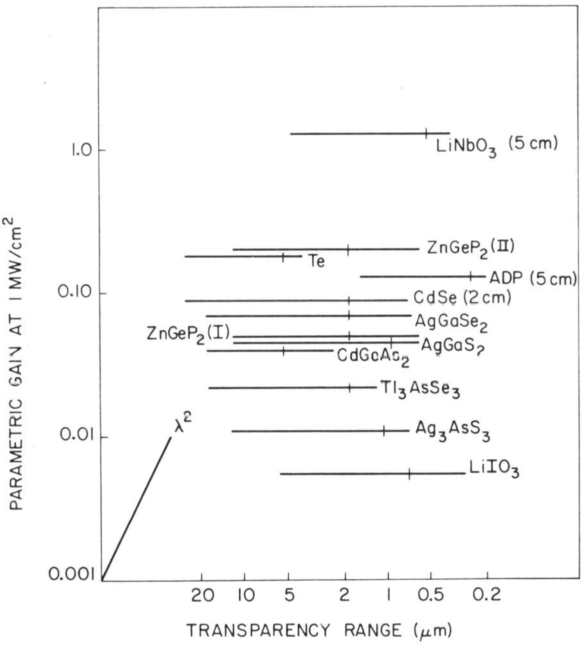

FIG. 26. Nonlinear conversion efficiency (parametric gain) at 1 MW/cm² pump intensity and transparency range of nonlinear crystals. The pump wavelength for parametric conversion (second harmonic wavelength for SHG) is shown by the vertical tick mark. The nonlinear efficiency varies as the square of the pump wavelength and crystal length. The crystal lengths are taken to be 1 cm unless specified.

sign and the coherence length of the interaction. At this time, absolute nonlinear coefficients have been measured for a number of crystals including ADP, KDP, $LiIO_3$, $LiNbO_3$, Ag_3SbS_3, CdSe, and $AgGaSe_2$. In the visible $LiIO_3$ and KDP and in the infrared GaAs are used as standard nonlinear crystals. There is interest in improving the accuracy of measurement for nonlinear coefficients and work is in progress toward that goal.

(b) *Transparency*

Known nonlinear materials have transparency ranges that extend from 2200 Å in ADP through the infrared in a number of semiconductor compounds and beyond the reststrahl band into the far infrared in both oxide and semiconductor crystals. In general, the transparency range in a single material is limited by the band-edge absorption at high frequencies and by two-phonon absorption at twice the reststrahlen frequency at low frequencies. Materials are also transparent in the very low frequency range between dc and the first crystal vibrational mode. Thus nonlinear interactions may extend from the ultraviolet through the visible and infrared to the far infrared. In fact, only a few crystals having the proper birefringence and overlapping transparency ranges are needed to cover the entire extended frequency range. This is important because the growth of high optical quality nonlinear crystals is, in general, a difficult task. In addition, nonlinear crystals that can be grown well, are transparent, and have proper birefringence for phasematching, are very rare among all known acentric crystals[147]. For example, among 13 000 surveyed crystals only 684, or 5.25%, are uniaxial and phasematchable, and of these less than half have a nonlinearity greater than KDP and far fewer are amenable to crystal growth in high optical quality centimeter sizes.

The transmission loss in a nonlinear crystal reduces the SHG conversion efficiency by $e^{-(\alpha_2/2+\alpha_1)l}$ where α_2 and α_1 are the loss per length at the second harmonic and fundamental waves. Thus a 0.05 cm^{-1} and 0.025 cm^{-1} loss at 2ω and ω in a 1 cm long crystal reduces the SHG efficiency by 0.95, which is negligible. However, in a 5 cm crystal the same losses reduce the efficiency by a factor of 0.77, which is significant. High optical quality oxide materials have losses in the 10^{-3} to 10^{-5} cm^{-1} range, whereas semiconductor materials show much higher losses in the 1 cm^{-1} to 10^{-2} cm^{-1} range. The reduction of optical loss in nonlinear crystals is important if the nonlinear interaction is to proceed efficiently. It becomes even more important for operation of a parametric oscillator where the threshold pumping power and operating efficiency directly depend on the crystal and cavity losses.

Recently intensity-dependent losses have been recognized as important factors in the transparency of a crystal. At the short wavelength end of a material's transparency range, two-photon absorption may become the dominant loss mechanism even in a frequency range that is normally transparent to low intensity radiation. For example, at 10 MW/cm^2 the two-photon absorption in GaAs at 1.32 μm is 0.3 cm^{-1}, which is significant compared to the 0.05 cm^{-1} normal absorption loss[148]. This nonlinear absorption mechanism has also been observed in oxide nonlinear materials for intense pump radiation within the two-photon frequency range of the band edge. Thus for maximum efficiency, pump wavelengths must be longer than the two-photon absorption edge of the nonlinear crystal, although operation near the band edge is possible at lower efficiencies and pump intensities. Gandrud and Abrams[45] have reported an additional loss mechanism in tellurium due to two-photon induced free carrier absorption. This mechanism reduced the SHG conversion efficiency from 15 to 5% for doubling a CO_2 laser source.

The two-photon absorption limit was in mind when pump wavelengths for the parametric gain calculations shown in Fig. 26 were chosen. Figure 26 does illustrate that with a few nonlinear materials, nonlinear interactions are possible over the entire 2200 Å to 30 μm spectral region. Thus there is the possibility of generating continuously tunable coherent radiation in nonlinear crystals over this extended spectral range.

(c) *Birefringence and Phasematching*

For efficient nonlinear interactions phasematching must be achieved in the nonlinear crystal. As an example, SHG in cubic crystals that lack birefringence and thus are not phasematchable occurs only over a distance of one coherence length between 10–100 μm. By using birefringence to offset dispersion, the phasematched interaction may proceed over the full crystal length, or approximately 1 cm. Since the SHG efficiency varies as l^2, phasematching increases the efficiency by at least 10^4. The enormous improvement in nonlinear conversion efficiency due to phasematching makes phasematching essential in nonlinear crystals. Thus if a crystal is nonlinear and transparent, it must still phasematch to be useful. Adequate birefringence for phasematching is the most restrictive requirement placed on a crystal and reduces the number of potential crystals to only a few hundred out of over 13 000 known crystals. A comparison of the crystal birefringence against its index of refraction and similar quantities, for known phasematching crystals, gives a quick indication of whether the potential nonlinear crystal has adequate birefringence for phasematching. Wemple[151], Wemple and DiDomenico[152], and Byer[147] have plotted birefringence versus index of refraction for known nonlinear crystals to determine the minimum required birefringence for phasematching. Such plots are useful since birefringence of a particular crystal cannot accurately be determined theoretically at this time.

Of course the above preliminary indications of possible phasematching in a new material must be carried a step further to determine the actual phasematching properties of a crystal. This is usually done by accurately measuring the crystal indices of refraction and fitting the results to an analytical expression which is useful for phasematching calculations. The measurement of crystal indices of refraction is tedious experimentally, especially in the infrared, so that complete measurements have been made on only a few crystals[35,95]. The measurements must be accurate to 0.01% for phasematching calculations to be made. For calculation purposes, the index values versus wavelength are usually expressed by a Sellmeier equation or modified Sellmeier equation. Once in this form, the phasematching expression for SHG, $n_o^{\omega} = n_e^{2\omega}(\theta)$, can be solved directly for the phasematching angle θ. For three-frequency phasematching conditions, a small computer program is usually written to solve simultaneously the two equations $\omega_3 = \omega_2 + \omega_1$ and $\mathbf{k}_3 = \mathbf{k}_2 + \mathbf{k}_1$ where $k = 2\pi n(\lambda)/\lambda$. Care must be taken in solving these equations since they may be double-valued. The $LiNbO_3$ parametric oscillator tuning curves, shown in Figure 7, were calculated in this manner using analytical expressions for the index of refraction given by Hobden and Warner[153].

Nonlinear crystals must also have a very uniform birefringence over the crystal length for efficient nonlinear conversion. Birefringence uniformity places one of the most stringent requirements on the growth of quality nonlinear crystals. Nash *et al.*[154] have discussed the reduction in conversion efficiency due to birefringent nonuniformities.

Another limitation to nonlinear conversion efficiency is the breaking of phasematching due to thermally-induced birefringence changes. This problem has been treated in detail for SHG by Okada and Ieiri[155]. They show that the optimum phasematching temperature shifts with increasing average laser power. In crystals with a large birefringence change with temperature, such as ADP and $LiNbO_3$, thermal breaking of phasematching becomes serious at average power levels near 1 W. On the other hand, crystals such as $LiIO_3$ which have a small temperature variation of birefringence can handle much higher average powers.

(d) *Damage Intensity*

An important limitation to the maximum nonlinear conversion efficiency is crystal damage due to the input laser intensity. Laser-induced damage may be the result of a number of interactions in the material. Ready[156] discusses possible damage mechanisms in a monograph on the effects of high-power laser radiation. Among the mechanisms considered are thermal heating, induced absorption due to multiphoton absorption which leads to heating or to breakdown, stimulated Brillouin scattering, self-focussing, surface preparation, and dielectric breakdown. Thermal heating, in which the temperature rise of the material is proportional to the input pulse duration or deposited energy is a common damage mechanism in absorbing materials. Semiconductors behave more like metals than dielectrics for this form of damage. Ready[156] lists semiconductor damage thresholds (see p. 305 of Ref. 156) as energy dependent near 5–10 J/cm^2. For a 100 nsec laser pulse the damage intensity is thus 50–100 MW/cm^2 which agrees with reported values for silicon and germanium[157]. For 1 nsec pulses the power density increases by 100 to greater than 1 GW/cm^2 if thermal damage remains the dominant mechanism. There is experimental evidence in GaAs that such high damage thresholds are reached for pulses near 1 nsec. This has important implications for infrared nonlinear optics since, in general, parametric gains are limited by the much lower 20–60 MW/cm^2 damage threshold for long pulses. In the limit of cw pumping, semiconductor materials have damage thresholds as low as 1 kW/cm^2 due to surface heating and melting.

Damage thresholds for dielectric materials are generally much higher than for semiconductors. For example, sapphire damages near 25 GW/cm^2 for Q-switch pulse irradiation. Glass and Guenther[158] have reviewed damage studies in dielectrics. They point out that nonlinear materials show a marked decrease in damage threshold for phasematched second harmonic generation. For example, $LiIO_3$ shows surface damage at 400 MW/cm^2 for 10 nsec pulses. When phasematched it damages at 30 MW/cm^2 for the fundamental and 15 MW/cm^2 for the second harmonic. Similar results have been noted in $LiNbO_3$ and $Ba_2NaNb_5O_{15}$.

In a series of very careful studies, Ammann[159] measured $LiNbO_3$ surface damage thresholds with various quarter-wave coatings. He found that uncoated $LiNbO_3$ damages at 40 MW/cm^2 at 1.08 μm for a 0.5 W, 4 kHz, 140 nsec repetitively Q-switch laser source. The damage threshold was increased to 150 MW/cm^2 for a $\lambda/4$ layer of Al_2O_3-coated $LiNbO_3$ crystal. Additional measurements on thin ZnS films showed the importance of overcoating and the lowering of the damage threshold with time upon exposure to air[160]. Similar lowering of damage threshold for ThF_2 overcoated CdSe crystals used in an infrared parametric oscillator has been noted[161]. In that case the damage threshold decreased from 60 MW/cm^2 to 10 MW/cm^2 over a period of a few days. In addition, damage occurred much more readily along surface scratches, indicating that surface preparation is important.

Bass and Barrett[162] proposed a probabilistic model for laser-induced damage based on an avalanche breakdown model. For this damage mechanism, the laser field acts as an ac analog to dc dielectric breakdown. A laser power density of 10 GW/cm^2 corresponds to 4×10^6 V/cm, which is close to the measured dc dielectric breakdown fields near 30×10^6 V/cm. Bass and Barrett[162] have presented the laser damage threshold in a probabilistic way such that the probability to induce surface damage is proportional to $\exp(-K/E)$ where K is a constant and E is the root mean square (rms) optical field strength. Measured dielectric breakdown intensities lie near 25 GW/cm^2 for glasses and fused silica and between 2–4 GW/cm^2 for nonlinear crystals. It appears that if other damage mechanisms do not limit the laser intensity at lower levels, then laser-induced dielectric breakdown determines the maximum incident intensity. Efforts to understand the mechanisms of laser induced damage have been increasing[163,164]. Hopefully these studies will lead to a better understanding of the material's damage intensity limits.

Fortunately the damage intensity for most useful nonlinear materials lies between 10 MW/cm^2 and 1 GW/cm^2. Figure 26 then shows that nonlinear conversion efficiencies or parametric gains are high for crystal lengths on the order of 1 cm. The damage intensity does place a limit on the energy handling capabilities of nonlinear crystals. Thus Q-switched laser pulse energies transmitted through 1 cm^2 area crystals are limited to 1–10 J. For most applications of nonlinear crystals this is not a serious limit since the laser host medium and the nonlinear material damage at similar energy densities.

B. OXIDE MATERIALS

(a) KDP *and its Isomorphs*

KDP, ADP, and their isomorphs are ferroelectric crystals[165] belonging to the $\bar{4}2m$ tetragonal point group above their Curie temperatures. KDP crystals are transparent from 0.21 to 1.4 μm for the non-deuterated and 0.21 to 1.7 μm for the deuterated materials[166]. ADP has a similar transparency range. KDP and its isomorphs are negative birefringent and are phasematchable over most of their transparency range. For a complete reference to crystal indices of refraction see Milek and Welles[166], Bechmann and Kurtz[95], and Singh[35].

Because of the availability of large, high optical quality crystals, KDP and ADP were the subject of early nonlinear optical experiments by Miller *et al.*[167] and Miller[21]. Later Francois[140] and Bjorkholm and Siegman[141] made accurate cw measurements of ADP's nonlinearity. In addition,

Bjorkholm[168] has made comparative nonlinear coefficient measurements of other crystals relative to KDP and ADP.

KDP and ADP have played a significant role as efficient second harmonic generators for both cw and pulsed sources. For cw doubling ADP and KDP can be temperature tuned to the 90° phasematching condition over a limited range of fundamental wavelengths between 0.54 μm and 0.49 μm at temperatures between $+60$°C and -80°C[169,170]. In particular, ADP and KDP 90° phasematch for doubling 5145 Å at -9.2°C and -11.0°C. Other isomorphs of KDP 90° phasematch over different wavelength regions. For example, rubidium dihydrogen arsenate (RDA) 90° phasematches for doubling the 0.694 μm Ruby laser[171] and cesium dihydrogen arsenate (CDA) 90° phasematches for doubling 1.06 μm[172].

Using 90° phasematched ADP, Dowley and Hodges[169] obtained up to 100 mW of 2573 Å in 1 msec pulses and 30–50 mW of cw power. The doubling was performed internal to an argon ion laser cavity to take advantage of the high circulating fields. The SHG efficiency was strongly dependent on crystal losses.

The ultraviolet transparency and phasematching characteristics of KDP isomorph crystals make them useful for ultraviolet generation by sum and second harmonic generation. Huth *et al.*[173] were the first to demonstrate this capability by externally doubling a dye laser source. Similarly Yeung and Moore[174] and Sato[175] have generated tunable ultraviolet between 3044–3272 Å by summing a Ruby pumped dye laser and a Ruby laser.

Wallace[176] reported an intracavity doubled dye laser that tunes between 2610 and 3150 Å. This source uses a Q-switched internally double Nd:YAG laser as a pump source for the rhodamine-6G and sodium fluorescein dye laser. The ADP intracavity doubled dye laser produced 32 mW of average power at 2900 Å in a 2–3 cm^{-1} bandwidth. The conversion efficiency from input doubled Nd:YAG to ultraviolet power was 4.3%. At the 65° phasematching angle used for doubling the rhodamine 6G dye laser, the ADP double refraction angle is $\rho = 0.025$ rad, which results in a conversion efficiency to the second harmonic of 1% per 100 W of input power.

Using a 1 J per pulse, 10 pulse per second Nd:YAG laser source, Yarborough[177] has obtained high ultraviolet pulse energies and average powers. For the 1.06 μm source Yarborough uses a flashlamp pumped 1/4″ diameter rod Nd:YAG oscillator followed by 1/4″ diameter and 3/8″ diameter rod amplifiers. This source is doubled in a 90° phasematched CD*A crystal at 45% efficiency yielding 450 mJ, 0.532 μm pulses. The CD*A phasematches at 103°C and the incident 1.06 μm power is unfocused into the 20 mm CD*A crystal. The green 0.532 μm source is doubled into the ultraviolet in a 2 cm long 90° phasematched ADP crystal. The conversion efficiency for this step is 22%, yielding 10 mJ pulses at 0.266 μm. The harmonic generation in both CD*A and ADP takes place without damage to the crystals. The power levels reported are limited by the laser source becoming unpolarized at high pumping levels. Simultaneously, thermal heating of the doubling crystals begins to limit the second harmonic generation efficiency. The ultraviolet source has been used as a pump for an ADP parametric oscillator[66]. In addition, the green output at 0.532 μm has been used to pump dye lasers with up to 60% conversion efficiency.

The high energy, high average power source reported by Yarborough illustrates the optical quality of KDP and its isomorphs. In addition, the use of 90° phasematching for efficient second harmonic generation demonstrates its advantage in these experiments. Although KDP-type crystals have not been utilized extensively in parametric oscillator studies, they play an important role in generating tunable ultraviolet radiation by second harmonic and sum generation of tunable visible sources.

(b) $LiNbO_3$ *and* $LiIO_3$

$LiNbO_3$ is a ferroelectric material[178] with a Curie temperature approximately 40°C below its melting point of 1253°C. Since the recognition of the unique electro-optical[179] and nonlinear optical[180] properties of $LiNbO_3$ in 1964 it has been extensively studied[181]. The growth and physical properties of $LiNbO_3$ have been discussed in a series of papers by Nassau *et al.*[182] and by Abrahams *et al.*[183–185]. The electro-optic coefficients for $LiNbO_3$ have been measured by a number of workers[186–189], as have the indices of refraction[153,190,191]. A particularly useful form for the refractive indices, including temperature dependence, is given by Hobden and Warner[153].

Early work with $LiNbO_3$ showed two potentially troublesome optical properties: optically-induced inhomogeneities in the refractive index[11,192–196] and growth-dependent birefringent variations[197–200]. The optically-induced index inhomogeneities were found to be self-annealing for crystal temperatures above approximately 180°C for visible radiation and 100°C for near infrared radiation. This fact led to attempts to grow "hot" $LiNbO_3$ by adjusting the composition[198] and by doping with MgO to increase the birefringence to achieve phasmatching for 1.06 μm at temperatures approaching 180°C. In addition, new but related crystals such as $BaNaNb_5O_{15}$[39,201] were grown in an attempt to overcome the index damage problem. Unfortunately, these attempts were not successful due to the poor crystal quality that resulted from doping $LiNbO_3$ and to the striations inherent in $Ba_2NaNb_5O_{15}$ crystals.

Although growth striations severely limit the size and quality of $Ba_2NaNb_5O_{15}$ crystals, they have been used to double cw Nd:YAG lasers[39,202] with good efficiency. The nonlinear coefficient, which is approximately three times that of $LiNbO_3$, partially offsets the disadvantage of smaller crystals of less optical quality. Serious attempts to grow striation-free $Ba_2NaNb_5O_{15}$ crystals have not been successful so that the use of this material is not widespread.

The growth-dependent birefringent variations in $LiNbO_3$ were less difficult to eliminate. The problem was solved by growth of $LiNbO_3$ from its congruent melting composition near a lithium to niobium ratio of 0.48 mole %[203–205]. The growth of $LiNbO_3$ crystals from a congruent melt plus improved optical quality tests[199,206] led to uniform high quality single crystals of over 5 cm in length.

Ferroelectric $LiNbO_3$ has a large variation of birefringence with temperature. This allows SHG at 90° phasematching for fundamental wavelengths between 1 and 3.8 μm at temperatures between 0°C and 550°C. Conversely, $LiNbO_3$ 90° phasematches for parametric oscillation for a number of pump wavelengths and can be temperature tuned over a broad spectral region. Figure 7 shows the parametric oscillator tuning curves for various pump wavelengths of an internally doubled Q-switched Nd:YAG laser. These $LiNbO_3$ phasematching curves were calculated by using the Hobden and Warner[153] index of refraction expressions. Additional $LiNbO_3$ tuning curves, including angular dependence, are given by Harris[79,207] and Ammann *et al.*[208].

The internally doubled Q-switched Nd:YAG pumped $LiNbO_3$ parametric oscillator is the only commercially available parametric oscillator at this time. This system is described by Wallace[209]. The Nd:YAG laser pump source operates with an internal $LiIO_3$ doubling crystal, acousto-optic Q-switch, and two Brewster angle prisms for wavelength selection. The combination of four doubled Nd:YAG pump wavelengths and temperature tuning allows the oscillator to cover the 0.54 to 3.65 μm spectral region. Table III gives the oscillator spectral output wavelengths, laser pump wavelengths, and output power levels. The values used in this table are useful for comparative purposes since they reflect the characteristics of the particular Q-switched Nd:YAG pump laser and are not $LiNbO_3$ oscillator limits.

Threshold for the $LiNbO_3$ SRO with a 5 cm length 90° phasematched crystal is typically between 300 and 600 W. The peak power conversion efficiency is near 50% for an oscillator operating a few times above the threshold. Due to the finite pump pulse width and oscillator build-up time, the energy conversion efficiency is approximately 30%. However, much higher conversion efficiencies have been reported.

TABLE III. $LiNbO_3$ parametric oscillator operating parameters

Tuning range (μm)	Peak power (W)	Average power (mW)	Pulse length (nsec)	Laser pump wavelength (μm)
3.50–2.50	80–150	5–10	70	0.532
0.975–2.50	250–350	30	80	0.659
0.725–0.975	250	20	200	0.562
0.623–0.760	250	50	150	0.532
0.546–0.593	300	3–5	200	0.473

$LiNbO_3$ is one of the highest quality nonlinear optical materials. Its use in parametric oscillators to generate tunable radiation over the 0.6 to 3.5 μm spectral range is a demonstration of its importance as a nonlinear material.

Recently Byer *et al.*[132] have grown improved optical quality $LiNbO_3$ crystals in an off-angle direction perpendicular to the (01.4) plane. The crystal growth axis lies in the y-z plane at 38° to the z-axis. Thus the crystal must be poled properly to maximize the effective nonlinear coefficient. The growth of large high quality (01.4) and very recently (01.3) boules allows fabrication of 5 cm long 15 mm diameter $LiNbO_3$ parametric oscillator crystals. Crystals of this size can handle over 4 J of 1.06 μm pump energy and still remain below the surface intensity damage limit. Fig. 27 shows a photograph of a $LiNbO_3$ boule and 1.06 μm pumped angle-tuned $LiNbO_3$ SRO.

In 1968 Kurtz and Perry[142] applied a technique based on the measurement of SHG of powders to the search for nonlinear materials. That search led to the evaluation of α-iodic acid (α-HIO_3) for phasematched second harmonic generation by Kurtz *et al.*[210]. Measurement of the nonlinearity of α-HIO_3 showed that it had a nonlinear coefficient approximately equal to that of $LiNbO_3$. The favourable nonlinear properties of α-HIO_3 led to consideration of other AIO_3 crystals[211]. One of the first crystals considered was lithium iodate, $LiIO_3$. Its optical and nonlinear optical properties were studied by Nath and Haussuhl[212] and Nash *et al.*[213]. The more favourable optical quality of $LiIO_3$ compared to α-HIO_3 has led to its use in a number of nonlinear applications spanning its entire transparency range from 0.35 to 5.5 μm.

$LiIO_3$ (point group 6) has a measured nonlinear coefficient slightly greater than that of $LiNbO_3$[168,171,214,215]. Its indices of refraction and large birefringence ($\Delta n = 0.15$) have been measured as have phasematching angles for various pump wavelengths. $LiIO_3$ phasematches at 52° for doubling 0.6943 μm of a Ruby laser[44,168,216] and at 30% for doubling a 1.06 μm Nd:YAG laser.

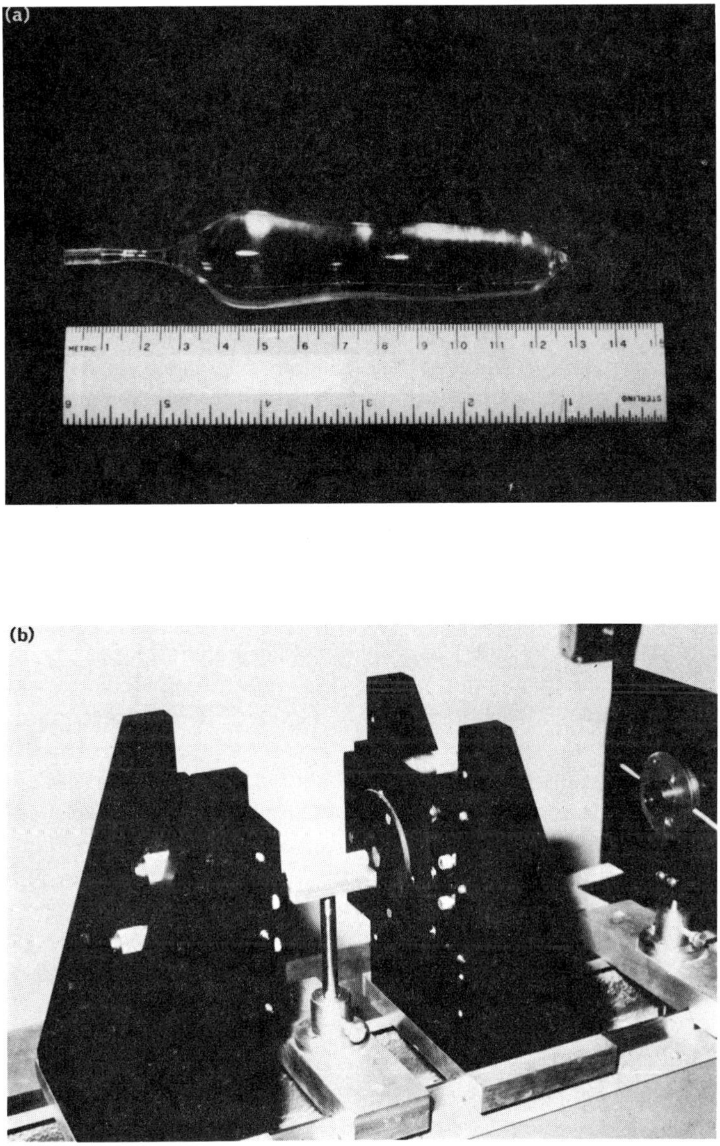

FIG 27. (a) Photograph of a (01.4) boule after annealing and prior to poling. (b) Photograph of a fabricated parametric oscillator crystal 10 mm diameter by 50 mm long long oriented at 46° to the optic axis. At room temperature the parametric oscillator tunes from degeneracy at 2.12 μm ($\theta = 45.5°$) to 1.39–4.4 μm ($\theta = 48.5°$).

$LiIO_3$ has proved particularly useful as a high optical quality nonlinear crystal for internal second harmonic generation of a Nd:YAG laser. For this application the laser mirrors are highly reflecting in the infrared but are transparent at the second harmonic. To obtain efficient doubling the laser operates Q-switched. In this mode, the $LiIO_3$ acts as a nonlinear output coupler and efficiently doubles the Nd:YAG. By operating with a prism in the laser cavity any one of 15 Nd:YAG laser lines can be selected and efficiently doubled by rotating the $LiIO_3$ to the phasematching angle. In this way wavelengths at 0.473, 0.532, 0.579, and 0.659 μm can be generated. For example, peak powers of over 10 kW and average powers of greater than 1 W have been obtained from the internally doubled Nd:YAG laser source[72].

In 1970 Campillo and Tang[217] studied spontaneous parametric scattering in $LiIO_3$ and Dobrzhanskii *et al.*[218] carried out similar measurement in α-HIO_3. Shortly afterward, Goldberg[81] constructed a $LiIO_3$ parametric oscillator pumped with a Ruby laser and Izailenko *et al.*[82] demonstrated an oscillator using $LiIO_3$ and α-HIO_3 pumped by a doubled Nd:Glass laser.

Campillo[219] also externally doubled the $LiIO_3$ oscillator using an 8 mm $LiIO_3$ crystal cut at 21.4°. Figure 28 shows the phasematching angles obtained in that experiment over a range of fundamental wavelengths between 1.1 and 1.8 μm. The second harmonic output at 100 W peak power-tuned between 0.560 and 0.915 μm. Figure 4 also shows phasematching data obtained by Herbst[161] in an internally doubled $LiNbO_3$ parametric oscillator and illustrates the phasematching properties of $LiIO_3$ for SHG over a broad spectral region.

The infrared transmission of $LiIO_3$ allows interaction out to 5.5 μm. Parametric oscillation is possible, but not useful for idler wavelengths this far in the infrared due to low gain. However, $LiIO_3$ does phasematch for mixing. Meltzer and Goldberg[220] have demonstrated mixing in $LiIO_3$ internal to a Ruby pumped dye laser. Output powers of 100 W were generated over the 4.1 to 5.2 μm region by mixing the Ruby source with the wavelengths from a DTTC dye laser. The spectral width of the dye laser was 6 Å in its 0.802 to 0.835 μm region. Although $LiIO_3$ is transparent and phasematchable in the near infrared spectral region, its low effective nonlinearity due to the 19° phasematching angle reduces its conversion efficiency to the point that other materials may prove more useful for infrared generation by mixing.

C. SEMICONDUCTOR NONLINEAR MATERIALS

$LiNbO_3$ and $LiIO_3$ parametric oscillators have been tuned to 3.7 and 4.5 μm in the infrared. Infrared oscillation is limited to these wavelengths due to the onset of crystal absorption. Use of semiconductor nonlinear materials, as

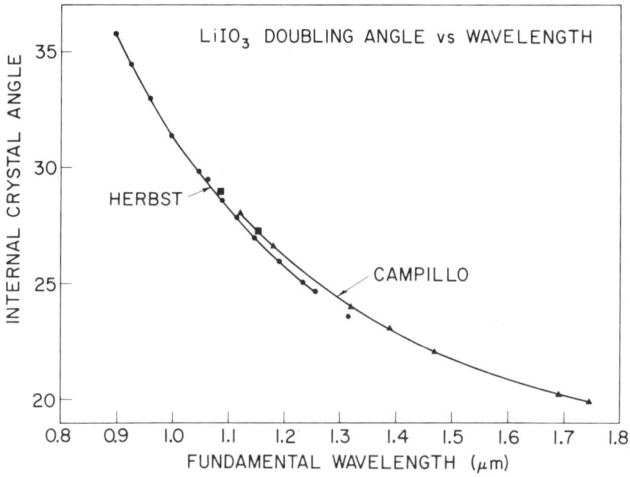

FIG. 28. Measured $LiIO_3$ phasematching angles versus fundamental wavelength. Data between 1.1 and 1.8 μm from Campillo[219] and data between 0.9 and 1.3 μm from Herbst[161].

discussed in this section, means the capability of extended infrared tuning from the two-photon absorption limit to beyond 10 μm. All materials that satisfy this extended infrared transparency range are semiconductors with bandgaps between 2 and 0.5 eV.

Tellurium was an early crystal used for SHG in the infrared[221–223]. However, problems in crystal growth and handling have held back its widespread use. Tellurium has the largest nonlinear coefficient $d_{11} = 649 \times 10^{-12}$ m/V of any known crystal, but its large birefringence of $\Delta n = -1.45$ and indices of refraction $n_o = 6.25$, $n_e = 4.80$ make it difficult to use and force phasematching to very small angles. In spite of these difficulties, tellurium has been used to double a 10.6 μm CO_2 laser with 5% efficiency[45]. Higher efficiencies were not possible due to induced absorption loss by free electron generation by the intense fundamental and second harmonic waves[45]. This problem is intensified in tellurium due to its 4.0 μm bandgap which permits two-photon absorption, thus generating the free carriers which lead to free-carrier absorption.

Proustite (Ag_3AsS_3) is an extensively studied infrared nonlinear crystal first considered by Hulme *et al.*[224]. Its large birefringence allows phase-matched nonlinear interactions over its entire 0.6 to 13 μm transparency range. Proustite has been used to double a 10.6 μm CO_2 laser[225] and to up-convert 10.6 μm into the visible using a Ruby pump source[49]. Its indices of refraction have been measured by Hobden[226]. Proustite's large birefringence ($\Delta n = 0.22$) allows off-angle phasematching near $\theta = 20°$. Propagation at such a phasematching angle results in a reduced effective nonlinear coefficient, $d_{\text{eff}} = d_{31} \sin\theta + d_{22} \cos\theta$, that is approximately twice d_{31} for $LiNbO_3$. In addition, the large birefringence results in a large double refraction angle ($\rho = 0.08$), giving a short effective interaction length. In spite of these drawbacks and the low crystal burn density near 25 MW/cm^2, proustite has operated as a parametric oscillator near 2.1 μm, pumped by a 1.06 μm Q-switched source[227,228]. Recently Hanna *et al.*[229] extended the operation of a proustite parametric oscillator to the singly resonant configuration. They were able to angle tune the oscillator over an extended 1.22 to 8.5 μm spectral range. However, operation of the infrared oscillator in proustite has operated as a parametric oscillator near 2.1 μm, pumped by a coating problems. These difficulties can be bypassed by mixing in proustite using a near-infrared source[56,57].

In addition to proustite, two closely related crystals have been investigated for infrared nonlinear optical applications. These materials are pyrargyrite (Ag_3SbS_3)[222,224,230] and Tl_3AsSe_3 described by Feichtner and Roland[231]. Pyrargyrite has properties very similar to proustite. It belongs to the same point group (3m), is transparent between 0.6 and 13 μm, and has a large enough birefringence to phasematch over its transparency region. The measured phasematching angle for doubling 10.6 μm is 29°. Its nonlinear coefficient is very close to that of proustite.

The material Tl_3AsSe_3 appears useful as an infrared nonlinear material. Its transparency range extends from 1.26 to 17 μm and the crystal has been grown in sizes up to 1.2 cm diameter and 3 cm in length. The crystal has large birefringence similar to proustite and phasematches for doubling 10.6 μm at 22°. The measured nonlinear coefficient is 3.3 times that of proustite and pyrargyrite. In addition, Feichtner and Roland measured the crystal burn density at 10.6 μm to be 32 MW/cm^2 compared to 20 MW/cm^2 for proustite and 14 MW/cm^2 for pyrargyrite.

Unlike proustite and its related crystals, cadmium selenide (CdSe) 90° phasematches over an extended infrared region. Wurtzite (6 mm) CdSe does not have adequate birefringence for second harmonic generation; however, it does phasematch for mixing and off-degenerate parametric oscillator operation[79]. In addition to its high optical quality, CdSe has a nonlinear coefficient three times that of $LiNbO_3$.

Herbst and Byer[58] performed the first phasematched interaction in CdSe by mixing 10.6 μm against the 1.833 μm Nd:YAG line[232] to generate 2.2 μm. The measured angle for the allowed Type II phasematching was 77°. The observed mixing efficiency was 35% in a 0.6 cm crystal at 48 MW/cm^2, which is less than the measured crystal burn density of 60 MW/cm^2.

The extended 0.75 to 25 μm transparency range of CdSe makes it useful as a tunable mixing source between 8 and 25 μm. In a recent experiment using a $LiNbO_3$ parametric oscillator source, Herbst and Byer[86] generated tunable radiation between 10 and 13.5 μm by mixing in CdSe. The measured peak output power of 2 W agreed with the expected conversion efficiency.

In an extension of the mixing experiment in CdSe, Herbst and Byer[83] demonstrated the first singly resonant infrared parametric oscillator in CdSe pumped by the 1.833 μm line of Nd:YAG. Davydov *et al.*[233] also reported achieving singly resonant oscillation in CdSe. The observed threshold for the Herbst-Byer oscillator was 550 W and pump depletions of up to 40% were obtained. Figure 29 is a photograph of the CdSe parametric oscillator showing the 2 cm CdSe crystal within the 5.7 cm confocal cavity. At this time CdSe crystals up to 3 cm in length are commercially available with longer (74 mm) higher-quality crystals having been reported[234]. CdSe singly resonant oscillators should operate reliably without crystal burning for 4 cm long crystals.

Considerable effort has gone into growing new nonlinear materials, expecially materials that are useful in the infrared. One group of materials with the chalcopyrite ($CuFeS_2$, point group $\bar{4}2m$) structure[235] has particularly useful nonlinear properties.

The chalcopyrite crystals form two groups which are ternary analogs of the II–VI and III–V binary semiconductors. The first group is the I–III–VI_2 compounds of which $AgGaS_2$ and $AgGaSe_2$ are examples. The second group is the II–IV–V_2 compounds which include $CdGeAs_2$ and $ZnGeP_2$. Goryunova *et al.*[236] were the first to investigate the semiconducting and nonlinear optical properties of chalcopyrite crystals. In 1971 Chemla *et al.*[237] reported on $AgGaS_2$ and later that year Boyd *et al.*[238] measured the properties of $ZnGeP_2$ and carried out a phasematched up-conversion experiment of 10.6 μm[52]. Byer *et al.*[239] measured the properties of $CdGeAs_2$. The properties of these chalcopyrites and other potentially useful crystals of the same class were extensively studied by Boyd *et al.*[146,240,241] in a series of papers. As a result of these and other measurements[125,242] four crystals belonging to the chalcopyrite semiconductors have been identified as phasematchable and potentially useful for infrared nonlinear optics. These crystals are $CdGeAs_2$, $ZnGeP_2$, $AgGaSe_2$, and $AgGaS_2$.

$AgGaS_2$ is transparent between 0.6 and 13 μm. It is negative birefringent and phasematchable over a wide range of the infrared[237,240,241]. At this time crystals up to 1 cm in size have been grown, but optical quality is not yet sufficient for use in parametric oscillators. The material has been used for infrared generation by mixing and for second harmonic generation[59].

FIG. 29. Photograph of a CdSe singly resonant parametric oscillator operating at 22 and 10 μm in the infrared. The pump radiation is incident from the right through a lens. The 2 cm long CdSe crystal within the 5.7 μm confocal cavity is rotated for tuning.

$AgGaSe_2$ is a close analogy to $AgGaS_2$. However, its transparency range (0.73–17 μm) and phasematching curves are shifted to longer wavelengths. Figure 30 shows the calculated $AgGaSe_2$ tuning curves based on the index of refraction data of Boyd *et al.*[146]. Crystals have been recently grown up to 4 cm in length with 0.02 cm^{-1} loss at 10.6 μm. Although $AgGaSe_2$ has not been used in a parametric oscillator, efficient SHG of 10.6 μm and 7–15 μm parametric mixing have been demonstrated[61]. $AgGaSe_2$ single crystals grow readily. This, along with its improved optical quality, high nonlinearity, and infrared phasematching, make it a very useful infrared nonlinear material[60].

Unlike the negative birefringence $AgGaS_2$ and $AgGaSe_2$, $ZnGeP_2$ and $CdGeAs_2$ are positive birefringent. Thus the Type II phasematching $(n_p^o\omega_p = n_s^e(\theta)\omega_s + n_i^o\omega_i)$ is allowed at 90° where Type I phasematching has a zero effective nonlinear coefficient. However, the effective nonlinearity is maximum for Type I phasematching $[n_p^o\omega_p = n_s^e(\theta)\omega_s + n_i^e(\theta)\omega_i]$ at $\theta = 45°$. Type II phasematching doubles the birefringence needed to achieve phasematching since it averages the birefringence at the longer wavelengths.

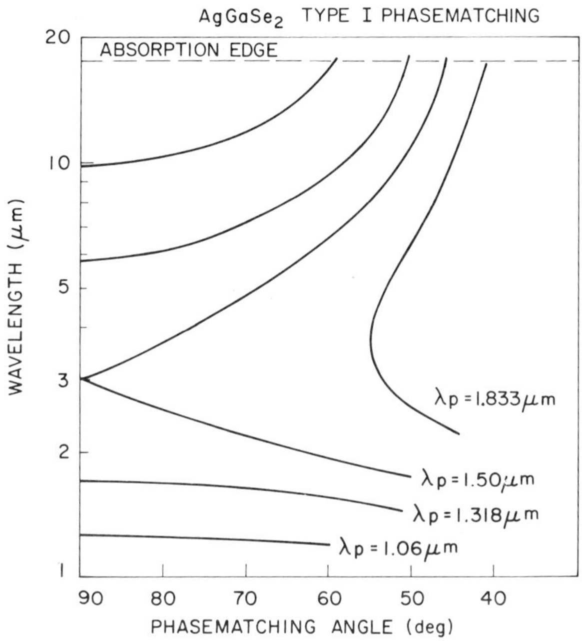

FIG. 30. Calculated parametric oscillator tuning curves for $AgGaSe_2$ for various pump wavelengths showing the extended 2–18 μm phasematching range.

$ZnGeP_2$ does not have enough birefringence to Type II phasematch for SHG. However, it does phasematch for Type I SHG. Complete Type I SHG phasematching curves for $ZnGeP_2$ are given by Boyd *et al.*[146]. $ZnGeP_2$ has been used for infrared up-conversion[52].

$CdGeAs_2$ has the highest parametric figure of merit of any known crystal except Tellurium. In addition, it is particularly useful for SHG of a 10.6 μm CO_2 laser or for parametric oscillation with a 5.3 μm pump source[239]. Recent work[46] has demonstrated the potential of $CdGeAs_2$ by SHG of a CO_2 laser with 15% efficiency in a 9 mm length crystal. Phasematched third harmonic generation is also possible in $CdGeAs_2$[243] and has been extensively studied both theoretically and experimentally[121]. $CdGeAs_2$ crystals have been grown up to 9 mm in length with a loss near 0.2 cm^{-1} at 10.6 μm. Crystal size and optical quality improvements are required before the full potential of this material can be realized.

At this time the chalcopyrite crystals satisfy all of the requirements for nonlinear materials except low absorption loss and crystal uniformity.

The crystals cannot yet be grown in adequate sizes of high optical quality. However, the very useful nonlinear properties, especially in the infrared, lend considerable importance to chalcopyrite crystal growth efforts.

VI. CONCLUSION

The properties of nonlinear materials are better understood when discussed with reference to nonlinear devices and the theory of nonlinear interactions. Therefore, in the first half of this chapter, the author discussed nonlinear phenomena including the origin of the nonlinear susceptibility and the theoretical models which describe crystal nonlinearity. Nonlinear interactions which were briefly reviewed included second harmonic generation and the three-frequency processes of sum- and difference-frequency generation and parametric generation and oscillation. The important aspects of focussing and the concept of phasematching were introduced.

With the above discussion as a frame of reference, the nonlinear material requirements including crystal nonlinearity, transparency, birefringence, and damage intensity were discussed. These properties were illustrated by describing useful nonlinear crystals. The properties of KDP and ADP, $LiNbO_3$, $LiIO_3$, and semiconductor materials including proustite, pyrargyrite, cadmium selenide, and four chalcopyrite materials were summarized. For reference Table IV lists the linear and nonlinear properties of selected nonlinear materials.

The first column of Table IV gives the presently accepted values for the crystal nonlinear coefficients. These values are based upon Levine and Bethea's[244] best value for GaAs. The parametric gain is proportional to the material parameters $d^2/n_{\omega}^2 n_{2\omega}$ where d is the effective nonlinear coefficient at the angle of phasematching and n_{ω} and $n_{2\omega}$ are the indices of refraction for the degenerate and pump waves. Column 8 of Table IV lists the material figure of merit which is also shown in Figure 25 along with the crystal transparency range.

In many applications the pump laser has more than adequate power. In this case the quantity of interest is the nonlinear conversion efficiency per input intensity and the crystal burn intensity. The calculated conversion efficiency $\Gamma^2 l^2$ given in Table IV assumes a 1 cm length crystal unless stated otherwise. Figure 26 shows the conversion efficiency at 1 MW/cm^2 pump intensity for the crystals listed in Table IV. The four chalcopyrite crystals and $LiNbO_3$ and ADP show good efficiencies. An important limitation to the maximum parametric gain is the material damage. Table IV lists approximate values for damage in nonlinear crystals based on reported damage intensity measurements. Since laser induced damage may be the result of a number of interactions in the material, the reported values vary widely. However, the listed values are useful for estimating the conversion efficiency of a nonlinear crystal as limited by the onset of damage.

The properties of a large number of nonlinear materials have been measured and useful nonlinear crystals for the ultraviolet, visible, and infrared regions have been characterized. Although the nonlinear crystals listed in Table IV allow nonlinear interactions over the spectral range from 0.25 to 25 μm, there are large classes of materials that have not been investigated. Among these are organic materials[245] and crystals potentially useful in the ultraviolet at wavelengths less than 0.2 μm.

There remains considerable work yet to be done for the full theoretical understanding of nonlinear crystals and the growth and improvement of presently known phasematchable crystals. However, progress since the first experiment of Franken *et al.*[1] has been rapid. I would like to conclude this review by suggesting what lies ahead in tunable coherent sources based on nonlinear interactions in crystals. The parametric oscillator is a unique tunable coherent source because its gain mechanism due to the crystal nonlinearity is independent of its tuning and bandwidth, which depend on the crystal birefringence and dispersion. Where tunable laser sources typically operate over a 10% bandwidth (a dye laser for example) the parametric oscillator tunes over a greater than two to one range. As an example, Fig. 19 shows the tuning curve of a 1.06 μm pumped $LiNbO_3$ parametric oscillator which angle tunes between 1.5 μm and 4.0 μm. This oscillator was recently demonstrated and is capable of rapid tuning and high output energies[132]

The $LiNbO_3$ parametric oscillator's basic tuning range can be extended toward the infrared by mixing the signal and idler in $AgGaSe_2$ to cover the 3–18 μm range and by mixing in CdSe to tune over the 10 30 μm range. Second harmonic generation in $LiNbO_3$ and in $LiIO_3$ extends the tuning to 0.3–1.5 μm. Finally, sum generation in ADP phasematches for generation of 0.22–0.3 μm in the ultraviolet. Figure 31 illustrates the phasematching angles and tuning ranges for these nonlinear interactions. Measurements[118] show that for 10 mJ pump energies available from a 1.06 μm Q-switched Nd:YAG laser source, the parametric oscillator and all following nonlinear interactions are 10–30% efficient. This widely tunable, high energy device should operate very much like a coherent spectrometer source. The spectrometer concept illustrates the unique capabilities of nonlinear interactions for generation of coherent radiation over an extended spectral range. The efficiency and high power capability of nonlinear interactions assure wider application of nonlinear devices as future tunable coherent sources.

TABLE IV. Nonlinear coefficient, figure of merit, conversion efficiency, burn intensity, and

Material (point group pump wavelength)	$d \times 10^{12}$ (m/V) (References)	n_o n_e	$n_e - n_o$	θ_m	ρ	$d_{\text{eff}} \times 10^{12}$
Te (32) $\lambda_p = 5.3\ \mu$m	$d_{11} = 649$ (222)	6.25 4.80	−1.45	14°	0.10	638 ($d \cos^2 \theta_m$)
$CdGeAs_2$ ($\bar{4}2m$) $\lambda_p = 5.3\ \mu$m	$d_{36} = 236$ (241, 239)	3.51 3.59	+0.086	II 55°	0.021	193 ($d \sin \theta$)
				I 35°	0.021	212 ($d \sin 2\theta$)
GaAs ($\bar{4}3m$)	$d_{36} = 90.1$ (244)	3.30	0	—	—	—
$ZnGeP_2$ ($\bar{4}2m$) $\lambda_p = 1.83\ \mu$m	$d_{36} = 75$ (238)	3.11 3.15	+0.038	II 90°	0.0	d_{36}
				I 62°	0.01	62.2 ($d \sin 2\theta$)
Tl_3AsSe_3 (3m) $\lambda_p = 1.83\ \mu$m	$d_+ = 40$ (231)	3.34 3.15	−0.182	36°	0.055	d_+
$AgGaSe_2$ ($\bar{4}2m$) $\lambda_p = 1.83\ \mu$m	$d_{36} = 33$ (146)	2.62 2.58	−0.32	I 55°	0.01	27 ($d \sin \theta$)
				I 90°	0.0	d_{36}
CdSe (6 mm) $\lambda_p = 1.83\ \mu$m	$d_{31} = 19$ (238, 58)	2.45 2.47	+0.019	90°	0.0	d_{31}
$AgGaS_2$ ($\bar{4}$2m) $\lambda_p = 0.946\ \mu$m	$d_{36} = 12$ (240)	2.42 2.36	−0.054	I 64°	0.17	10.8 ($d \sin \theta$)
				I 90°	0.0	d_{36}
Ag_3SbS_3 (3m) $\lambda_p = 1.06\ \mu$m	$d_+ = 12$ (222)	2.86 2.67	−0.19	—	—	d_+
Ag_3AsS_3 (3m) $\lambda_p = 1.06\ \mu$m	$d_+ = 11.6$ (231)	2.76 2.54	−0.223	30°	0.078	d_+
$LiIO_3$ (6) $\lambda_p = 0.694\ \mu$m	$d_{31} = 7.5$ (214)	1.85 1.72	−0.135	23°	0.071	3.04 ($d \sin \theta$)
$LiNbO_3$ (3m) $\lambda_p = 0.532$	$d_{31} = 6.25$ (137)	2.24 2.16	−0.081	90°	0.0	d_{31}
ADP ($\bar{4}2m$) $\lambda_p = 0.266$	$d_{36} = 0.57$ (140, 141)	1.53 1.48	−0.0458	90°	0.0	d_{36}
KDP ($\bar{4}2m$) $\lambda_p = 0.266$	$d_{36} = 0.50$ (143)	1.51 1.47	−0.0417	90°	0.0	d_{36}
SiO_2 (32)	$d_{11} = 0.33$ (143)	1.55 1.56	+0.0095	—	—	d_{11}

transmission range for nonlinear crystals

$d^2_{eff}(n^2_\omega n_{2\omega} \times 10^{24}$	$l(\rho)_{eff}$(cm)	$\Gamma^2 l^2$ (1 W)	$\Gamma^2 l^2$ (1 MW/cm^2)	I_{burn} (MW/cm^2)	Transmission range (μm)
3634	0.011	0.95×10^{-4}	0.18	40–60	4–25
861	0.34	3.8×10^{-4}	0.033	20–40	2.4–17
1039		4.6×10^{-4}	0.040		
226	$l_{coh} = 104\,\mu$m	—	—	60	0.9/1.4 – 17
187	$l = 1$ cm	7.1×10^{-3}	0.21	>4	0.7–12
127	0.59	1.8×10^{-3}	0.05		
51	0.019	3.7×10^{-3}	0.022	32	1.2–18
42	0.71	5.52×10^{-4}	0.02	>10	0.73–17
63.4	$l = 1$ cm	1.98×10^{-3}	0.07		
24	$l = 2$ cm	1.3×10^{-3}	0.09	60	0.75–25
9.2	0.14	2.3×10^{-4}	0.013	12–25	0.60–13
11	$l = 1$ cm	2.3×10^{-3}	0.045		
7.6	—	8.2×10^{-6}	0.01	14–50	0.60–14
8.2	0.007	9.0×10^{-6}	0.011	12–40	0.60–13
1.88	0.008	5.6×10^{-6}	5.5×10^{-3}	125	0.31–5.5
3.88	$l = 5$ cm	2.1×10^{-2}	1.28	50–140	0.35–4.5
0.100	$l = 5$ cm	2.9×10^{-3}	0.131	>1000	0.20–1.1
0.079	$l = 5$ cm	2.30×10^{-3}	0.103	>1000	0.22–1.1
0.029	$l_{coh} = 14\,\mu$m	—	—	>1000	0.18–3.5

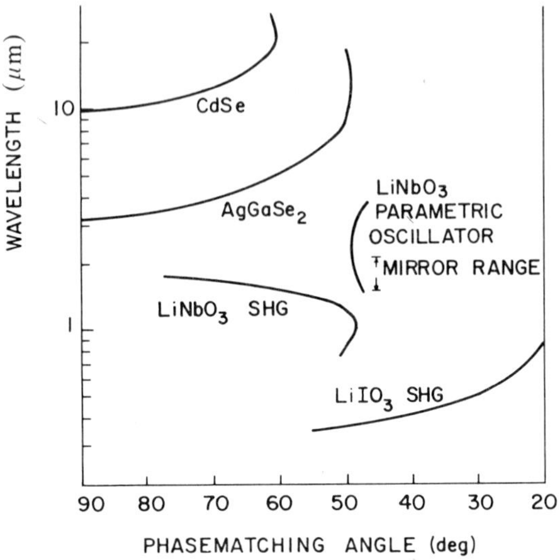

FIG. 31. Spectral range versus crystal phasematching angle for the 1.06 μm Nd:YAG pumped $LiNbO_3$ parametric oscillator and following nonlinear crystal generators. CdSe and $AgGaSe_2$ phasematch for infrared generation by mixing the $LiNbO_3$ oscillator's signal and idler frequencies. $LiNbO_3$ and $LiIO_3$ phasematch for doubling the primary oscillator frequency range into the visible and ultraviolet.

APPENDIX

(d_{eff} example–GaAs ($\bar{4}3m$))

A cubic but asymmetric crystal class is $\bar{4}3m$ to which many II–IV and III–V semiconductor compounds belong. As an example of a calculation of d_{eff} we consider GaAs which belongs to point group $\bar{4}3m$. The polarization is

$$\begin{pmatrix} P_x \\ P_y \\ P_z \end{pmatrix} = \begin{pmatrix} 0 & 0 & 0 & d_{14} & 0 & 0 \\ 0 & 0 & 0 & 0 & d_{25} & 0 \\ 0 & 0 & 0 & 0 & 0 & d_{36} \end{pmatrix} \begin{pmatrix} E_x^2 \\ E_y^2 \\ E_z^2 \\ 2E_yE_z \\ 2E_xE_z \\ 2E_xE_y \end{pmatrix}$$

Therefore

$$P_x = 2d_{14}E_yE_z$$
$$P_y = 2d_{25}E_xE_z$$
$$P_z = 2d_{36}E_xE_y$$

By Kleinman's symmetry

$$d_{14} = d_{25} = d_{36} \equiv d_{14}.$$

The components of polarization are

$$P_o = P_x \sin\varphi - P_y \cos\varphi$$
$$P_e = -P_x \cos\theta \cos\varphi - P_y \cos\theta \sin\varphi + P_z \sin\theta$$

Substituting for P_x, P_y and P_z we find

$$P_o = 2d_{14}E_yE_z \sin\varphi - 2d_{14}E_xE_z \cos\varphi$$
$$P_e = -2d_{14}E_yE_z \cos\theta\cos\varphi - 2d_{14}E_xE_z \cos\theta \sin\varphi + 2d_{14}E_xE_y \sin\theta$$

Since

$$E^o = E_x^o \sin\varphi - E_y^o \cos\varphi$$
$$E^e = -E_x^e \cos\varphi\cos\theta - E_y^e \sin\varphi\cos\theta + E_z^e \sin\theta$$

we substitute and find

$$P_o = 2d_{14}[E^e \sin\theta \sin\varphi\,(-E^o \cos\varphi - E^e \sin\varphi\cos\theta) \\ -E^e \sin\theta\cos\varphi\,(E^o \sin\varphi - E^e \cos\varphi\cos\theta)]$$

$$P_e = 2d_{14}[-E^e \cos\theta\sin\theta\cos\varphi\,(-E^o \cos\varphi - E^e \sin\varphi\cos\theta) \\ -E^e \cos\theta\sin\theta\sin\varphi\,(E^o \sin\varphi - E^e \cos\varphi\cos\theta) \\ +\sin\theta\,(E^o \sin\varphi - E^e\cos\varphi\cos\theta)(-E^o\cos\varphi - E^e\sin\varphi\cos\theta)]$$

$$P_o = 2d_{14}[-E^oE^e \sin\theta\sin\varphi\cos\varphi - (E^e)^2 \sin\varphi^2 \sin\theta\cos\theta \\ -(E^eE^o \sin\theta\sin\varphi\cos - (E^e)^2\cos^2\varphi\sin\theta\cos\theta)]$$

$$P_e = 2d_{14}[E^eE^o \cos\theta\sin\theta\cos^2\varphi + (E^e)^2\cos^2\theta\sin\theta\sin\varphi\cos\varphi \\ -E^eE^o\cos\theta\sin\theta\sin^2\varphi + (E^e)^2\cos^2\theta\sin\theta\sin\varphi\cos\varphi \\ -(E^o)^2\sin\theta\sin\varphi\cos\varphi - E^oE^e\sin^2\varphi\sin\theta\cos\theta \\ +E^eE^o\cos^2\varphi\sin\theta\cos\theta + (E^e)^2\cos^2\theta\sin\theta\sin\varphi\cos\varphi]$$

Since $\sin 2x = 2 \sin x \cos x$ and $\cos 2x = \cos^2 x - \sin^2 x$ the above expressions simplify to

$$P_o = d_{14}[-2E^oE^e \sin\theta \sin 2\varphi + (E^e)^2 \sin 2\theta \cos 2\varphi]$$

$$\begin{aligned} P_e = d_{14}[&-(E^o)^2 \sin\theta \sin 2\varphi \\ &+ 2E^oE^e(\cos\theta \sin\theta \cos^2\varphi - \cos\theta \sin\theta \sin^2\varphi - \sin^2\varphi \sin\theta \cos\theta \\ &+ \cos^2\varphi \sin\theta \cos\theta) \\ &+ 2(E^e)^2(\cos^2\theta \sin\theta \sin\varphi \cos\varphi + \cos^2\theta \sin\theta \sin\varphi \\ &+ \cos^2\theta \sin\theta \sin\varphi \cos\varphi)] \end{aligned}$$

$$\begin{aligned} P_e = d_{14}[&-(E^o)^2 \sin\theta \sin 2\varphi \\ &+ E^oE^e(\sin 2\theta \cos^2\varphi - \sin 2\varphi \sin^2\varphi - \sin^2\varphi \sin 2\theta + \cos^2\varphi \sin\varphi) \\ &+ \tfrac{3}{2}(E^e)^2 \cos\theta \sin 2\theta \sin 2\varphi] \\ = d_{14}[&-(E^o)^2 \sin\theta \sin 2\varphi + 2E^oE^e \sin 2\theta \cos 2\varphi \\ &+ 3(E^e)^2 \sin\theta \cos^2\theta \sin 2\varphi] \end{aligned}$$

If the $\bar{4}3m$ medium has negative birefringence then phasematching could occur by Type I process and

$$\text{Type I(neg)} \quad d_{\text{eff}} = -d_{14} \sin 2\varphi \sin\theta$$

or by Type II process where

$$\text{Type II(neg)} \quad d_{\text{eff}} = 2d_{14} \cos 2\varphi \sin 2\theta$$

For positive birefringence $\bar{4}3m$ symmetry crystals we find

$$\begin{aligned} \text{Type I(pos)} \quad d_{\text{eff}} &= d_{14} \cos 2\varphi \sin 2\theta \\ \text{Type II(pos)} \quad d_{\text{eff}} &= -2d_{14} \sin 2\varphi \sin\theta \end{aligned}$$

For a non-birefringent crystal there is also the possibility of polarizing all waves in the same direction in a non-phasematched interaction. For this case

$$d_{\text{eff}} = 3d_{14} \sin\theta \cos^2\theta \sin 2\varphi$$

which maximizes to

$$d_{\text{eff}} = 2d_{14}$$

for propagation along the body diagonal or (111) direction.

References

1. P. A. Franken, A. E. Hill, C. W. Peters and G. Weinreich. *Phys. Rev. Lett.* **7**, 118 (1961).
2. D. A. Kleinman. *Phys. Rev.* **126**, 1977 (1962).

3. J. A. Giordmaine. *Phys. Rev. Lett.* **8**, 19 (1962).
4. P. D. Maker, R. W. Terhune, N. Nisenoff and C. M. Savage. *Phys. Rev. Lett.* **8**, 21 (1962).
5. S. A. Akhmanov, A. I. Kovrigin, R. V. Khokhlov and O. N. Chunaev. *Zh. Eksp. Teor. Fiz.* **45**, 1336 (1963). Transl. *Sov. Phys. JETP* **18**, 919 (1964).
6. J. E. Bjorkholm. *Phys. Rev.* **142**, 126 (1966).
7. D. A. Kleinman, A. Ashkin and G. D. Boyd. *Phys. Rev.* **145**, 338 (1966).
8. N. Bloembergen and P. S. Pershan. *Phys. Rev.* **128**, 606 (1962).
9. D. A. Kleinman. *Phys. Rev.* **128**, 1761 (1962).
10. G. D. Boyd, A. Ashkin, J. M. Dziedzic and D. A. Kleinman. *Phys. Rev.* **A137**, 1305 (1965).
11. A. Ashkin *et al. Appl. Phys. Lett.* **9**, 72 (1966).
12. S. A. Akhmanov, V. G. Dmitriev and V. P. Modenov. *Radiotekh. Elektron.* **10**, 649 (1965). Transl. *Radio Eng. Electron. Phys.* **10**, 552 (1965).
13. J. K. Wright. *Proc IEEE* **51**, 1663 (1963).
14. R. G. Smith, K. Nassau and M. F. Galvin. *Appl. Phys. Lett.* **7**, 256 (1965).
15. C. C. Wang and G. W. Racette. *Appl. Phys. Lett.* **6**, 169 (1965).
16. R. H. Kingston. *Proc. IRE* **50**, 472 (1962).
17. N. M. Kroll. *Phys. Rev.* **127**, 1207 (1962).
18. S. A. Akhmanov and R. V. Khokhlov. *Zh. Eksp. Teor. Fiz.* **43**, 351 (1962). Transl. *Sov. Phys. JETP* **16**, 252 (1963).
19. J. A. Armstrong, N. Bloembergen, J. Ducuing and P. S. Pershan. *Phys. Rev.* **127**, 1918 (1962).
20. J. A. Giordmaine and R. C. Miller. *Phys. Rev. Lett.* **14**, 973 (1965).
21. R. C. Miller. *Appl. Phys. Lett.* **5**, 17 (1964).
22. S. A. Akhmanov and R. V. Khokhlov. "Problems in Nonlinear Optics", Akad. Nauk SSR. Moscow (1964). English edn 1973. Gordon and Breach, New York (1973).
23. N. Bloembergen. "Nonlinear Optics", Benjamin, New York (1965).
24. P. N. Butcher, "Nonlinear Optical Phenomena", Ohio State-University, Columbus, Ohio (1965).
25. P. A. Franken and J. F. Ward, 1963. *Rev. Mod. Phys.* **35**, 23 (1963).
26. L. N. Ovander. *Sov. Phys. Usp.* **8**, 337 (1965).
27. A. M. Bonch-Bruevich and V. A. Khodovoi. *Sov. Phys. Usp.* **8**, 1 (1965).
28. R. W. Minck, R. W. Terhune and C. C. Wang. *Appl. Opt.* **5**, 1595 (1966).
28a. P. S. Pershan. 1966. *In* "Progress in Optics", (E. Wolf, Ed.), Interscience, New York. North-Holland, Amsterdam, **5**, 85–114 (1966).
29. S. A. Akhmanov, *et al.* "Parametric Generators of Light". Presented at the Symp. Mod. Opt., Polytechnic Institute of Brooklyn, Brooklyn, New York (March 1967).
30. R. W. Terhune and P. D. Maker. *In* "Advances in Lasers", (A. K. Levine, Ed.), Dekker, New York, **2**, 295–370 (1968).
31. S. A. Akhmanov and R. V. Khokhlov. *Sov. Phys. Usp.* **11**, 394 (1968).
32. S. Kielich. *Opt. Electron.* **2**, 125 (1970).
33. V. S. Suvorov and S. A. Sonin. *Sov. Phys. Crystallogr.* **11**, 711–23 (1967).
34. I. S. Rez. *Sov. Phys. Usp.* **10**, 759–82 (1968).
34a. K. F. Hulme. *Rep. Progr. Phys.* **36**, 497–540 (1973).
35. S. Singh. *In* "Handbook of Lasers", (R. J. Pressley, Ed.), Chem. Rubber Co. Press, p. 489 (1971).

36. G. C. Baldwin. "An Introduction to Nonlinear Optics", Plenum, New York (1969).
37. F. Zernike and J. E. Midwinter. "Applied Nonlinear Optics", Academic Press, New York (1973).
38. H. Rabin and C. L. Tang. "Treatise in Quantum Electronics", Academic Press, New York (1975).
39. J. E. Geusic, H. J. Levinstein, J. J. Rubin, S. Singh and L. G. van Uitert. *Appl. Phys. Lett.* **11**, 269 (1968). *See also* J. E. Geusic, H. J. Levinstein, S. Singh, R. G. Smith and L. G. van Uitert. *Appl. Phys. Lett.* **12**, 306 (1968).
40. M. W. Dowley. *Appl. Phys. Lett.* **13**, 395 (1968).
41. W. F. Hagen and P. C. Magnante. *J. Appl. Phys.* **40**, 219 (1969).
42. R. B. Chesler, M. A. Karr and J. E. Geusic. *J. Appl. Phys.* **41**, 4125 (1970).
43. R. W. Wallace. *IEEE J. Quantum Electron.* **8**, 819 (1972).
44. G. Nath, H. Mehmanesch and M. Gsaenger. *Appl. Phys. Lett.* **17**, 286 (1970).
45. W. B. Gandrud and R. L. Abrams. *Appl. Phys. Lett.* **7**, 302 (1970).
46. H. Kildal and J. C. Mikkelsen. *Opt. Comm.* **10**, 306 (1974).
47. H. A. Smith and H. Mahr. "An Infrared Detector for Astronomy Using Up-Conversion Techniques". Presented at the Int. Quantum Electron. Conf., Kyoto, Japan (September, 1970).
48. G. D. Boyd, T. J. Bridges and E. G. Burkhardt. *IEEE J. Quantum Electron.* **4**, 515 (1968).
49. J. Warner. *Appl. Phys. Lett.* **12**, 222 (1968).
50. W. B. Gandrud and G. D. Boyd. *Opt. Comm.* **1**, 187 (1969).
51. Y. Klinger and F. R. Abrams. *Proc. IEEE.* **57**, 1797 (1969).
52. G. D. Boyd, W. B. Gandrud and E. Buehler. *Appl. Phys. Lett.* **18**, 446 (1971).
52a. G. D. Boyd, T. J. Bridges, C. K. N. Patel and E. Buchler. *Appl. Phys. Lett.* **21**, 553 (1972).
53. J. E. Midwinter. *IEEE J. Quantum Electron.* **4**, 716 (1968).
54. R. A. Andrews, *IEEE J. Quantum Electron.* **6**, 68 (1970).
55. F. Zernike and P. R. Berman. *Phys. Rev. Lett.* **15**, 999 (1965).
56. D. C. Hanna, R. C. Smith and C. R. Stanley. *Opt. Comm.* **4**, 300 (1971).
57. C. D. Decker and F. K. Tittel. *Opt. Comm.* **8**, 244 (1973).
58. R. L. Herbst and R. L. Byer. *Appl. Phys. Lett.* **19**, 527 (1971).
59. D. C. Hanna, V. V. Rampal and R. C. Smith. *Opt. Comm.* **8**, **151** (1973).
60. H. Kildal and J. C. Mikkelsen. *Opt. Comm.* **9**, 315 (1973).
61. R. L. Byer, M. M. Choy, R. L. Herbst, D. S. Chemla and R. S. Feigelson. *Appl. Phys. Lett.* **24**, 65 (1974).
62. W. H. Louisell. "Coupled Mode and Parametric Electronics". Wiley, New York (1960).
63. A. Yariv and J. E. Pearson. *In* "Progress in Quantum Electronics", (J. H. Sanders and K. W. H. Stevens, Eds), Pergamon, New York, **1**, 1–49 (1969).
64. S. A. Akhmanov *et al. JETP Lett.* **3**, 241 (1966).
65. A. G. Akmanov *et al. IEEE J. Quantum Electron.* **4**, 828 (1968).
66. J. M. Yarborough and G. A. Massey. *Appl. Phys. Lett.* **18**, 438 (1971).
67. J. A. Giordmaine and R. C. Miller. "Physics of Quantum Electronics", (P. L. Kelley, B. Lax and P. E. Tannenwald, Eds), McGraw, New York, 21–42 (1966). *See also* "Proc. Phys. Quantum Electron Conf.". San Juan, Puerto Rico, June 28–30 (1965).

68. J. A. Giordmaine and R. C. Miller. *Appl. Phys. Lett.* **9**, 298 (1960).
69. J. E. Bjorkholm. *Appl. Phys. Lett.* **13**, 53 (1968).
70. J. Falk and J. E. Murray. *Appl. Phys. Lett.* **14**, 245 (1969).
71. E. O. Ammann, J. D. Foster, M. K. Oshman and J. M. Yarborough. *Appl. Phys. Lett.* **15**, 131 (1969).
72. R. W. Wallace and S. E. Harris. "Laser Focus", p. 42 (1970).
73. R. G. Smith, J. E. Geusic, H. J. Levinstein, S. Singh and L. G. van Uitert. *J. Appl. Phys.* **39**, 4030 (1968).
74. R. G. Smith, *et al. Appl. Phys. Lett.* **12**, 308 (1968).
75. C. Laurence and F. Tittel. *J. Appl. Phys.* **42**, 2137 (1971).
76. J. F. Weller, T. G. Giallorenzi and R. A. Andrews. *J. Appl. Phys.* **43**, 4650 (1972).
77. R. L. Byer, M. K. Oshman, J. F. Young and S. E. Harris. *Appl. Phys. Lett.* **13**, 109 (1968).
78. R. L. Byer, A. I. Kovrigin and J. F. Young. *Appl. Phys. Lett.* **15**, 136 (1969).
79. S. E. Harris. *Proc. IEEE* **57**, 2096 (1969).
80. R. W. Wallace. "IEEE Conference on Laser Applications. Washington, D.C." (1971).
81. L. S. Goldberg. *Appl. Phys. Lett.* **17**, 489 (1970).
82. A. I. Izrailenko, A. I. Kovrigin and P. V. Nikles. *JETP Lett.* **12**, 331 (1970).
83. R. L. Herbst and R. L. Byer. *Appl. Phys. Lett.* **21**, 189 (1972).
84. R. G. Smith. *In* "Laser Handbook", (F. T. Arecchi and E. O. Schultz-DuBois, Eds), North Holland, Amsterdam, 837 (1972).
85. R. L. Byer. *In* "Treatise in Quantum Electronics", (H. Rabin and C. L. Tang, Eds), Academic Press, New York (1973).
86. R. L. Byer. "Parametric Oscillators". *In* "Laser Spectroscopy", R. G. Brewer and A. Mooradian, Eds), Plenum, New York (1973).
87. M. Bass, P. A. Franken, J. F. Ward and G. Weinreich. *Phys. Rev. Lett.* **9**, 446 (1962).
88. R. W. Terhune, P. D. Maker and C. M. Savage. *Phys. Rev. Lett.* **8**, 404 (1962).
89. W. Kaiser and C. G. B. Garrett. *Phys. Rev. Lett.* **7**, 229 (1961).
90. G. Eckhart *et al. Phys. Rev. Lett.* **9**, 455 (1962).
91. R. Y. Chiao, C. H. Townes and B. P. Stoicheff. *Phys. Rev. Lett.* **12**, 592 (1964).
92. D. I. Mash, V. V. Morozov, V. S. Starunov and I. L. Fabelinskii. *JETP Lett.* **2**, 22 (1965).
93. S. H. Wemple and M. DiDomenico, Jr. *In* "Applied Solid State Science", (R. Wolfe, Ed.), Academic Press, New York, Vol. 3 (1972).
93a. J. Ducuing and C. Flytzanis. *In* "Optical Properties of Solids", (F. Abeles, Ed.), North-Holland, Amserdam (1970).
94. J. F. Nye. "Physical Properties of Crystals", Oxford, London, Ch. VII (1960).
95. R. Bechmann and S. K. Kurtz. "Landolt-Bornstin Numerical Data and Functional Relationships in Science and Technology New Series Group III", (K. H. Hellwege and A. M. Hellwege, Eds), Springer-Verlag, Berlin, **2**, 167–209 (1969).
96. P. S. Pershan. *Phys. Rev.* **130**, 919 (1963).
97. G. D. Boyd and D. A. Kleinman. *J. Appl. Phys.* **39**, 3597 (1968).
98. S. Bhagvantam. "Crystal Symmetry and Physical Properties", Academic Press, London and New York (1966).
99. "Standards on Piezoelectric Crystals", *Proc. IRE*, **37**, 1378 (1949).
100. M. V. Hobden. *J. Appl. Phys.* **38**, 4365 (1967).

101. G. D. Boyd, A. Ashkin, J. M. Dziedic and D. A. Kleinman. *Phys. Rev.* **B7**, A1305 (1965).
102. G. D. Boyd and A. Ashkin. *Phys. Rev.* **146**, 187 (1966).
103. S. A. Akhmanov, A. P. Sukhorukov and R. V. Khokhlov. *Usp. Fiz. Nauk* **93**, 19 (1967). Transl. *Sov. Phys. Usp.* **10**, 609 (1968).
104. T. A. Rabson, H. J. Ruiz, P. L. Shah and F. K. Tittle. *Appl. Phys. Letts.* **21**, 129 (1972).
105. S. E. Harris, M. K. Oshman and R. L. Byer. *Phys. Rev. Lett.* **18**, 732 (1967).
106. D. Magde and H. Mahr. *Phys. Rev. Lett.* **18**, 905 (1967).
107. R. L. Byer. Ph.D. Thesis, Stanford University, Stanford, California (1968).
108. J. A. Giordmaine. *Phys. Today* **22**, 38 (1969).
109. R. L. Byer and S. E. Harris. *Phys. Rev.* **168**, 1064 (1968).
110. T. G. Giallorenzi and C. L. Tang. *Phys. Rev.* **166**, 225 (1968).
111. D. A. Kleinman. *Phys. Rev.* **174**, 1027 (1968).
112. S. E. Harris. *IEEE J. Quantum Electron.* **QE-2**, 701 (1966).
113. R. G. Smith. *J. Appl. Phys.* **41**, 4121 (1970).
114. J. E. Bjorkholm, A. Ashkin and R. G. Smith. *IEEE J. Quantum Electron.* **QE-6**, 797 (1970).
115. L. B. Kreuzer. *Appl. Phys. Lett.* **13**, 57 (1968).
116. J. E. Pearson, U. Ganiel and A. Yariv. *IEEE J. Quantum Electron.* **QE-8**, 433 (1972).
117. H. Kildal. "$CdGeAs_2$ and $CdGeP_2$ Chalcopyrite Crystals for Infrared Nonlinear Optics", Ph.D. dissertation, Stanford University, (ML report 2118).
118. R. L. Byer, R. L. Herbst and R. N. Fleming. "A Broadly Tunable IR Source". *In* "Laser Spectroscopy" (S. Haroche, J. C. Pebay-Peyroula, T. W. Hansch and S. E. Harris, Eds), Springer-Verlag, Berlin, 207 (1975).
119. J. Falk. *IEEE J. Quantum Electron.* **QE-7**, 230 (1971).
120. A. E. Siegman. *Appl. Optics* **1**, 739 (1962).
121. J. E. Bjorkholm. *IEEE J. Quantum Electron.* **QE-5**, 293 (1969).
122. R. Fischer. Zum Wirkungsgrad parametrisch abstimmbarer Laser. *In* "Laser und ihre Anwendungen", Internationale Tagung, Dresden, 10.6–17.6 **Teil 13**, 773–856 (1970). *See also* R. Fischer. *Exp. Technik der Physik* **XIX**, 193 (1971).
123. J. E. Bjorkholm. *Appl. Phys. Lett.* **13**, 399 (1968).
124. L. B. Kreuzer. "Single and Multimode Oscillation of the Singly Resonant Optical Parametric Oscillator", Proc, Joint Conf. on Lasers and Opto-Electronics, University of Southampton, Southampton, England (1969). *See also Appl. Phys. Lett.* **15**, 263 (1969).
125. J. E. Bjorkholm. *IEEE J. Quantum Electron.* **QE-7**, 109 (1971).
126. P. P. Bey and C. L. Tang. *IEEE J. Quantum Electron.* **QE-8**, 361 (1972).
127. L. B. Kreuzer. *Appl. Phys. Lett.* **13**, 57 (1968).
128. Y. N. Belyaev, A. M. Kiselev and G. I. Friedman. *JETP Letts.* **9**, 263 (1969).
129. R. W. Wallace. *Appl. Phys. Lett.* **17**, 497 (1970).
130. J. M. Yarborough and G. A. Massey. *Appl. Phys. Lett.* **18**, 438 (1971).
131. A. I. Kovrigin and P. V. Nikles. *JETP Lett.* **13**, 313 (1971).
132. R. L. Herbst, R. N. Fleming and R. L. Byer. *Appl. Phys. Lett.* **25**, 520 (1974). *See also* R. L. Byer, R. L. Herbst, R. S. Feigelson and W. L. Kway. *Optics Comm.* **12**, 427 (1974).
133. N. I. Adams III and J. J. Barrett. *IEEE J. Quantum Electron.* **QE-2**, 430 (1966).

134. (a) L. B. Kreuzer. *Appl. Phys. Lett.* **10**, 336 (1967); (b) G. V. Krivoshchekov, S. V. Kruglov, S. I. Marennikov and Yu. N. Polivanov. *JETP Lett.* **7**, 63 (1968).
135. A. J. Campillo and C. L. Tang. *Appl. Phys. Lett.* **16**, 242 (1970); *Appl. Phys. Lett.* **12**, 376 (1970).
136. A. Horkvik, H. R. Schlossberg and C. M. Stickley. *Appl. Phys. Lett.* **18**, 448 (1971).
137. J. Pinard and J. F. Young. *Opt. Comm.* **4**, 425 (1972).
138. (a) R. Damaschini. *Compt. Rend. Acad. Sci.* (*Paris*) **268**, 1169 (1969). (b) P. W. Smith. *IEEE J. Quantum Electron.* **QE-1**, 343 (1965).
139. (a) W. R. Leeb. *Appl. Phys.* **6**, 267 (1975); (b) G. Holton and O. Teschke. *IEEE. J. Quantum Electron.* **QE-10**, 577 (1974).
140. G. E. Francois. *Phys. Rev.* **143**, 597 (1966).
141. J. E. Bjorkholm and A. E. Siegman. *Phys. Rev.* **154**, 851 (1967).
142. S. K. Kurtz and T. T. Perry. *J. Appl. Phys.* **39**, 3798 (1968).
143. J. Jerphagnon and S. K. Kurtz. *J. Appl. Phys.* **41**, 1667 (1970). *See also* J. Jerphagnon and S. K. Kurtz. *Phys. Rev.* **B1**, 1738 (1970).
144. J. J. Wynne and N. Bloembergen. *Phys. Rev.* **188**, 1211 (1969).
145. D. S. Chemla and P. Kupecek. *Rev. Phys. Appl.* **6**, 31 (1971).
146. G. D. Boyd, H. M. Kasper, J. H. McFee and F. G. Storz. *IEEE J. Quantum Electron.* **8**, 900 (1972).
147. R. L. Byer. *Opt. Spectra* **42** (1970).
148. D. A. Kleinman, R. C. Miller and W. A. Nordland. *Appl. Phys. Lett.* **23**, 243 (1973).
149. J. D. H. Donnay and G. Donnay. "Crystal Data", American Crystallographic Assoc. 2nd edn (1963).
150. A. N. Winchell and H. Winchell. "Optical Properties of Artificial Minerals", Academic Press, New York (1964).
151. S. H. Wemple. *Phys. Rev. Lett.* **23**, 1156 (1969).
152. S. H. Wemple and M. DiDomenico, Jr. *J. Appl. Phys.* **40**, 735 (1969).
153. M. V. Hobden and J. Warner. *Phys. Lett.* **22**, 243 (1964).
154. F. R. Nash, G. D. Boyd, M. Sargent, III and P. M. Bridenbaugh. *J. Appl. Phys.* **41**, 2564 (1970).
155. M. Okada and S. Ieiri. *IEEE J. Quantum Electron.* **7**, 560 (1971).
156. J. F. Ready. "Effects of High Power Laser Radiation", Academic Press, New York (1971).
157. D. C. Hanna, B. Luther-Davis, H. N. Rutt, R. C. Smith and C. R. Stanley. *IEEE J. Quantum Electron.* **8**, 317 (1972).
158. A. J. Glass and A. H. Guenther, *Appl. Opt.* **11**, 832 (1972).
159. E. O. Ammann. "Technical Report AFAL-TR-72-177. Air Force Avionics Laboratory", Wright Patterson AFB, Ohio (1972).
160. E. O. Ammann and J. D. Wintemute. *J. Opt. Soc. Am.* **63**, 965 (1973).
161. R. L. Herbst. Private communication (1973).
162. M. Bass and H. H. Barrett. *IEEE J. Quantum Electron.* **8**, 338 (1972).
163. C. R. Giuliano. *IEEE J. Quantum Electron.* **8**, 749 (1972).
164. M. Bass and D. W. Fradin. *IEEE J. Quantum Electron.* **9**, 890 (1973).
165. F. Jona and G. Shirane. "Ferroelectric Crystals", McMillan, New York (1962).
166. J. T. Milek and S. J. Welles. "Linear Electro-Optic Modulator Materials", Electronic Properties Information Center, Hughes Aircraft Company (1970).
167. R. C. Miller, D. A. Kleinman and A. Savage. *Phys. Rev. Lett.* **11**, 146 (1963).

168. J. E. Bjorkholm. *IEEE J. Quantum Electron.* **4**, 970 (1968) and correction to above, *IEEE J. Quantum Electron.* **5**, 260 (1969).
169. M. W. Dowley and E. B. Hodges. *IEEE J. Quantum Electron* **4**, 552 (1968).
170. B. G. Huth and Y. C. Kiang. *J. Appl. Phys.* **40**, 4976 (1969).
171. J. E. Pearson, G. A. Evans and A. Yariv. *Opt. Comm.* **4**, 366 (1972).
172. R. S. Adhav and R. W. Wallace. *IEEE J. Quantum Electron.* **9**, 854 (1973).
173. B. G. Huth, G. I. Farmer, L. M. Taylor and M. R. Kagan. *Spectrosc. Lett.* **1**, 425 (1968).
174. E. S. Yeung and C. B. Moore. *J. Am. Chem. Soc.* **93**, 2059 (1971).
175. T. Sato. *J. Appl. Phys.* **43**, 1837 (1972).
176. R. W. Wallace. *Opt. Comm.* **4**, 316 (1971).
177. J. M. Yarborough. Private communication (1972).
178. B. T. Matthias, and J. P. Remeika. *Phys. Rev.* **76**, 1886 (1949).
179. G. E. Peterson, A. A. Ballman, P. V. Lenzo and P. M. Bridenbaugh. *Appl. Phys. Lett.* **5**, 62 (1964).
180. G. D. Boyd, R. C. Miller, K. Nassau, W. L. Bond and A. Savage. *Appl. Phys. Lett.* **5**, 234 (1964).
181. R. C. Miller, G. D. Boyd and A. Savage. *Appl. Phys. Lett.* **6**, 77 (1965).
182. K. Nassau, H. J. Levinstein and G. M. Loiacono. *J. Phys. Chem. Solids* **27**, 983, 989 (1966).
183. S. C. Abrahams, J. M. Reddy and J. L. Bernstein. *J. Phys. Chem. Solids* **27**, 997 (1966).
184. S. C. Abrahams, W. C. Hamilton and J. M. Ready. *J. Phys. Chem. Solids* **27**, 1013 (1966).
185. S. C. Abrahams, H. J. Levinstein and J. M. Ready. *J. Phys. Chem. Solids* **27**, 1019 (1966).
186. P. V. Lenzo, E. G. Spencer and K. Nassau. *J. Opt. Soc. Am.* **56**, 633 (1966).
187. E. H. Turner. *Appl. Phys. Lett.* **8**, 303 (1966).
188. J. D. Zook, D. Chen and G. N. Otto. *Appl. Phys. Lett.* **11**, 159 (1967).
189. K. F. Hulme, P. H. Davies and V. M. Cound. *J. Phys. C.* **2**, 855 (1969).
190. R. C. Miller and A. Savage. *Appl. Phys. Lett.* **9**, 169 (1966).
191. G. D. Boyd, W. L. Bond and H. L. Carter. *J. Appl. Phys.* **48**, 1941 (1967).
192. F. S. Chen. *J. Appl. Phys.* **40**, 3389 (1969).
193. W. D. Johnston, Jr. *J. Appl. Phys.* **41**, 3279 (1970).
194. N. B. Angert, V. A. Pisskov and N. M. Solov'eva. *Sov. Phys. JETP* **35**, 867 (1972).
195. V. E. Wood. *J. Appl. Phys.* **44**, 1391 (1973).
196. H. B. Serreze and R. B. Goldner. *Appl. Phys. Lett.* **22**, 629 (1973).
197. J. G. Bergman *et al. Appl. Phys. Lett.* **12**, 92 (1968).
198. H. Fay, W. J. Alford and H. M. Dess. *Appl. Phys. Lett.* **12**, 89 (1968).
199. J. E. Midwinter. *Appl. Phys. Lett.* **11**, 128 (1968a).
200. R. C. Miller, W. A. Nordland and P. M. Bridenbaugh. *J. Appl. Phys.* **42**, 4145 (1971).
201. R. L. Byer, S. E. Harris, D. J. Kuizenga, J. F. Young and R. S. Feigelson. *J. Appl. Phys.* **40**, 444 (1969).
202. M. A. Kerr. *J. Appl. Phys.* **42**, 4517 (1971).
203. P. Lerner, C. Legras and J. P. Dumas. "Crystal Growth". Proc. Second Int. Conf. Crystal Growth, Birmingham, U.K. July 1968 (F. C. Frank, J. B. Mullin and H. S. Peiser, Eds), North-Holland, Amsterdam, 231 (1968).

204. J. R. Carruthers, G. E. Peterson, M. Grasso and P. M. Bridenbaugh. *J. Appl. Phys.* **42**, 1846 (1971).
205. J. E. Midwinter. *J. Appl. Phys.* **39**, 3033 (1968).
206. R. L. Byer, J. F. Young and R. S. Feigelson. *J. Appl. Phys.* **41**, 2320 (1970).
207. S. E. Harris. *Chromatix Tech. Lett. No. 1* (1969).
208. E. O. Ammann, J. M. Yarborough and J. Falk. *J. Appl. Phys.* **42**, 5618 (1971).
209. R. W. Wallace. *Appl. Phys. Lett.* **17**, 497 (1970).
210. S. K. Kurtz, T. T. Perry and J. G. Bergman, Jr. *Appl. Phys. Lett.* **12**, 186 (1968).
211. J. G. Bergman, G. D. Boyd, A. Ashkin and S. K. Kurtz. *J. Appl. Phys.* **40**, 2860 (1969).
212. G. Nath and S. Haussuhl. *Appl. Phys. Lett.* **14**, 154 (1969).
213. F. R. Nash, J. G. Bergman, Jr., G. D. Boyd and E. H. Turner. *J. Appl. Phys.* **40**, 5201 (1969).
214. A. J. Campillo and C. L. Tang. *Appl. Phys. Lett.* **19**, 36 (1971).
215. J. Jerphagnon. *Appl. Phys. Lett.* **16**, 298 (1970).
216. G. Nath and S. Haussuehl. *Phys. Lett.* **A29**, 91 (1969).
217. A. J. Campillo and C. L. Tang. *Appl. Phys. Lett.* **16**, 242 (1970). *See also Appl. Phys. Lett.* **12**, 376 (1968).
218. G. F. Dobrzhanskii *et al. JETP Lett.* **12**, 353 (1970).
219. A. J. Campillo. *IEEE J. Quantum Electron.* **8**, 809.
220. D. W. Meltzer and L. S. Goldberg. *Opt. Comm.* **5**, 209 (1972).
221. C. K. N. Patel. *Phys. Rev. Lett.* **16**, 613 (1966).
222. J. H. McFee, G. D. Boyd and P. H. Schmidt. *Appl. Phys. Lett.* **17**, 57 (1970).
223. G. H. Sherman and P. D. Coleman. *IEEE J, Quantum Electron.* **9**, 403 (1973).
224. K. F. Hulme, O. Jones, P. H. Davies and M. V. Hobden. *Appl. Phys. Lett.* **10**, 133 (1967).
225. D. M. Boggett and A. F. Gibson. *Phys. Lett.* **A28**, 33 (1968).
226. M. V. Hobden. *Opt. Electron.* **1**, 159 (1969).
227. E. O. Ammann and J. M. Yarborough. *Appl. Phys. Lett.* **17**, 233 (1970).
228. D. C. Hanna, B. Luther-Davies, H. N. Rutt and R. C. Smith. *Appl. Phys. Lett.* **20**, 34 (1972).
229. D. C. Hanna, B. Luther-Davies and R. C. Smith. *Appl. Phys. Lett.* **22**, 440 (1973).
230. W. B. Gandrud, G. D. Boyd, J. H. McFee and F. H. Wehmeier. *Appl. Phys. Lett.* **16**, 59 (1970).
231. J. D. Feichtner and G. W. Roland. *Appl. Opt.* **11**, 993 (1972).
232. R. W. Wallace. *IEEE J. Quantum Electron.* **7**, 203 (1971).
233. A. A. Davydov, L. A. Kulevskii, A. M. Prokhorov, A. D. Savel'ev and V. V. Smirnov. *JETP Lett.* **15**, 513 (1972).
234. S. A. Abagyan, G. A. Ivanov, A. A. Kartushina and G. A. Koroleva. *Sov. Phys. Semicond.* **5**, 1452 (1972).
235. E. Parthe. "Crystal Chemistry of Tetrahedral Structures", Gordon and Breach, New York (1964).
236. N. A. Goryunova, S. M. Ryvkin, I. M. Fishman, G. P. Shpen'kov and I. D. Yaroshetskii. *Sov. Phys. Semicond.* **2**, 1272 (1965). *See also* N. A. Goryunova. "The Chemistry of Diamond Like Semiconductors", M.I.T. Press, Cambridge, Mass. (1965).
237. D. S. Chemla, P. J. Kupecek, D. S. Robertson and R. C. Smith. *Opt. Comm.* **1**, 29 (1971).

238. G. D. Boyd, E. Buehler and F. G. Storz. *Appl. Phys. Lett.* **18**, 301 (1971).
239. R. L. Byer, H. Kildal, and R. S. Feigelson. *Appl. Phys. Lett.* **19**, 237 (1971).
240. G. D. Boyd, H. Kasper and J. H. McFee. *IEEE J. Quantum Electron.* **7**, 563 (1971).
241. G. D. Boyd, E. Buehler, F. G. Storz and J. H. Wernick. *IEEE J. Quantum Electron.* **8**, 419 (1972).
242. G. C. Bahr and R. C. Smith. *Phys. Status Solidi* **13**, 157 (1972).
243. H. Kildal, R. F. Begley, M. M. Choy and R. L. Byer. *J. Opt. Soc. Am.* **62**, 1398 (1972).
244. B. F. Levine and C. G. Bethea. *Appl. Phys. Lett.* **20**, 272 (1972).
245. J. Jerphagnon. Private communication (1972).

3. The Laser Revolution in Spectroscopy

B. P. STOICHEFF

Department of Physics, University of Toronto, Canada

The following is a summary of the four lectures given by Professor Stoicheff.

The laser, with its remarkable properties of high intensity, monochromaticity, and directionality, has stimulated a surge of activity in spectroscopy. At the present time, there are lasers which operate at specific wavelengths, or a range of wavelengths, from $\sim 400\ \mu m$ in the infrared to 1200 Å in the ultraviolet. Coherent radiation has been obtained by various nonlinear processes from $\sim 3000\ \mu m$ to ~ 800 Å. This radiation is tunable over limited regions; linewidths as narrow as a few Hz have been realized, and frequency stability of a part in 10^{13}, has been achieved.

While the emphasis during these lectures will be on light scattering, the author would like to put this work in proper prespective from the outset. Therefore, it is my purpose in this chapter to review very briefly the recent achievements in spectroscopy, and to discuss the main ideas and developments which have led to these improvements.

I. LIGHT SCATTERING

When electromagnetic radiation traverses a gas, or solid, and is not absorbed, a small fraction ($\sim 10^{-3}$) of the incident radiation is scattered in all directions, some at the incident frequency, and some with changed frequency. It is customary to distinguish the several features which occur in the scattered spectrum as Rayleigh, Brillouin, and Raman components.

A. RAYLEIGH SCATTERING

The Rayleigh line is unshifted in frequency, and is very narrow: it originates from free particle scattering in gases, and from non-propagating entropy fluctuations (or diffusive motions) in liquids. It is often accompanied by a broad background caused by orientational motions of anisotropic molecules.

Investigations of Rayleigh scattering have come to the fore in recent years, with the use of lasers and the technique of photon-beating spectroscopy. In this technique, the wave fronts of the laser reference field

$$E_1(r, t) = E_1(r)\cos \omega_1 t$$

and of the scattered field ($E_2(r, t) = E_2(r)\cos \omega_2 t$) arrive in parallel alignment over the surface of a photo-detector. The detector has a non-linear response, that is, the current density $j(r, t) \propto |E(r, t)|^2$, thus the time-average over a period long compared to ω_1^{-1} or ω_2^{-1} is

$$\begin{aligned} j(r, t) &\propto |E_1(r)\cos \omega_1 t + E_2(r)\cos \omega_2 t|^2_{\text{av}} \\ &= \tfrac{1}{2}|E_1|^2 + \tfrac{1}{2}|E_2|^2 + E_1 E_2 \cos(\omega_1 - \omega_2)t \end{aligned}$$

and produces a beat note at frequency $(\omega_1 - \omega_2)$. In an experiment, a signal wave analyzer is swept over a range of frequencies (i.e. beat frequencies) corresponding to the Rayleigh linewidth, and widths as narrow as a few Hz have been measured. This technique is limited in frequency range to ~ 3 GHz, and to relatively high scattered intensity. Consequently, most of the research has been concentrated on investigations of scattering at critical temperatures of fluids, of critical mixtures of fluids, and of biological molecules in suitable solvents. Experimental techniques and applications have been reviewed by several authors[1].

B. BRILLOUIN SCATTERING

This type of scattering is caused by propagating density-fluctuations, or acoustic phonons. Brillouin components occur with small frequency shifts, less than ~ 10 GHz, which vary with scattering angle θ, according to the equation

$$\Delta \nu_B = \pm 2\nu_0 \frac{V}{c} n \sin\left(\frac{\theta}{2}\right).$$

Here, ν_0 is the incident frequency, V the phonon velocity, c the velocity of light, and n the refractive index. A measurement of $\Delta \nu_B$ gives the phonon velocity, and the Brillouin linewidth gives the phonon attenuation. By varying the angle θ from the forward to the backward direction, the velocity and attenuation of phonons in a frequency range of $\sim 10^7$ to 10^{10} Hz may be investigated—in effect, one can study "phonon spectroscopy" by light scattering.

While experimental work in Brillouin spectroscopy had been carried out in the pre-laser period[2], observations were limited to a few liquids and solids, and linewidth could not be measured. With the use of laser sources, and recent improvements in interferometry and detection techniques, Brillouin spectroscopy has become a flourishing field of research. Spectra of gases have been investigated for the first time, and used to test recent theories. Studies in the neighbourhood of critical points have given information on thermal and transport properties, and of their dependence on $(T - T_c)$. In superfluid helium, the velocity and attenuation of thermal waves, and of the second sound, have been measured. Spectra of surface waves have been observed. Depolarized Brillouin spectra have been studied and attributed to orientational fluctuations of anisotropic molecules, caused by heavily dampled shear modes in liquids. Also, considerable work on the spectra of solids has been reported, dealing with determinations of elastic and photo-elastic constants, phonon damping, ferroelectric phase transitions, and the phonon bottleneck. Several reviews are available[3]. Finally, it should be mentioned that stimulated Brillouin scattering has been observed in gases, liquids and solids[4].

C. RAMAN SCATTERING

This type of scattering occurs with relatively large frequency shifts (independent of scattering angle), and is caused by rotational and vibrational transitions in individual molecules in fluids, and by optical phonons and other excitations in solids. It is perhaps the best known field of light scattering[5].

Since the discovery of the Raman effect in 1928, investigations of the spectra of gases, liquids and solids have been carried out, although the majority have been concerned with liquids. These have given useful information on the vibrational levels of molecules, including the nature and number of vibrational modes, their frequencies, isotope shifts, Raman activity, and state of polarization. Together with the corresponding data from the infrared spectrum, such information has been widely used for the determination of the geometrical configuration of many molecules. Rotational spectra of gases have been investigated and used to obtain internuclear distances and bond angles in simple non-polar molecules[6].

Several review articles[7] and books[8] have recently appeared on Raman studies with laser excitation. The use of laser excitation has paved the way for great improvements in high-resolution Raman spectroscopy, not only because of the monochromaticity of laser radiation, but also because scattering in the forward direction effectively minimizes Doppler breadths. This has been demonstrated in the rotational spectrum of H_2. The high intensity of laser radiation in the focal region has permitted the observation of rotation-vibration spectra of O_2, N_2 and CO_2 from only $\sim 10^{11}$ molecules in volumes of $\sim 10^{-7}$ cm^3. One of the most exciting observations has been the Raman scattering from rotons in super-fluid He^4.

To date, laser excitation has had the most profound impact on investigations of Raman scattering in solids[9], and many new spectra have been observed for the first time. For example: spin-flip scattering, transitions between Landau levels in semiconductors, scattering from metal surfaces, from magnons, polaritons and F-centres, electric-field induced scattering and resonance scattering. The spectra of many crystals have been obtained, and polarization experiments have given unambiguous designations of the symmetry species of vibrational modes. Ferroelectric phase transitions have been studied, and electronic spectra of ionic crystals observed.

Nonlinear or hyper-Raman scattering[10] is produced when pulsed radiation at high peak power is focussed in gases, liquids or solids. This occurs at nearly twice the laser frequency; its intensity is only 10^{-13} of the incident intensity and varies as the square of the incident intensity. Nonlinear scattering originates from the second and higher order polarizabilities and their derivatives. The induced dipole moment of a molecule in an electric field $E = E_D \sin \omega t$, is given by $P = \alpha E + \beta E^2/2 + \gamma E^3/6 + \cdots$, and it will have components oscillating at frequencies ω, 2ω, 3ω etc, as follows: $P(\omega) = \alpha E + \gamma E^3/8$; $P(2\omega) = \beta E^2/4$; $P(3\omega) = \gamma E^3/24$. Thus, it is now possible to determine β and γ, and their derivatives. Also, the selection rules generally differ from those for normal Raman spectra, leading to new information on molecular structure, and on orientational relaxation times.

Stimulated Raman scattering[4,11] has been produced, and intense coherent radiation has been generated at Stokes and anti-Stokes frequencies of many orders. The characteristics of this radiation, including growth of intensity, directionality, bandwidth and duration have been thoroughly investigated, and the relevant processes are understood. The latter include competition between stimulated Brillouin scattering and transient scattering observed with picosecond pulses. Recent applications include: generation of intense pulses of short duration at prescibed frequencies; measurement of lifetimes and dephasing times[12] of energy levels which are highly populated during the stimulated scattering process; and development of coherent anti-Stokes spectroscopy[13] which offers advantages of spectral and spatial discrimination against fluorescence and does not require a spectrometer.

II. Doppler-free Spectroscopy

High resolution spectroscopy[14] has made significant advances with the development of tunable lasers, and with their use in nonlinear or saturation spectroscopy and in two-photon absorption spectroscopy. Both techniques dispense with the necessity of a dispersing element or spectrometer, and both probe transitions which are Doppler-free, the observed widths being limited by the frequency stability of the lasers used.

A. Saturation Spectroscopy[15]

In saturation spectroscopy, two beams from the same laser are incident on a medium from opposite directions. Only atoms (or molecules) which have zero velocity components in the beam directions can interact with radiation from both beams, and thus only those atoms which have velocities at right angles to the beam or observation direction take part in the resonance transition. Under these conditions the absorption intensity at the laser frequency ω is given by

$$I(\omega) = g(\omega)\left\{1 - \frac{2|\mu_{01}E/(2\hbar)|^2}{(\omega - \omega_{10})^2 + \gamma^2}\right\}$$

Here $g(\omega)$ is the usual absorption intensity, and the nonlinear term is dependent on the incident electric field E, on the natural linewidth γ, and on the dipole moment μ_{01} for the transition between the states 0 and 1. With laser power density of $\sim 10\ \text{mW/cm}^2$, and assuming $\mu_{01} \sim 0.01$ debye, and $\gamma \sim 1$ MHz, the nonlinear term is ~ 0.1. Signals as small as $1:10^8$ of the incident laser can be detected and pressures of a few Torr are adequate.

Initially experiments were limited to gas laser transitions, or to atomic and molecular transitions which were in accidental resonance or could be tuned into resonance (by electric or magnetic fields) with available laser radiation. The recent availability of tunable lasers, however, has made the technique of saturation spectroscopy widely applicable[16]. Extremely high resolution, with linewidths $\sim$ a few hundred kHz, has been achieved with several atomic and molecular systems. The P(7) line in the ν_3 band of CH_4 has been investigated by magnetically tuning the He–Ne laser emission at 3.39 μm, and the 13 Stark components have been observed over a frequency range of ~ 15 MHz. The fine-structure components of the hydrogen Balmer α line have been resolved, and the Lamb-shift between the levels $2P_{1/2}$ and $2S_{1/2}$ determined to be 1052.7 MHz, in good agreement with theory. Precision measurements of the absolute wavelength of the $2P_{3/2} - 3D_{5/2}$ components of the H_α and D_α lines have been made, giving an improved value (by a factor of 10) of the Rydberg constant, R = 109737.3143(4)cm^{-1}.

B. TWO-PHOTON ABSORPTION SPECTROSCOPY

When an atom (or molecule) makes a transition by absorbing simultaneously two photons, one from each of two opposed beams, the Doppler effect is eliminated[18]. Moreover, this effect is independent of the velocities of the atoms, and thus every atom in the laser beam may take part in the process. For example: an atom with velocity component v_x along the beam axis "sees" photons of frequency $\omega(1 - v_x/c)$ in one beam, and of frequency $\omega(1 + v_x/c)$ in the other, and makes the transition $\omega(1 - v_x/c) + \omega(1 + v_x/c)$ which equals 2ω, independent of the velocity.

Experiments with sodium atoms have been made using pulsed, tunable lasers and c.w. dye lasers[19]. Atoms raised from the $3^2S_{1/2}$ ground state to the $4^2D_{3/2}$ and $D_{5/2}$ states (or $4^2S_{1/2}$ state) make radiative transitions to the 4^2P states and then to the ground state with emission of radiation at 3303 Å. Thus as the laser is tuned through the region of 5787.3 Å, the observation of radiation at 3303 Å is a measure of two-photon absorption to the 4^2D states. In this way, the hyperfine components of the ground state and fine structure of excited states have been clearly resolved, with linewidths as small as ~ 10 MHz (laser limited). The Zeeman and Stark effects have also been observed. It has been shown that the sensitivity increases by several orders of magnitude when two different lasers are used, one tuned close to the intermediate levels $3P_{1/2}$, $3P_{3/2}$, and the other tuned to complete the two-photon transition[20] (although a residual Doppler linewidth is obtained because the two opposed beams have different frequencies). Experiments on the 1S–2S transitions of H and D have given values of the Lamb-shift of the 1S state for the first time, 8.6 ± 0.8 and 8.3 ± 0.3 GHz for H and D, respectively.

In this brief account of spectroscopy with lasers, I have stressed more the measurement of frequencies, and have not discussed several other important achievements, namely, quantum-beat spectroscopy[22], the measurement of transient phenomena[23], and lifetime and dephasing times[12] of electronic and vibrational levels and of various processes in gases, liquids and solids: The two latter are discussed in this volume.

However, mention should be made of the recent determination[24] of the speed of light, c = 299,792,458.0 ± 1.2 m/sec. This is based on the comparison of wavelength and frequency of a stabilized He–Ne laser (at 3.39 μm) against the Kr^{86} wavelength and Cs frequency standards. The frequency comparison was made in several stages of mixing microwave and optical radiation on a nonlinear diode capable of responding to frequencies of $\sim 10^{14}$ Hz, and developed at the Massachusetts Institute of Technology and National Bureau of Standards. The final result for the velocity of light gives an improvement by a factor of 100 over the earlier value.

REFERENCES

1. G. B. Benedek. *In* "Polarization Matière et Rayonnement", Livre de Jubilé en l'honneur du Professeur A. Kastler, Presses Universitaires de France, Paris, pp. 49–84 (1969); B. Chu. *Ann. Rev. Phys. Chem.* **21**, 145 (1970); H. Z. Cummins and H. L. Swinney. *In* "Progress in Optics VIII" (E. Wolff, Ed.), North-Holland, Amsterdam, pp. 135–197 (1971).
2. I. L. Fabelinskii. "Molecular Scattering of Light", Nauka Press, Moscow (1965). Transl. Plenum Press, New York (1968).
3. G. B. Benedek and K. Fritsch. *Phys. Rev.* **149**, 647 (1966); P. Fleury. *In* "Physical Acoustics", (W. P. Masov, Ed.), Academic Press, New York, Vol. VI, pp. 1–64 (1970); B. P. Stoicheff. *In* "Fundamental and Applied Laser Physics", (M. S. Feld, A. Javan and N. Kurnit, Eds), John Wiley, New York, pp. 573–611 (1973).
4. W. Kaiser and M. Maier. *In* "Laser Handbook", (F. T. Arecchi and E. O. Schulz-DuBois, Eds), North-Holland, Amsterdam, pp. 1077–1150 (1972); N. Bloembergen. *Amer. J. Phys.* **35**, 989 (1967).
5. G. Herzberg. "Infrared and Raman Spectra of Polyatomic Molecules", Van Nostrand, Princeton (1945); G. Placzek. *In* "Handbuch der Radiologie", (E. Marx, Ed.), Akademische Verlag, Leipzig, Vol. 6, Part 2, p. 205 (1934); B. P. Stoicheff. *In* "Methods of Experimental Physics", (D. Williams, Ed.), Academic Press, New York, Vol. 3, pp. 111–155 (1962).
6. B. P. Stoicheff. *In* "Advances in Spectroscopy", (H. W. Thompson, Ed.), Wiley-Interscience, New York, Vol. 1, pp. 91–174 (1957); A. Weber. *In* "The Raman Effect", (A. Anderson, Ed.), M. Dekker, Inc. New York, Vol. 2, pp. 543–757 (1973).
7. B. P. Stoicheff, Ref. 3; A. Weber, Ref. 6.
8. P. J. Hendra and T. R. Gibson. "Laser Raman Spectroscopy", Wiley-Interscience, London, 1970; "The Raman Effect", (A. Anderson, Ed.), M. Dekker, Inc., New York, Vols. 1 and 2 (1971, 1973).
9. G. B. Wright. "Light Scattering Spectra of Solids", Springer-Verlag, New York (1969); M. Balkanski. "Light Scattering in Solids", Flammarion Sciences, Paris (1971); R. Loudon. *Adv. Phys.* **13**, 423 (1964), **14**, 621 (1965); R. A. Cowley, Ref. 8. "The Raman Effect", Vol. 1 (1971).
10. P. D. Maker. *In* "Physics of Quantum Electronics", (P. L. Kelley, B. Lax and P. E. Tannenwald, Eds), McGraw-Hill, New York, p. 60 (1966); *Phys. Rev.* **A1**, 923 (1970).
11. G. Eckhardt, R. W. Hellwarth, F. J. McClung, S. E. Schwarz, D. Weiner and E. J. Woodbury. *Phys. Rev. Lett.* **9**, 455 (1962); E. Garmire, F. Pandarese and C. H. Townes. *Phys. Rev. Lett.* **11**, 160 (1963); B. P. Stoicheff. *Phys. Lett.* **7**, 186 (1963); R. Y. Chiao, C. H. Townes and B. P. Stoicheff. *Phys. Rev. Lett.* **12**, 592 (1964); C. W. Cho, N. D. Foltz, D. H. Rank and T. A. Wiggins. *Phys. Rev. Lett.* **18**, 107 (1967); R. M. Herman and M. A. Gray. *Phys. Rev. Lett.* **19**, 824 (1967); D. H.

Rank, C. W. Cho, N. D. Foltz and T. A. Wiggins. *Phys. Rev. Lett.* **19**, 928 (1967); S. L. Shapiro, J. A. Giordmaine and K. W. Wecht. *Phys. Rev. Lett.* **19**, 1093 (1967); K. Darée and W. Kaiser. *Phys. Rev. Lett.* **26**, 816 (1971).

12. W. Kaiser and A. Laubereau. "Ultra-short Pulses in Nonlinear Optics", Chap. 7, this volume (1977).
13. P. R. Regnier and J. P. E. Taran. *Appl. Phys. Lett.* **23**, 240 (1973); R. F. Begley, A. B. Harvey and R. L. Byer. *Appl. Phys. Lett.* **25**, 387 (1975).
14. R. G. Brewer and A. Mooradian. "Laser Spectroscopy", Plenum Press, New York (1974).
15. R. G. Brewer. *Science* **178**, 247 (1972).
16. T. W. Hänsch, I. S. Shahin and A. L. Schawlow. *Phys. Rev. Lett.* **27**, 707 (1971); *Nature (London) Phys. Sci.* **235**, 63 (1972).
17. T. W. Hänsch, M. H. Hayeh, S. A. Lee, S. M. Curry and I. S. Shahin. *Phys. Rev. Lett.* **32**, 1336 (1974).
18. L. S. Vasilenko, V. P. Chebataev and A. V. Shishaev. *Pis'ma Zh. Eksp. Teor. Fiz.* **12**, 161 (1970); *JETP Lett.* **12**, 113 (1970); B. Cagnac, G. Grynberg and F. Birabeu. *J. Phys. (Paris)* **34**, 845 (1973).
19. F. Birabeu, B. Cagnac and G. Grynberg. *Phys. Rev. Lett.* **32**, 643 (1974); M. D. Levenson and N. Bloembergen. *ibid.* **32**, 645 (1974); T. W. Hansch, K. C. Harvey, G. Meisel and A. L. Schawlow. *Opt. Comm.* **11**, 50 (1974).
20. P. F. Liao and J. E. Bjorkholm. *Phys. Rev. Lett.* **34**, 1 (1975).
21. T. W. Hänsch, S. A. Lee, R. Wallenstein and C. Wieman. *Phys. Rev. Lett.* **34**, 307 (1975).
22. S. Haroche, M. Gross and M. P. Silverman. *Phys. Rev. Lett.* **33**, 1063 (1974).
23. R. G. Brewer. "Coherent Optical Spectroscopy", Chap. 8, this volume (1977).
24. K. M. Evenson, B. L. Danielson, G. W. Day, R. L. Barger and J. L. Hall. *Phys. Rev. Lett.* **29**, 1346 (1972).

4. Free-Electron Nonlinear Optical Processes in Semiconductors†

P. A. WOLFF

Department of Physics, Massachusetts Institute of Technology, Cambridge, Massachusetts 02139, U.S.A.

I. INTRODUCTION

Nonlinear optics, as a sub-discipline of physics, began with the observation of optical second harmonic generation by Franken *et al.*[1]. This experiment was performed in an insulating crystal (quartz), as were other, early measurements in the field. It was soon recognized that these nonlinear optical processes are caused by valence electrons in the solid. Such processes are now often termed *bound-electron nonlinearities*. Soon thereafter, scientists speculated that *mobile electrons* in solids could also give rise to nonlinear optical processes. Semiconductors are an especially advantageous host for experiments of this type, because they are transparent to photons of energy below the band gap, yet still may contain a relatively high density of free electrons (concentrations in the 10^{15}–10^{16}/cc range are typical). Thus, one can envisage measurements in which an intense laser beam traverses a

† Research sponsored by the U.S. Air Force Office of Scientific Research Air Force Systems Command under Contract/Grant AFOSR-71-2010.

semiconductor crystal, interacting as it does so with the electron gas. These electrons cause a variety of nonlinear optical processes such as Raman scattering[2], multiphoton mixing[3], far-infrared (FIR) generation[4], and self-focussing[5], which have been intensively studied, both theoretically and experimentally.

In recent years (since about 1970) research concerning free electron nonlinearities has focussed on processes involving spin degrees of freedom. There are several reasons for this fact. In the first place, spin-induced processes are observed in many semiconductors. Spontaneous spin-flip Raman scattering, for example, has been seen in a far wider range of materials[6] than any of the other electronic Raman processes referred to above. Moreover, under favourable circumstances, the spin-flip cross-section is large, far bigger than the classical Thomson cross-section. Finally, spin-flip processes often have sharp lines[7] which can be orders of magnitude narrower than those of other electronic transitions in semiconductors. Sharp lines are a great advantage in any application of the spin-flip process. It is hardly surprising, therefore, that the most important nonlinear optical device whose operation depends upon mobile electrons in semiconductors involves spin degrees of freedom[8].

The recent emphasis on spin-dependent, nonlinear processes has tended to suppress interest (at least among solid state physicists) in other sorts of optical nonlinearities due to mobile electrons. Nevertheless, there are still challenging problems in these areas. Certain collective modes—whose existence is predicted by theory—have not yet been observed in solids. Light scattering from semiconductor plasmas is the most promising method for study of these waves. In addition, with recent developments in laser technology, there is a definite possibility of achieving stimulated scattering from collective modes. Such experiments would create large amplitude excitations in the solid state plasma, and give some insight into the nonlinear behaviour of this medium.

In this chapter we will review the nonlinear optical properties of mobile electrons in semiconductors. Our discussion will begin with the subject of light scattering, since that process is fundamental to many of the other nonlinear processes mentioned above. In studying light scattering, we will learn how light couples to electrons moving through crystals. Generally speaking, this interaction is considerably more complicated than that of light with a *classical*, free electron. As a consequence, electrons in semiconductors give rise to important, nonlinear optical processes that have no classical counterparts. Spin-flip scattering is a well-known example. We will also investigate the effects of many-body interactions on the light scattering spectrum and cross-section. In some instances, collective effects dramatically alter the spectrum; in others they are relatively unimportant. The theory of these phenomena has been worked out in detail, agrees with experiment, and is an excellent illustration of the effects of many-body interactions on a typical scattering experiment.

The later lectures of this series will deal with a variety of nonlinear optical phenomena—most of them involving spin degrees of freedom. We will consider stimulated spin-flip scattering, multi-photon mixing and far-infrared generation, line-shape problems, phasematching, self-focussing, and the use of spin-flip scattering to study the Mott transition in semiconductors.

Generally speaking, the most striking spin-dependent, nonlinear optical phenomena are found in narrow-gap semiconductors. Electrons in such materials have large g-values and large spin-photon interactions—which favour nonlinear optical interactions. To date, much of the experimental work has been done in n-InSb since it can easily be grown in large, perfect crystals. In principle, however, the concepts under consideration here could equally well be applied to other narrow-gap semiconductors, provided crystals of sufficient size and quality were available. This is a field in which materials research is sorely needed.

II. Scattering by a Single Mobile Electron

The process of light scattering by mobile, Bloch electrons was first analyzed by Wolff[9], Yafet[10] and Kelley and Wright[11]. Their calculations preceded experiments in this field because, at the time they were performed, there were no infrared lasers suitable as sources for scattering work in semiconductors. Soon thereafter, however, two powerful IR lasers were developed—the Nd:YAG laser at 1.06 μm and the CO_2 laser at 10.6 μm. These new sources made possible a variety of light scattering (and other nonlinear optical) experiments in semiconductors. Subsequent work has confirmed that the Nd:YAG laser is an excellent tool for probing moderate-gap materials (GaAs, InP, CdTe, etc.), whereas the CO_2 laser is equally well matched to the narrower-gap crystals, such as InSb, InAs and PbTe. Most of the experiments which tested the theory described below were performed with these devices.

To study the light scattering cross-section of an electron in a semiconductor, we consider the idealized situation in which there is a *single* electron in the conduction band of an otherwise intrinsic crystal. We also neglect electron–electron interactions for the time being. This system is described by the Hamiltonian:

$$H = \frac{\left(\mathbf{p} - \frac{e}{c}\mathbf{A}\right)^2}{2m} + V + \frac{\hbar}{4m^2c^2}\left[\nabla V \times \left(\mathbf{p} - \frac{e}{c}\mathbf{A}\right)\cdot\boldsymbol{\sigma}\right]$$

$$\equiv H_0 - \frac{e}{c}(\mathbf{v}\cdot\mathbf{A}) + \frac{e^2}{2mc^2}A^2, \qquad (1)$$

where H_0 is the unperturbed Hamiltonian of an electron in the crystal, including the crystal potential, V, and the spin-orbit interaction; $\mathbf{v} \equiv \partial H_0/\partial \mathbf{p}$ is the velocity operator; and $\mathbf{A}$ is the vector potential of the light field. Light scattering is a two-photon process, so its matrix element is second-order in the perturbation, $\mathbf{A}$. Thus, since Equation (1) contains both linear and quadratic terms in the vector potential, there are two different contributions to the matrix element—those arising in second-order perturbation theory from the $(\mathbf{v} \cdot \mathbf{A})$ interaction, and those due to the A^2 operating in first order. These contributions are combined in the complex matrix element. For classical, free electrons (those with a Hamiltonian $H_0 \equiv p^2/2m$) it can be shown that the major contribution to the light scattering cross-section arises (via first-order perturbation theory) from the interaction, $e^2A^2/2mc^2$. Second-order contributions of the $(\mathbf{v} \cdot \mathbf{A})$ interaction almost exactly cancel†. The free-electron matrix element turns out to be twice the coefficient of the A^2 term in the electron–photon interaction which, when squared, gives the well-known Thomson cross-section:

$$\frac{d\sigma}{d\Omega} = \left(\frac{e^2}{mc^2}\right)^2. \tag{2}$$

In the Bloch-electron problem, the cancellation described above persists as far as the *intraband* matrix elements of the $(\mathbf{v} \cdot \mathbf{A})$ interaction are concerned. However, the operator $\mathbf{v}$ now also has large *interband* matrix elements, which do not cancel and give an important contribution to the cross-section. This takes the form (ignoring the small momentum of the light waves)

$$\cdot\left(\frac{e}{c}\right)^2 \frac{2\pi\hbar c^2}{(\omega_0\omega_1)^{1/2}} \sum_{\substack{\mu' \\ (\mu' \neq \mu)}} \left\{\frac{[\boldsymbol{\varepsilon}_0 \cdot \langle\mu\mathbf{p}|\mathbf{v}|\mu'\mathbf{p}\rangle][\langle\mu'\mathbf{p}|\mathbf{v}|\mu\mathbf{p}\rangle \cdot \boldsymbol{\varepsilon}_1]}{[E_\mu(\mathbf{p}) - E_{\mu'}(\mathbf{p}) - \hbar\omega_0]} + \frac{[\boldsymbol{\varepsilon}_1 \cdot \langle\mu\mathbf{p}|\mathbf{v}|\mu'\mathbf{p}\rangle][\langle\mu'\mathbf{p}|\mathbf{v}|\mu\mathbf{p}\rangle \cdot \boldsymbol{\varepsilon}_0]}{[E_\mu(\mathbf{p}) - E_{\mu'}(\mathbf{p}) + \hbar\omega_1]}\right\}. \tag{3}$$

Here the indices $(\mathbf{p}, \mu)$ label Bloch states of crystal momentum $\mathbf{p}$, and band index μ; ω_0 and ω_1 are the (angular) frequencies of the incident and scattered light waves; and $\boldsymbol{\varepsilon}_0$ and $\boldsymbol{\varepsilon}_1$ are the corresponding polarization vectors. The sum is carried over all bands—filled as well as empty.

Equation (3) describes a process in which an electron, initially in state $(\mathbf{p}, \mu)$, scatters light via a virtual, interband transition. We will most often be

† This cancellation will be discussed in the next section. *See also* P. M. Platzman and P. A. Wolff, "Waves and Interactions in Solid State Plasmas", Academic Press, New York, Chs III and V (1973).

concerned with the case in which this electron starts in the conduction band ($\mu = c$). Hole scattering has also been studied theoretically[10], but relatively little experimental work has been done on p-type crystals. In many semiconductors, the most important transition contributing to Equation (3) is that between the valence and conduction bands ($\mu = c$, $\mu' = v$). When this is the case, the second-order process has the form shown schematically in Fig. 1. This illustration portrays the energy bands of an intrinsic semiconductor (such as InSb) with a single electron in the conduction band.

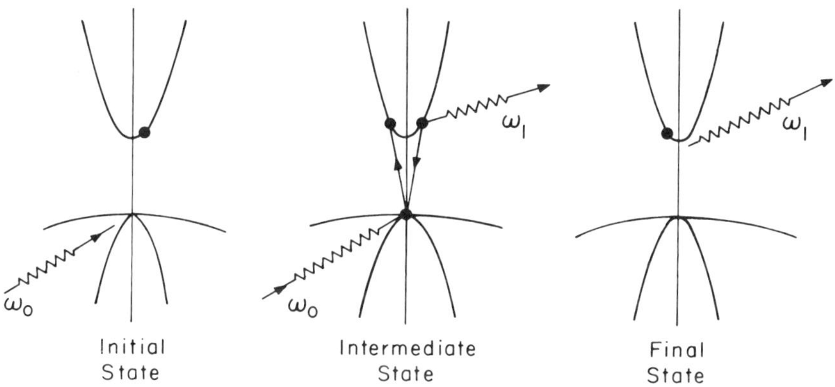

FIG. 1. Schematic of light scattering process.

A photon strikes the system and is absorbed, creating a virtual electron-hole pair. The intermediate state contains two electrons in the conduction band, and a hole in the valence band. In the second step of this two-step process, the original electron recombines with the hole, emitting a photon. The net result is light scattering, with a small recoil of the electron in the conduction band.

To obtain the total matrix element for light scattering, one must add to Equation (3) the contribution, in first order, of the A^2 interaction. The result is

$$M = \left(\frac{e}{c}\right)^2 \frac{2\pi\hbar c^2}{(\omega_0 \omega_1)^{1/2}} \left\{ \frac{\boldsymbol{\varepsilon}_0 \cdot \boldsymbol{\varepsilon}_1}{m} + \sum_{\substack{\mu' \\ (\mu' \neq \mu)}} \left(\frac{[\boldsymbol{\varepsilon}_0 \cdot \langle \mu\mathbf{p}|\mathbf{v}|\mu'\mathbf{p}\rangle][\langle \mu'\mathbf{p}|\mathbf{v}|\mu\mathbf{p}\rangle \cdot \boldsymbol{\varepsilon}_1]}{[E_\mu(\mathbf{p}) - E_{\mu'}(\mathbf{p}) - \hbar\omega_0]} + \frac{[\boldsymbol{\varepsilon}_1 \cdot \langle \mu\mathbf{p}|\mathbf{v}|\mu'\mathbf{p}\rangle][\langle \mu'\mathbf{p}|\mathbf{v}|\mu\mathbf{p}\rangle \cdot \boldsymbol{\varepsilon}_0]}{[E_\mu(\mathbf{p}) - E_{\mu'}(\mathbf{p}) + \hbar\omega_1]} \right) \right\}. \tag{4}$$

In the zero-frequency limit ($\omega_0, \omega_1 \to 0$), the expression in brackets reduces to the familiar sum rule:

$$\frac{\partial^2 E_\mu(\mathbf{p})}{\partial p_i\,\partial p_j} \equiv \frac{\delta_{ij}}{m} + \sum_{\mu'} \left\{ \frac{\langle \mu\mathbf{p}| v_i |\mu'\mathbf{p}\rangle\langle \mu'\mathbf{p}| v_j |\mu\mathbf{p}\rangle}{[E_\mu(\mathbf{p}) - E_{\mu'}(\mathbf{p})]} + \frac{\langle \mu\mathbf{p}| v_j |\mu'\mathbf{p}\rangle\langle \mu'\mathbf{p}| v_i |\mu\mathbf{p}\rangle}{[E_\mu(\mathbf{p}) - E_{\mu'}(\mathbf{p})]} \right\}. \tag{5}$$

Thus, as $\omega_0 \to 0$, the matrix element becomes

$$M \simeq \left(\frac{e}{c}\right)^2 \frac{2\pi\hbar c^2}{\sqrt{\omega_0\omega_1}} \left[\boldsymbol{\varepsilon}_0 \cdot \frac{\partial^2 E_\mu(\mathbf{p})}{\partial \mathbf{p}\,\partial \mathbf{p}} \cdot \boldsymbol{\varepsilon}_1 \right]. \tag{6}$$

For a single, isotropic conduction band, Equation (6) yields the effective-mass formula:

$$M = \left(\frac{e^2}{m^* c^2}\right) \frac{2\pi\hbar c^2}{\sqrt{\omega_0\omega_1}} (\boldsymbol{\varepsilon}_0 \cdot \boldsymbol{\varepsilon}_1), \tag{7}$$

and differential cross-section;

$$\left(\frac{d\sigma}{d\Omega}\right) = \left(\frac{e^2}{m^* c^2}\right)^2 (\boldsymbol{\varepsilon}_0 \cdot \boldsymbol{\varepsilon}_1)^2. \tag{8}$$

Equations (6–8) are the justification for the use of the effective-mass approximation[12,13] in the low frequency ($\hbar\omega_0 \ll E_G$) limit.

In several semiconductors, the effective mass is small compared to that of the free electron. Carriers in such materials have cross-sections much larger than the Thomson cross-section. For example, in n-InSb $m^* = 0.015m$ and the cross-section is more than three orders of magnitude bigger than that of a classical electron.

It is clear from Equation (4) that, at finite frequencies, the light scattering cross-section must differ from the effective mass value (Equation (8)). There are numerical differences, but in addition new physical effects appear. To study these phenomena, it is convenient to symmetrize Equation (4) as follows, assuming that $\omega_1 \simeq \omega_0$:

$$\begin{aligned} M \simeq \left(\frac{e}{c}\right)^2 \frac{2\pi\hbar c^2}{\sqrt{\omega_0\omega_1}} &\left\{ \left(\frac{\boldsymbol{\varepsilon}_0 \cdot \boldsymbol{\varepsilon}_1}{m}\right) \right. \\ &+ \sum_{\substack{\mu' \\ (\mu' \neq \mu)}} \left[\frac{[\boldsymbol{\varepsilon}_0 \cdot \mathbf{v}_{\mu\mu'}(\mathbf{p})\mathbf{v}_{\mu'\mu}(\mathbf{p}) \cdot \boldsymbol{\varepsilon}_1 + \boldsymbol{\varepsilon}_1 \cdot \mathbf{v}_{\mu\mu'}(\mathbf{p})\mathbf{v}_{\mu'\mu}(\mathbf{p}) \cdot \boldsymbol{\varepsilon}_0][E_\mu(\mathbf{p}) - E_{\mu'}(\mathbf{p})]}{[E_\mu(\mathbf{p}) - E_{\mu'}(\mathbf{p})]^2 - (\hbar\omega_0)^2} \right. \\ &\left.\left. + \frac{[\boldsymbol{\varepsilon}_0 \cdot \mathbf{v}_{\mu\mu'}(\mathbf{p})\mathbf{v}_{\mu'\mu}(\mathbf{p}) \cdot \boldsymbol{\varepsilon}_1 - \boldsymbol{\varepsilon}_1 \cdot \mathbf{v}_{\mu\mu'}(\mathbf{p})\mathbf{v}_{\mu'\mu}(\mathbf{p}) \cdot \boldsymbol{\varepsilon}_0]}{[E_\mu(\mathbf{p}) - E_{\mu'}(\mathbf{p})]^2 - (\hbar\omega_0)^2} \hbar\omega_0 \right] \right\}, \end{aligned} \tag{9}$$

where $\mathbf{v}_{\mu\mu'}(\mathbf{p}) \equiv \langle \mu\mathbf{p}|\mathbf{v}|\mu'\mathbf{p}\rangle$. The first sum in Equation (9) is closely related to that which appears in Equation (5) and determines the effective mass, whereas the second is related to the magnetic moment[14]. To evaluate these sums in detail requires, of course, considerable knowledge of the band structure and interband matrix elements. There are several important cases, however, in which the properties of the conduction band of a semiconductor are largely determined by its interaction with a single lower-lying valence band (well-known examples are InSb, InAs and PbTe). Within this simple, two-band model, there is only a single term in each of the sums appearing in Equation (9). One then finds

$$M \simeq \left(\frac{e}{c}\right)^2 \frac{2\pi\hbar c^2}{\sqrt{\omega_0\omega_1}} \left\{ \left(\frac{\boldsymbol{\varepsilon}_0 \cdot \boldsymbol{\varepsilon}_1}{m^*}\right) \left[\left(\frac{E_G^2}{E_G^2 - (\hbar\omega_0)^2}\right) \left(\frac{1}{m^*} - \frac{1}{m}\right) + \frac{1}{m} \right] + i\left(1 - \frac{m_s}{m}\right)\left(\frac{\hbar\omega_0 E_G}{E_G^2 - (\hbar\omega_0)^2}\right)\left(\frac{\boldsymbol{\sigma} \cdot \boldsymbol{\varepsilon}_0 \times \boldsymbol{\varepsilon}_1}{m_s}\right) \right\}, \tag{10}$$

where m_s is the spin mass, defined via the relation

$$\left(\frac{m}{m_s}\right) \equiv \frac{|g|}{2}. \tag{11}$$

Equation (10) has two interesting new features. The matrix element is enhanced by a factor $E_G^2/[E_G^2 - (\hbar\omega_0)^2]$ as the light frequency approaches the (direct) gap. This phenomenon is the solid state analogue of resonance fluorescence. In addition, the last term of Equation (10) is spin-dependent, implying the possibility of spin-flip light scattering. This process does, indeed, occur and has a cross-section

$$\frac{d\sigma}{d\Omega}(\text{spin-flip}) = \left(\frac{e^2}{m_s c^2}\right)^2 \left[\frac{\hbar\omega_0 E_G}{E_G^2 - (\hbar\omega_0)^2}\right]^2 \left(1 - \frac{m_s}{m}\right)^2 |\langle\alpha|\boldsymbol{\sigma} \cdot \boldsymbol{\varepsilon}_0 \times \boldsymbol{\varepsilon}_1|\alpha'\rangle|^2, \tag{12}$$

where the labels α, α' refer to spin states of the electron.

Spin-flip light scattering was first predicted by Yafet[10]. It is best studied as a Raman process by applying a dc magnetic field to the sample. One then observes scattered light at frequencies ($\omega_1 = \omega_0 \pm \omega_s$), where $\omega_s \equiv \mu g B$ is the spin-resonance frequency. This process has been seen in about ten different semiconductors.

The analysis leading to Equation (12) is not, strictly speaking, applicable to a situation in which the crystal is in a finite, dc magnetic field. When a magnetic field is applied, the Hamiltonian is modified via the replacement $\mathbf{p} \rightarrow (\mathbf{p} - (e/c)\mathbf{A}_0)$, and Bloch states are converted into Landau levels. In calculating the light scattering cross-section it is then necessary to evaluate velocity matrix elements between Landau levels in different bands, rather than Bloch states. The formalism is similar, in principle, to that leading to Equation (4), but in practice leads to quite lengthy and complicated calculations. Such computations have been performed by Yafet[10], and by Kelley and Wright[11] for III–V semiconductors. Their results for elastic and spin-flip scattering essentially agree with those predicted by Equation (9). In addition, they find the possibility of Raman processes involving a change of Landau-level quantum number and, in *p*-type materials, intervalence band transitions. Landau-Raman scattering had also, earlier, been predicted by Wolff[9] from a nonparabolic, effective-mass Hamiltonian.

Neither spin-flip nor Landau-Raman scattering can be caused by classical, free electrons (those with Hamiltonian $H_0 = p^2/2m^*$). The spin-flip process requires spin-orbit interaction; the Landau-Raman process requires that the electron dynamics be more complicated than those of classical electrons. To illustrate these points, we consider the III–V semiconductors (Fig. 2). In

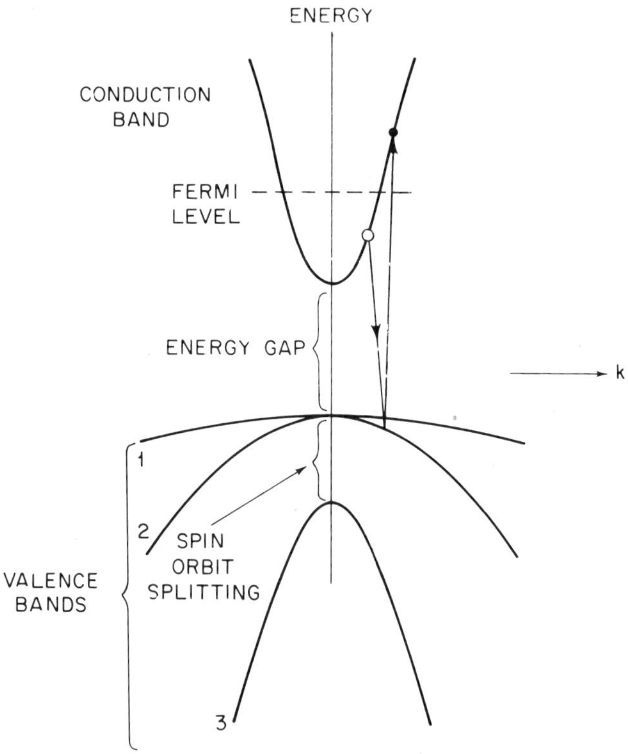

FIG. 2. Bond structure of III–V semiconductors at $\mathbf{k} = 0$.

these crystals the $k = 0$ valence band edge is split, by the spin-orbit interaction, into $J = \frac{3}{2}$ and $J = \frac{1}{2}$ multiplets. The band-edge wave functions contain an admixture of spin-up and spin-down states. For example, the $J = \frac{3}{2}$, $m_J = \frac{1}{2}$ and $J = \frac{1}{2}$, $m_J = \frac{1}{2}$ eigenfunctions have the form:

$$\psi_{1/2}^{3/2} = -\sqrt{2/3}|z\uparrow\rangle + \sqrt{1/6}|(x + iy)\downarrow\rangle$$
$$\psi_{1/2}^{1/2} = \sqrt{1/3}[|(x + iy)\downarrow\rangle + |z\uparrow\rangle]. \quad (13)$$

Each of these states can give rise to spin-reversing interband transitions via the channels:

$$|s\uparrow\rangle \rightarrow \psi_{1/2}^{3/2} \rightarrow |s\downarrow\rangle$$

and

$$|s\uparrow\rangle \rightarrow \psi_{1/2}^{1/2} \rightarrow |s\downarrow\rangle.$$

The velocity matrix elements for the processes are equal, and of opposite sign. Hence, if the $J = \frac{3}{2}$ and $J = \frac{1}{2}$ states had the same energies—giving equal energy denominators in Equation (9)—their contributions to M would cancel, and there would be no spin-flip scattering. In practice, the $J = \frac{3}{2}$, $J = \frac{1}{2}$ degeneracy is lifted by spin-orbit interaction, and the spin-flip cross-section is finite. In a material such as InSb, the spin-orbit splitting ($\Delta \simeq 0.9$ eV) is actually large compared to the band gap ($E_G \simeq 0.25$ eV). Under these circumstances, only the

$$|s\uparrow\rangle \rightarrow \begin{matrix} J = \frac{3}{2} \\ m_J = \pm\frac{1}{2} \end{matrix} \rightarrow |s\downarrow\rangle$$

channel contributes to Equation (9). In that limit ($\Lambda \gg E_G$) the elastic (Rayleigh) and spin-flip cross-sections are comparable.

To understand the origin of Landau-Raman scattering, we must ask why this process does *not* occur for classical electrons. Though the $(\mathbf{v} \cdot \mathbf{A})$ interaction has matrix elements between Landau levels, when *all* terms are summed, the total matrix element for Landau-Raman scattering by free electrons vanishes. This cancellation requires that the Landau levels be equally spaced in energy. With hindsight, the result is hardly surprising. A classical electron behaves as a simple harmonic oscillator (of frequency ω_c) when placed in a magnetic field. The incoming photon is another harmonic oscillator. When these oscillators are coupled (through the $\mathbf{v} \cdot \mathbf{A}$ interaction) there are no mixed frequencies, such as $(\omega_0 \pm \omega_c)$ or $(\omega_0 \pm 2\omega_c)$, in the electron's response. But mixed-frequency currents are precisely what is required to generate Raman scattered radiation at frequencies $(\omega_0 - \omega_c)$ or $(\omega_0 - 2\omega_c)$. This argument depends crucially upon the fact that the free electron has a Hamiltonian of the form $p^2/2m^*$. In many III–V semiconductors (especially narrow-gap crystals such as InSb or InAs) there are important corrections to this Hamiltonian, resulting from band-nonparabolicity. In InSb, for example, the effective-mass Hamiltonian is more accurately written[15] as

$$H = \sqrt{\left(\frac{E_G}{2}\right)^2 + E_G\left(\frac{p^2}{2m^*}\right)} - \left(\frac{E_G}{2}\right)$$
$$\simeq \left(\frac{p^2}{2m^*}\right) - \left(\frac{1}{E_G}\right)\left(\frac{p^2}{2m^*}\right)^2. \tag{14}$$

In a magnetic field the Landau levels of such an electron are *not* equally spaced in energy but crowd together as the electron energy increases. The system no longer is perfectly harmonic, hence can cause orbital, magnetic Raman scattering. The cross-section for the $\Delta n = \pm 2$ process ($\omega_1 = \omega_0 \pm 2\omega_c$) is estimated[9–11] to be

$$\left.\frac{d\sigma}{d\Omega}\right|_{\Delta n = 2} \simeq \left(\frac{e^2}{m^*c^2}\right)^2\left(\frac{\hbar\omega_c}{E_G}\right)^2. \tag{15}$$

Here, the factor $(\hbar\omega_c/E_G)^2$ is a measure of the band nonparabolicity. Both $\Delta n = 1$ and $\Delta n = 2$ Landau-Raman scattering peaks are observed in n-InSb[16] and n-InAs[17]. The $\Delta n = 1$ transition was overlooked in the early theoretical work. It has zero matrix element for electrons with $p_z = 0$ (momentum parallel to field), but can occur in doped samples, because the electrons than have (via the exclusion principle) an appreciable value of p_z, This question has been discussed by Makarov[18].

Experimentally, spin-flip and Landau-Raman scattering were first observed[16] in InSb, where the cross-sections and the frequency shifts are large. The observed scattering intensities were in reasonable agreement with those predicted by Equations (12) and (15), though the $\Delta n = 2$ Landau-Raman process did not have the B^2 field dependence implied by Equation (15). This discrepancy remains unexplained. The frequency shifts (ω_s, ω_c, $2\omega_c$) of the $\Delta s = 1$, $\Delta n = 1$ and $\Delta n = 2$ processes were those expected from the known-mass ($m^* \simeq 0.015$) and g-value ($g \simeq -50$) of n-InSb. As implied by Equations (12) and (15), the spin-flip process is considerably stronger in n-InSb (and other semiconductors) than the orbital Raman process. It also has a much sharper line. Slusher *et al.*[16] found Landau-Raman linewidths of 30 cm^{-1}, as compared to 5 cm^{-1} for the spin-flip transition. Subsequent work has shown that the spin-flip transition is often far narrower than this. Linewidths in the 0.1 cm^{-1} range are common in n-InSb. We will return later, in some detail, to the problem of spin-flip linewidths, since they play a crucial role in determining the threshold pump power of the spin-flip Raman laser (SFRL). For the time being, it suffices to say that the linewidths of the Landau-Raman processes are determined by electron scattering times measured via mobility ($\Delta\omega \simeq 1/\tau_s$). The spin-flip linewidth, on the other hand, is much smaller than the orbital collision rate. A considerably more subtle theory is required to discuss this question.

Soon after the InSb work, magnetic Raman scattering was also observed[17] in n-InAs. Again, there are three lines in the Raman spectrum—corresponding to $\Delta s = 1$, $\Delta n = 1$ and $\Delta n = 2$ processes. As before, the $\Delta n = 1$ and $\Delta n = 2$ lines are broad. In other respects, there is nothing exceptional about this work; the results generally support the predictions of Equations (12) and (15).

To our knowledge, the experiments described above (on n-InSb and n-InAs) constitute the only observation of Landau-Raman scattering—and these measurements have never been repeated. Orbital Raman processes have not been of great interest to experimentalists since they are weak, diffuse, and have only been seen in two crystals. Spin-flip scattering, on the other hand, seems to be an almost universal phenomenon. It has now been seen in n-InSb, n-InAs, n-PbTe[19], n-CdS[20], n-ZnSe[20], n-CdTe[21], n-CdSe[21], n-GaAs[22], n-SiC[23] and p-ZnTe[24]. The spin-flip lines are generally narrow (at least compared to electron collision rates) and in several cases they are also very strong. These features, needless to say, are of great value in applications of spin-flip scattering. We will return to such questions later.

III. Many-Body and Mode-Coupling Effects in Light Scattering

The discussion given above deals with a particularly simple problem—that in which light scatters from a semiconductor containing one (or, at most, a few) free carriers. Under such conditions, carrier–carrier interactions are unimportant. In practice, however, the semiconductor crystals used in scattering experiments contain a large electron density. Values in the $n \simeq 10^{15}$–10^{18}/cc range are typical. Such an electron gas is a dense "plasma", whose properties are largely determined by collective interactions between its constituents. Under these conditions, the light scattering spectrum is drastically modified, and can only be understood from a viewpoint which properly includes many-body effects. Such a description is the aim of this section.

In studying the effects of Coulomb interactions on the spectrum, we will simplify the problem, in other respects, as much as possible. We will assume an isotropic, nondegenerate conduction band edge (such as that of GaAs, InSb, or InAs), and further assume that the optical frequencies are low enough that the effective-mass approximation applies. Under these conditions (essentially those leading to Equations (7) and (8)), the problem can be described by a free electron-like Hamiltonian:

$$H = \sum_i \left[\frac{\left(\mathbf{p}_i - \frac{e}{c}\mathbf{A}_i\right)^2}{2m^*} \right] + \frac{1}{2} \sum_{i \neq j} \left[\frac{e^2}{\varepsilon r_{ij}} \right]. \tag{16}$$

Here the sums on "i" and "j" run over the various carriers in the conduction band, $\mathbf{A}_i \equiv \mathbf{A}(\mathbf{r}_i)$ is the vector potential evaluated at the position of the ith particle; m^* is the effective mass; and ε the static dielectric constant of the crystal. Note that Equation (16) includes the Coulomb interactions between carriers, which are responsible for its plasma-like properties.

To proceed, we rewrite Equation (16) in the form

$$H = H_0 - \left(\frac{e}{m^*c}\right) \sum_i (\mathbf{p}_i \cdot \mathbf{A}_i) + \left(\frac{e^2}{2m^*c^2}\right) \sum_i (A_i)^2, \tag{17}$$

where

$$H_0 \equiv \left[\sum_i \left(\frac{p_i^2}{2m^*}\right) + \frac{1}{2} \sum_{i \neq j} \left(\frac{e^2}{\varepsilon r_{ij}}\right) \right]. \tag{18}$$

We recall (see discussion preceding Equation (2)) that in the scattering of light by a single, classical, free electron the A^2 term in the electron-photon interaction is dominant. This statement remains true in the many-electron, plasma case. In fact, one can use Equation (18) to give a plausibility argument for this fact. To do so, we recall that the wavelength of visible light is large compared to atomic dimensions (or even the mean-free-path of an electron in a crystal). As a consequence, the vector potential, $\mathbf{A}(\mathbf{r})$, varies relatively slowly in space. Therefore one can approximate the $(\mathbf{p} \cdot \mathbf{A})$ interaction in Equation (17) as follows:

$$\frac{e}{m^*c} \sum (\mathbf{p}_i \cdot \mathbf{A}) \simeq \left(\frac{e\bar{\mathbf{A}}}{m^*c}\right) \cdot \sum_i (\mathbf{p}_i) = \left(\frac{e\bar{\mathbf{A}} \cdot \mathbf{P}}{m^*c}\right). \tag{19}$$

Here $\mathbf{P} = \sum_i (\mathbf{p}_i)$ is the total electron momentum. If one ignores electron-lattice collisions, (usually a small perturbation), $\mathbf{P}$ is a constant of the motion whose value is zero (or nearly so) for an electron gas in a semiconductor. Though oversimplified, this argument is essentially correct, and can be quantitatively extended to show that the $(\mathbf{p} \cdot \mathbf{A})$ terms play practically no role in determining the spectrum of light scattered from a classical plasma[25].

We now use the A^2 interaction to determine the spectrum. For this purpose it is important to consider the geometry of the experiment. Figure 3 defines the important variables in such a measurement. Light, of wave vector $\mathbf{k}_0$ and

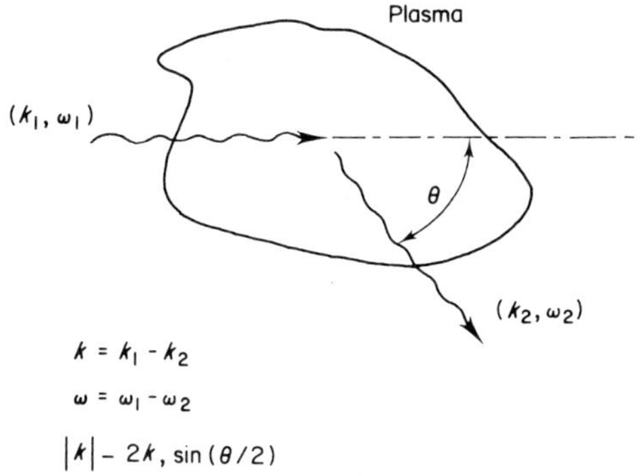

FIG. 3. Schematic scattering by a plasma.

relatively large wave vector transfer, $\mathbf{k} \equiv (\mathbf{k}_0 - \mathbf{k}_1)$ to the plasma. For near frequency ω_0, strikes the plasma. A detector observes the scattered spectrum at angle θ. We denote the wave vector and frequency of the scattered light by symbols $\mathbf{k}_1$ and ω_1. Generally speaking, $\omega_1 \simeq \omega_0$. However, there may be a relatively large vector transfer, $\mathbf{k} \equiv (\mathbf{k}_0 - \mathbf{k}_1)$ to the plasma. For near-elastic scattering,

$$|\mathbf{k}| \simeq 2k_0 \sin\left(\frac{\theta}{2}\right). \tag{20}$$

In scattering from $(\mathbf{k}_0, \omega_0)$ to $(\mathbf{k}_1, \omega_1)$, the electron-photon interaction takes the form:

$$\left(\frac{e^2}{m^*c^2}\right)\frac{2\pi\hbar c^2}{\sqrt{\omega_0\omega_1}}(\boldsymbol{\varepsilon}_0 \cdot \boldsymbol{\varepsilon}_1)\sum_i (e^{i\mathbf{k}\cdot\mathbf{r}_i}). \tag{21}$$

The operator appearing in this expression is the Fourier transform of the electron density operator:

$$\begin{aligned}\sum_i (e^{i\mathbf{k}\cdot\mathbf{r}_i}) &= \sum_i \int \delta(\mathbf{r} - \mathbf{r}_i)e^{i\mathbf{k}\cdot\mathbf{r}}\, d^3\mathbf{r} \\ &= \int n(\mathbf{r})e^{i\mathbf{k}\cdot\mathbf{r}}\, d^3\mathbf{r} \equiv n_{-\mathbf{k}}.\end{aligned} \tag{22}$$

The matrix element for scattering is

$$M_{if} = \frac{2\pi\hbar c^2}{\sqrt{\omega_0\omega_1}}\left(\frac{e^2}{m^*c^2}\right)(\boldsymbol{\varepsilon}_0 \cdot \boldsymbol{\varepsilon}_1)\langle\psi_f|n_{-\mathbf{k}}|\psi_i\rangle, \tag{23}$$

where ψ_i and ψ_f are *exact, many-body* eigenfunctions of the Hamiltonian H_0. Finally, the scattering rate is

$$R_{if} = \left(\frac{2\pi}{\hbar}\right)|M_{if}|^2\delta(E_i + \hbar\omega_0 - E_f - \hbar\omega_1). \tag{24}$$

From this point on, the analysis parallels that of van Hove[26], in his classic paper on scattering theory. Since his work is well known, we omit all details. Van Hove calculates the *total* scattering rate,

$$R = \sum_f (R_{if}), \tag{25}$$

consistent with a given frequency (or energy) transfer. This quantity determines the spectrum. Standard manipulations† give the following result for the differential cross-section

$$\frac{d^2\sigma}{d\Omega\, d\omega} = \left(\frac{e^2}{m^*c^2}\right)^2\left(\frac{\omega_1}{\omega_0}\right)(\boldsymbol{\varepsilon}_0 \cdot \boldsymbol{\varepsilon}_1)^2 S(\mathbf{k}, \omega), \tag{26}$$

where

$$S(\mathbf{k}, \omega) \equiv \left(\frac{1}{2\pi}\right)\int_{-\infty}^{\infty} \langle n_{\mathbf{k}}(t) n_{-\mathbf{k}}(0)\rangle e^{i\omega t}\, dt \tag{27}$$

is the *dynamic structure factor* or *density-density correlation function*. In these formulae

$$\omega \equiv (\omega_0 - \omega_1)$$

$$\mathbf{k} \equiv (\mathbf{k}_0 - \mathbf{k}_1),$$

and

$$n_{\mathbf{k}}(t) \equiv e^{iH_0 t} n_{\mathbf{k}} e^{-iH_0 t}. \tag{28}$$

Equations (26) and (27), which are the starting point for all discussions of light scattering by classical plasmas‡, imply that scattering is caused by electron density fluctuations. Physically, we may view the fluctuations as modulating the plasma dielectric constant:

$$\varepsilon = \left(1 - \frac{\omega_p^2}{\omega^2}\right) = 1 - \left(\frac{4\pi n e^2}{\varepsilon m^* \omega^2}\right). \tag{29}$$

† P. M. Platzman and P. A. Wolff. "Waves and Interactions in Solid State Plasmas", Academic Press, New York, ch. III (1973).

‡ The problem of light scattering by classical (i.e. gas) plasmas has been thoroughly studied. See, for example, E. E. Salpeter. *Phys. Rev.* **120**, 1528 (1960); M. N. Rosenbluth and N. Rostoker. *Phys. Fluids* **5**, 776 (1962).

The modulation then produces the scattering described by Equation (26). This viewpoint was originally discussed by Einstein[27] in 1910. It should also be noted that the *total* scattering rate (at a given scattering angle) is related to the *static structure factor*, or equal-time correlation function:

$$\left(\frac{d\sigma}{d\Omega}\right) = \int \frac{d^2\sigma}{d\Omega\, d\omega}\, d\omega$$
$$= \left(\frac{e^2}{m^*c^2}\right)^2 (\boldsymbol{\varepsilon}_0 \cdot \boldsymbol{\varepsilon}_1)^2 S(\mathbf{k}), \tag{30}$$

where

$$S(\mathbf{k}) \equiv \int S(\mathbf{k}, \omega) d\omega$$
$$= \langle n_{\mathbf{k}} n_{-\mathbf{k}} \rangle. \tag{31}$$

Finally, $S(\mathbf{k}, \omega)$ obeys an important sum rule:

$$\int_{-\infty}^{\infty} \omega S(\mathbf{k}, \omega) d\omega = \left(\frac{N\hbar k^2}{2m}\right). \tag{32}$$

The central, theoretical problem in analyzing the spectrum of light scattered from a plasma (or any other many-body system) is that of calculating the correlation function, $S(\mathbf{k}, \omega)$. This is a complicated, technical problem which has been extensively studied in a variety of cases. We will not attempt to reproduce that analysis here, but instead will use physical arguments to develop the spectrum. Our discussion is intuitive, but correct. Readers interested in details are referred to the literature.

We first consider the spectrum of radiation scattered from a non-interacting electron gas. In this case, the electrons move as independent particles, in straight-line trajectories. Hence

$$n_{\mathbf{k}}(t) = \sum_i [e^{-i\mathbf{k}\cdot\mathbf{r}_i(t)}],$$
$$= \sum_i [e^{-i\mathbf{k}\cdot(\mathbf{r}_i(0) + \mathbf{v}_i t)}]. \tag{33}$$

Substituting this expression into Equation (27) gives the result

$$S^{(0)}(\mathbf{k}, \omega) = \sum_i [\delta(\omega - \mathbf{k}\cdot\mathbf{v}_i)]$$
$$= \int \delta(\omega - \mathbf{k}\cdot\mathbf{v}) f(\mathbf{v}) d^3\mathbf{v}, \tag{34}$$

where $f(\mathbf{v})$ is the electron velocity distribution. Equation (34) has a simple interpretation. It can easily be seen that the frequency shift, $\omega = \mathbf{k}\cdot\mathbf{v}$, is a Doppler shift caused by electron motion relative to the observer. Thus, the complete spectrum is no more than the sum of all possible Doppler shifts. This result implies that the spectrum is determined by—and can be used to measure—the electron velocity distribution. Light scattering experiments are now, in fact, a routine tool for determining velocity distributions in gas plasmas. In the Maxwellian case, Equation (34) yields a spectrum of the form

$$S(\mathbf{k}, \omega) = \left(\frac{m}{2\pi k_B T}\right)^{1/2}\left(\frac{1}{k}\right)\exp\left(-\frac{m\omega^2}{2k^2(k_B T)}\right), \tag{35}$$

where T is the electron temperature. Note that this spectrum is Gaussian in frequency, centred at $\omega = 0$, with a width proportional to $T^{1/2}$.

We now consider how the preceding picture must be modified when electron-electron interactions are included. At long wavelengths, the response of a plasma to perturbations becomes entirely "collective" in nature, and is best described by normal modes of the entire medium. A classical equation of motion for the density fluctuation follows from Poisson's equation, the continuity equation, and Newton's law describing the acceleration of the carriers by the electric field. These equations have the form:

$$\begin{aligned} \nabla\cdot\mathbf{E} &= 4\pi e n \\ \frac{\partial n}{\partial t} + \nabla\cdot\mathbf{\Gamma} &= 0 \\ \frac{\partial \Gamma}{\partial t} &= \frac{n_0 e\mathbf{E}}{m}, \end{aligned} \tag{36}$$

where $\mathbf{\Gamma}$ is the particle current, and n the fluctuation in electron density. Upon eliminating $\mathbf{\Gamma}$ and $\mathbf{E}$, we find the equation

$$\frac{\partial^2 n}{\partial t^2} + \omega_p^2 n = 0 \tag{37}$$

Here n oscillates harmonically in time, at frequency ω_p. A more accurate approximation gives the dispersion relation

$$\omega^2(\mathbf{k}) = [\omega_p^2 + k^2\langle v^2\rangle]. \tag{38}$$

Thus, in the "collective" limit, the system can be viewed as a set of independent, harmonic oscillators—one for each value of $\mathbf{k}$. The displacement

($x_{\mathbf{k}}$) of each oscillator is proportional to $n_{\mathbf{k}}$. To calculate the correlation function, we note that

$$S(\mathbf{k}, \omega) = \int \langle n_{\mathbf{k}}(t) n_{-\mathbf{k}}(0) \rangle e^{i\omega t} \frac{dt}{2\pi}$$
$$= A \int \langle x_{\mathbf{k}}(t) x_{\mathbf{k}}(0) \rangle e^{i\omega t} \frac{dt}{2\pi}, \tag{39}$$

where A is a constant that we will later evaluate from the sum rule (Equation (32)). Since $x_{\mathbf{k}}$ is the amplitude of a harmonic oscillator, of frequency $\omega(\mathbf{k})$, we find (with the aid of the sum rule):

$$S(\mathbf{k}, \omega) = \left(\frac{N\hbar k^2}{2m}\right)\left[\left(\frac{e^{\beta\hbar\omega_{\mathbf{k}}}}{e^{\beta\hbar\omega_{\mathbf{k}}} - 1}\right)\delta(\omega - \omega_{\mathbf{k}}) + \left(\frac{1}{e^{\beta\hbar\omega_{\mathbf{k}}} - 1}\right)\delta(\omega + \omega_{\mathbf{k}})\right]. \tag{40}$$

This spectrum is quite different from that predicted by Equations (34) or (35). It consists of two peaks, shifted in frequency by amounts $\omega = \pm\omega_{\mathbf{k}}$; there is no scattering at $\omega = 0$. Moreover, the total intensity of the scattering is drastically reduced compared to that predicted by Equations (30) and (34). In the latter case (which is usually called the single-particle limit)

$$\frac{d\sigma}{d\Omega} = N\left(\frac{e^2}{m^*c^2}\right)^2 (\boldsymbol{\varepsilon}_0 \cdot \boldsymbol{\varepsilon}_1)^2. \tag{41}$$

When collective effects are dominant (so-called collective régime) we find

$$\frac{d\sigma}{d\Omega} \simeq N\left(\frac{e^2}{m^*c^2}\right)^2 (\boldsymbol{\varepsilon}_0 \cdot \boldsymbol{\varepsilon}_1)^2 \left[\frac{k^2(k_B T)}{\left(\dfrac{4\pi n e^2}{\varepsilon}\right)}\right]$$
$$= N\left(\frac{e^2}{m^*c^2}\right)^2 (\boldsymbol{\varepsilon}_0 \cdot \boldsymbol{\varepsilon}_1)^2 \left(\frac{k^2}{\kappa_D^2}\right), \tag{42}$$

where

$$\kappa_D = \sqrt{\frac{4\pi n e^2}{\varepsilon(k_B T)}} \equiv \lambda_D^{-1} \tag{43}$$

is the reciprocal of the Debye length. The peaks at frequencies $\omega = \pm\omega_{\mathbf{k}}$ in Equation (40) describe Raman scattering from the collective (plasma) modes of the electron gas. The integrated intensity is reduced (compared to that in the single-particle régime) by correlation effects. In an interacting electron gas, a density fluctuation, $n_{\mathbf{k}}$, has a Coulomb energy (per unit volume)

$$V_{\mathbf{k}} = \left[\frac{4\pi n_{\mathbf{k}}^2 e^2}{\varepsilon k^2}\right]. \tag{44}$$

This energy becomes large in the limit $k \to 0$. The net result is a drastic suppression of density fluctuations at long wavelengths. This suppression is responsible for the factor (k^2/κ_D^2) in Equation (42).

In an actual plasma, the light scattering spectrum can show either collective (Equation (40)) or single particle (Equation (34)) behaviour, depending upon the wave vector (**k**) with which it is probed. Single-particle scattering is observed when $k > \kappa_D$; collective-mode scattering when $k < \kappa_D$. Physically speaking, the transition from single-particle to collective scattering (as $k \to 0$) can be viewed as an interference effect. In the large-k limit ($k \gg \kappa_D$), waves scattered from different electrons in the plasma do not interfere with one another. The net result is a formula (Equation (34)) in which one sums the intensities of these waves. On the other hand, as k becomes smaller than κ_D, waves scattered from different electrons interfere (destructively). In the long-wavelength limit ($k \ll \kappa_D$) the total cross-section is obtained by summing the amplitudes (weighted with appropriate correlation function) of waves scattered from each electron.

We next consider how these ideas may be applied to light scattering experiments in solids. Collective effects in light scattering were first observed by Mooradian and Wright[28] in n-GaAs. They scattered 1.06 μm (Nd:YAG laser) radiation from the plasma in such crystals, and measured spectra for various doping densities. In their initial work, the carrier densities were sufficiently large that all spectra were in the collective regime ($k \ll \kappa_D$). Thus one expects to observe Raman scattering from plasma modes, rather than single-particle behaviour. To understand their results, however, we must recall that a crystal supports other collective modes besides those due to the plasma. In particular, in an acentric material such as GaAs, both the transverse optic (TO) and longitudinal optic (LO) modes at $k = 0$ are Raman-active. Moreover, since both the plasmon and the LO phonon are longitudinal waves, with large macroscopic electric fields, these modes interact to give hybrid excitations—normal modes that are partially plasmon-like and partially phonon-like. This mode-coupling effect is best described with a dielectric constant formalism[29]. In the long wavelength limit, the system has the dielectric constant

$$\varepsilon(\omega) = \varepsilon_\infty + \frac{(\varepsilon_0 - \varepsilon_\infty)\omega_T^2}{(\omega_T^2 - \omega^2)} - \left(\frac{4\pi n e^2}{m^* \omega^2}\right), \tag{45}$$

where ε_∞ is the high-frequency dielectric constant (well above the reststrahl frequency); ε_0 is the low-frequency dielectric constant (below the reststrahl); and ω_T is the TO frequency. The second term in Equation (45) is due to lattice polarization; the third is the free carrier (plasma) contribution. Normal mode frequencies are determined by the condition

$$\varepsilon(\omega) = 0,$$

with roots (often called L_+ and L_-)

$$\omega_\pm^2 = \tfrac{1}{2}(\omega_L^2 + \overline{\omega}_p^2) \pm \tfrac{1}{2}[(\omega_L^2 + \omega_p^2)^2 - 4\overline{\omega}_p^2\omega_T^2]^{1/2}. \tag{46}$$

Here

$$\overline{\omega}_p^2 = \left(\frac{4\pi ne^2}{\varepsilon_\infty m^*}\right) \tag{47}$$

is the plasma frequency in the absence of phonon effects, and

$$\omega_L = \left(\frac{\varepsilon_0}{\varepsilon_\infty}\right)^{1/2} \omega_T \tag{48}$$

is the LO frequency of an undoped crystal. Figure 4 shows a comparison of the predictions of Equation (46) with data of Mooradian and Wright[28]. The agreement is excellent. For small n, the lower root ω_- (or L_-) starts out as a plasmon of frequency

$$\left(\frac{4\pi ne^2}{\varepsilon_0 m^*}\right)^{1/2},$$

then at higher concentrations becomes phonon-like. In the limit of large n_e, it is the longitudinal optic mode of the crystal. Conversely, the root ω_+ (or L_+) starts as the LO and turns into a plasmon (now at frequency $\overline{\omega}_p$) as n increases.

As Fig. 4 indicates, the variation of $\omega_\pm$ with carrier concentration conforms exactly to theory. It is less easy to account for the intensities of these lines because, in addition to the electron scattering, the phonons in GaAs are also Raman-active. The modes are mixtures of phonon and plasmon, so both couplings contribute to the scattering. A good deal of work has been done on this problem and it is now fairly well understood.

To date, the most extensive work on plasmon light scattering has been done in n-GaAs. However, the effect has also been seen in other semiconductors, such as CdTe, InP, InAs, InSb and CdS. Plasmon-phonon

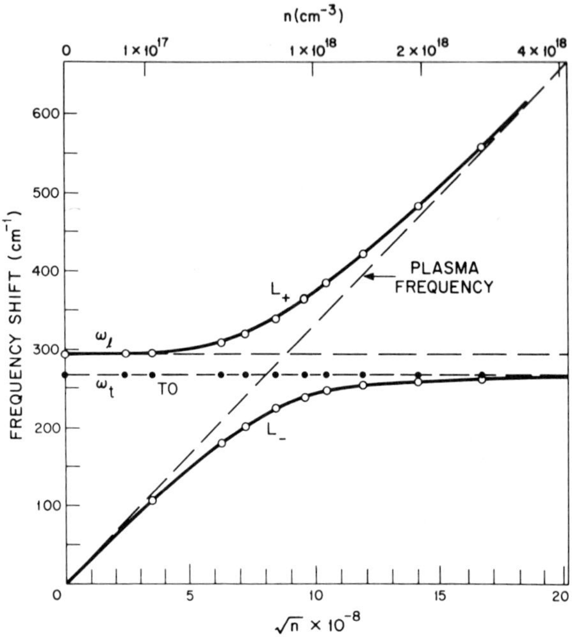

FIG. 4. Coupled plasmon—phonon modes in GaAs, 300° K.

coupling is observed in all of these materials, except InAs. In cases where it is seen, the data fit Equation (46) well.

The early experiments on plasmon light scattering in semiconductors were performed on fairly heavily doped materials, i.e. in the collective régime ($k \ll \kappa_D$ or $k \ll \kappa_{FT}$ = Fermi-Thomas wave vector). As techniques improved, the experiments were extended to lightly doped samples. The portion of the spectrum near the laser line—single-particle scattering—was also studied[30]. This work produced a major surprise. Even in the collective régime, the single-particle scattering was found to be very strong—several orders of magnitude more intense than the classical theory would permit. This scattering is now known to be caused by spin-density fluctuations. The polarization of the scattered radiation was the clue to the resolution of this problem. Mooradian[30] found that in the more heavily doped *n*-GaAs samples, the single-particle scattering is polarized *perpendicular* to the laser beam. Such behaviour is the reverse of that expected for density fluctuation scattering (Equations (41) and (42)), but is in agreement with the selection rules for spin-flip scattering (Equation (12)). This observation lead Hamilton and McWhorter[31] to suggest that spin interactions are responsible for the strong single-particle scattering when $k \ll \kappa_{FT}$. Detailed calculations confirm this view. The essential point is that spin-density fluctuations have no charge, hence are not suppressed by many-body interactions. The calculation of spin-density fluctuation scattering is performed as if one were dealing with a noninteracting electron gas. The results of such calculations are in quantitative agreement with Mooradian's work.

IV. STIMULATED SPIN-FLIP SCATTERING AND SPIN-FLIP LASERS

Before discussing stimulated spin-flip scattering, we briefly review the predictions of the theory (Equation (12)) and some of the important experimental facts concerning spontaneous spin-flip scattering in n-InSb. Equation (12) implies that:

1. In narrow-gap materials, $|g|/2 \gg 1$. As a consequence, the basic cross-section for the spin-flip process, $(e^2/m_s c^2)^2$, is much larger than the free-electron Thomson cross-section. For InSb the ratio is 625. The large g-value also provides a large tuning range for spin-flip scattering, since $\omega_1 = \omega_0 - \omega_s \equiv \omega_0 - \mu g B_0$. For example, if InSb is pumped with a CO_2 laser ($\omega_0 \simeq 10.6\ \mu m$), the Stokes radiation can be tuned from 10.6 μm to 13 μm with fields in the 0–100 kG range.

2. As $\hbar\omega_0 \to E_G$, there is resonant enhancement of the spin-flip cross-section. This effect is the solid state analogue of resonance fluorescence. In InSb resonance is achieved by pumping with CO lasers, whose frequencies closely match the gap of the crystal.

3. For pump frequencies well below the gap, the spin-flip cross-section (which is proportional to $|M|^2$) varies as ω_0^2. Phonon Raman cross-sections, on the other hand, have an ω_0^4 frequency dependence. As a consequence, the Raman gain for stimulated spin-flip scattering (which is essentially the scattering rate per mode of the radiation field) actually *increases* as the pump frequency decreases. To our knowledge, this situation only occurs in mobile electron scattering. In all other cases, the cross-section decreases more rapidly than the density of states as $\omega_0 \to 0$, and the Raman gain vanishes in the low frequency limit. As a consequence, conventional (phonon) Raman lasers cannot operate with IR pumps. The spin-flip laser is the only IR pumped, Raman oscillator.

4. The polarization selection rules for spin-flip scattering are quite different from those for elastic (Rayleigh) scattering. Equation (12) implies a maximum cross-section when the incident and scattered radiation have

orthogonal polarizations, whereas in Rayleigh scattering parallel polarization is favoured. One can also show from Equation (12), that the spin-flip process has its largest cross-section when one beam is polarized parallel to the dc magnetic, and the other orthogonal to it. This is the standard configuration for spin-flip Raman lasers.

Equation (12) has been thoroughly tested in the n-InSb case. Brueck and Mooradian[32] find a strong resonance enhancement as $\hbar\omega_0 \to E_G$, and also confirm the polarization selection rules implied by this formula. Under optimal conditions, the spin-flip cross-section is almost 10^5 times larger than the Thomson cross-section. They[33] have also studied the spontaneous spin-flip linewidth and lineshape in considerable detail. Several important conclusions emerge from that work:

1. The spin-flip line is quite sharp in InSb. Linewidths between 0.1–1 cm^{-1} are achieved under conditions of interest for spin-flip laser operation. The corresponding spin relaxation time (T_2) is in the 10^{-10}–10^{-11} sec range. Such times are considerably longer than the scattering time (τ) which determines the mobility of electrons in InSb; τ is usually about 10^{-12} sec. Hence, an electron in InSb scatters many times before the phase of its spin precession is destroyed. In this sense, spin-flip scattering is unique. All other mobile-electron, Raman processes have linewidths of order $(1/\tau)$.
2. The spin-flip linewidth depends upon the geometry of the experiment. If $\mathbf{k}_0$ and $\mathbf{k}_1$ are wave vectors of the incident and scattered light waves, $\hbar\mathbf{k} = \hbar(\mathbf{k}_0 - \mathbf{k}_1)$ is the recoil momentum imparted to the electron in the scattering process. Broad spin-flip lines are observed[33] in the $(\mathbf{k}\cdot\mathbf{B}_0) \neq 0$ geometry, whereas the line is considerably sharper in the $(\mathbf{k}\cdot\mathbf{B}_0) = 0$ configuration (here $\mathbf{B}_0$ is the dc magnetic field). This effect is caused by electron motion. In a magnetic field, the electrons are free to move *along* field lines, but cannot move *across* them. Motion along field lines gives a Doppler shift to the scattered wave in the $(\mathbf{k}\cdot\mathbf{B}_0) \neq 0$ geometry, but does not in $(\mathbf{k}\cdot\mathbf{B}_0) = 0$ situation. The upshot is a considerably larger linewidth in the $(\mathbf{k}\cdot\mathbf{B}_0) \neq 0$ case. Spin-flip laser experiments are usually done in the $(\mathbf{k}\cdot\mathbf{B}_0) = 0$ configuration.
3. At least two different processes contribute to the spin-flip linewidth in InSb. We have seen above that, in the $(\mathbf{k}\cdot\mathbf{B}_0) \neq 0$ case, the linewidth is due to electron motion. This effect is nearly absent in the $(\mathbf{k}\cdot\mathbf{B}_0) = 0$ configuration. Band nonparabolicity, which gives rise to a spread of g-values, is believed responsible for the linewidth in this geometry. Several theoretical papers have been devoted to this problem[34], but the situation is still not entirely clear.

It is often convenient[35] to view Equation (9) as the matrix element of a "Raman interaction" of the form

$$H_R = \left(\frac{e^2}{m_s c^2}\right)\left[\frac{E_G \hbar\omega_1}{E_G^2 - (\hbar\omega_1)^2}\right](\boldsymbol{\sigma} \cdot \mathbf{A}_0 \times \mathbf{A}_1^*). \tag{49}$$

This coupling results, of course, from the virtual, interband process described earlier. For many purposes, however, the details of this virtual process are unimportant and one may equally well use the direct Raman coupling of Equation (49). Equation (49) can, in fact, be derived by an effective-mass-type transformation which removes the interband $(\mathbf{p} \cdot \mathbf{A})$ terms from our basic Hamiltonian. Some mathematical simplifications result from this transformation, but the major advantage is a conceptual one. Equation (49) is similar to the Zeeman interaction;

$$H_Z = \mu g(\boldsymbol{\sigma} \cdot \mathbf{B}). \tag{50}$$

In H_Z the spin is coupled to a magnetic field; in Equation (49), on the other hand, the spin couples to a nonlinear combination $(\mathbf{A}_0 \times \mathbf{A}_1^*)$ of two electromagnetic fields. Nevertheless, as far as the electron is concerned, the interactions are similar—in each case an external force exerts a torque on its spin. The spin-flip process can be viewed, therefore, as a sort of "nonlinear spin resonance" in which the light waves combine, via the Raman interaction, to excite spin precession[35]. This observation is useful because electron spin motion has been extensively studied in connection with the EPR problem. Much of this analysis directly applies to the present case as well.

A. STIMULATED SPIN-FLIP SCATTERING

Bloembergen and Shen[36] first pointed out that the Raman process can be described by a third-order, nonlinear optical coefficient. If a Raman-active crystal is subjected to a pump field $(\mathbf{E}_0)$ and a field $(\mathbf{E}_1)$ near the Raman frequency, the combination induces a nonlinear polarization at frequency ω_1 given by

$$\mathbf{P}^{NL}(\omega_1) = \chi_R |E_0|^2 \mathbf{E}_1. \tag{51}$$

The nonlinear coefficient (χ_R) has a sharp resonance when the Raman condition, $\omega_1 = \omega_0 - \omega_s$, is satisfied. At resonance, χ_R is purely imaginary and negative (ignoring background contributions due to other, nonresonant processes). This fact implies a negative absorption, or "gain", at frequency ω_1.

To relate the gain to χ_R, we consider the wave equation for the field $\mathbf{E}_1$:

$$\frac{\partial^2 \mathbf{E}_1}{\partial z^2} + \frac{\varepsilon_0 \omega_1^2}{c^2} \mathbf{E}_1 = -\frac{4\pi\omega_1^2}{c^2} \mathbf{P}^{NL}$$

$$= -\frac{4\pi\omega_1^2}{c^2} \chi_R |E_0|^2 \mathbf{E}_1. \tag{52}$$

This equation has a solution of the form

$$\mathbf{E}_1 = \mathbf{E}_1(z) e^{ik_1 z}, \tag{53}$$

where $\mathbf{E}_1(z)$ is a slowly varying function, and $k_1 = \sqrt{\varepsilon_0}\,\omega_1/c \equiv n\omega_1/c$. $\mathbf{E}_1(z)$ is determined by the equation

$$2ik_1 \frac{\partial \mathbf{E}_1(z)}{\partial z} \simeq -\left(\frac{4\pi\omega_1^2}{c^2}\right) \chi_R |E_0|^2 \mathbf{E}_1(z) \tag{54}$$

whence

$$\mathbf{E}_1(z) \sim e^{i\beta z}, \tag{55}$$

with

$$\beta = \left(\frac{2\pi\omega_1}{nc}\right) \chi_R |E_0|^2. \tag{56}$$

The energy density varies as follows

$$|E_1(z)|^2 \sim e^{gz},$$

where g is the gain:

$$g = -\left(\frac{4\pi\omega_1}{nc}\right) |E_0|^2 \operatorname{Im}(\chi_R). \tag{57}$$

Physically, the gain represents coherent, stimulated scattering of radiation from the pump field ($\mathbf{E}_0$) into the Stokes field ($\mathbf{E}_1$). This statement implies a relation between the gain (g) and the spin-flip cross-section ($d\sigma/d\Omega$), which has the form

$$g(\mathrm{cm}^{-1}) = \frac{16\pi^2 c^2 N_e \left(\dfrac{d\sigma}{d\Omega}\right) I_0}{\hbar\omega_0 \omega_1^2 n^2 \Gamma}. \tag{58}$$

Here N_e is the electron density, I_0 the pump power density, and Γ the (spontaneous) spin-flip linewidth. This formula follows from a simple rate-equation argument[37], which we will not reproduce here. To understand its meaning, we rewrite Equation (58) as follows:

$$g = \left[\frac{nN_e(\text{vol})\left(\dfrac{d\sigma}{d\Omega}\right)I_0}{\hbar\omega_0 c}\right] \Big/ \left[\frac{n^3\omega_1^2\Gamma}{(2\pi c)^3}\right]. \tag{59}$$

In this expression, g can be seen to be the ratio of the total scattering rate per unit solid angle, to the density of photon states per unit solid angle. We conclude, therefore, that g is the *scattering rate per mode* of the field.

Equation (58) is an important result, central to the theory of the spin-flip laser (SFRL). It implies that g is proportional to $(d\sigma/d\Omega)$, and varies inversely with linewidth and frequency cubed. As mentioned earlier, the spin-flip process in InSb is favourable in all these respects. Its cross-section is large, the lines are narrow and, most important of all, the laser can be pumped at relatively low frequencies. A numerical example, discussed by Patel and Shaw[38], illustrates the importance of the final condition. In CS_2 the spontaneous Raman efficiency (fraction of incident light scattered per cm of travel) is 10^{-6} cm^{-1} ster.$^{-1}$ at 5000 Å, which is somewhat larger than that of the CO_2 pumped spin-flip Raman process in n-InSb. In addition, the linewidth Γ for CS_2 Raman scattering is $\sim 1.4\ \text{cm}^{-1}$, again comparable to that of the spin-flip process in InSb. Nevertheless, the calculated Raman gain for CS_2 is only $g \sim 10^{-8}\ \text{cm}^{-1}/\text{W cm}^{-2}$, compared to $g \sim 10^{-5}\ \text{cm}^{-1}/\text{W cm}^{-2}$ for spin-flip in InSb. This enormous difference is a direct consequence of the ω_0^3 factor in the denominator of Equation (58) (we should emphasize, again, that it does no good to decrease the pumping frequency in the CS_2 case, because the cross-section decreases faster (ω_0^4) than the photon state density).

The spin-flip process in n-InSb has, by far, the largest known Raman gain. With CO_2 laser pumping

$$g \sim 10^{-5}\ \text{cm}^{-1}/\text{watts cm}^{-2}.$$

This figure can be enhanced with resonant pumping ($\hbar\omega_0 \sim E_G$, see Equation (12)) by CO laser beams. Gains as large as $10^{-2}\ \text{cm}^{-1}/\text{W cm}^{-2}$ have been achieved with this technique. The Raman laser oscillates when the gain (g) exceeds the loss (α) due to absorption and mirror loss. In InSb, α is usually of order 1 cm^{-1}. Thus, pump power densities in the 10^5 watts/cm^2 range (with CO_2 laser pumping) or the 10^3 watts/cm range (with CO pumping) are required to achieve net gain ($g > \alpha$). These requirements are well within reach of present laser technology.

B. SPIN-FLIP RAMAN LASERS (SFRL)

(a) *Pulsed Operation*

Spin-flip laser operation was first demonstrated by Patel and Shaw[39] in 1970. The device consisted of a liquid He cooled, n-InSb crystal ($n = 1.6 \times 10^{16}$/cc) pumped by a Q-switched CO_2 laser. The pump pulses had powers in the 3–5 kW range, and were about 200 nsec long. Power densities exceeding 1MW/cm^2 were achieved by focussing these pulses into the InSb crystal. Laser operation (as opposed to spontaneous spin-flip scattering) was confirmed by two types of experiments: measurements of Stokes output versus pump power, and studies of the spectral distribution of the Raman radiation. For pump powers below threshold for laser action, the scattered Raman (Stokes) power is low, and varies linearly with pump power. At a certain critical pump power density, the Raman output abruptly increases by two to three orders of magnitude. Simultaneously, there is a dramatic narrowing of the spectral distribution of the Raman radiation. These features indicate the onset of laser action. Beyond threshold, the output power continues to increase, but eventually saturates. Output powers were in the 10W range, corresponding to a conversion efficiency of 1%. This value is typical of CO_2-pumped SFRL's. With a 10.59 5 μm CO_2 laser pump, the tuning range of the Raman radiation was 10.9 μm–13 μm. The upper limit was set by the available field (100 kG); the lower by a quantum condition in the InSb crystal. For fields above 18 kG (corresponding to Stokes radiation at 10.9 μm) all electrons in the n-InSb crystal were in the lowest, $n = 0$, spin-down quantum level. This situation is called the "quantum limit". For fields below 18 kG (in the particular sample studied by Patel and Shaw) electrons began to occupy the $n = 0$, spin-up state. As a consequence, there were electrons in the final state for the Raman process, which blocked (via the exclusion principle) transitions which could otherwise take place. The upshot was a significant reduction in gain. This blocking effect has been studied in detail by Wherrett and Harper[40], and by Patel and Shaw[38].

High-power CO_2-pumped, SFRL's have been investigated by Aggarwal *et al.*[41]. These authors pumped n-InSb crystals with CO_2 TEA lasers, having powers in the 25–100 kW range. The spin-flip power saturated at 1 kW—presumably because of spin saturation. The experiments imply a spin relaxation time, $T_1 \sim 6 \times 10^{-8}$ sec in cold InSb.

Pulsed SFRL operation in n-InSb has also been demonstrated[42] with high power ($\sim$3 kW) 5.3 μm pump pulses, obtained by doubling CO_2 TEA laser pulses in Te. In this case, conversion efficiencies exceeded 50%.

(b) CW *Operation*

CW SFRL operation was first demonstrated by Mooradian *et al.*[43]. Their device also operated in cooled *n*-InSb crystals, but was pumped with a CO laser whose photon energy (5–6 μm) closely matched the gap of the InSb crystal. As a consequence (see Equation (3)), there was a dramatic enhancement of the Raman cross-section, and a corresponding reduction in threshold power. CO-pumped SFR lasers have operated with pump powers as low as a few milliwatts[44]. One the other hand, cw output powers as high as 1 W have also been achieved. Conversion efficiencies generally exceed 50 %.

Much effort has been devoted to the frequency stabilization of cw, SFR lasers. Patel[45] demonstrated that the inherent linewidth of the Raman laser radiation is less than 1 kHz. Single-axial, single-transverse mode operation was achieved by Brueck and Mooradian[46]. More recently, these authors[47] have made a complete study of the frequency stabilization and fine-tuning characteristics of cw SFR lasers. In low carrier concentration samples ($n < 10^{15}$(cc), operating in the low field range, the spin-flip linewidths are exceedingly small [$\Gamma \simeq 0.02$ cm^{-1} at 5 kG]. The output of such devices can be continuously tuned over most (>90 %) of its frequency range. As magnetic field is varied, the device "hops" from one cavity mode to another. However, with a narrow spin-flip line the individual mode frequencies "pull" with applied field to give nearly complete coverage. The tuning rate is about -75 MHz/Gauss. Absolute frequencies of the SFRL output are determined by comparison (via heterodyning) with a frequency stabilized CO laser. Stable output, with linewidth less than 30 kHz was obtained. This device provides the ease and accuracy of microwave tuning in the 5–6 μm region.

V. Spin-Induced Far-Infrared Generation

One of the important goals of quantum optics is the development of a tunable, coherent, reasonably intense source of far-infrared (FIR) radiation. Several experiments have been reported[48] in which visible or near infrared laser beams were mixed in nonlinear materials to generate 100 μm–1000 μm radiation. To date, however, these processes have not been sufficiently efficient or controllable to be used as spectroscopic sources. The problem is tantalizing, because the recent development of the SFRL has made available an excellent, tunable source of near-infrared radiation.

In this device, the difference-frequency, between the pump and Raman fields, spans the FIR region. It is natural, therefore, to envisage using such a system to generate tunable FIR. The principal problem in such an application is that of finding an efficient mixing crystal. Recently, it has been demonstrated[35] that the spin system of n-InSb, itself, is an excellent medium for this purpose. Microwatt, tunable, FIR powers (in the 100 μm range) have been achieved in pulsed experiments via this technique. There is a good chance that comparable, cw powers can also be realized. Such a device—a tunable cw source of 100 μm–1000 μm radiation—could have a considerable impact on research in solid state physics, chemistry, and astronomy.

To investigate spin-induced FIR generation, we consider the "Raman Hamiltonian" discussed earlier (Equation (49)):

$$H_R = \left(\frac{e^2}{m_s c^2}\right)\left[\frac{E_G \hbar\omega_1}{E_G^2 - (\hbar\omega_1)^2}\right](\boldsymbol{\sigma} \cdot \mathbf{A}_0 \times \mathbf{A}_1^*). \tag{60}$$

If two laser fields ($\mathbf{A}_0$ and $\mathbf{A}_1^*$) of appropriate polarization are simultaneously present, they exert, via this interaction, a torque on the electron spin similar to that produced by a transverse RF magnetic field in a conventional EPR experiment. As a consequence, the spins precess about the dc field at the driving frequency. For the case of the Raman interaction, this driving force is at the difference frequency, $\omega = \omega_0 - \omega_1$, of the two light beams. The force has a maximum effect when its frequency equals the natural precessional frequency, $\omega_s = \mu g B_0$, of electrons in the dc magnetic field. For the InSb case, the electron response is sharply resonant because the spin-flip (or spin resonance) linewidth is small.

The preceding argument implies that the spin-flip Raman interaction can be used to excite a form of nonlinear spin resonance in a material such as n-InSb. This effect is detected by observing radiation, at the difference frequency, generated by the precessing spins. Since the excitation is coherent, all spins in the sample precess together, giving rise to a macroscopic, oscillating, magnetic dipole moment. The signal radiated by this dipole moment is somewhat analogous to the free induction decay signal studied in NMR experiments. In the present case, however, the radiation is in the FIR, and excited via a nonlinear, difference-frequency mechanism.

We emphasize that the process described above is caused by magnetic dipole rather than electric-dipole radiation. Electric dipole-induced, three-photon mixing is forbidden in materials having a centre of inversion. Though InSb is acentric, this fact has little effect on conduction electron dynamics, The Hamiltonians used in all light scattering calculations are centro-symmetric, and thus do *not* permit the three-photon process, $\omega_1 \to \omega_2 + \omega_3$ via an electric dipole transition. Electrons can, however, emit FIR via a *magnetic-dipole* transition.

To estimate the strength of this two-step, magnetic-dipole process, we return to Equation (60)—the Raman interaction. As emphasized earlier, this interaction produces a torque on the electron spins which causes them to precess, at frequency $\omega_3 = \omega_1 - \omega_2$, about the dc magnetic field. The problem is formally identical to that encountered in the standard EPR experiment. Using this analogy, one can easily calculate the transverse spin motion and the induced magnetization. The result is[35]

$$M^{NL}(\omega_3) = C(\omega_3) E_{\omega_1} E^*_{\omega_2}, \tag{61}$$

where

$$C(\omega_3) = \left\{ N\mu^* \left(\frac{e^2}{m_s c^2} \right) \left[\frac{\hbar\omega_1 E_G}{E_G^2 - (\hbar\omega_1)^2} \right] \left(\frac{e^2}{\omega_1 \omega_2} \right) \frac{1}{2\hbar\left(\omega_3 - \omega_s + \dfrac{i\Gamma}{2} \right)} \right\}. \tag{62}$$

Here N is the electron density, μ^* the magnetic moment of an electron in InSb. This formula has a sharp resonance—the usual EPR resonance—when $\omega_3 = \omega_s$. In addition, there is further enhancement as $\hbar\omega_1 \to E_G$. The nonlinear magnetization, $M^{NL}(\omega_3)$, is the source of the FIR field, whose intensity can now be calculated from Maxwell's equation:

$$\frac{\partial^2 H_x(\omega_3)}{\partial y^2} + \frac{\varepsilon_3 \omega_3^2}{c^2} H_x(\omega_3) = \frac{-4\pi\varepsilon_3 \omega_3^2}{c^2} M_x^{NL}(\omega_3). \tag{63}$$

To solve, we write $H_x(\omega_3) = h(y)e^{ik_3 y}$, with $h(y)$ a slowly varying envelope function. Equation (63) then takes the form

$$2ik_3 \frac{dh}{dy} + \frac{i\,\mathrm{Im}(\varepsilon_3)\omega_3^2 h}{c^2} = -\left(\frac{4\pi\varepsilon_3 \omega_3^2}{c^2} \right) C(\omega_3) E_{\omega_1} E_{\omega_2} e^{i\Delta k y}, \tag{64}$$

where $\Delta k = k_1 - k_2 - k_3$ is the phase mismatch. With the boundary condition $h(0) = 0$, the solution to Equation (64) is

$$h(y) = 4\pi \left[\frac{k_3}{(2\Delta k - i\alpha)} \right] C(\omega_3) E_{\omega_1} E_{\omega_2} (e^{i\Delta k y} - e^{-\alpha y/2}). \tag{65}$$

Here α is the absorption coefficient of the InSb crystal at frequency ω_3. In the limit $\Delta k y$, $\alpha y \ll 1$ (perfect phase matching and low loss), one finds

$$h(y) \simeq 2\pi i (k_3 y) C(\omega_3) E_{\omega_1} E_{\omega_2}. \tag{66}$$

Mixing, via the spin-dependent process described above, has been studied by Nguyen and Bridges[49]. They pumped an n-InSb crystal ($n = 2 \times 10^{15}$/cc) with beams from two CO_2 lasers, operating at 9.6 μm(ω_1) and 10.6 μm(ω_2), respectively. The power generated at the difference frequency ($P(\omega_3)$) was monitored as a function of applied magnetic field. This radiation was at 90 cm^{-1}. A typical experimental trace is illustrated in Fig. 5. Note the sharp resonance in the far-infrared power when the spin resonance condition ($\omega_3 = \omega_s$) is satisfied. The strength of this resonance was estimated by comparing the peak intensity to the known, off-resonance signal due to valence electrons. The ratio was measured to be 88; the theory sketched above gives the value 200.

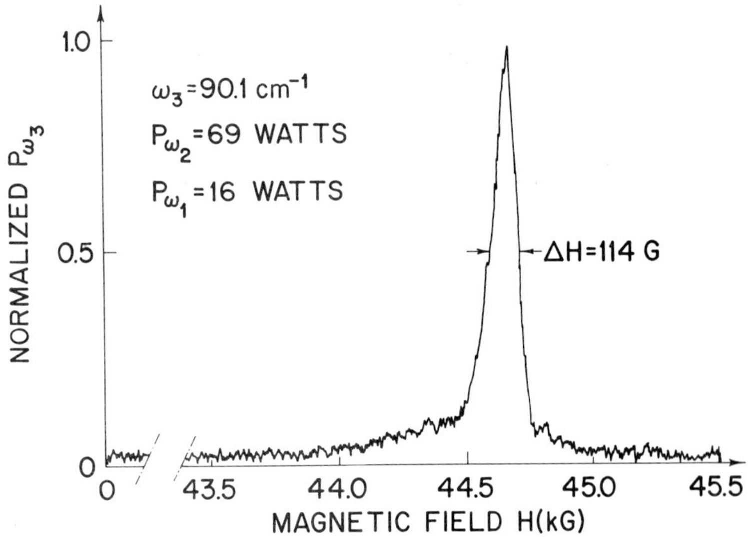

FIG. 5. Far-infrared power generated by resonant difference-frequency mixing in InSb.

The spin-dependent mixing coefficient in n-InSb is the largest known. Moreover, one can anticipate that it will be further enhanced by pumping with near band-gap radiation (see Equation (9)). To date, this spin-induced mixing process has been used to generate fixed-frequency[49] and tunable[50] FIR at the microwatt level. In the latter experiment, one of the near IR pump beams was a pulsed, CO_2 pumped, spin-flip laser whose frequency could be adjusted with magnetic field—thus tuning the FIR output. Note that the magnetic field required to achieve spin resonance is exactly the same as that used in the spin-flip laser [$\omega_3 = \omega_1 - \omega_2 = \omega_1 - (\omega_1 - \omega_s) \equiv \omega_s$]. Thus, the whole experiment could be done in a single field coil and, in principle, might even be performed in one InSb crystal, which would simultaneously serve as the spin-flip oscillator (to generate ω_2 from ω_1) and the FIR mixing medium.

The theory of the spin-induced mixing process (Equation (62)) predicts a large enhancement of the mixing coefficient as the driving frequencies (ω_1, ω_2) approach the band gap. This enhancement has been thoroughly documented in the spin-flip cross-section; there is every reason to believe it will also be present in $C(\omega_3)$. In InSb, resonance is best achieved with CO laser pumping. This combination is also attractive because it is the system used in the cw SFRL. In principle, therefore, one could simultaneously pump *n*-InSb with CO and SFR lasers—both nearly resonant with the band gap—to generate tunable, cw FIR. Under optimal conditions, the mixing coefficient in such an arrangement could be as large as 100 times that of the conventional, valence electron nonlinearity. This system is clearly an attractive one for FIR generation.

Unfortunately, to date, phasematching problems have prevented the development of a CO-pumped, FIR source of the type described above. To understand the difficulty, we briefly consider the general question of phasematching in three wave mixing. As indicated in Equation (64), the nonlinear magnetization which drives the FIR field at frequency ω_3, contains a phase factor $e^{i\Delta ky}$, where

$$\begin{aligned}\Delta k &\equiv [k_1 - k_2 - k_3] \\ &= \frac{1}{c}[n(\omega_1)\omega_1 - n(\omega_2)\omega_2 - n(\omega_3)\omega_3] \end{aligned} \tag{67}$$

Here $n(\omega)$ is the index of refraction, at frequency ω, of the mixing crystal. Efficient FIR generation (see Equation (65)) can only be achieved when $\Delta k \simeq 0$. Otherwise, there is destructive interference (and almost complete cancellation) between waves excited at different positions along the interaction path. Since $\omega_1 - \omega_2 = \omega_3$, the condition $\Delta k = 0$ implies

$$[n(\omega_1) - n(\omega_2)]\omega_2 + [n(\omega_1) - n(\omega_3)]\omega_3 = 0. \tag{68}$$

This is the basic equation for analyzing collinear phasematching problems in cubic (optically isotropic) media.

If the pump frequencies ω_1 and ω_2 are in the 10 μm range, i.e. well below the band gap of InSb, $n(\omega_1) \simeq n(\omega_2)$ and the phasematching condition takes the form:

$$n(\omega_1) = n(\omega_3) \tag{69}$$

Here ω_3 is generally below the reststrahl frequency, whereas ω_1 is well above it, so this equation can also be written as

$$n_\infty = n_0$$

or

$$\varepsilon_\infty = \varepsilon_0. \tag{70}$$

ε_0 is the low-frequency dielectric constant, whereas ε_∞ is the dielectric constant at a frequency well above the lattice resonance (but below the band gap). The condition $\varepsilon_0 = \varepsilon_\infty$ can never be satisfied in undoped, ionic materials, such as pure InSb. The contribution of the lattice resonance to the dielectric constant invariably makes $\varepsilon_0 > \varepsilon_\infty$. In fact, for crystals such as InSb, the Lyddane-Sachs-Teller relation implies that

$$\varepsilon_0 = \left(\frac{\omega_{LO}^2}{\omega_{TO}^2}\right)\varepsilon_\infty, \tag{71}$$

where ω_{LO} and ω_{TO} are the frequencies of longitudinal and transverse optic modes. On the other hand, in *doped* crystals the plasma contribution to the dielectric constant reduces the low-frequency dielectric constant, but has little effect on that at frequency ω_1. This trick permits phasematching for FIR generation via mixing of CO_2 laser beams[51]. The explicit condition for phasematching is

$$\varepsilon_0\left(1 - \frac{\omega_p^2}{\omega_3^2}\right) \simeq \varepsilon_\infty. \tag{72}$$

In n-InSb ,Equation (72) is satisfied (for ω_3 near 100 μm) if $n \simeq 2 \times 10^{15}$/cc. Thus, in this technique, the electrons whose spins cause the mixing are also responsible for phasematching. The experiments of Ref. (50) were phasematched, via the plasma contribution to $n(\omega_3)$, in this way.

One must now ask why this phasematching trick cannot be used in the case of CO laser pumping of InSb. The problem is one of dispersion. As the pump frequencies (ω_1 and ω_2) approach the band gap, the index of refraction becomes frequency dependent. The total change is not large (about 1% overall in InSb), but is concentrated in the region $\hbar\omega_1 \simeq E_G$. In fact, the simplest theory predicts that $\partial n/\partial\omega \to \infty$ as $\hbar\omega \to E_G$ (though n, itself, is finite at the band edge). Because of the large frequency variation of the refractive index, it is no longer possible to make the approximation $n(\omega_1) \simeq n(\omega_2)$ in Equation (68). Instead, one must include a first-order correction due to the frequency difference ($\omega_1 - \omega_2 \equiv \omega_3$):

$$\begin{aligned} n(\omega_1) &= n(\omega_2 + \omega_3) \\ &\simeq n(\omega_2) + \omega_3 \frac{\partial n(\omega_2)}{\partial\omega}. \end{aligned} \tag{73}$$

The phasematching condition now takes the form:

$$\left[n(\omega_2) + \omega_2 \frac{\partial n(\omega_2)}{\partial\omega}\right] = n(\omega_3). \tag{74}$$

If $\hbar\omega_2 \simeq E_G$, the left-hand side of this equation is invariably *larger* than the low-frequency index, $n(\omega_3) \simeq \sqrt{\varepsilon_0}$. Under these conditions, the plasma effect *cannot* be used for phasematching. To date, two methods for circumventing this problem have been proposed. One uses uniaxial strain to make the InSb crystal birefringent—and thus permit phasematching. The stresses required are not exceedingly large, but the overall system is a clumsy one, requiring the simultaneous application of strain and magnetic field, low temperatures and good optical access. Experimentally, it is difficult to combine all these features in a single setup. Another possible technique for phasematching is that of inhomogeneous doping. By modulating the carrier concentration of an *n*-InSb crystal, one can arrange the carrier distribution in such a way as to prevent the phase cancellation mentioned above. This technique involves a difficult materials problem, but otherwise is a seemingly ideal way of achieving phasematching.

VI. Theory of Spin-Flip Line Shape in CdS

A. INTRODUCTION

As we have seen earlier, spin-flip scattering is a process in which an electron in a crystal scatters light with simultaneous change of its spin direction. Usually such experiments are performed in a static magnetic field. The scattered frequency is then shifted from the incident frequency by the electron spin resonance frequency—hence the process is magnetically tunable. This feature provided much of the impetus for early research on spin-flip scattering. The spin-flip laser resulted from this work.

So far, it has been less well appreciated that the spin-flip process can also be used to deduce fundamental information concerning electron dynamics in semiconductors—dynamics both of the electron spin motion, and of its translational motion through the crystal lattice. To exploit this fact, however, one must know how to relate the features of the spin-flip line (such as its position, intensity, linewidth, line shape, etc.) to parameters which characterize electron motion. The purpose of this section is to discuss these relationships and, in particular, to express the spin-flip line shape in terms of microscopic parameters. The most important of these is the *spin diffusion constant*, which, to our knowledge, cannot be measured in any other way. In particular, practically nothing is known about spin diffusion in the vicinity of the metal-insulator transition in semiconductors.

We will test our ideas concerning the spin-flip line shape by comparing the theory to experiments on *n*-CdS. CdS has an enormous spin-flip cross-section, and the line shape has been carefully measured. For this case, the theory works remarkably well. It accounts in detail, and without adjustable parameters, for most of the data concerning angular and temperature variation of the spin-flip linewidth. In the low temperature range, where discrepancies are found, they are understandable in terms of either the effects of electron–electron interaction or impurity banding. Neither the data nor the theory are yet good enough to characterize this region well, but it is clearly an area for further study.

In all spin-flip measurements, the magnetic tunablility of the scattering aids the experimentalist by giving a unique signature to the spin-flip signals—a resonance whose position in frequency is known, and varies linearly with magnetic field. In this respect, spin-flip scattering is akin to EPR. This analogy is, actually, a far reaching one. Spin-flip scattering and EPR both measure the same basic property of the conduction electron system—the transverse spin susceptibility, $\chi^+(\mathbf{k}, \omega)$. EPR, however, measures essentially the $\mathbf{k} = 0$ value of $\chi^+(\mathbf{k}, \omega)$, whereas spin-flip scattering can be used to determine $\chi^+(\mathbf{k}, \omega)$ over a range of $\mathbf{k}$. In this sense, spin-flip scattering is a more powerful technique than EPR.

B. FORMULATION OF THEORY

For n-type III–V semiconductors, the spin-flip matrix element has the form (see Equation (10))

$$M_{\text{spin-flip}} = D\left(\frac{e^2}{mc^2}\right)\langle f|(\boldsymbol{\sigma} \cdot \boldsymbol{\varepsilon}_0 \times \boldsymbol{\varepsilon}_1)e^{i\mathbf{k}\cdot\mathbf{r}}|i\rangle, \tag{75}$$

where $\mathbf{k} = (\mathbf{k}_0 - \mathbf{k}_1)$, $\mathbf{k}_0$ and $\mathbf{k}_1$ are wave vectors of the incident and scattered light waves; $\boldsymbol{\varepsilon}_0$ and $\boldsymbol{\varepsilon}_1$ are the corresponding polarization vectors; and D is a dimensionless constant whose value is determined by a sum of interband matrix elements. The states "i" and "f" are both in the conduction band. We will regard D in Equation (75) as a known parameter—evaluated either from theory or experiment. D has a resonance (see Equation (12)) as the photon energy approaches the direct band-gap, which can be used to enhance spin-flip cross-sections. In CdS there is a particularly fortunate coincidence of the 4880 Å line of the Ar^+ laser with a strong, interband transition. Huge spin-flip cross-sections—of order 10^{-18} cm^2/ster.—have been reported[52] under these conditions.

The theory of spin-flip scattering has not yet been worked out in detail for II–VI semiconductors. These crystals are more complicated—for the present purpose—than III–V semiconductors, because they are hexagonal. As a consequence, there is a uniaxial crystal field which lifts the degeneracy (required by symmetry in cubic crystals) at the $\mathbf{k} = 0$ point of the valence band. In CdS the splitting is small—about .05 eV. Hence, if the photon frequencies are well away from resonance, the energy denominators in the virtual interband transitions which determine the matrix element are nearly the same for all valence-to-conduction band transitions. Under these conditions, the effects of the uniaxial crystal field can be ignored, and the spin-flip matrix element has the same form as for a III–V semiconductor. Spin-flip polarization selection rules consistent with this interpretation are observed with green (5145 Å) Ar^+ laser light scattering in CdS.

On the other hand, if the incident light is resonant with a particular interband transition, this channel dominates the spin-flip matrix and gives it a different form from that indicated by Equation (75). Such a situation occurs with blue light (4880 Å line of the Ar^+ laser) illumination of CdS. As was first pointed out by Thomas and Hopfield[52], this frequency coincides with that of a bound exciton transition, of the exciton associated with the uppermost ($J = \frac{3}{2}$, $m_J = \pm\frac{3}{2}$) valence band. The spin-flip process is then dominated by the virtual transition passing through this level, and the spin-flip matrix element takes the fc ·m:

$$M_{\text{spin-flip}} = D\left(\frac{e^2}{mc^2}\right)\langle f\,|(\boldsymbol{\sigma}\cdot\hat{\mathbf{c}})(\hat{\mathbf{c}}\cdot\boldsymbol{\varepsilon}_0 \times \boldsymbol{\varepsilon}_1)e^{i\mathbf{k}\cdot\mathbf{r}}|\,i\rangle, \tag{76}$$

where $\hat{\mathbf{c}}$ is a unit vector in the c-axis direction. Spin-flip selection rules consistent with Equation (76) were observed by Thomas and Hopfield with 4880 Å radiation.

Thus, depending upon the laser frequency, the spin-flip cross-section in CdS can take a variety of forms. For our purpose, it is sufficient to note that the matrix element has the general form:

$$M_{\text{spin-flip}} = D\left(\frac{e^2}{mc^2}\right)\langle f\,|(\boldsymbol{\sigma}\cdot\boldsymbol{\alpha})e^{i\mathbf{k}\cdot\mathbf{r}}|\,i\rangle, \tag{77}$$

where $\boldsymbol{\alpha}$ is a vector that does not involve the electronic variables. We can anticipate that spin-flip matrix elements will generally have this form. The essential point is that, in spin-flip scattering with momentum transfer $\mathbf{k} = (\mathbf{k}_0 - \mathbf{k}_1)$, the light couples to the electronic variable $\boldsymbol{\sigma}e^{i\mathbf{k}\cdot\mathbf{r}}$ which is the kth Fourier component of the spin density. Our subsequent considerations will be based on this fact and will not require a knowledge of the vector $\boldsymbol{\alpha}$.

So far, the discussion has been concerned with single-particle spin-flip scattering. In deriving Equation (75), for instance, one imagines a semiconductor crystal containing a single conduction band electron in otherwise intrinsic material. Equation (77) determines the spin-flip scattering cross-section for such a carrier. Practically speaking, however, a semiconductor always contains many carriers—often sufficiently many that carrier-carrier interactions play a central role in determining its properties. These interactions are particularly important at low temperatures, and can lead to a Mott (metal-insulator) transition in appropriately doped crystals. We believe that spin-flip scattering will prove to be an especially powerful tool for studying such transitions. It is important, therefore, to extend our analysis to the case of a sample containing a large density of interacting conduction electrons.

Many body effects in light scattering have been discussed by Hamilton and McWhorter[31]. Their prescription for treating the problem is a straightforward extension of Equation (77). In an interacting electron system they assume that light couples to the conduction electron spin density via a spin-flip interaction of the form

$$H_{sf} = D\left(\frac{e^2}{mc^2}\right)\sum_j [(\boldsymbol{\sigma}_j \cdot \boldsymbol{\alpha})e^{i\mathbf{k}\cdot\mathbf{r}_j}]. \tag{78}$$

Here the j-sum runs over all conduction band electrons. The matrix elements of H_{sf}, taken between many-body states for the conduction electrons, then determine the scattering rate. This prescription is correct for a non-interacting electron system. When interactions are present, it includes their effect in the initial and final states, but ignores their influence on the virtual, intermediate states which appear in the sum which determines D. Generally speaking, however, it is a good approximation to ignore many-body effects in virtual states, since electrons spend only a short time in them. It should be recognized, however, that the argument which permits one to neglect many-body effects in intermediate states fails at resonance. In CdS, 4880 Å light is near resonance, though probably not sufficiently close to invalidate the Hamilton-McWhorter procedure. Nevertheless, it is clear that a question of principle remains. The behaviour of carrier cross-sections, as the photon frequency approaches resonance, deserves further study.

Henceforth, we will use Equation (78) to describe spin-flip scattering. This interaction involves the operator

$$\boldsymbol{\sigma}_{-\mathbf{k}} = \sum_j (\boldsymbol{\sigma}_j e^{i\mathbf{k}\cdot\mathbf{r}_j}), \tag{79}$$

which is the Fourier transform of the electron spin density. A straightforward calculation, paralleling van Hove's[26] well-known analysis, yields the following expression for the Stokes portion of the spin-flip spectrum:

$$\left.\frac{d^2\sigma}{d\omega\, d\Omega}\right|_{\text{Stokes}} = 2\pi|D|^2\left(\frac{e^2}{mc^2}\right)^2\left(\frac{\omega_1}{\omega_0}\right)S^+(\mathbf{k}, \omega), \tag{80}$$

where

$$S^+(\mathbf{k}, \omega) \equiv \int_{-\infty}^{\infty} e^{i\omega t} \langle \boldsymbol{\sigma}_{\mathbf{k}}(t) \boldsymbol{\sigma}_{\mathbf{k}}^+(0) \rangle \frac{dt}{2\pi} \tag{81}$$

is the Fourier transform of the spin-spin correlation function. The angular brackets in this expression indicate a thermal average over the exact, many-body states of the interacting, conduction electron system. At equilibrium, the spin-spin correlation function can be related to the transverse spin susceptibility by the fluctuation-dissipation theorem[53]. One finds:

$$S^+(\mathbf{k}, \omega)\Bigg|_{\text{Equil.}} = \frac{-4 \operatorname{Im} \chi^+[(\mathbf{k}, \omega)]}{[1 - e^{-\beta\hbar\omega}]}, \tag{82}$$

where

$$\chi^+(\mathbf{k}, \omega) = -\frac{i}{4} \int_0^{\infty} e^{i\omega t} \langle [\boldsymbol{\sigma}_{\mathbf{k}}(t), \boldsymbol{\sigma}_{\mathbf{k}}^+(0)] \rangle \, dt \tag{83}$$

is the transverse spin susceptibility.

The manipulations leading from Equation (77) to Equation (80), (81), (82) and (83) are quite standard in scattering theory, so we have not discussed them in detail. The final result is important, however, since it demonstrates that spin-flip light scattering experiments measure the k-dependent spin susceptibility. This conclusion is already implicit in several analyses of the spin-flip problem[34], but has not yet been fully exploited to study electron dynamics in semiconductors.

C. THE SPIN SUSCEPTIBILITY

The central problem in calculating the spin-flip spectrum, is that of determining the transverse spin susceptibility, $\chi^+(\mathbf{k}, \omega)$. We are interested in this function for k-values typical of light scattering experiments. In the CdS work referred to above[52,54] the wave vector transfer (k) was generally of order 5×10^5 cm^{-1}. The corresponding length, $l = (2\pi/k) \simeq 10^{-5}$ cm, is large compared to the average interelectron spacing, the effective Bohr radius of electrons in CdS, and the mean free path. In all these experiments, the electron system was perturbed by a relatively long-wavelength disturbance compared to the characteristic dimensions of the system. Under such conditions, the spin magnetization (even in the presence of strong many-body effects) can be described by a Bloch equation of the form:

$$\frac{\partial M^+}{\partial t} + i\omega_s M^+ + \frac{(M^+ - \chi_0 H^+)}{T_2} - D_s \nabla^2 M^+ = i\mu g M_0 H^+. \tag{84}$$

Here T_2 is the transverse spin-relaxation time, D_s the spin diffusion constant, and H^+ the perturbing field that excites the spin system. If H^+ is a plane wave,

$$H^+(\mathbf{r}, t) = H_0^+ e^{i(\mathbf{k}\cdot\mathbf{r} - \omega t)}, \tag{85}$$

Equation (84) has the solution:

$$M^+ = \frac{\left(-\omega_s + \dfrac{i}{T_2}\right)\chi_0 H^+}{\left[(\omega - \omega_s) + i\left(\dfrac{1}{T_2}\right) + D_s k^2\right]}, \tag{86}$$

implying that

$$\chi^+(\mathbf{k}, \omega) \equiv \frac{M^+(\mathbf{k}, \omega)}{H^+(\mathbf{k}, \omega)} = \frac{\left(-\omega_s + \dfrac{i}{T_2}\right)\chi_0}{\left[(\omega - \omega_s) + i\left(\dfrac{1}{T_2}\right) + D_s k^2\right]} \tag{87}$$

The structure factor then becomes

$$S^+(\mathbf{k}, \omega) = \frac{4}{\pi(1 - e^{-\beta\omega})}\left[\frac{\gamma\chi_0}{(\omega - \omega_s)^2 + \gamma^2}\right], \tag{88}$$

where

$$\gamma \equiv \left[\frac{1}{T_2} + D_s k^2\right]. \tag{89}$$

The main features of this spectrum have been confirmed by experiments of Scott *et al.*[54]. Their line shapes are Lorentzian, centred at the (known) spin resonance frequency for electrons in Cds. The k^2 variation of the linewidth was tested by measuring the dependence of linewidth on scattering angle (θ). For elastic (or near elastic) scattering

$$k = 2k_0 \sin\left(\frac{\theta}{2}\right). \tag{90}$$

Thus, the linewidth γ is expected to have angular variation

$$\gamma = \left[\frac{1}{T_2} + 4D_s k_0^2 \sin^2\left(\frac{\theta}{2}\right)\right]. \tag{91}$$

The data fit Equation (91) to within experimental error. The excellent agreement between theory and experiment suggests that our general picture of the spin-flip line shape is correct, and that spin diffusion indeed is responsible for the k^2 variation of the linewidth. We believe that this feature of spin-flip scattering is particularly important since it enables one, even in an interacting electron system, to make a direct measurement of the spin diffusion constant. Thus, we next consider how the measured values of the spin diffusion constant (D_s) compare with those expected on theoretical grounds.

D. SPIN DIFFUSION—COMPARISON WITH EXPERIMENT

Electron motion in semiconductors is most often described as a single-particle process, in which electron-electron interactions are ignored. This picture is valid at elevated temperatures, or in heavily doped (degenerate) samples even near $T = 0$. When the single-particle picture is applicable, the spin diffusion (D_s) and particle diffusion (D) constants are equal. The particle diffusion constant, moreover, is related to the mobility (μ) via the Einstein relation. Thus, whenever many-body effects are unimportant, there is a direct relationship between the spin-flip linewidth and a standard transport coefficient.

The preceding argument implies that mobility data can be used to estimate high temperature spin-flip linewidths in CdS. At low temperature, however, the single-particle picture fails in n-CdS samples of the doping level ($n \simeq 5 \times 10^{17}$/cc) studied by Scott *et al.*[54]. This sample has, in fact, a doping level only slightly higher than that at which a transition from metallic to insulating behaviour takes place at low temperatures (Mott transition)[55]. Nevertheless, in the absence of a theory for this transition, we have attempted to analyze the SDF data by assuming $D = D_s$, and relating D to the measured mobility (μ) via the Einstein relation. This procedure is suspect in two ways—D_s need not equal D in an interacting system; there can also be many-body corrections to the Einstein relation[56]. Despite these possible shortcomings, the procedure works surprisingly well in predicting the observed, temperature variation of the linewidth.

We now compare the measured spin-flip linewidths, with those calculated on the assumption $D_s = D$. To estimate D, we use the Einstein relation and mobility data[55] for n-CdS. The classical Einstein relation has the form $D = (kT/e)\mu$. This result has been generalized to degenerate and interacting systems by Kubo[56]. Kubo's analysis cannot immediately be applied to the present problem because his definition of the diffusion constant (for which we will use the symbol D^*) differs from that we need, in the case of a many-body system interacting via Coulomb forces. Kubo's diffusion constant (D^*) is defined by the relation

$$k^2 D^*(\mathbf{k}) = \gamma_{\mathbf{k}}, \tag{92}$$

where $\gamma_{\mathbf{k}}$ is the decay rate of an electron density fluctuation of wave vector $\mathbf{k}$. In a charged system, such fluctuations generate an electrostatic field. Thus, as a consequence, the diffusion constant defined by Kubo describes the decay of a fluctuation *under the combined action of two, separate forces—the particle density gradient, and the induced electrostatic field.* The latter depends upon the wave vector of the perturbation. It is not surprising, therefore, that Kubo finds D^* to be wave vector dependent. In the spin-flip problem, on the other hand, we are concerned with spin-density fluctuations, which are neutral. Thus, we are only interested in that portion of the diffusion current which is driven by the particle density gradient. To separate this component from the total diffusion current, we use Kubo's relation:

$$D^* = \frac{n\mu}{\chi[\mathbf{k}, 0]} \tag{93}$$

where $\chi[\mathbf{k}, 0]$ is the static density susceptibility:

$$\chi[\mathbf{k}, 0] \equiv -\frac{\partial\langle n_{\mathbf{k}}\rangle}{\partial\varphi_{\mathbf{k}}^{\text{ext}}} \tag{94}$$

In this definition, $\langle n_{\mathbf{k}}\rangle$ is an electron density fluctuation of wave vector $\mathbf{k}$; $\varphi_{\mathbf{k}}^{\text{ext}}$ an external potential at the same wave vector. For a Coulomb system, $\chi[\mathbf{k}, 0]$ can be written in the form:

$$\begin{aligned} \chi[\mathbf{k}, 0] &= -\frac{\partial\langle n_{\mathbf{k}}\rangle}{\partial\varphi_{\mathbf{k}}^{\text{total}}}\frac{\partial\varphi_{\mathbf{k}}^{\text{total}}}{\partial\varphi_{\mathbf{k}}^{\text{ext}}} = -\frac{1}{(k)}\frac{\partial\langle n_{\mathbf{k}}\rangle}{\partial\varphi_{\mathbf{k}}^{\text{total}}} \\ &= \frac{-1}{\left[1 + \dfrac{1}{(kl)^2}\right]}\frac{\partial\langle n_{\mathbf{k}}\rangle}{\partial\varphi_{\mathbf{k}}^{\text{total}}}, \end{aligned} \tag{95}$$

where l is the screening length. Thus

$$D^* = -n\left[\frac{\partial\langle n_{\mathbf{k}}\rangle}{\partial\varphi_{\mathbf{k}}^{\text{total}}}\right]^{-1}\mu\left[1 + \left(\frac{1}{kl}\right)^2\right]. \tag{96}$$

It can be shown that the first term in this expression represents current flowing in response to the electron density gradient $(\partial n/\partial\zeta)$; the second is caused by the induced electrostatic field. Thus, for spin diffusion, the diffusion constant must be calculated from the first part of Equation (96), i.e.

$$D = -n\left[\frac{\partial\langle n_{\mathbf{k}}\rangle}{\partial\varphi_{\mathbf{k}}^{\text{total}}}\right]^{-1}\mu = \frac{n\mu}{e}\left(\frac{\partial n}{\partial\zeta}\right)^{-1} \tag{97}$$

This formula can also be rewritten in terms of a compressibility (κ) via the relation

$$n^2\kappa = \left(\frac{\partial n}{\partial \zeta}\right). \tag{98}$$

It is clear from the procedure used to derive Equation (97), that the compressibility defined by Equation (98) is that for a process in which both the electrons and the background positive charge (donors in the n-CdS case) are compressed together.

To calculate κ (Equation (98)), we now assume a non-interacting, free-electron gas. κ can then be evaluated, as a function of temperature, from tabulated integrals of Fermi functions. The mobility data of Toyotomi and Morigki then enable us to determine D via Equation (97), and the linewidth via Equation (89) and the assumption $D = D_s$. Figure 6 shows a comparison of measured and calculated linewidths determined in this way for an $n = 5 \times 10^{17}$/cc CdS crystal. The agreement is good. Both μ and $\partial n/\partial \zeta$ for this sample are temperature dependent. At higher temperatures, μ is constant and

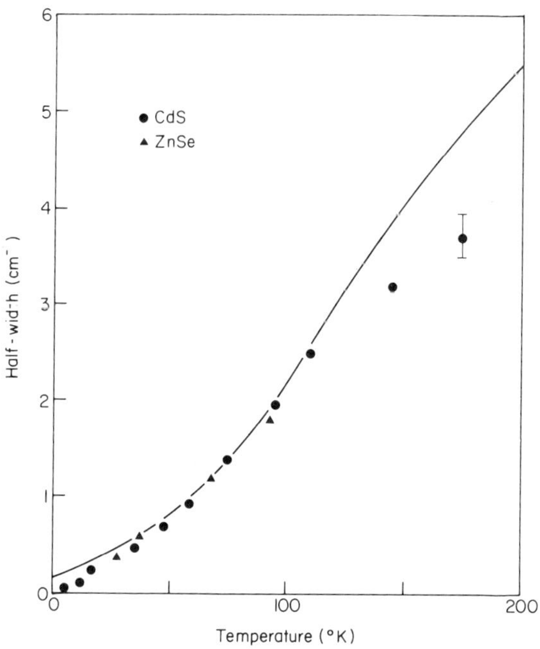

FIG. 6. Comparison of measured and calculated (full line) linewidths for spin-flip scattering in CdS and ZnSe.

the nearly linear temperature variation of the linewidth follows from the classical Einstein relation $D = (kT\mu/e)$. On the other hand, at low temperatures the Einstein coefficient saturates and the linewidth variation is caused by the temperature variation of μ, which is quite strong.

At the lowest temperatures ($T < 20°K$) the theory outlined predicts values of D (in the 5×10^{17} sample) larger than those observed. The discrepancy, at 4°K, is about a factor 7—well outside experimental error. The source of this discrepancy is not yet known, though there are two possibilities:

1. Impurity banding effects, which modify $\partial n/\partial \zeta$ (see Equation (98)). This amounts to a modification of the single-electron dynamics by the impurity potential.
2. Many-body effects. Electron-electron interactions can modify the Einstein relation (Equation (97)), or cause the assumption $D = D_s$ to fail. In either case, the analysis leading to Fig. 6 would be incorrect.

Further analysis will be required to resolve the questions raised above. Nevertheless, it is clear that spin-flip scattering gives fundamental information concerning electron dynamics in semiconductors. We expect that this technique will play an important role in future studies of the metal-insulator transition in such crystals.

References

1. P. Franken, A. E. Hill, C. W. Peters and G. Weinreich. *Phys. Rev. Lett.* **7**, 118 (1961).
2. There is a large literature on this subject. A summary, with a bibliography, is given by P. M. Platzman and P. A. Wolff. *In* "Waves and Interactions in Solid State Plasmas", Solid State Physics, Supplement 13, Academic Press, New York (1973).
3. C. K. N. Patel, R. E. Slusher and P. A. Fleury. *Phys. Rev. Lett.* **17**, 1011 (1966); P. A. Wolff and G. A. Pearson. *ibid.* **17**, 1015 (1966).
4. V. T. Nguyen and T. J. Bridges. *Phys. Rev. Lett.* **29**, 359 (1972); T. L. Brown and P. A. Wolff, *ibid.* **29**, 362 (1972).
5. N. Tzoar and J. I. Gersten. *Phys. Rev. Lett.* **26**, 1634 (1971); N. Tzoar and J. I. Gersten. *Phys. Rev.* **B4**, 3540 (1971).
6. To date, spin-flip scattering by mobile electrons has been observed in InSb (R. E. Slusher, C. K. N. Patel and P. A. Fleury. *Phys. Rev. Lett.* **18**, 530 (1967)); InAs (C. K. N. Patel and R. E. Slusher. *Phys. Rev.* **167**, 413 (1968)); PbTe (C. K. N. Patel and R. E. Slusher. *Phys. Rev.* **177**, 1200 (1969)); CdS and ZnSe (P. A. Fleury and J. F. Scott.); *Phys. Rev.* **B3**, 1979 (1971)); CdTe and CdSe (T. W. Walker, *et al.*, "Proceedings 11th Intl. Conf. Phys. of Semiconductors", Warsaw (1972)); GaAs (T. W. Walker. *et al.*, *Bull. Am. Phys. Soc.* **18**, 300 (1973)); SiC (R. L. Hollis, *et al.*, *ibid.* **18**, 300 (1973)); ZnTe (R. L. Hollis. *et al.*, to be published).
7. S. R. J. Brueck and F. A. Blum. *Phys. Rev. Lett.* **28**, 1458 (1972).

8. A good review of research on spin-flip lasers is given in an article by M. J. Colles and C. R. Pidgeon. *Reports on Prog. in Physics* **38** (1975).
9. P. A. Wolff. *Phys. Rev. Lett.* **16**, 225 (1966).
10. Y. Yafet. *Phys. Rev.* **152**, 858 (1966).
11. P. L. Kelley and G. B. Wright. *Bull. Am. Phys. Soc.* **11**, 812 (1966); G. B. Wright, D. L. Kelley and S. H. Groves. "Procs. of Intl. Conf. on Light Scattering Spectra of Solids", Springer-Verlag, New York (1969).
12. A. L. McWhorter. "Procs. of Intl. Conf. on the Physics of Quantum Electronics", McGraw-Hill, New York (1965).
13. P. M. Platzman. *Phys. Rev.* **139**, A379 (1965).
14. Y. Yafet. "Solid State Physics", Academic Press, New York, Vol. 14 (1963).
15. E. O. Kane. *J. Phys. Chem. Solids* **1**, 249 (1957).
16. R. E. Slusher, C. K. N. Patel and P. A. Fleury. *Phys. Rev. Lett.* **18**, 530 (1967).
17. C. K. N. Patel and R. E. Slusher. *Phys. Rev.* **167**, 413 (1968).
18. V. P. Makarov. *Soviet Physics—JETP* **28**, 336 (1969).
19. C. K. N. Patel and R. E. Slusher. *Phys. Rev.* **177**, 1200 (1969).
20. P. A. Fleury and J. F. Scott. *Phys. Rev.* **B3**, 1979 (1971).
21. T. W. Walker *et al.* "Procs. 11th Intl. Conf. Phys. of Semiconductors", Warsaw (1972).
22. T. W. Walker. *Bull. Am. Phys. Soc.* **18**, 300 (1973).
23. R. L. Hollis *et al.* **18**, 300 (1973).
24. R. L. Hollis, J. F. Ryan, D. J. Toms and J. F. Scott. *Phys. Rev. Lett.* **31**, 1004 (1973).
25. P. M. Platzman and N. Tzoar. *Phys. Rev.* **136**, A11 (1964).
26. L. van Hove. *Phys. Rev.* **95**, 249 (1954).
27. A. Einstein. *Ann. Phys. (Leipzig)* **33**, 1275 (1910).
28. A. Mooradian and G. B. Wright. *Phys. Rev. Lett.* **16**, 999 (1966).
29. B. B. Varga. *Phys. Rev.* **A137**, 1896 (1965).
30. A. Mooradian. "Procs. of Intl. Conf. on Light Scattering Spectra of Solids", (Springer-Verlag, New York, p. 285 (1968).
31. D. C. Hamilton and A. L. McShorter. *ibid.* p. 297.
32. S. R. J. Brueck and A. Mooradian. *Phys. Rev. Lett.* **28**, 161 (1972).
33. S. R. J. Brueck and A. Mooradian. *Opt. Comm.* **8**, 236 (1973).
34. R. W. Davies and F. A. Blum. *Phys. Rev.* **B3**, 3321 (1971); R. W. Davies. *Phys. Rev.* **B7**, 3731 (1972); S. Y. Yuen, P. A. Wolff and B. Lax. *Phys. Rev.* **B9**, 3394 (1974); *see also* Ref. 33.
35. T. L. Brown and P. A. Wolff. *Phys. Rev. Lett.* **29**, 362 (1972).
36. N. Bloembergen and Y. R. Shen. *Phys. Rev.* **133**, A37 (1964).
37. R. B. Dennis, C. R. Pidgeon, S. D. Smith, B. S. Wherrett and R. A. Wood. *Proc. Roy. Soc.* **A331**, 203 (1972).
38. C. K. N. Patel and E. D. Shaw. *Phys. Rev.* **B3**, 1279 (1971).
39. C. K. N. Patel and E. D. Shaw. *Phys. Rev. Lett.* **24**, 451 (1970).
40. B. S. Wherrett and P. G. Harper. *Phys. Rev.* **183**, 692 (1969).
41. R. L. Aggarwal, B. Lax, C. E. Chase, C. R. Pidgeon, D. Linibert and F. Brown. *Appl. Phys. Lett.* **18**, 383 (1971).
42. R. A. Wood, A. McNeish, N. L. Brignall and C. R. Pidgeon. *Opt. Comm.* **8**, 248 (1973); S. D. Smith, C. R. Pidgeon, R. A. Wood, A. McNeish and N. Brignall. "Proc. Intl. Conf. on Tunable Laser Spectroscopy", Plenum Press, New York (1973).
43. A. Mooradian, S. R. J. Brueck and F. A. Blum. *Appl. Phys. Lett.* **17**, 481 (1970).
44. A. Mooradian. Private communication.

45. C. K. N. Patel. *Phys. Rev. Lett.* **28**, 649 (1972).
46. S. R. J. Brueck and A. Mooradian. *Appl. Phys. Lett.* **18**, 229 (1971).
47. S. R. J. Brueck and A. Mooradian. *IEEE J. Quantum Electron.* (To be published)
48. F. Zernicke. *Phys. Rev. Lett.* **22**, 931 (1969); T. Yazima and K. Inoue. *IEEE J. Quantum Electron.* **5**, 140, (1969); V. T. Nguyen and C. K. N. Patel. *Phys. Rev. Lett.* **22**, 463 (1969); C. K. N. Patel and V. T. Nguyen, *Appl. Phys. Lett.* **15**, 189 (1969); D. W. Faries, K. A. Gehring, P. L. Richards and Y. R. Shen. *Phys. Rev.* **180**, 363 (1969); D. W. Faries, P. L. Richards, Y. R. Shen and K. H. Yang. *Phys. Rev.* **A3**, 2148 (1971); J. Morris and Y. R. Shen. *Opt. Comm.* **3**, 81 (1971); K. H. Yang, P. L. Richards and Y. R. Shen. *Appl. Phys. Lett.* **19**, 320 (1971); T. J. Bridges and A. R. Strand. *Appl. Phys. Lett.* **20**, 382 (1972); G. D. Boyd, H. M. Kasper, J. H. McFee and F. G. Stonz. *Appl. Phys. Lett.* **21**, 553 (1972); R. L. Aggarwal, B. Lax and G. Favrot. *Appl. Phys. Lett.* **22**, 329 (1973); B. Lax, R. L. Aggarwal, and G. Favrot. *Appl. Phys. Lett.* **23**, 669 (1973); K. H. Yang, J. R. Morris, P. L. Richards and Y. R. Shen. *Appl. Phys. Lett.* **23**, 669 (1973); R. L. Aggarwal, *et al. J. Appl. Phys.* **45**, 3972 (1974).
49. V. T. Nguyen and T. J. Bridges. *Phys. Rev. Lett.* **29**, 359 (1972).
50. V. T. Nguyen and T. J. Bridges. "Procs. of Intl. Conf. on Laser Spectroscopy", Plenum Press, New York (1974); T. J. Bridges and V. T. Nguyen. *Appl. Phys. Lett.* **23**, 107 (1973).
51. V. T. Nguyen and C. K. N. Patel. *Phys. Rev. Lett.* **22**, 463 (1963).
52. D. G. Thomas and J. J. Hopfield. *Phys. Rev.* **175**, 1021 (1968).
53. D. N. Zubarev. *Soviet Physics—Uspekhi* **3**, 320 (1960).
54. P. A. Fleury and J. F. Scott. *Phys. Rev.* **B3**, 1979 (1971). J. F. Scott, T. C. Damen and P. A. Fleury. *Phys. Rev.* **B6**, 3856 (1972).
55. S. Toyotomi and K. Morigaki. *J. Phys. Soc. Japan* **25**, 807 (1968); *see also* N. F. Mott and E. A. Davis. "Electronic Processes in Non-crystalline Materials", Clarendon Press, Oxford (1971).
56. R. Kubo. *Reports on Progress in Physics*, Vol. XXIX, p. 255 (1966).

5. Tunable Semiconductor Lasers†

A. MOORADIAN

Lincoln Laboratory, Massachusetts Institute of Technology, Lexington, Massachusetts 02173, U.S.A.

I. INJECTION LASERS

A. INTRODUCTION

Recent development of the use of tunable lasers for high resolution spectroscopy, photochemistry, pollution monitoring, etc. has spurred both fundamental and applied research in tunable laser device technology. This section describes the characteristics of spontaneous and stimulated recombination processes in semiconductors that have bandgaps in the infrared. Of particular importance are the III–V and lead salt binary compounds as well as the ternary lead salt alloys. The infrared region in the wavelength range between 2 and 25 μm has particular significance because the vibrational-rotational absorption bands for most molecules occur there.

B. SPONTANEOUS EMISSION

In order to understand better the operation of infrared semiconductor lasers, some of the spontaneous emission processes and the mechanisms that govern

† This work was sponsored by the Department of the Air Force, NSF/RANN, and the U.S. Energy Research and Development Administration.

their lineshapes will be described here. Study of lineshapes has been most easily done by use of optical pumping where one can deal with the bulk properties of a crystal. Excitation sources usually are a high pressure arc lamp or a laser where the exciting wavelength is shorter than the optical bandgap. Figure 1 shows the energy band structure that is typical of III–V compounds. A broadband source excites electrons from the valence band high into the conduction band within the penetration layer of the exciting light of about 1 μm on the surface of the crystal. Excited carriers then rapidly lose energy within the band by emission of phonons. The emission of an optical phonon usually occurs in about 10^{-13} sec as the excited carriers cascade down to within one optical phonon energy of the bottom of the band. Relaxation then proceeds via acoustic phonon emission with the relaxation time being dependent upon the energy of the carrier as well as the lattice temperature. At low temperatures, the acoustic phonon relaxation times can be as long as 10^{-7} sec which is longer than the interband recombination time. After the photoexcited carriers have relaxed via optical phonon emission within the band, they then diffuse from the surface into the bulk of the crystal. Most of the energy in such an optical pumping process is deposited in a 1 μm region at the surface and can be efficiently removed by appropriate heat sinking.

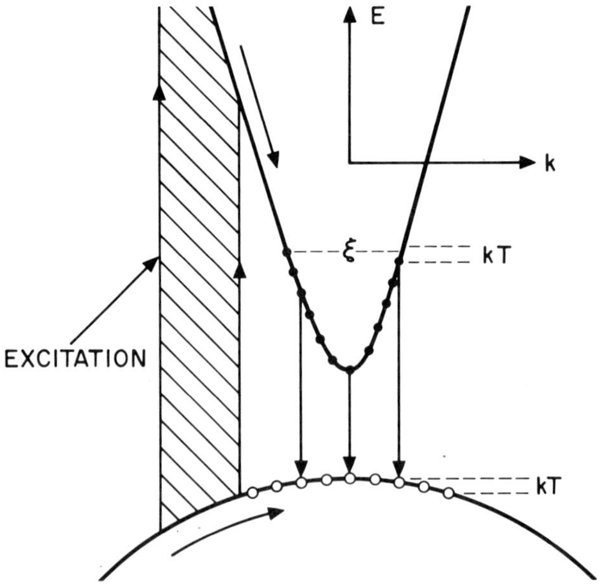

FIG. 1. Excitation and recombination shown using a simplified plot of energy versus momentum for a typical III-V semiconductor (only the heavy-hole band is shown).

Nonradiative recombination processes can be important loss mechanisms in semiconductor lasers. Excited carriers can recombine nonradiatively via impurity or defect levels within the bandgap or via Auger processes in which an electron in the conduction band can recombine directly with a hole in the valence band and impart the energy to a carrier in either band. Surfaces and interfaces also provide a large number of nonradiative recombination centres. In most high quality semiconductor lasers, the stimulated emission rate well above threshold is much greater than nonradiative recombination rates.

Radiative recombination processes in most infrared bandgap semiconductors proceed via recombination of electrons in the conduction band with holes in the valence band or recombination of excited carriers with an impurity level in the bandgap. Exciton and phonon-assisted transitions are not important for the infrared bandgap semiconductor lasers. While laser action from impurity levels has been observed in a few of the III–V compounds, the dominant stimulated emission usually occurs from free electron-hole recombinations. This is because the radiative transition rate is proportional to the product of the electron and hole densities and for $\Delta n \approx \Delta p \gg n_0, p_0$, increases quadratically with pump intensity. In the case of lead salts where the dielectric constants are high and the effective masses are low, bound impurity states in general do not exist because of dielectric screening and only band-to-band recombinations are observed.

Optical transitions in which the sum of electron and photon wave vectors is conserved are called direct transitions as against indirect transitions which violate the wave vector conservation. In dealing with optical radiation, the photon momentum may be neglected since it is much less than the electron or hole momentum in the band. Only direct transitions are allowed when the effects of lattice imperfections and lattice vibrations are negligible.

The emission spectrum for direct recombination of holes and electrons can be expressed by[1]

$$I(h\nu) \propto \nu^2 \int_S |M(\mathbf{k})|^2 f_e(\mathbf{k}) f_h(\mathbf{k}) \frac{dS}{|\nabla_k(\varepsilon_e + \varepsilon_h)|}, \tag{1}$$

where $I(h\nu)$ is the intensity per unit range of photon energy, M is the transition matrix element, $f_e(\mathbf{k})$ and $f_h(\mathbf{k})$ are, respectively, the electron and hole distribution functions, and S is the surface corresponding to the energy $\varepsilon_e(\mathbf{k}) + \varepsilon_h(\mathbf{k}) = (h\nu - E_g)$ in $\mathbf{k}$-space. In the case where the extrema of the conduction and valence bands coincide at $\mathbf{k} = 0$ and the surfaces of constant energy are spherical, Equation (1) reduces to

$$I(h\nu) \propto \nu^2 |M|^2 f_e f_h \rho \tag{2}$$

where ρ is the density of states per unit range of energy $h\nu$, and M, f_e, f_h and ρ are functions of $h\nu$. For energy bands with constant effective masses, $\rho \propto (h\nu - E_g)^{1/2}$. In most cases, $|M|^2$ can be taken to be constant, and we get:

$$I(h\nu) \propto \nu^2(h\nu - E_g)^{1/2} f_e f_h. \tag{3}$$

The distribution function, f_e or f_h, is the distribution of carriers in the bands that occurs under excitation. For classical statistics, we get

$$I(h\nu) \propto \nu^2(h\nu - E_g)^{1/2} \exp\left[-\frac{h\nu - E_g}{kT}\right], \tag{4}$$

where T is the electron and hole gas temperature which is usually the same as the lattice temperature. In n-type III–V compounds, the electron distribution is highly degenerate at low temperatures, but the heavy-hole distribution which dominates the emission spectra may be considered nondegenerate in most experiments. Under such conditions, we have:

$$I(h\nu) \propto \nu^2(h\nu - E_g)^{1/2} \exp\left[-\frac{m_e}{m_e + m_h}\frac{h\nu - E_g}{kT}\right] \times \left\{\exp\left(\left[\frac{m_h}{m_e + m_h}(h\nu - E_g) - \zeta\right]\Big/kT\right) + 1\right\}^{-1}, \tag{5}$$

where ζ is the quasi Fermi level of the electrons. Since $m_h \gg m_e$, the degenerate electron distribution gives at low temperatures a cut-off for the emission at $(h\nu - E_g) \sim \zeta$. The hole distribution does not influence the emission spectrum strongly unless the electron concentration is so high that $\zeta \gtrsim [(m_h + m_e)/m_e]kT$. This is not the case for the lead salt compounds and their alloys because the hole mass is comparable to the electron mass.

In the case of intrinsic emission, emitted photons can be strongly absorbed on their way out of the crystal, and correction has to be made in order to obtain the true emission spectrum. With a sample thickness large compared to the diffusion length, the ratio of the observed and true intensities is given by:

$$\frac{[I(h\nu)]_{\text{obs}}}{[I(h\nu)]_{\text{true}}} = \int_0^\infty [\Delta n(x)]^p \exp[-\alpha(h\nu)x]dx, \tag{6}$$

where $\Delta n(x)$ is the concentration of excess carriers as a function of distance x from the surface and $\alpha(h\nu)$ is the absorption coefficient for emitted photons of energy $h\nu$. $p = 1$ if Δn is small compared to the majority carrier concentration, and $p = 2$ otherwise. It can be shown that

$$\Delta n(x) \propto \left[\frac{\alpha' L^2 + S\tau}{L + S\tau} e^{-x/L} - e^{-\alpha' x}\right] \tag{7}$$

where α' is the absorption coefficient for the exciting light, L is the diffusion length of the excited carriers, S is the surface recombination velocity, and τ is the lifetime of excess carriers. The expression applies under the condition $\alpha' L \gg 1$, with $\alpha' \gtrsim 10^4\ \mathrm{cm}^{-1}$ and $L \sim 10^{-3}$ cm. Substituting (7) in (6) and using the condition $\alpha' L \gg 1$, we get

$$\frac{[I(h\nu)]_{\mathrm{obs}}}{[I(h\nu)]_{\mathrm{true}}} \propto [\alpha(h\nu)L + 1]^{-1}. \tag{8}$$

Figure 2 shows the relative positions of the absorption and emission of an InAs crystal at 77 and 4.2 K while Fig. 3 shows the 77 K spectrum corrected for self-absorption effects[2]. Both of these figures demonstrate the above arguments relating to lineshape. The experimental curve does not agree with the calculated shape near the bandgap in Fig. 3. This is generally true for heavily doped n-type crystals and is a result of impurity banding near the

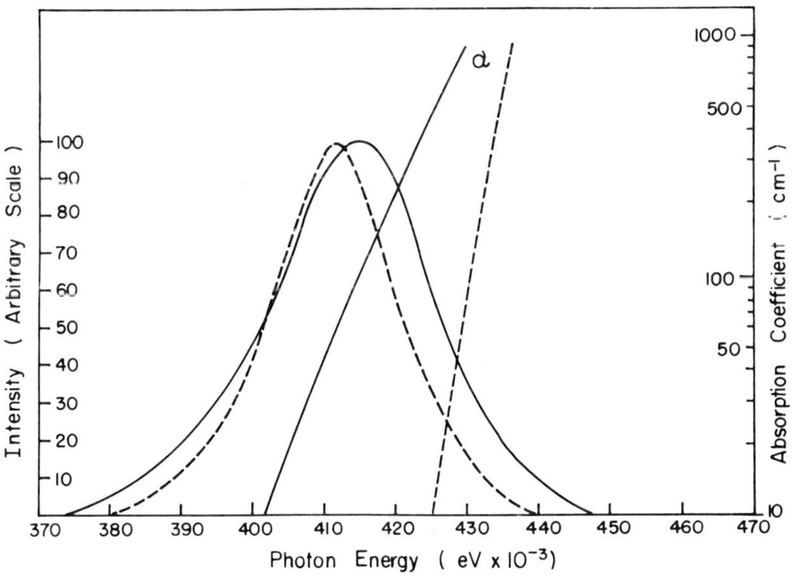

FIG. 2. Photoluminescence and absorption data for InAs ($n = 9 \times 10^{16}\ \mathrm{cm}^{-3}$). Dashed curves are 4.2 K and solid curves are 77 K.

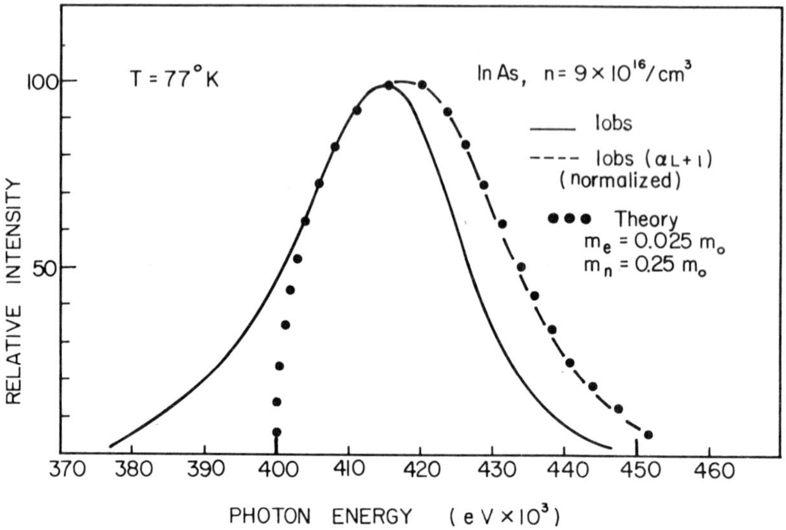

FIG. 3. Experimental and theoretical spectra for InAs corrected for self-absorption effects.

bottom of the conduction band that produces a tailing of the density of states into the bandgap as well as a violation of the $\Delta k = 0$ selection rule. For intense excitation, a dynamic shift in the Fermi levels of electrons and holes can produce substantial bandwidths for laser emission.

C. TUNING MECHANISMS

(a) *Temperature Tuning*

This section describes the gross bandgap and fine mode-tuning mechanisms for semiconductor lasers[3]. The wavelength ranges covered by compound semiconductors in which lasers have been fabricated to date are shown in Fig. 4. By choosing the appropriate composition of one of the compounds, lasers can be produced at any desired wavelength in the broad range from 0.6 to 32 μm. However, the level of materials and device development and hence of laser performance varies considerably among the various compounds. An individual laser can be tuned by a magnetic field, hydrostatic pressure, or by temperature. Figure 5 shows the variation of bandgap for some binary semiconductor compounds with temperature. A specific operating temperature is usually chosen using a fixed temperature bath of liquid helium or liquid nitrogen, although variable temperature refrigerators presently have sufficient temperature stability for most applications.

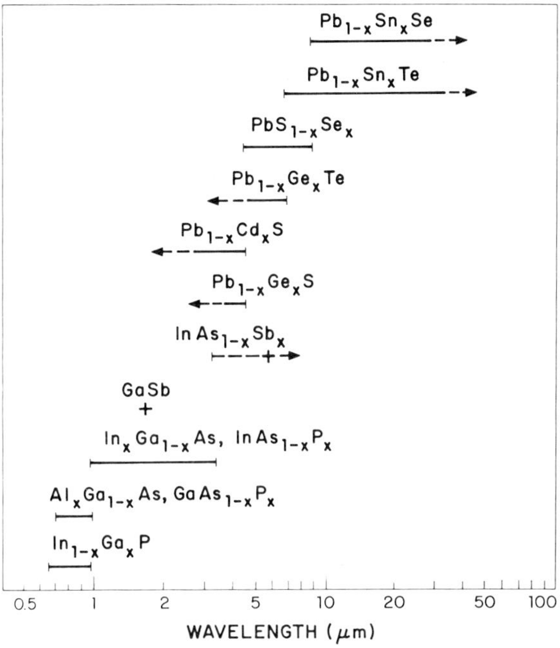

FIG. 4. Wavelength ranges covered by various semiconductor diode laser materials. The solid lines represent ranges covered at present. The dotted portions and the arrows indicate regions where lasers may be developed in the future.

Most lasers have a spectral output consisting of several modes which increase in number and shift in frequency with increased drive current. For example, the mode tuning of a diode laser with injection current is shown in Fig. 6. Heating of the laser due to losses causes a change in refractive index of the mode volume which changes the effective cavity length. Continuous tuning of a single mode from this effect can be several cm^{-1} before a mode jump occurs. The jumping of modes results from a shift of peak frequency of the gain spectrum as the energy gap of the material changes.

The output frequency of a laser in the case of a homogeneously broadened Lorentzian line is given by[4]

$$\nu = \frac{\nu_c \Gamma_s + \nu_s \Gamma_c}{\Gamma_c + \Gamma_s} \tag{9}$$

where ν_s is the frequency at the gain peak, Γ_s the linewidth (full width at half height) of the gain, ν_c and Γ_c the corresponding cavity parameters. The cavity

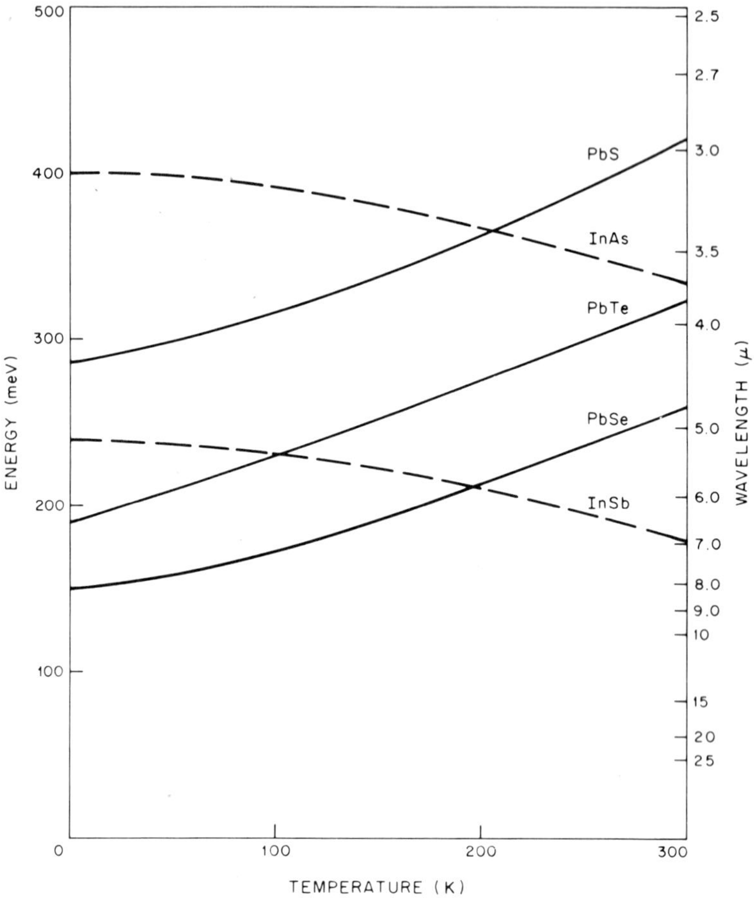

FIG. 5. Variation of bandgap for some binary semiconductors as a function of temperature.

linewidth and frequency are given by

$$\Gamma_c = \left[\alpha - \frac{\ln R_1 R_2}{2l}\right] \Big/ \left(n + \nu \frac{\partial n}{\partial \nu}\right)$$
$$\nu_c = \frac{cM}{2l}\left(n + \nu \frac{\partial n}{\partial \nu}\right) \tag{10}$$

where n is the refractive index, α is an effective loss parameter, R_1 and R_2 are the end mirror reflectivities, l is the cavity length, M is a large integer, $n \approx n_0 + (dn/dT)\Delta T$ is the index of refraction, and T is the operating temperature.

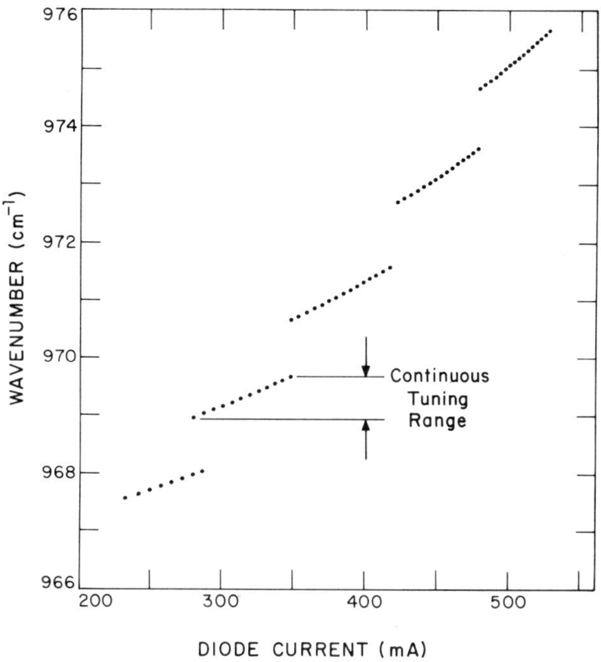

FIG. 6. Fine tuning of modes from a $Pb_xSn_{1-x}Te$ diode laser as a function of injection current. Tuning is continuous within a mode. From Ref. 32.

In the case of semiconductor lasers, the spontaneous linewidth is not Lorentzian, but Equation (9) is still a valid approximation.

Cavity tuning rates for typical semiconductor lasers are usually about one-third the bandgap tuning rates with temperature. Increasing the pumping level can not only shift the gain peak but also increase its magnitude and change its lineshape, which would complicate any detailed mode-tuning analysis.

(b) *Magnetic Field Tuning*

When a magnetic field is applied to a semiconductor, the density-of-states in both the conduction and valence bands are modified in the following manner

$$n(E) = \frac{\hbar\omega_c}{8\pi^2}\left(\frac{2m_c}{\hbar^2}\right)^{3/2} \sum_{n=0}^{n_{\max}} \sum_{m} [E + m\beta g H - (n + \tfrac{1}{2})\hbar\omega_c]^{-1/2}, \qquad (11)$$

where the summation over $m = \pm\frac{1}{2}$ is for the two spin states of each Landau level denoted by n. Here, $\omega_c = eH/m^*c$ is the cyclotron frequency, g is the effective g-factor, and β is the Bohr magneton. This density of states has a singularity at $k_z = 0$ in contrast to zero magnetic field. In fact, however, broadening mechanisms that would normally remove this singularity have not been treated in any great detail but are certainly operative. This simple description is quite useful in understanding the operating characteristics of tunable semiconductor lasers.

The gross tuning with magnetic field in lead-salt diode lasers[5] has been investigated in the past. The tuning of a $Pb_{0.79}Sn_{0.21}Te$ diode laser with magnetic field is shown[6] in Fig. 7 and the magnetic energy levels in the conduction and valence bands, together with the electronic transitions that occur between them, are schematically represented in Fig. 8. Because of the large mass anisotropy in the lead salts, the magnetic energy levels depend on the orientation of the applied magnetic field with the crystallographic axes. Both the conduction and valence bands have their energies quantized into Landau levels, each one of which is split into two spin energy states. Usually, electrons injected into the conduction band thermalize to the lowest available energy state, while holes injected into the valence band thermalize to the highest

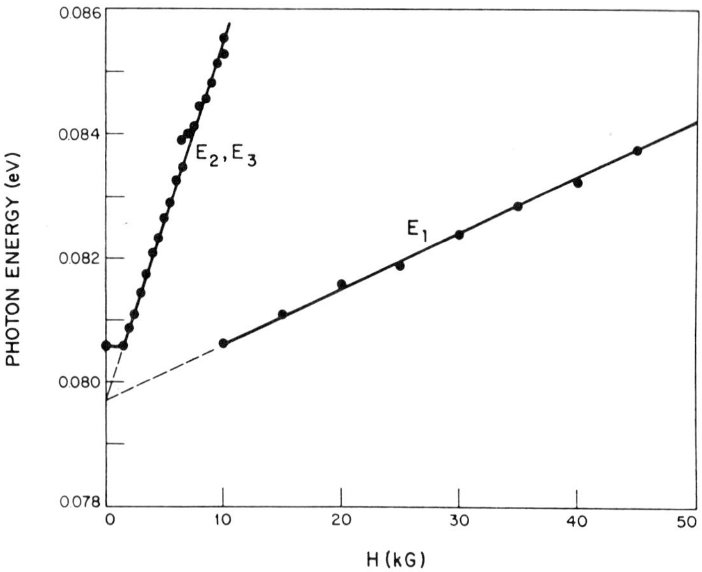

FIG. 7. Magnetic tuning of $Pb_{0.79}Sn_{0.21}Te$ diode laser at 7 K in a <100> magnetic field. From Ref. 6.

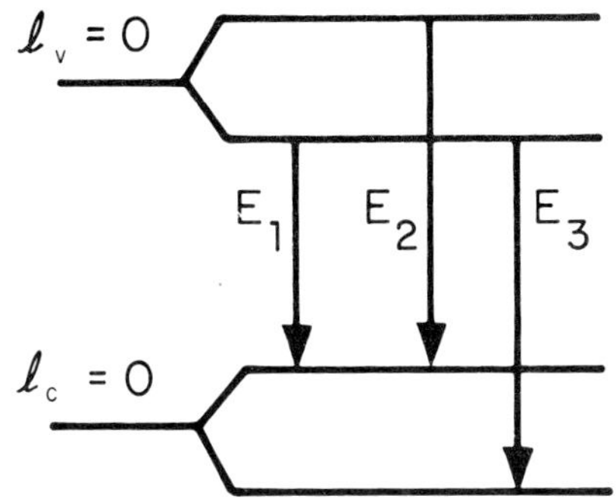

FIG. 8. Electronic energy level scheme for lead-salts in a magnetic field, for the lowest angular momentum states, $l_v = 0$ and $l_c = 0$.

available energy states before recombining. The energies of the respective transitions in Fig. 8 are given by

$$\begin{aligned} E_1 &= E_g + \left[\frac{m_0}{m_r} - \frac{1}{2}(g_c + g_v)\right]\beta H \\ E_2 &= E_g + \left[\frac{m_0}{m_r} + \frac{1}{2}(g_c - g_v)\right]\beta H \\ E_3 &= E_g + \left[\frac{m_0}{m_r} - \frac{1}{2}(g_c - g_v)\right]\beta H \end{aligned} \tag{12}$$

where β is the Bohr magneton, m_r is the reduced effective mass in the appropriate crystallographic direction and is defined as $1/m_r = 1/m_c + 1/m_v$, with m_v and m_c being the valence and conduction band effective masses, and g_c and g_v are the absolute values of the appropriate electron and hole g-factors. The average tuning in Fig. 7 is nearly linear over most of the magnetic field range with the small variations due to the magnetic field tuning of individual modes. In the lead salts, the energies E_2 and E_3 are very nearly equal because the conduction and valence bands are nearly mirror images of each other. At magnetic fields below 10 kG, the spin-conserving transitions E_2 and/or E_3 dominate over the spin-reversing transition E_1. As the field is increased, the highest energy level in the conduction band passes through the electron quasi-Fermi level and becomes empty, thus preventing recombination in transition E_2. Similarly, E_3 can become impossible as the lowest level in the valence band passes through the hole quasi-Fermi level. Consequently, at high magnetic fields emission is observed in transition E_1. The extrapolation of the data to zero magnetic field is lower in energy than the observed zero-field laser emission because the peak of the gain does not occur at the band edge.

The gross magnetic field tuning of Fig. 7 is qualitatively similar for most semiconductor lasers including the III–V compounds which have a negligible contribution to the tuning from the heavy-hole bands. Fine tuning of the axial modes with magnetic field occurs in a similar fashion to that of temperature tuning. The tuning rate within a mode and for different modes can vary with magnetic field, and for a $PbS_{0.82}Se_{0.18}$ diode laser has been measured to vary from 0.4 to 2.0 MHz/G.

(c) *Pressure Tuning*

The application of hydrostatic pressure to a semiconductor laser can provide a very broad tuning range for a single device. Figure 9 shows bandgap tuning characteristics for a number of binary lead-salt and III–V compound semiconductors at 77 K. A lead selenide diode laser at 77 K has been tuned[7] from 7.5 to 22 μm using hydrostatic pressure up to 14 kilobars. More recently, the stability of the hydrostatic pressure has been controlled sufficiently to perform Doppler-limited spectroscopy[8] using a pressure-tuned laser of GaAs and PbS_xSe_{1-x}.

The semiconductor laser is coarsely tuned to the absorption lines of interest by setting the hydrostatic pressure, and the fine tuning is done by varying the temperature of the junction with dc current in the case of cw operation, or by the frequency chirp which occurs in pulsed operation. Figure 10 shows an example[8] of the latter case where a GaAs diode laser is used to look at the real time, Doppler-limited absorption spectrum of the $6s^2S_{1/2} \rightarrow 6p^2P_{3/2}$ transition in cesium. Since helium freezes at these high pressures for temperatures less than 77 K, most of the diodes to date have been operated at 77 K in the pulsed mode with chirp rates low enough to sweep through a Doppler linewidth in a time longer than the response time of the detector. Uniaxial pressure tuning of diode lasers can avoid the limitations of low temperature operation with helium as the pressure transmitting fluid; however, in practice it is difficult to keep from crushing the lasers. Uniaxial pressure tuning rates are about one-third the hydrostatic rates. The recent development[9] of the infrared diodes which operate continuously at 77 K or higher will greatly facilitate the usefulness of hydrostatic pressure tuning. Because of the broadband wavelength coverage using hydrostatic pressure tuning, only the binary semiconductor compounds and one or two alloy semiconductors would be necessary to cover the wavelength range from 2 to 35 μm.

(d) *External Cavity Tuning*

Initial studies of the mode control properties of a semiconductor laser have been made with the use of a grating controlled-cavity GaAs laser[10–12] in

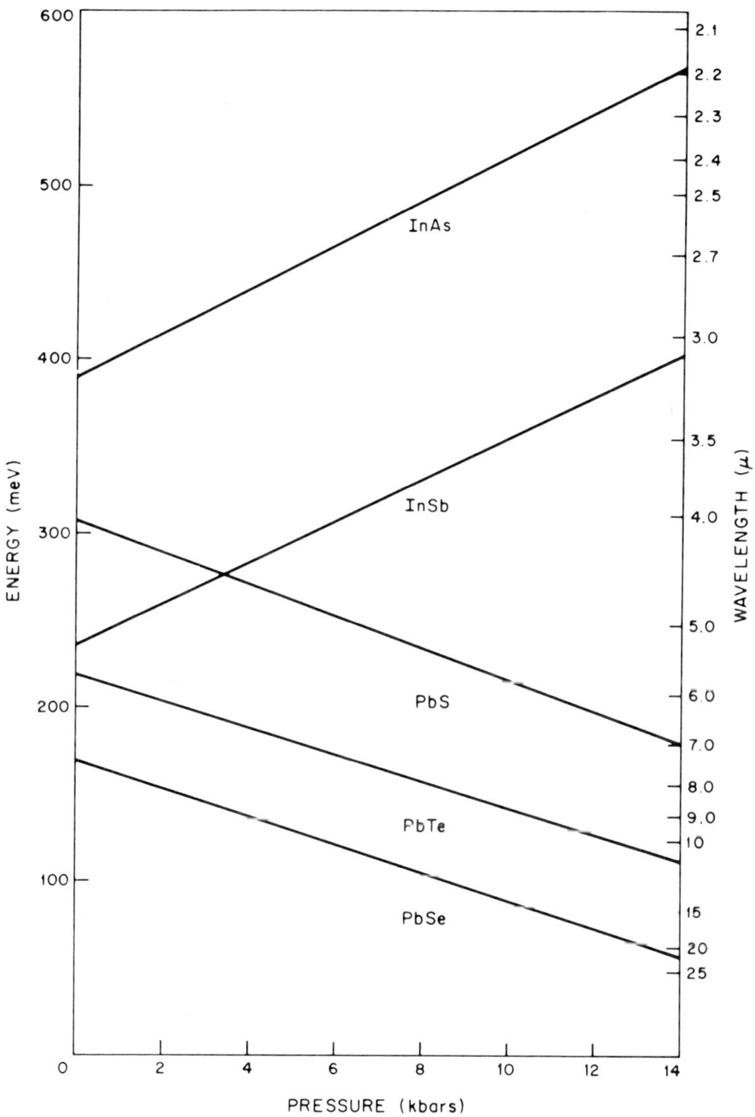

FIG. 9. Bandgap variation with hydrostatic pressure for some binary semiconductor compounds at 77 K. From Ref. 8.

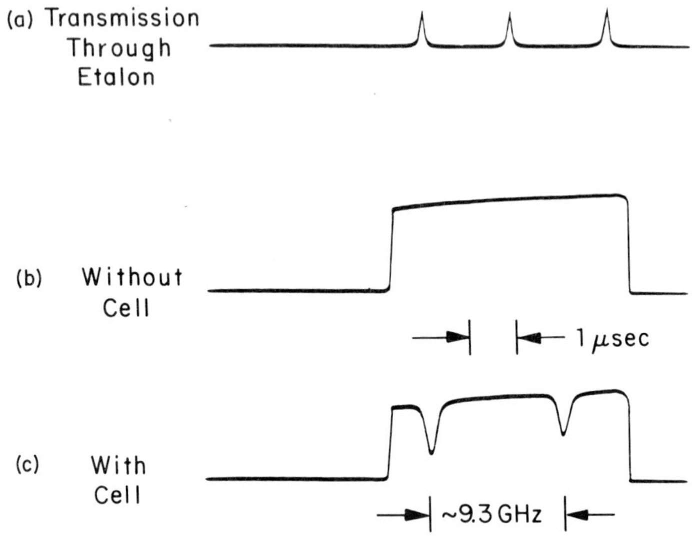

FIG. 10. Absorption spectrum of the Doppler limited $6s^2S_{1/2} \rightarrow 6p^2P_{3/2}$ transition in cesium using a 77 K chirped GaAs diode laser (λ = 8521.1 Å, hydrostatic pressure = 160 bars). (a) transmission of chirped axial mode of GaAs diode through the free spectral range of a Fabry–Perot interferometer; (b) mode intensity without absorption cell; (c) mode intensity with cesium cell. From Ref. 8.

order to improve the usefulness of the spectral output of these devices. A grating-controlled room temperature GaAs diode laser has operated[12] with a spectral width of less than 0.2 Å and a peak output power of nearly 3 W. It is interesting to note that the same diode operating with a 100% reflector in place of the grating produced 3.5 W in a bandwidth of over 40 Å. Figure 11 shows a photograph of this device. Only one end of the diode is antireflection coated. Figure 12 shows the output power as a function of wavelength for a number of GaAs and $GaAs_xP_{1-x}$ external cavity-grating controlled diode lasers at room temperature[13]. Tuning has been achieved from 8300 to 9300 Å with a spectral width of less than 0.4 Å. For temperatures below 77 K, output powers of up to 1 W cw in a single frequency should be possible from GaAs.

The spectral output of these grating-controlled semiconductor diode lasers occurs in a number of axial modes for pumping levels much above threshold. Additional intercavity mode-control elements would be necessary to limit the output to a single mode. One of the advantages of an external cavity-controlled semiconductor laser is the reduced frequency chirp for a given mode during pulsed operation. Compared to a conventional diode laser, the frequency chirp is reduced by the ratio of the optical length of the diode to the

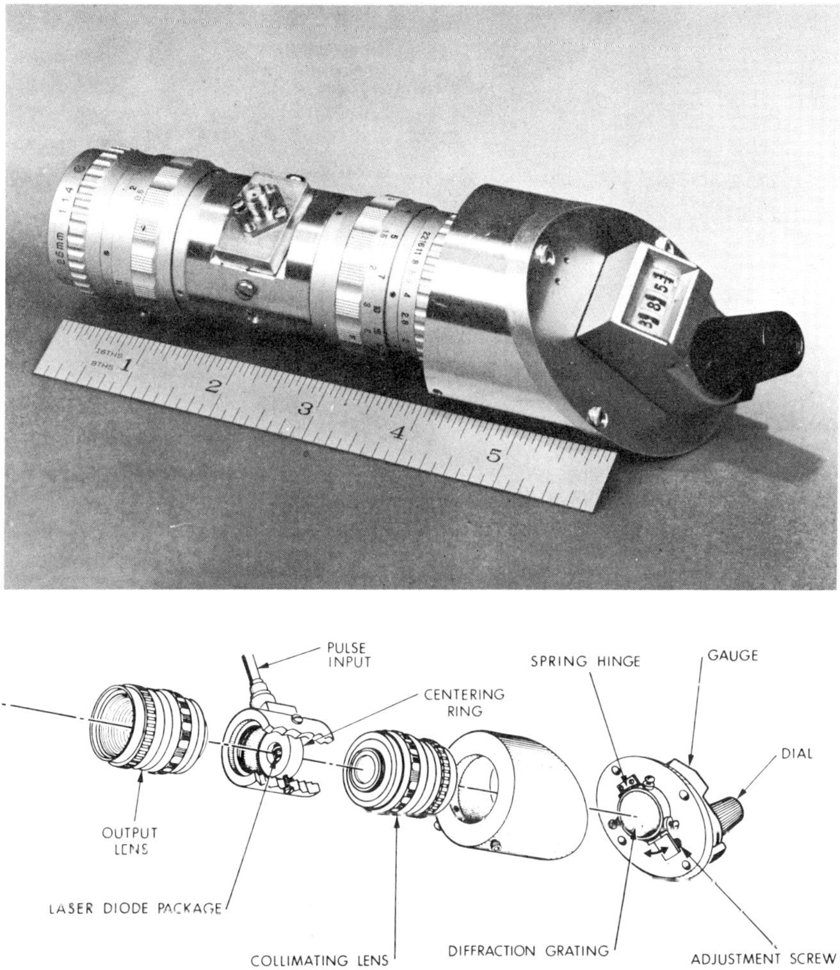

FIG. 11. Photograph of a room temperature external cavity gallium arsenide diode laser.

total length of the cavity. This reduction can be as much as a factor of one hundred for typical devices; however, the smaller mode spacing in long external cavities makes it more difficult to achieve single-frequency operation. For a number of spectroscopic applications, single-frequency operation may not be necessary or desired. An appropriate external-cavity device operated pulsed could have its modes chirped enough to fill in the spectral envelope defined by the external cavity grating dispersion and the mode width defined by the junction. Such an envelope width could be significantly less than

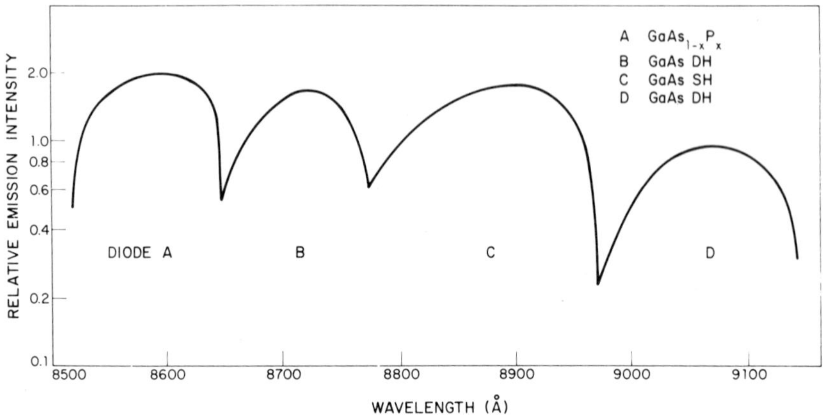

FIG. 12. Peak output power as a function of wavelength for a $GaAs_xP_{1-x}$ and three different GaAs cavity-grating controlled diode lasers at room temperature. Different amounts of silicon doping extend the GaAs laser wavelengths. From Ref. 13.

0.1 cm^{-1} if the diode were not driven very far above threshold. Tuning would effectively be continuous for spectroscopic lines that were broader than the envelope width.

(e) *Distributed-Feedback Lasers*

The external cavity-controlled diode laser is a hybrid device. A distributed-feedback grating fabricated along the active gain region of a diode can provide oscillation at a single wavelength given by

$$\lambda = \frac{2n(T)d}{M} \tag{13}$$

where d is the spacing of the grating, $n(T)$ is the temperature-dependent refractive index, and M is an integer. Optically-pumped distributed-feedback laser action has been demonstrated in GaAs at 77 K[14], while a distributed-feedback GaAs single-heterostructure diode laser has operated[15] pulsed at 77 K. An important feature of Equation (13) is that the temperature-dependent refractive index provides a mechanism for continuously tuning the resonant condition without any mode jumps. A temperature change in the grating region of several degrees is sufficient to tune the output frequency from 20 to 50 cm^{-1}. The reproducibility of such grating fabrication using holographic techniques eliminates uncertainties in the random cavity lengths obtained in conventional cleaved diode lasers. The fabrication procedure for an optically-pumped distributed-feedback semiconductor

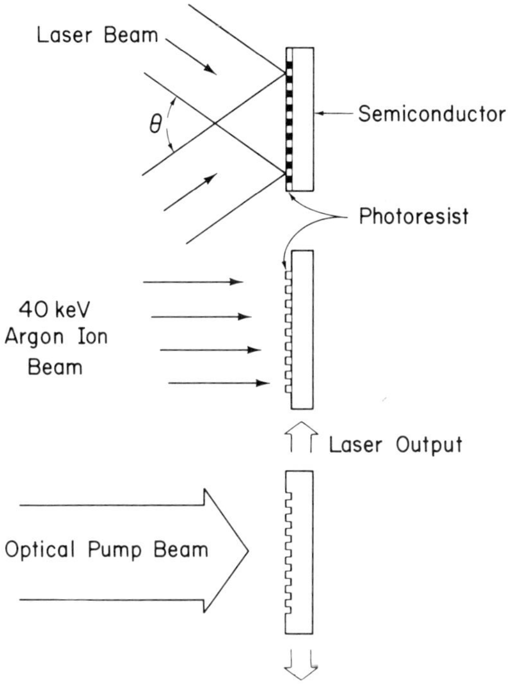

FIG. 13. Procedure for fabricating a distributed-feedback grating on a semiconductor.

laser is shown in Fig. 13. Exposure of a photoresist on the semiconductor is made with two interfering beams for visible or U.V. laser with the grating spacing being determined by the wavelength of this laser and the angle between them. After processing of the photoresist, a 40 keV argon ion beam is used to etch away the exposed semiconductor which can then be optically pumped. Regrowth of GaAs-$Ga_xAl_{1-x}As$ over the gratings with the inclusion of a *p-n* junction has successfully produced a GaAs distributed-feedback diode laser[15]. Extension of these techniques to infrared bandgap lasers could eliminate the multimode problem in these devices.

D. OPTICALLY-PUMPED SEMICONDUCTOR LASERS

One of the simplest ways of studying stimulated emission in semiconductors is by optical excitation using a laser. Figure 14 shows the schematic arrangements used for both transverse and collinear excitation of semiconductor platelets using a laser source. For the excitation of infrared bandgap semiconductor laser crystals for use in high resolution spectroscopy, the most

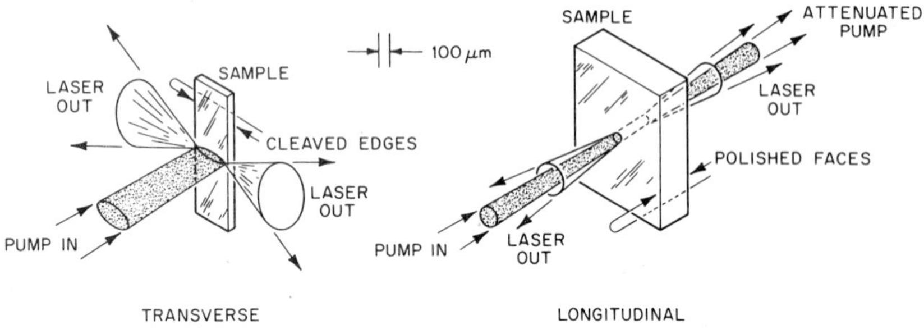

FIG. 14. Transverse and longitudinal geometries for optical pumping of semiconductor lasers. From Ref. 16.

successful[16] pump sources have been GaAs injection lasers, Nd:YAG, carbon monoxide and carbon dioxide lasers. Gallium arsenide injection lasers have the advantage of being small, efficient and inexpensive and have been used to excite a substantial number of semiconductors. Figure 15 shows a particular mounting arrangement for using GaAs diode lasers as pumps at low temperature. Cleaved platelets of several binary and ternary semiconductors have been optically pumped to produce laser sources stable enough to conduct high resolution spectroscopic experiments. Figure 16 shows the emission spectra of a GaSb laser that is pumped using a GaAs

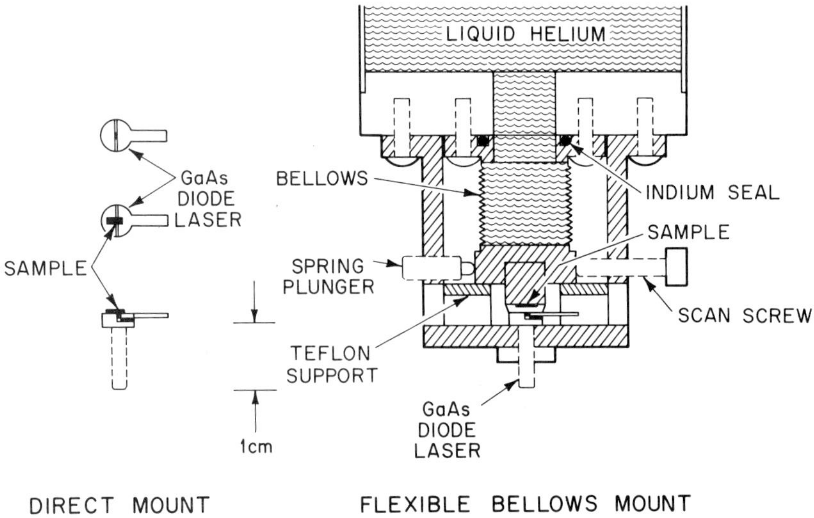

FIG. 15. Direct and scanning bellows mounts for transverse optical pumping of semiconductor lasers with a GaAs diode laser. From Ref. 16.

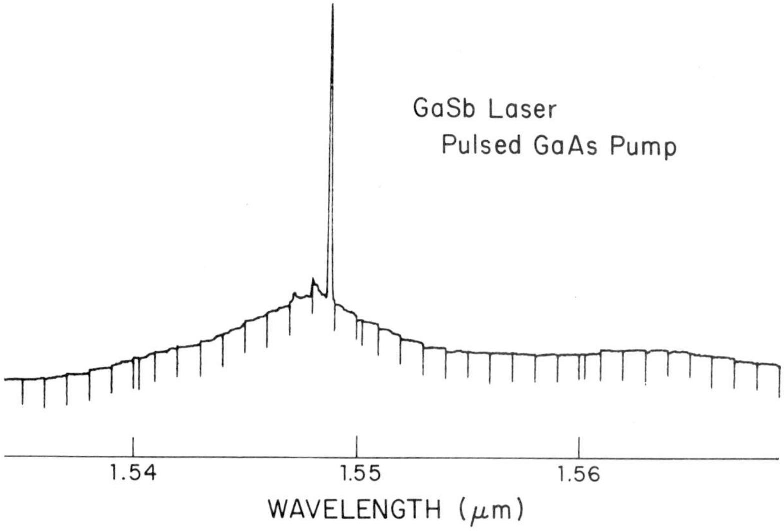

FIG. 16. Spectrum of GaSb laser transverse-pumped with a pulsed GaAs diode laser. From Ref. 16.

diode laser source. Both the spontaneous and stimulated emission can be seen in this trace. Figure 17 shows the output power trace in real time (top) of a single mode from a cw InAs laser that is optically-pumped by a GaAs diode laser. The bottom trace in Fig. 17 shows the transmission of the InAs laser through a low finesse germanium etalon used as a relative frequency marker while the middle trace shows an absorption line of NH_3 at 2 Torr pressure. This figure demonstrates the utility of using optically-pumped semiconductor lasers for high resolution spectroscopy.

An InSb laser has been optically excited using resonant pumping with a carbon monoxide laser[17]. Because of the good energy match between the pump photons and the gap, power conversion efficiencies of 20% were achieved using collinear pumping geometries with single longitudinal mode output powers of up to 20 mW. Using a wedged sample of InSb. continuous tuning across the gain bandwidth was obtained by moving the pump beam along the sample. The useful linewidth for such tuning was limited by the variation of cavity length over the mode diameter. Transverse pumping of InSb with a CO laser produced substantially lower output powers than collinear pumping due to the mode matching geometry. The external output quantum efficiency for a semiconductor laser can be expressed as

$$\eta_{ext} = \eta_{int} \frac{\ln(R_1 R_2)}{2\alpha l + \ln(R_1 R_2)}, \tag{14}$$

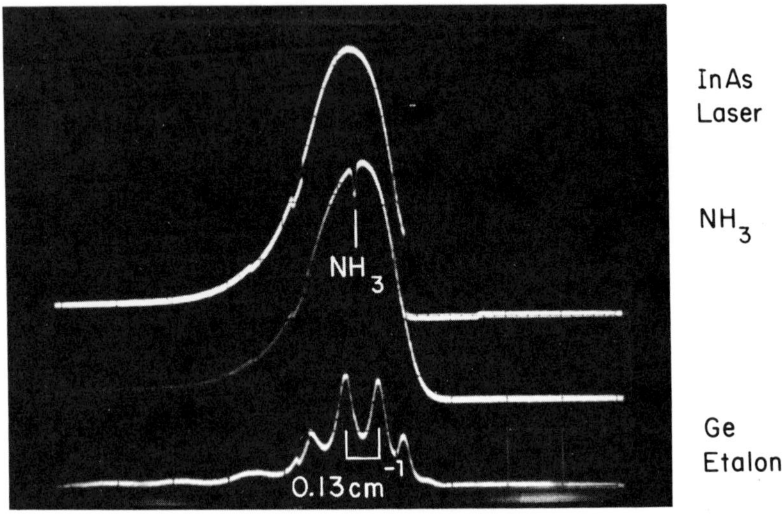

FIG. 17. Oscilloscope traces of (top) output of a thermally-tuned single mode from a cw optically-pumped (using a GaAs diode laser) InAs laser near liquid helium temperature; (middle) absorption spectrum of NH_3 at 2 Torr pressure; (bottom) transmission of laser mode through a low finesse, fixed Ge etalon. From Ref. 6.

where R_1 and R_2 are the mirror reflectivities, l is the crystal length, η_{int} is the internal quantum efficiency, and α is the background losses. In most semiconductor lasers, the external quantum efficiencies are limited by the background loss. Typical values for α of 10–20 cm^{-1} have been determined for diode lasers by measuring the laser threshold current density as a function of cavity length or mirror reflectivity. Such high losses effectively limit the length of diode lasers to less than one millimeter. Perhaps one of the more important potential features of optical pumping is that of scalability where large area or volume excitation can be achieved without the associated problems of such high-gain systems.

E. DIODE LASERS

Diode lasers operate by the electrical injection of carriers across a *p-n* junction[3]. The schematic energy band structure as a function of distance for a *p-n* junction with and without an electrical bias is shown in Fig. 18. Under conditions of forward bias, carriers are injected across the depletion region and diffuse into the bulk regions and recombine. One of the advantages of this type of excitation is that the energy required for injection is

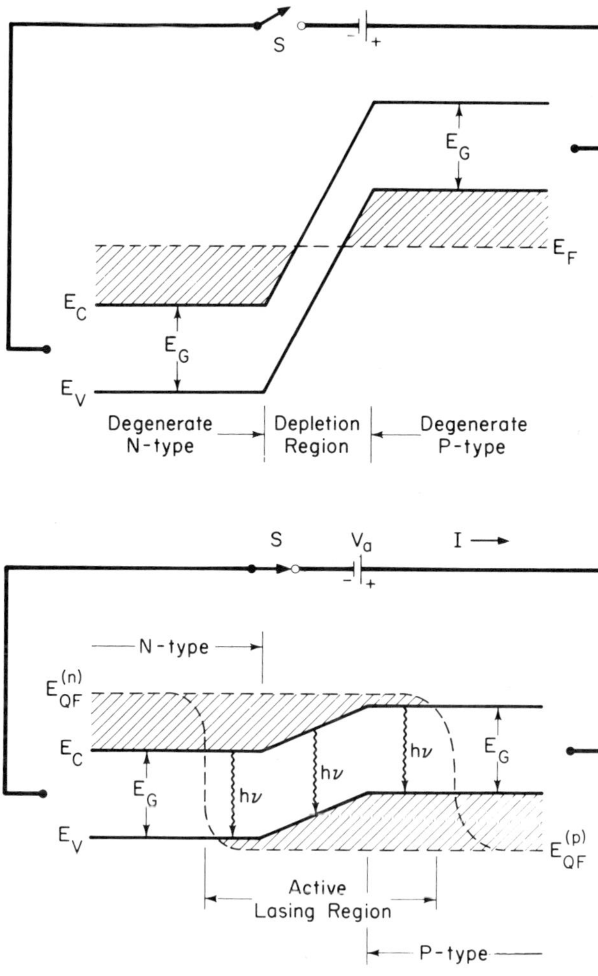

FIG. 18. Energy band diagram versus distance for a p-n junction both with and without forward bias. From Ref. 5.

comparable to the bandgap energy which is why these devices can be so efficient. The *p-n* junctions are usually formed by the diffusion of impurities into the surface of a doped semiconductor while a cavity is formed by cleaving the crystal along appropriate crystallographic axes.

Earlier lasers in which the junction extended across the entire device frequently operated in parasitic bounce-modes involving all four sides of the laser. The result was an undesirable beam pattern and low external efficiency.

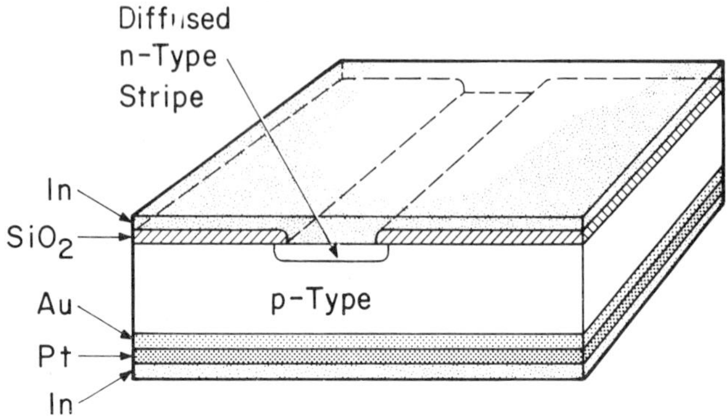

FIG. 19. Stripe geometry lead-salt laser structure. The stripe width is typically 50 μm and the length is about 500 μm. From Ref. 18.

In the stripe-geometry structure where bounce-modes are suppressed by the lossy bulk regions along the sides of the active region, operation in a fundamental spatial mode has been observed. The structure of a typical stripe-geometry diode laser[18] is shown in Fig. 19. The narrow geometry allows usually only the fundamental mode to propagate with the lowest loss in the cavity. Because of the compactness of diode lasers and their simplicity of operation, they have been used extensively for high resolution spectroscopy studies.

II. Spin-Flip Raman Lasers

A. Introduction

The spin-flip Raman laser can be a very useful device for use in high resolution spectroscopy and photochemistry. As a primary tunable laser device in the infrared, it has the broadest tuning range and the highest cw output power compared to any other laser presently available. The first operation of an InSb spin-flip laser was demonstrated by Patel and Shaw[19] using a pulsed CO_2 laser and the first continuously operating device was demonstrated by Mooradian *et al.*[20] using a CO laser. The spin-flip laser operates by scattering pump laser light from electrons whose energy levels have been magnetically quantized in a semiconductor crystal. Electrons in the lowest orbital energy state shown in Fig. 20 can flip their spin and reduce the energy of the pump photon by an amount $g\beta H$, where g is the magnitude of the electron g-factor, β is the Bohr magneton, and H is the magnitude of the applied magnetic

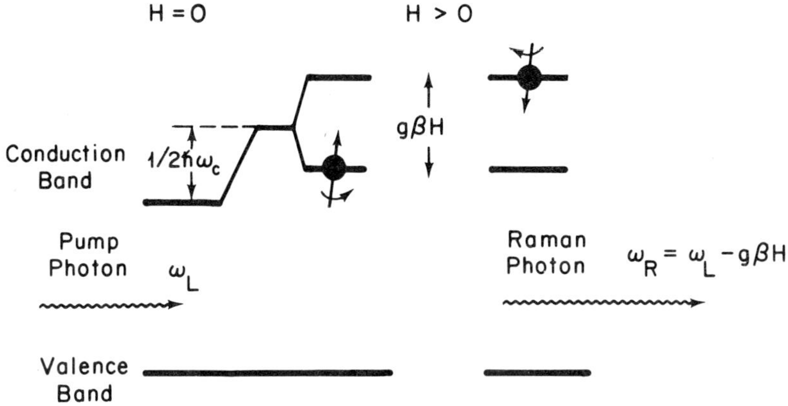

FIG. 20. Simplified energy diagram of electrons in a semiconductor both with and without an applied magnetic field.

field. For sufficient pump intensity, the spin-flip scattering can be stimulated with efficient conversion of pump energy into Stokes components. In the following sections, the operating characteristics of spin-flip lasers will be discussed with emphasis on its use as a tunable source.

B. SPONTANEOUS SCATTERING

In order to understand the operating characteristics of the spin-flip laser, it is important to discuss the spontaneous scattering mechanisms for spin-flip scattering. The gain for stimulated spin-flip Raman scattering can be expressed as

$$g = \frac{8\pi c(n_2 - n_1)(d\sigma/d\Omega)I}{\hbar\omega_s^3 n_p^2 \Gamma} \tag{15}$$

where n_1 and n_2 are the number of electrons in the upper and lower spin states, respectively, ω_s is the Stokes frequency, n_p is the refractive index at the pump frequency, Γ is the spontaneous spin-flip linewidth, I is the pump intensity, and $(d\sigma/d\Omega)$ is the frequency-dependent spontaneous scattering cross-section given by

$$\frac{d\sigma}{d\Omega} \approx \frac{9}{16}\sigma_T\left(\frac{E_g}{E_g - \hbar\omega_l}\right)^2 \tag{16}$$

where E_g is the optical energy gap, σ_T is the effective Thomson cross-section, and ω_l is the laser frequency. From Equation (16) it can be seen there is a strong resonant enhancement of the scattering cross-section as the pump frequency approaches the optical bandgap. Gains can vary from 10^{-5} to greater than 10^{-2} cm/W depending upon the pump frequency as well as electron concentration, sample temperature, and magnetic field. The specific parameters entering into Equations (15) and (16) have been carefully studied and provide a very good understanding of not only the threshold pump intensities but tuning characteristics of the InSb spin-flip Raman laser.

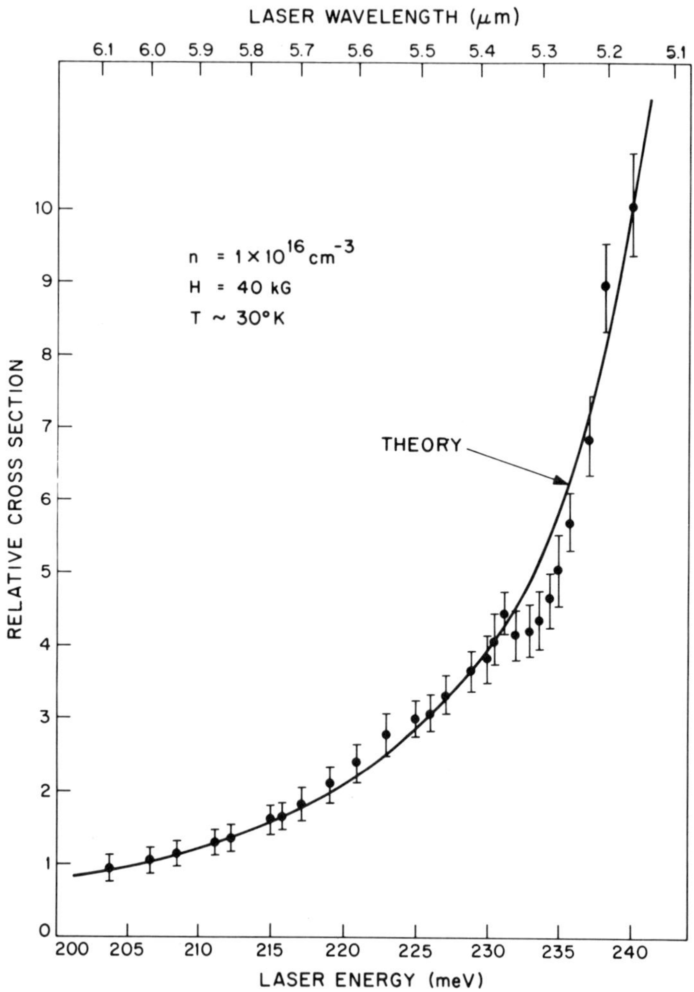

FIG. 21. Relative spontaneous spin-flip scattering cross-section in InSb. From Ref. 21.

The relative spontaneous scattering cross-section has been measured in InSb[21] by using a grating-tuned carbon monoxide laser as an excitation source. A liquid nitrogen cooled CO laser can operate with a line every few wave numbers between 5 and 6.5 μm. Figure 21 shows the experimental results for near-resonant spin-flip scattering in InSb compared to the theoretical predictions of Equation (16). It is interesting to note that there are basically no adjustable parameters necessary for the fit. The absolute value of the scattering cross-section is quite difficult to measure accurately and is not included in Fig. 21. By pumping close to resonance with a CO laser, the spin-flip laser has operated continuously.

The linewidth, Γ, in Equation (15) has also been measured[22] carefully in the case of collinear scattering. Initial linewidth measurements made by using a grating monochromator were limited by the weak signals and the resolution limit of about 0.1 cm^{-1} for a one meter instrument. A particularly clever method of determining the spontaneous linewidth is to use the small signal gain technique. In this technique, one CO laser at frequency ω_1 is used to pump an InSb crystal below threshold for stimulated scattering while a second, weak probe, CO laser at frequency $\omega_2 = \omega_1 - g\beta H$ is used to measure gain. The intensity of the pump laser is kept low enough so the peak gain is 10% or less in order to measure the true lineshape. The resolution is determined by the frequency jitter between the two lasers and was better than 100 kHz. Figure 22 shows the gain linewidth for an InSb sample at 2 K to be about 200 MHz. Such narrow linewidths together with the strong resonant enhancement to the scattering cross-section around 5 μm has produced stimulated scattering threshold of only a few milliwatts. The spontaneous linewidth, in general, is quite dependent upon magnetic field, temperature, and electron concentration. This complex variation as well as the line shape has not been theoretically treated in sufficient detail to fully understand the observed dependencies.

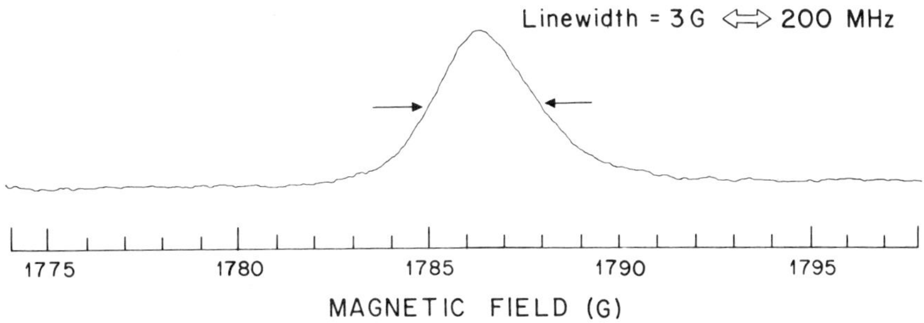

FIG. 22. Spin-flip Raman gain signal taken using small signal gain technique. From Ref. 22.

C. STIMULATED SCATTERING

Most of the stimulated spin-flip Raman scattering work performed so far has been on InSb because of the *relative* simplicity as well as the commercial availability of high quality material. Stimulated spin-flip scattering has been observed also in InAs, CdS, and HgCdTe; however, the present discussion will be limited to the results on InSb. The InSb spin-flip laser, pumped both with a CO or a CO_2 laser, is presently the most useful of any spin-flip laser and is already a commercially available device.

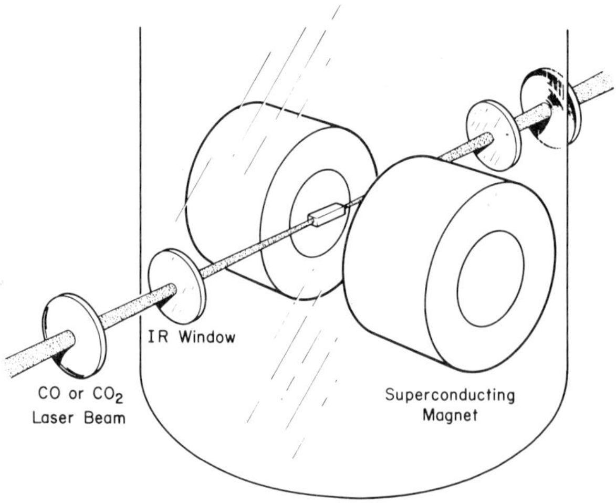

FIG. 23. Schematic of a spin-flip laser set up using a split superconducting solenoid.

Figure 23 shows the schematic arrangement of a spin-flip Raman laser using a split-solenoid superconducting magnet. The samples are usually heat sunk to a copper stud in contact with liquid helium. An alternate scheme for use with a cw CO laser is to use a low field electromagnet or a permanent magnet. In the case of permanent magnet operation, a tapered pole piece was used to successfully[23] provide gradient field tuning with sufficient frequency stability to perform high resolution spectroscopy. Low field, low concentration devices are particularly simple and useful, but a specially designed superconducting magnet system can offer the most general utility as well as simplicity. Most spin-flip crystals used with these magnet configurations have been in the form of unoriented, rectangular, parallelopipeds with polished ends acting as the Fabry-Perot resonators having the natural 36% reflectivities.

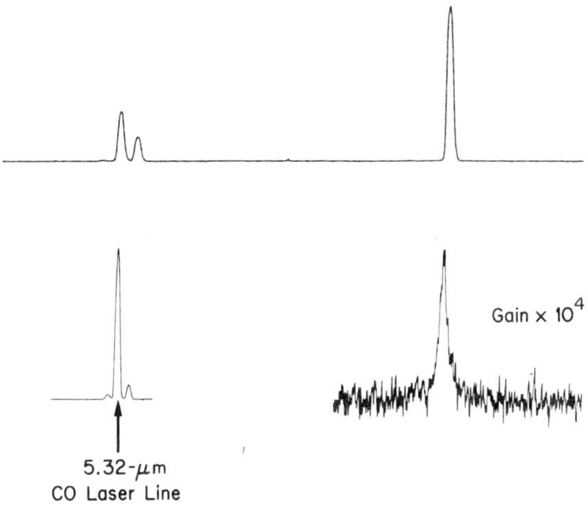

FIG. 24. Continuous stimulated and spontaneous spin-flip scattering in InSb ($T = 30$ K, $n = 1 \times 10^{16}$ cm^{-3}, $H = 35$ kG).

Recorder traces of the spontaneous and stimulated spin-flip scattering are shown in Fig. 24. The lower trace shows the spontaneous scattering with the gain increased by 10^4 from the pump laser signal. The top trace taken above threshold shows a dramatic increase in the intensity of the pump signal. The conversion efficiency in this case is so high, in fact, that one can clearly see the effects of pump depletion where the main CO pump line has been decreased in intensity relative to the weaker CO lines which have insufficient intensity to pump the spin-flip. There is sufficient intensity in the first stimulated Stokes wave to act as a pump for a second stimulated spin-flip process. In low concentration samples, for near-resonant pumping, as many as six Stokes and three anti-Stokes lines have been observed. Not all of these are due to stimulated spin-flip scattering but rather to a four-wave mixing process which uses the resonant spin nonlinearity to generate efficiently these components. The results of a degenerate four-wave frequency mixing experiment[22] are shown in Fig. 25 for a magnetic field of 535 G, with a frequency separation $\omega_1 - \omega_2 = 1.25$ cm^{-1}, in a sample of concentration $n = 1 \times 10^{14}$ cm^{-3}. Stimulated spin-flip scattering does not have to occur in order for this process to be efficient. In this experiment, ω_1 and ω_2, which are generated from a single pump laser, are mixed to produce outputs at $\omega_3 = 2\omega_1 - \omega_2$ and $\omega_4 = 2\omega_2 - \omega_1$. Resonance occurs for $\omega_1 - \omega_2 = g\beta H$. The top trace shows the transmitted power on resonance with the two new frequencies ω_3 and ω_4. The second trace was taken with $H = 0$ and shows no mixing. The bottom trace was again for $H = 0$ at a lower gain to show the

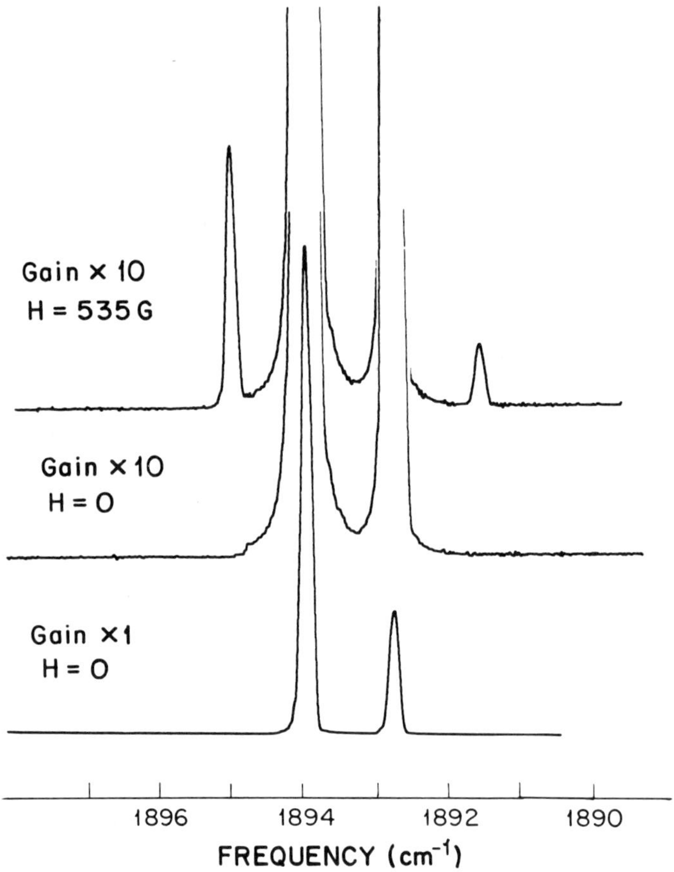

FIG. 25. Degenerate four-wave mixing in InSb ($n = 1 \times 10^{14}$ cm^{-3}, $T = 2$ K). From Ref. 22.

relative intensities of ω_1 and ω_2. These four-wave, resonant processes occur also during the operation of a spin-flip laser and account for some of the multiple Stokes and anti-Stokes components that are observed.

At the onset of stimulated emission, there is a sharp nonlinear increase in the spin-flip output that continues until limited by a number of effects. The limiting output from a spin-flip laser is dependent primarily upon spin saturation and heating effects. Figure 26 shows how a particular InSb spin-flip laser output might be expected to behave if pumped near 5 μm. Because of the low threshold for 5 μm excitation, there is high conversion efficiency well above threshold which causes a linear dependency of output

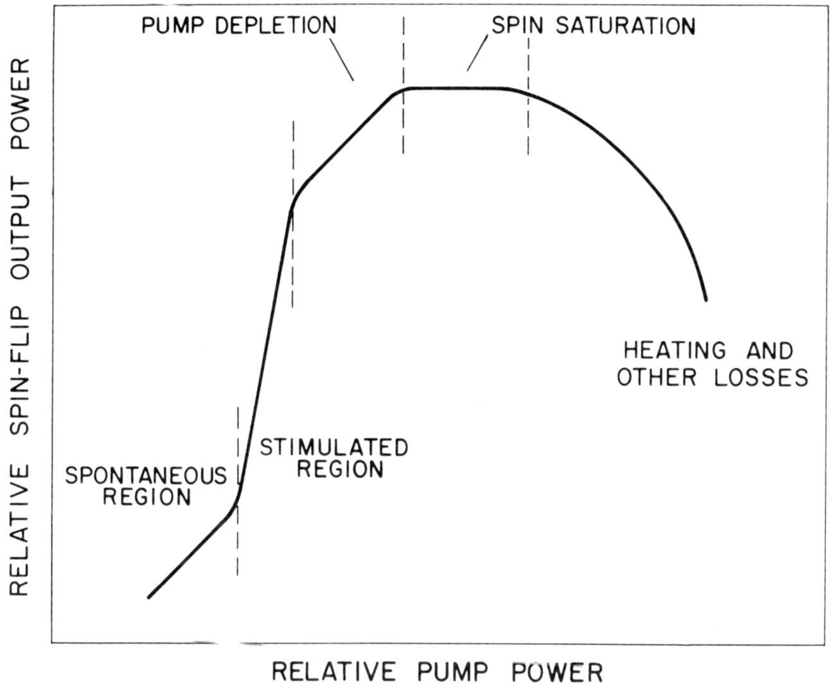

FIG. 26. Spin-flip output as a function of pump power input showing different regions of operation that can be obtained using cw excitation. Curve does not correspond to any actual conditions.

power on input power. As the pump intensity is further increased, the rate at which spins are flipped into the upper spin state becomes equal to the rate at which they spontaneously decay back to the lower spin state. This regime is called spin saturation and the output power from a fixed mode volume remains constant as the pump power is increased. However, the active mode volume increases with pump intensity. For very high pumping intensities the spin-flip output actually decreases due to heating of the lattice and the electron gas as well as from other intensity dependent losses. The order in which these power limiting effects can occur depends upon many parameters including pump wavelength, electron concentration, magnetic field, and temperature. As an example of spin saturation, for a sample having 10^{16} electrons/cm^3, a spin lifetime of 10^{-8} sec, and a mode volume of 1 cm by 200 μm diameter, a few watts of output power could be achieved.

D. TUNING CHARACTERISTICS

The gross tuning of an InSb spin-flip laser has been experimentally measured for both a CO and a CO_2 laser pumped device. Figure 27 shows the calculated tuning range for the first- and second-Stokes components using a 5.3 and a 6.0 μm pump. The spin-flip laser has in fact only been operated cw from 5 to 6.5 μm using a CO laser[24] and pulsed from 9 to 14.6 μm using a CO_2 laser[25]. The useful limit for superconducting magnets is about 140 kG while Bitter type magnets can exceed 200 kG. At present, a very compact and practical superconducting magnet system can be obtained with a split solenoid having fields in excess of 70 kG.

The fine tuning characteristics of the spin-flip laser have been studied[26] and the effects of mode pulling are well understood. Figure 28 shows the

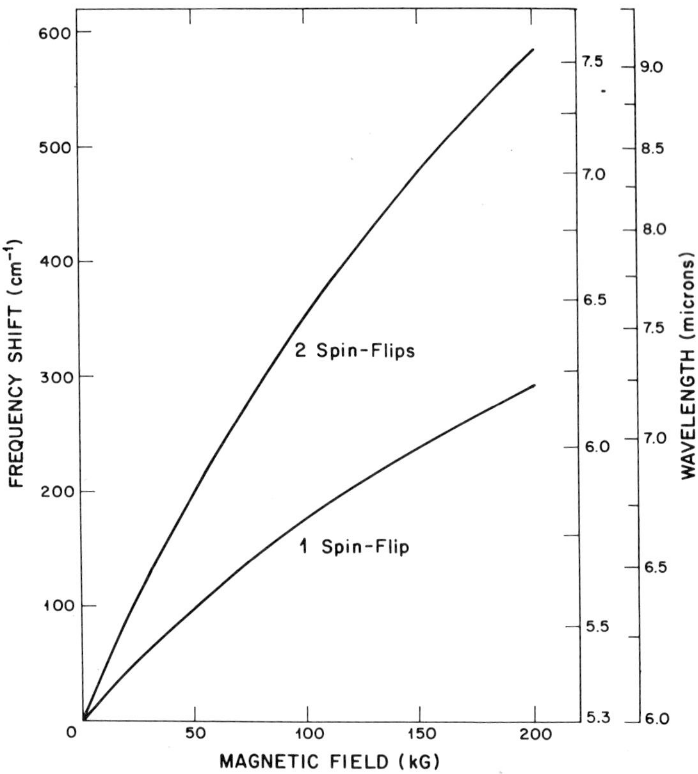

FIG. 27. Calculated tuning range of InSb spin-flip Raman laser using CO laser excitation. Actual tuning range demonstrated is 5.0 to 6.5 μm.

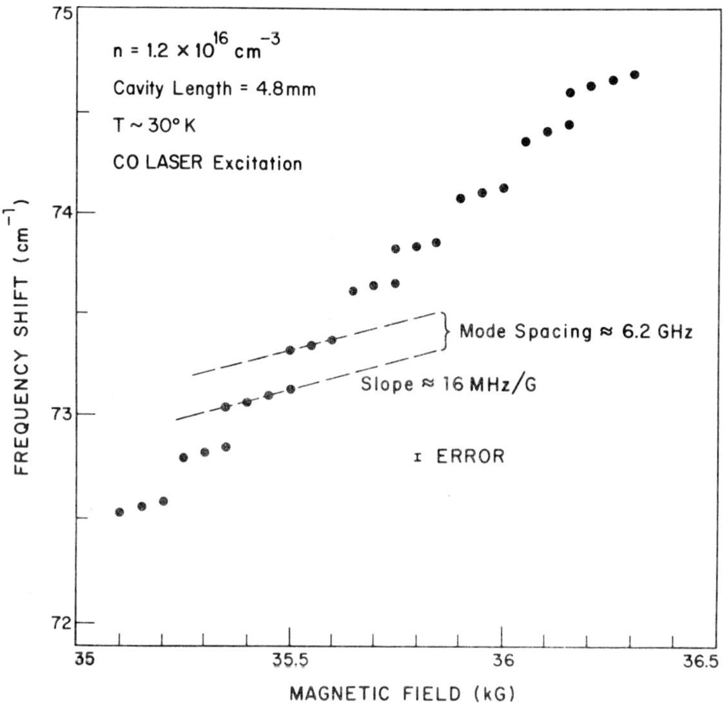

FIG. 28. Fine mode tuning of a cw spin-flip Raman laser. From Ref. 30.

experimentally measured fine tuning of a spin-flip laser using a high resolution spectrometer. The pulling and jumping of the individual longitudinal modes can be calculated from Equation (9–11). The mode spacing is

$$\Delta\nu = \frac{c\Gamma_s}{2l(\Gamma_s + \Gamma_c)} \frac{1}{n + \nu \dfrac{dn}{d\nu} \dfrac{\Gamma_s}{\Gamma_s + \Gamma_c}} \tag{17}$$

and is of the order of 6 GHz, or 0.2 cm^{-1}, for $l \sim 5$ mm.

The tuning rate of a mode, with magnetic field, is given by

$$\frac{d\nu}{dH} = \frac{\Gamma_c \dfrac{d\nu_s}{dH} - \dfrac{\nu\Gamma_s}{n} \dfrac{\partial n(\nu, H)}{\partial H}}{\Gamma_c + \Gamma_s\left[1 + \dfrac{\nu}{n} \dfrac{\partial n(\nu, H)}{\partial \nu}\right]} \tag{18}$$

$d\nu_s/dH$ is about -2 cm^{-1}/kG (-60 GHz/kG), while, for the experimental conditions of Fig. 28, the mode tuning rate is about -0.5 cm^{-1}/kG

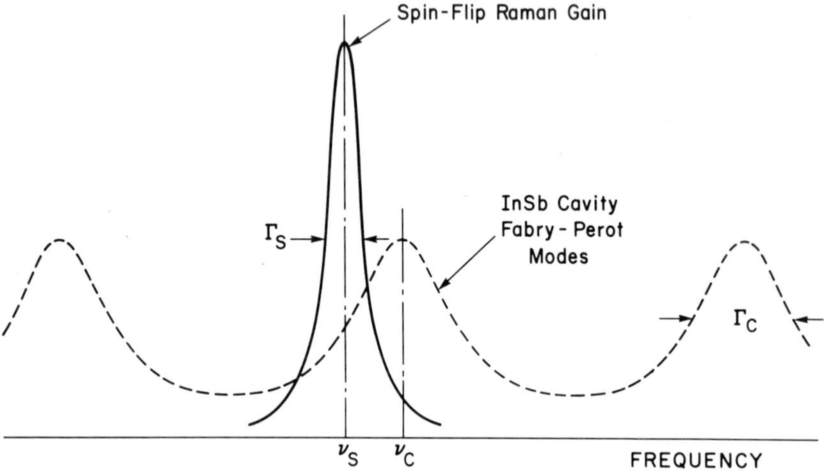

FIG. 29. Mode pulling between Raman gain and the InSb resonant cavity.

(−15 GHz/kG). The difference between $d\nu_s/dH$ and $d\nu/dH$ results in mode hopping and consequently gaps in the spectral range.

The coupling between the gain linewidth and the cold cavity linewidth is demonstrated in Fig. 29. In contrast to the case for recombination radiation lasers, the spontaneous spin-flip Raman linewidth discussed above is usually narrower than the cavity linewidths. From Equation (18), $d\nu/dH \rightarrow g\beta$ as the spontaneous linewidth becomes very small as occurs in the case of low concentration samples at low (less than 10 kG) magnetic fields. Table 1 lists the calculated variation of longitudinal mode tuning rate for different values of spontaneous linewidth. These results have been experimentally verified

TABLE I. Calculated mode-tuning characteristics of a spin-flip laser

Γ_0 (cm^{-1})	$d\nu/dH$ (MHz/G)	$(d\nu/dH)\Delta H$ (GHz)	$\Delta\nu$ (GHz)	Frequency coverage (%)
0.01	63	5.4	0.6	91
0.04	48	4.1	1.8	69
0.10	33	2.8	3.1	47
0.30	15	1.3	4.5	22
0.50	10	0.8	5.0	13
1.00	5	0.4	5.4	7

over most of the range. In addition to the mode tuning rate approaching the spontaneous rate as the linewidth decreases, the frequency range over which a mode tunes before a mode jump occurs increases.

In the case of multimode operation in a low concentration sample, the envelope width can be comparable to the mode jump which will result, in effect, to a continuous tuning. A spectrum analyser photograph of the spin-flip output operated in such a mode is shown in Fig. 30. This data was taken by heterodyning the output[26] of a cw spin-flip laser with a second CO laser in a high speed infrared detector. The heterodyning technique is an excellent method of measuring and controlling the output frequency of any tunable laser. The same InSb crystal could operate in a single axial mode by carefully adjusting the focus of the pump beam. The tuning characteristics for such a mode of operation is shown in Fig. 31. Here, the mode jumps are very small. As a demonstration of the utility of the spin-flip laser as a spectroscopic source, Fig. 32 shows the Doppler limit absorption spectrum of NO taken

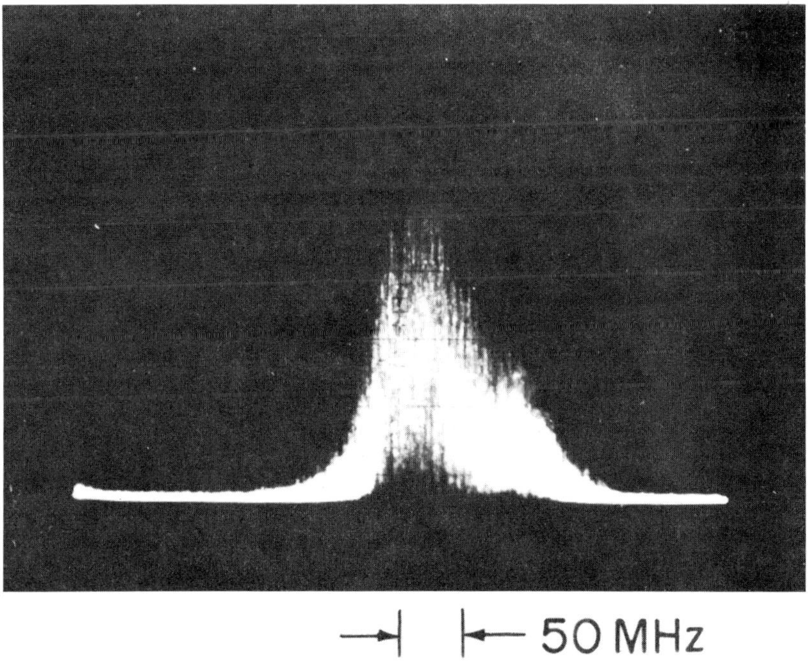

FIG. 30. Heterodyne signal of a multimode cw spin-flip laser ($n = 1 \times 10^{15}$ cm^{-3}, $H =$ 3770 G). Bandwidth = 300 kHz. From Ref. 26.

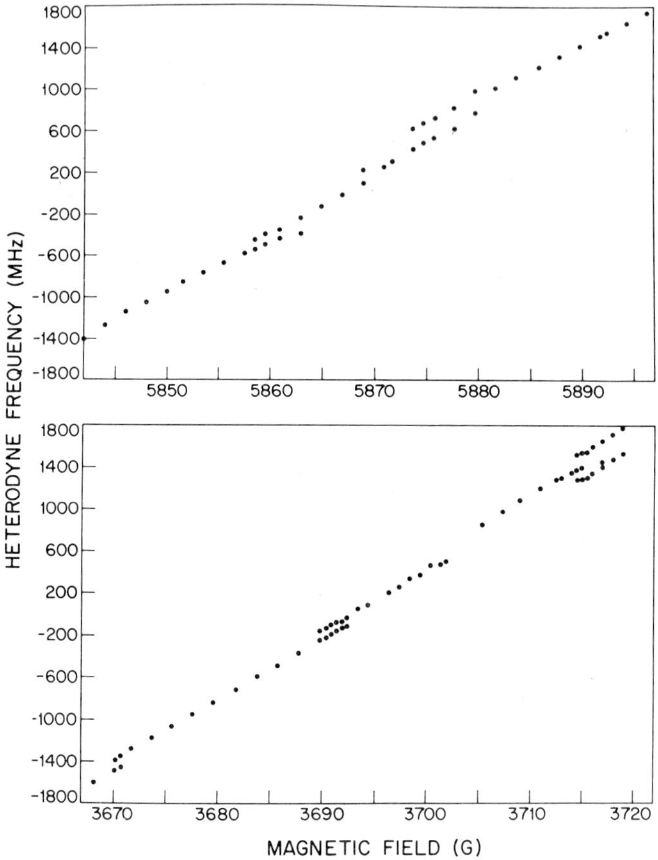

FIG. 31. Mode tuning of a cw spin-flip Raman laser ($n = 1 \times 10^{15}$, $T = 2$ K) taken using heterodyning techniques for two different magnetic field ranges. From Ref. 26.

using a cw spin-flip laser operating in a single longitudinal mode. The lambda-doubled splitting of 360 MHz is clearly resolved. The spin-flip laser has been used extensively for molecular absorption spectroscopy[24,27]. For use in very high resolution spectroscopy, precise frequency control of the spin-flip device has to be achieved. This has been done by using a feedback loop to drive a small magnetic coil around the InSb crystal. The feedback loop uses an error signal derived by heterodyning the spin-flip output with a stable CO laser together with a stable microwave oscillator in a HgCdTe photodetector to achieve stabilities of better than 30 kHz.

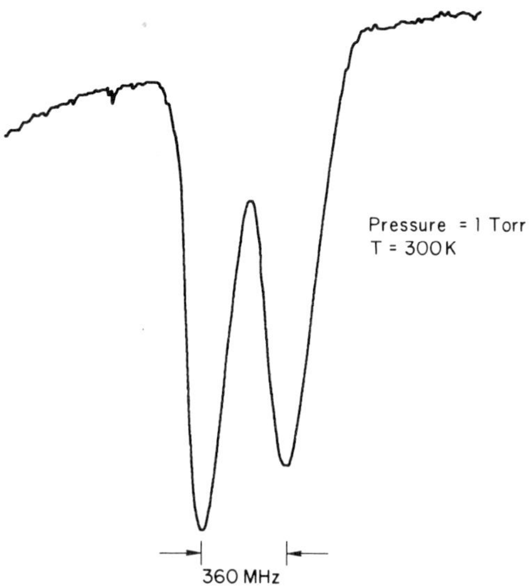

FIG. 32. Absorption spectrum of NO taken using a cw single frequency InSb spin-flip laser ($\nu \sim 1881.0\ \mathrm{cm}^{-1}$) showing the Lambda splitting of the $R(\frac{1}{2})_{1/2}$ line.

E. TRANSIENT SPIN-FLIP LASER

When a spin-flip laser in the form of a polished rectangular parallelopiped is excited with a pulsed laser that has a smooth (unmodelocked) intensity profile, the output usually has a complicated temporal behaviour that is dependent in part upon the intensity of the pump laser. Previous results using pulsed CO_2 lasers to excite InSb spin-flip lasers reported external conversion efficiencies limited to about 1 % because of spin saturation, but no detailed studies of the temporal characteristics of the spin-flip laser were made. Figure 33 shows a series of oscilloscope traces[28] of the output of an InAs spin-flip Raman laser operating at 2 K. The pump was a HF TEA laser with a smooth intensity profile and the crystals were 3 × 3 × 5 polished rectangular parallelopipeds. The traces were taken for increasing values of magnetic field from 70 to 85 kG. For this InAs crystal with 1.6×10^{16} electrons per cm^3 the quantum limit is reached for magnetic fields above 70 kG thereby reducing the threshold for stimulated spin-flip Raman scattering. Starting with the lowest magnetic fields, the spin-flip output is an initial transient pulse which decays rapidly to near-zero intensity and is followed by a broader peak that grows and shifts to longer times for increasing magnetic fields. Similar temporal behaviour was[29] observed for InSb spin-flip laser crystals when pumped by a CO_2 TEA

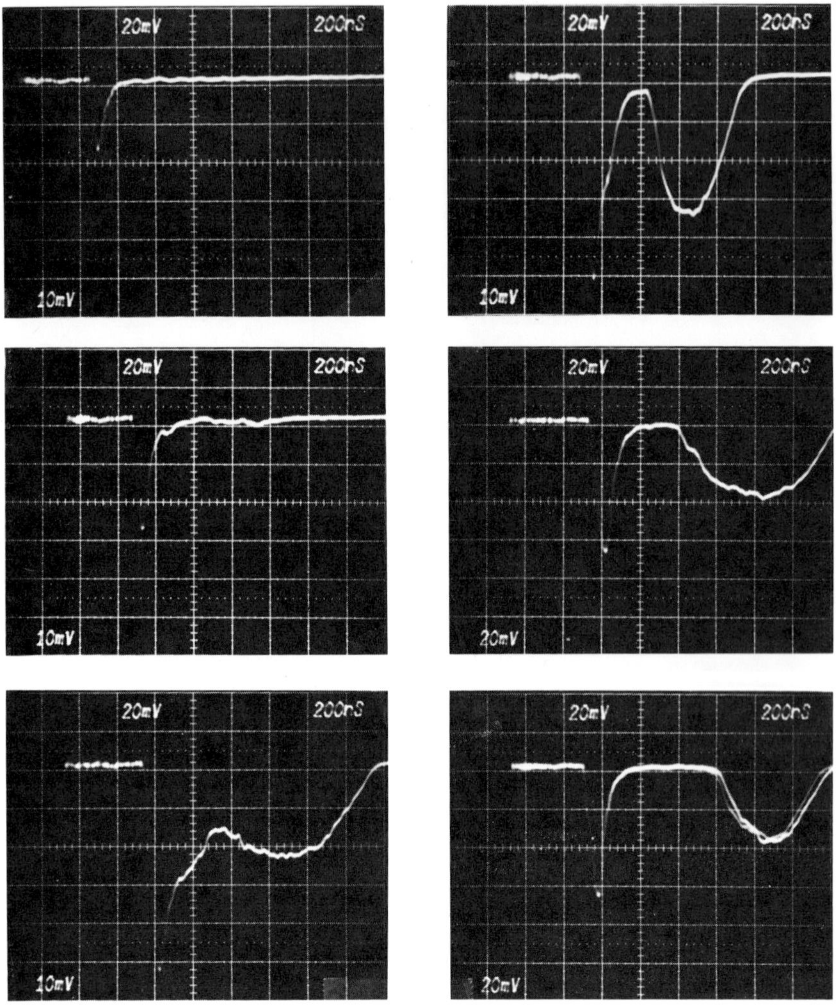

FIG. 33. Temporal output of an InAs spin-flip Raman laser at 2 K. Input pump intensity is held constant. Traces from upper left downward are for increasing magnetic fields from 70 to 85 kG.

laser with the exception that the secondary peak increased in amplitude and delay for increasing pump intensity. Figure 34 shows the output of an InSb ($n \sim 1 \times 10^{16}\ cm^{-3}$) spin-flip Raman laser for different values of CO_2 TEA laser pump intensity. The temporal variation of the pump pulse in all cases was the same. The spin-flip output pulse, as in the case for InAs, starts as just a single short pulse which is followed by a long pulse at the higher

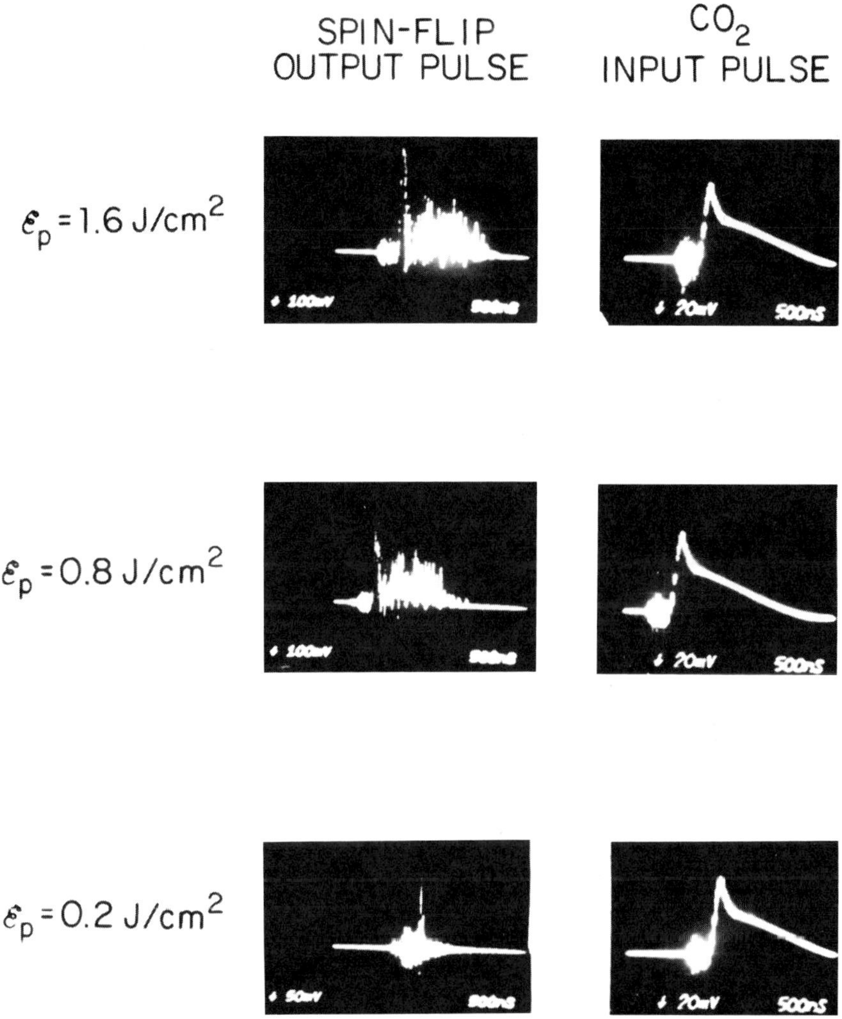

FIG. 34. Temporal output of an InSb spin-flip Raman laser for different values of pump energy from a CO_2 TEA laser. From Ref. 31.

excitation intensities. No mode-locking of the CO_2 TEA pump laser occurred because a cw gain cell was placed in the cavity.

The initial transient is interpreted as normal laser relaxation oscillations that occur when a laser is switched on. While one would expect steady-state laser oscillation to occur after the relaxation oscillations have died out, in both InAs and InSb, complete quenching of the spin-flip output could occur before buildup of the second peak. This quenching is attributed to the substantial amount of free carrier absorption present at the pump wavelength ($\sim 1\ cm^{-1}$ in the case of InSb) which results in scattering the free electrons from the bottom of the conduction band to high momentum states where their spins are scrambled and they are effectively removed from participating in the spin-flip process. The electrons which then relax back to the bottom of the band occupy either spin state with equal probability. This quenching occurs along the focus of the TEM_{00} mode pump beam within the crystal where the intensity is the greatest. In the wings of the Gaussian pump beam where the intensity is insufficient for collinear oscillation, build-up of parasitic bounce modes within the crystal occurs with a slower cavity build-up time. Special care in alignment can minimize but not eliminate these bounce modes. Scanning the intensity of the spin-flip output across the exit face of the crystal clearly indicated the difference between the collinear and bounce mode operation. The annular region of the crystal where the bounce modes build up occupy a much larger volume than the central area where the collinear mode oscillates. It is clear from Figs 33 and 34 that a substantial fraction of the spin-flip output energy can be obtained in the secondary pulse which is dominantly a multiple bounce-mode. The output coupling for bounce-modes normally is quite low and the external power conversion efficiency should be significantly greater than the observed 1% if an appropriately coupled cavity is used.

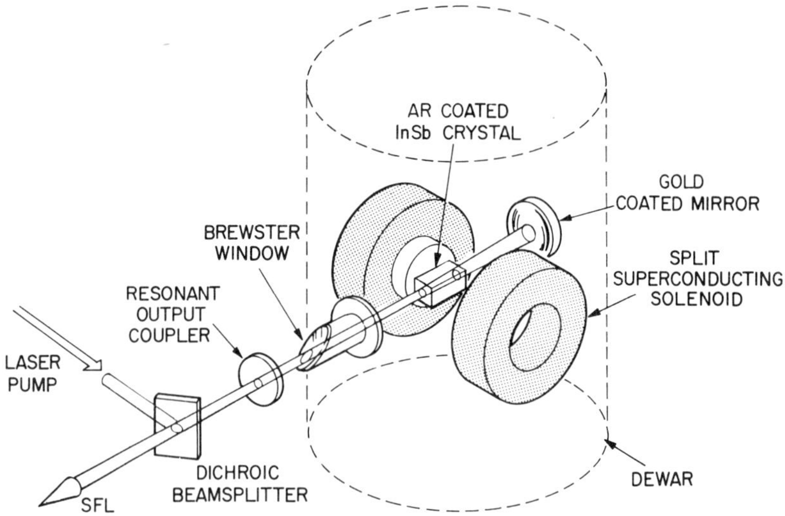

FIG. 35. Schematic of external cavity spin-flip laser.

Bounce-mode operation was effectively eliminated by the use of an external cavity on an antireflection coated InSb crystal where only the initial relaxation oscillations were observed with no subsequent buildup of a secondary peak. Figure 35 shows the schematic arrangement for the external-cavity InSb spin-flip Raman laser. The InSb crystal ($n = 1 \times 10^{16}$ cm^3) was heat sunk near liquid helium temperature in a split superconducting magnet. The output coupler was a resonant reflector made from either ZnSe or germanium which limited the number of longitudinal modes to one or two. Figure 36 shows both the unmodelocked CO_2TEA laser pump pulse and the spin-flip output pulse from the external cavity device. The saturated output energy of 70 microjoules measured from the external cavity spin-flip laser is what one would expect if half the spins in the mode volume were flipped. It is clear that most of the pump laser intensity was unnecessary to extract the 70 microjoules in this experiment. In this case, the real power conversion efficiency is

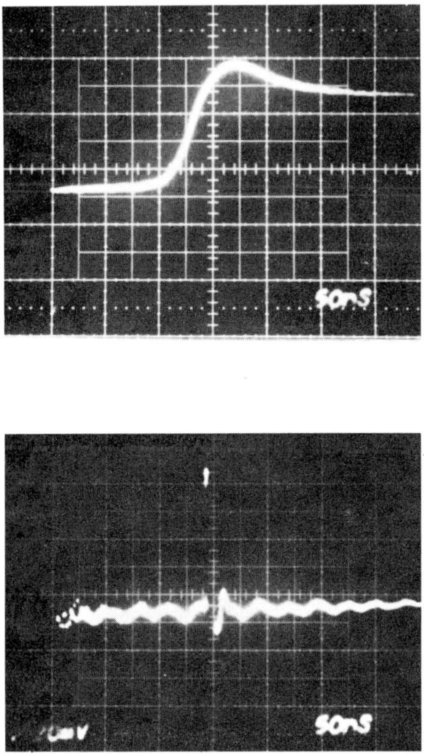

FIG. 36. Transient output of an external cavity InSb spin-flip laser using CO_2 TEA laser excitation. From Ref. 31.

very high and the implication is that a high average power, high efficiency spin-flip laser could be constructed by using a short pulse, high repetition rate pump laser. The pulse repetition rate would be limited by the spin relaxation time and the intraband relaxation time for electrons. The average power output would still be limited by the lattice, but several watts is not an unreasonable estimate. It is clear that external cavity operation is necessary to control effectively the output of either a pulsed or a cw spin-flip Raman laser, for maximum energy output as well as for spectral purity.

Acknowledgements

Many of the pioneering efforts described here were carried out by my colleagues at Lincoln Laboratory. In particular, I would like to thank Dr. S. R. J. Brueck with whom most of the spin-flip experiments were performed.

References

1. A. Mooradian and H. Y. Fan. *Phys. Rev.* **148**, 873 (1966).
2. A. Mooradian and H. Y. Fan. *In* "Proceedings of the Symposium on Radiative Recombination in Semiconductors", Academic Press, New York (1964).
3. For a review on tunable semiconductor lasers, see I. Melngailis and A. Mooradian. *In* "Laser Applications to Optics and Spectroscopy" (S. Jacobs, M. Sargent, M. Scully and J. Scott, Eds), Addison Wesley Company (1975).
4. W. R. Bennett, Jr. *Appl. Optics Suppl.* **1**, 24 (1962).
5. T. C. Harman. *In* "The Physics of Semimetals and Narrow-Gap Semiconductors", Pergamon Press, New York, p. 363 (1970).
6. J. F. Butler. *Solid State Commun.* **7**, 909 (1969).
7. J. M. Besson, J. F. Butler, A. R. Calawa, W. Paul and R. H. Rediker. *Appl. Phys. Lett.* **7**, 206 (1965).
8. A. S. Pine, C. J. Glassbrenner and J. A. Kafalas. *J. Quantum Electron.* **9**, 800 (1973).
9. S. H. Groves, K. W. Nill and A. J. Strauss. *Appl. Phys. Lett.* **25**, 331 (1974).
10. H. D. Edmunds and A. W. Smith. *IEEE J. Quantum Electron.* **6**, 356 (1970).
11. R. Ludeke and E. P. Harris. *Appl. Phys. Lett.* **20**, 499 (1972).
12. J. A. Rossi, S. R. Chinn and H. Heckscher. *Appl. Phys. Lett.* **23**, 25 (1973).
13. H. Heckscher and J. A. Rossi. *Appl. Optics* **14**, 94 (1975).
14. M. Nakamura, A. Yariv, H. W. Yen, S. Somekh and H. L. Garvin. *Appl. Phys. Lett.* **22**, 515 (1973); M. Nakamura, H. W. Yen, A. Yariv, E. Garmire, S. Somekh and H. L. Garvin. *Appl. Phys. Lett.* **23**, 224 (1973); H. W. Yen. M. Nakamura, E. Garmire, S. Somekh and A. Yariv. *Opt. Comm.* **9**, 35 (1973).
15. D. R. Scifres, R. D. Burnham and W. Streifer. *Appl. Phys. Lett.* **25**, 203 (1974).
16. A. Mooradian, A. S. Pine and N. Menyuk. In final report "Development of Tunable Semiconductor Lasers for Application to Air Pollution Measurements", NSF/RANN/IT/GI-34990/RR/75/4, (May, 1975).

17. N. Menyuk, A. S. Pine and A. Mooradian, *IEEE J. Quantum Electron.* **11**, 477 (1975).
18. R. W. Ralston, J. N. Walpole, A. R. Calawa, T. C. Harman and J. P. McVittie. *J. Appl. Phys.* **45**, 1323 (1974).
19. C. K. N. Patel and E. D. Shaw. *Phys. Rev. Lett.* **24**, 451 (1970).
20. A. Mooradian, S. R. J. Brueck and F. A. Blum. *Appl. Phys. Lett.* **17**, 481 (1970).
21. S. R. J. Brueck and A. Mooradian. *Phys. Rev. Lett.* **28**, 161 (1972).
22. S. R. J. Brueck and A. Mooradian. *Opt. Comm.* **8**, 263 (1973).
23. M. A. Guerra, S. R. J. Brueck and A. Mooradian. *IEEE J. Quantum Electron.* **QE-9**, 1157 (1973).
24. C. K. N. Patel. *In* "Proceedings of Laser Spectroscopy Conference" (R. G. Brewer and A. Mooradian, Eds), Plenum Press, New York, p. 471 (1974).
25. R. L. Aggarwal, B. Lax, C. E. Chase, C. R. Pidgeon and D. Lambert. *Appl. Phys. Lett.* **18**, 383 (1971).
26. S. R. J. Brueck and A. Mooradian. *IEEE J. Quantum Electron.* **QE-10**, 634 (1974).
27. S. D. Smith, C. R. Pidgeon, R. A. Wood, A. McNeish and N. Brignall. *In* "Proceedings of Laser Spectroscopy Conference" (R. G. Brewer and A. Mooradian, Eds), Plenum Press, New York, p. 523 (1974).
28. Data taken in collaboration with R. S. Eng and H. R. Fetterman.
29. S. R. J. Brueck and A. Mooradian. *Opt. Comm.* **18**, 539 (1976).
30. S. R. J. Brueck and A. Mooradian. *Appl. Phys. Lett.* **18**, 229 (1971).
31. Data taken in collaboration with S. R. J. Brueck.
32. E. D. Hinkley, K. W. Nill and F. A. Blum. *In* "Laser Spectroscopy of Atoms and Molecules" (H. Walther, Ed.), Springer-Verlag (1976).

6. Dye Lasers

H. WALTHER

Physik-Department, Universität München, München, West Germany

The following topics were discussed in lectures given by Professor Walther.

I PRINCIPLES OF THE DYE LASER
- A. The Laser Process and Critical Inversion
- B. Tuning
- C. C.W. Operation
- D. Frequency Stabilization

II APPLICATIONS TO ATOMIC SPECTROSCOPY
- A. Selective Quenching, Stepwise Excitation, Excited State Spectroscopy
- B. High Resolution Spectroscopy
  - (a) *Optical Pumping, Double Resonance, Level Crossing, Quantum Beats*
  - (b) *Fluorescence Line Narrowing, Saturation Spectroscopy, Two Photon Absorption*

III ISOTOPE SEPARATION USING LASERS
- A. Beam Deflection
- B. Photochemical Separation
- C. Two-step Photoionization and Dissociation

IV ATMOSPHERIC MEASUREMENTS BY MEANS OF LASERS
- A. Light Scattering Processes in the Atmosphere
  - (a) *Rayleigh Scattering*
  - (b) *Mie Scattering*
  - (c) *Raman Scattering*
- B. Absorption Measurements using Lasers
- C. Lidar

7. Ultrashort Pulses in Nonlinear Optics

W. KAISER and A. LAUBEREAU

Physik-Department, Technische Universität München, München, West Germany

I. INTRODUCTION

During the past years ultrashort light pulses have received increasing attention. A substantial number of papers have been published on the generation and the analysis of mode-locked picosecond light pulses. Special attention has been given to the study of important pulse parameters such as pulse duration and pulse shape. Several new experimental techniques have been developed for these investigations. It is now well established that the generation of well-defined single picosecond pulses requires careful experimentalists. The reproducible production of ultrashort pulses turned out to be more difficult than envisioned in early publications[1].

Ultrashort light pulses offer various new possibilities and a series of advantages for work in nonlinear optics. Most important is the potential to study dynamical processes on a time scale of 10^{-12} sec. It is now possible to measure important material parameters (e.g. certain relaxation times in the condensed phase) for which experimental techniques did not exist previously. Short light pulses can be generated at very high peak intensities.

Since the efficiency of nonlinear optical processes increases strongly with light intensity, the observation of a variety of nonlinear optical processes is readily possible with pulsed excitation. Of practical interest is the high conversion efficiency of pulses of fixed laser frequency to new pulses with tunable frequencies. In a number of cases, ultrashort light pulses allow the discrimination between different nonlinear processes, favouring those with short time constants. For instance, stimulated Raman scattering has a larger effective gain than stimulated Brillouin scattering for picosecond pulses while the reverse holds for nanosecond pulses[2]. Furthermore, ultrashort light pulses are advantageous when solids are subjected to high light intensities. The shorter build-up time for optical breakdown reduces surface and bulk damage, permitting the application of still higher electromagnetic fields.

Nonlinear optics and ultrashort pulses are linked in a variety of ways. On the one hand, nonlinear optical techniques are essential for pulse generation, for pulse shaping and for different ways of pulse analysis; on the other hand the high intensities of ultrashort light pulses give rise to strong light-matter interaction with resulting nonlinear phenomena. In fact, it is frequently difficult to avoid nonlinear optical processes when working with intense short light pulses. The knowledge of the different nonlinear effects is necessary for a full understanding of the specific experimental situation. (See for instance the severe intensity distortion in self-focussed light beams.)

In this chapter, we discuss a series of nonlinear processes which represent interesting examples of what can be done with ultrashort light pulses. For additional publications the reader is referred to a recent more extensive review article on part of the subject[3].

The generation of picosecond pulses is discussed first (Section II). The following Section III gives a short account of various techniques of pulse analysis. Several remarks are made concerning the state of the art of single picosecond pulses. In Section IV, results of the nonlinear generation of new frequencies are presented. Section V is concerned with a broad scope of applications in physics and chemistry. Some concluding remarks are given in Section VI.

II. Generation of Picosecond Pulses

In recent years, solid state mode-locked lasers have found wide applications. In particular Nd:glass has proven to be an attractive amplifier material where peak power values of 10^8 W are readily obtained from the oscillator. Multiple amplifier configurations provide power values exceeding 10^{12} W.

We first discuss a model of the pulse generation in a passively mode-locked laser system. This model is due to Letokhov[4] and has been extended by several authors[5]. In Figure 1 the set-up of a mode-locked laser is depicted schematically. The essential components are: the laser cavity, the amplifying laser rod, and the dye cell containing the nonlinear absorber. The dye cell must be contacted to one mirror in order to eliminate multiple pulse trains[6]. Concave mirrors are frequently used; they reduce the stringent requirements of laser stability and facilitate optical alignment. A small aperture suppresses higher order transverse cavity modes and generates TEM_{00} mode emission. It is important that longitudinal mode selection is carefully avoided. Brewster angles at the laser rod and wedged plates for the dye cell and optical components are strongly recommended.

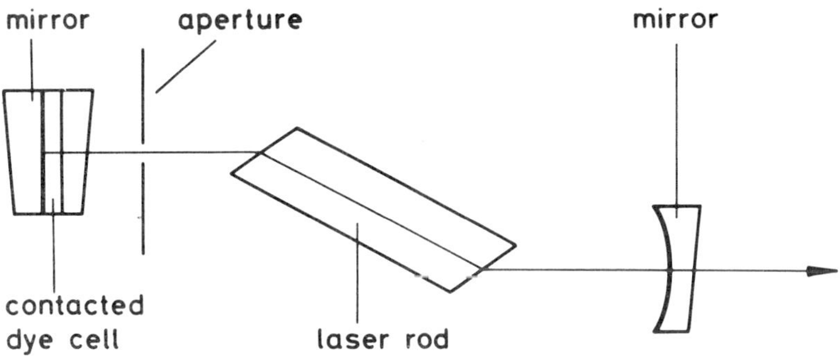

FIG. 1. Schematic of a passively mode-locked laser.

The generation of a train of picosecond pulses is readily discussed in the following steps: a linear amplification stage, the nonlinear pulse formation and the final pulse destruction. At the beginning, spontaneous fluorescence of the laser transition is linearly amplified by the optically-pumped laser rod. On account of the large fluorescence bandwidth $\Delta\omega$ a large number of longitudinal modes are initially excited. For a typical cavity length of ~ 1.5 m one estimates $\sim 10^5$ cavity modes in a laser system with $\Delta\omega \simeq 200$ cm^{-1}. During this initial stage, the population inversion of the laser transition and the transmission of the bleachable dye are not affected by the radiation.

The interference of the laser modes with random phase relations leads to fluctuations of the light intensity. The light intensity is periodic with respect to the cavity round-trip time $T = 2L/c \simeq 10^{-8}$ sec. While the total number of fluctuations is large, there are a small number of intensity peaks which exceed the average intensity significantly. The most intense radiation peak will be selected in the subsequent nonlinear stage. The following numbers are typical for the linear amplification processes of a Nd:glass laser. At a net gain of several per cent per cavity round-trip the linear stage comprises $\sim 10^3$ cavity transits corresponding to a build-up time of 10^{-5} sec. The light intensity rises by many orders of ten to approximately 10^7 W/cm^2.

At this intensity level the nonlinear pulse formation starts. The intensity peaks in the laser cavity approach values of the saturation intensity I_s of the saturable absorber and the development of the laser radiation is determined by intensity-dependent cavity losses. In Fig. 2 the nonlinear transmission of the dye cell is plotted as a function of light intensity. In the vicinity of I_s the transmission increases with intensity. Under favourable conditions the final transmission is close to one, i.e. the dye is completely transparent.

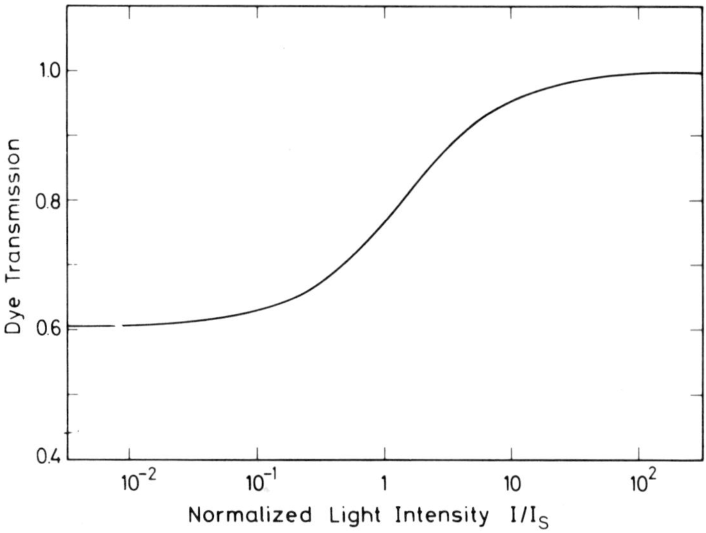

FIG. 2. Nonlinear transmission of the saturable absorber verus light intensity in units of the saturation intensity I_s of the dye. An initial transmission of 61% is assumed.

A second important parameter of the bleachable absorber is the recovery time τ_D of the ground state population which is equal to the lifetime of the excited state in a two level system. In solid state laser systems very short relaxation times τ_D are listed for four commonly used dyes, (Table I). In Nd:glass laser systems one works with saturation intensities of $I_s \simeq 5 \times 10^7$ W/cm^2. The short relaxation time of several picoseconds returns the dye very quickly to the high absorption state after the intense pulse has passed the dye cell.

TABLE I. Dye parameters

Dye	Kodak No. 9740[7]	Kodak No. 9860[7]	DDI [8]	Crypto-cyanine [9]
I_s(W cm^{-2})	4×10^7	5.6×10^7	2×10^7	5×10^6
τ_D(ps)	8.3	9.3	14	110
laser	Nd	Nd	ruby	ruby

The situation is somewhat different in flashlamp pumped dye lasers. Detailed investigations are available of the pulsed Rhodamine 6G dye laser which uses a dilute solution of 3,3′-diethyloxadicarbo-cyanine iodide (DODCI) for reproducible mode-locking[10,11]. In this laser system the combined action of amplifier and absorber saturation produces trains of ultrashort pulses[12] in spite of a long recovery time τ_D of the nonlinear absorber (for DODCI $\tau_D \simeq 10^{-9}$ sec[13]). The main pulse saturates the amplifier system to such an extent that the following satellite pulses find no gain for continuing growth.

In solid state systems the amplifier saturation is not significant. High peak intensities of the order of 10^9 W/cm^2 are obtained at the maximum of the pulse train. As a result of these high intensity values various nonlinear effects occur which strongly effect the generated short pulse. The nonlinear refractive index of the laser medium and the dye solvent gives rise to substantial frequency broadening (self-phase modulation). During the following passes of the pulse within the cavity, the frequency modulated pulse is decomposed on account of the colour dispersion of the laser components. It is now well established that in high-gain laser systems the important pulse parameters, pulse duration, pulse shape, peak intensity, and frequency bandwidth vary from the beginning to the end of the pulse train[14–17]. For this reason the application of the whole mode-locked pulse train should be avoided. It was found that pulses positioned several hundred nanoseconds before the maximum of the pulse train display optimum properties. Because of the small intensity of several 10^8 W/cm^2 one finds negligible frequency modulation and dispersion effects[16,18]. For selection of a single pulse we have used successfully a fast Kerr shutter in conjunction with a laser-triggered spark gap[19]. Adjusting the threshold of the spark gap, a light pulse of definite intensity is cut from the leading part of the pulse train. Figure 3 shows an oscilloscope display of the pulse train and the selected pulse. Pictures of this sort do not give the required time resolution, they represent necessary but not sufficient conditions for useful picosecond pulses.

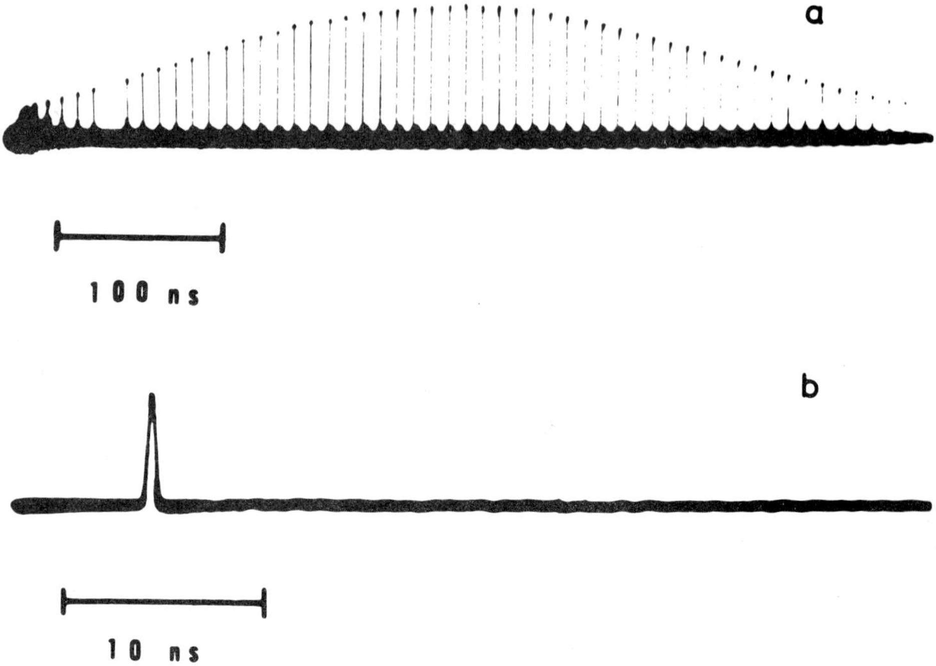

FIG. 3. Oscilloscope display of the pulse train (a) and of the selected pulse (b) rise time of the detection system is not sufficient to resolve the picosecond pulses.

III. PULSE ANALYSIS

There are a number of diagnostic techniques which provide information of various pulse parameters. In this chapter, three experimental systems will be briefly discussed: (1) the commonly used two-photon fluorescence technique (TPF) which gives values of the pulse duration, (2) a probing method using stimulated Raman scattering which provides information of the pulse shape, and (3) streak camera measurements which give direct pictures of the individual pulse.

1. In the two-photon fluorescence technique[20] two identical replicas of the pulse overlap in an organic dye solution which exhibits two-photon absorption (see left side of Fig. 5). A photograph is taken of the generated fluorescence from the side; it shows enhanced fluorescence where the individual pulses overlap with each other. The spatial distribution of the fluorescence track is directly related to the second-order correlation function

allowing an estimate of the pulse width with a single laser shot. While the experimental technique appears to be very simple at a first glance, serious pitfalls were accounted at the beginning in the interpretation of the observed photographs. It is now well established that careful alignment and quantitative evaluation of the photographs is required to obtain meaningful information. A smooth fluorescence peak, with a contrast ratio of 3:1 in relation to the background, indicates good mode-locking of the system[21]. The fluorescence profile is very insensitive to the form of the incident pulse, it gives no information of a possible asymmetry of the pulses investigated. In Fig. 4 a microdensitometer trace of a TPF photograph is presented[16]. The calibration steps on the left suggest an excellent contrast ratio of 3:1. The width of the fluorescence peak indicates a pulse duration of 6 psec.

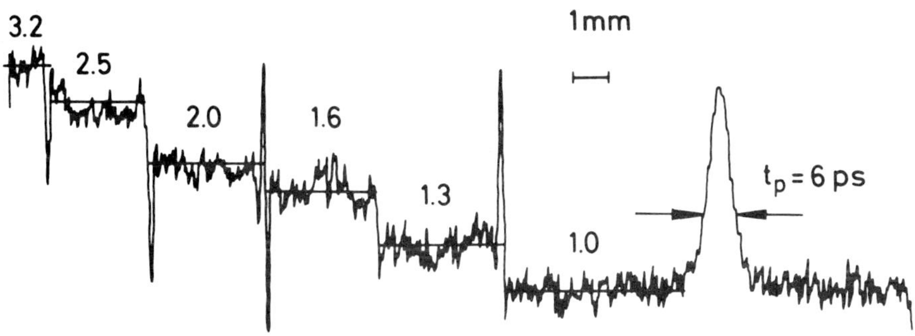

FIG. 4. Two-photon fluorescence trace of a single psec pulse. Calibration steps of the intensity scale are indicated on the left.

2. Higher-order nonlinear techniques have been devised to analyse the shape of the picosecond pulses[22–24]. In one method[22,23] a first pulse generates a rapidly rising and decaying material excitation (a phonon pulse) via stimulated Raman scattering. This shortlived phonon excitation interacts with a second pulse giving rise to a measured (anti-Stokes) scattered signal. With careful time delays between the phonon pulse and the second light pulse the momentary intensity of the second pulse can be measured over an intensity range of several orders of magnitude. This phonon probing technique requires a reproducible generation of pulses for an accurate analysis of the pulse shape. The major advantage of this technique is the possibility to obtain information on the pulse shape several orders of ten below the peak. It was found that our single picosecond pulses are well approximated by a Gaussian function over three orders of magnitude. The trailing edge indicates a somewhat slower decay as expected from the pulse generation process.

3. The most direct way of looking at ultrashort pulses is made possible by the development of high speed streak cameras[25,26]. Streak tubes are now commercially available with a time resolution of 3 to 5 psec (depending upon the wavelength). Figure 5 shows an experimental set-up schematically[25]. The pulses travel through an optical delay line consisting of an array of beam-splitters to produce from each incident pulse a series of subpulses with known delay between them. This delay time allows an accurate time calibration and a ready check of the linearity of the camera streak. A high gain (10^6) image intensifier tube is applied after the streak tube in order to avoid image distortion and loss of spatial resolution arising from high photocurrent densities. In Fig. 6 a microdensitometer trace of an enlarged streak photograph shows structure with a separation of 2.5 psec within each subpulse indicating the high time resolution of the system. Extensive work with streak cameras has provided valuable information on ultrashort light pulses[11]. Streak photographs revealed satellite pulses in poorly designed mode-locked laser systems, they showed clearly the pulse break-up during the later part of the pulse train and they demonstrated the evolution of picosecond pulses at an early time of formation.

As an example of the present state of the art of single ultrashort pulses a number of important pulse parameters are listed in Table II. These data were obtained after a detailed study of our single pulses switched out approximately 300 nsec before the peak of the pulse train of a Nd:glass laser system emitting at $\nu = 9455\ \text{cm}^{-1}$ (see Fig. 3). We have tested our pulses by various

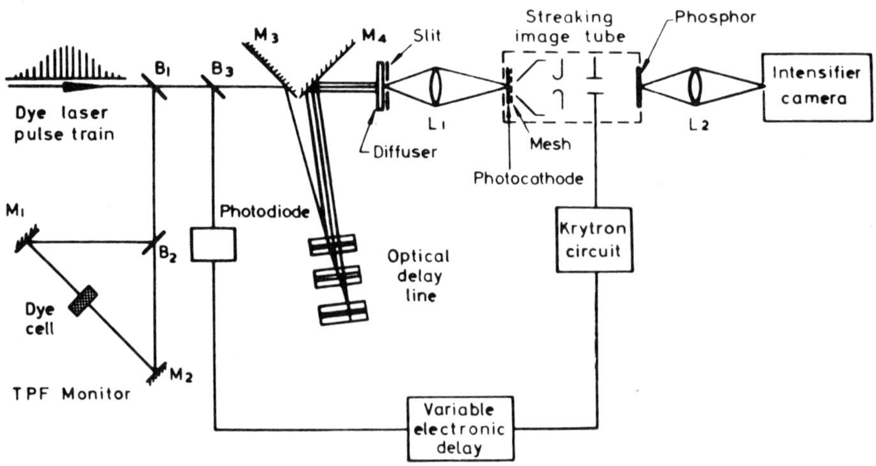

FIG. 5. Experimental set-up for the direct measurement of streak camera resolution employing a mode-locked dye laser (from Ref. 25).

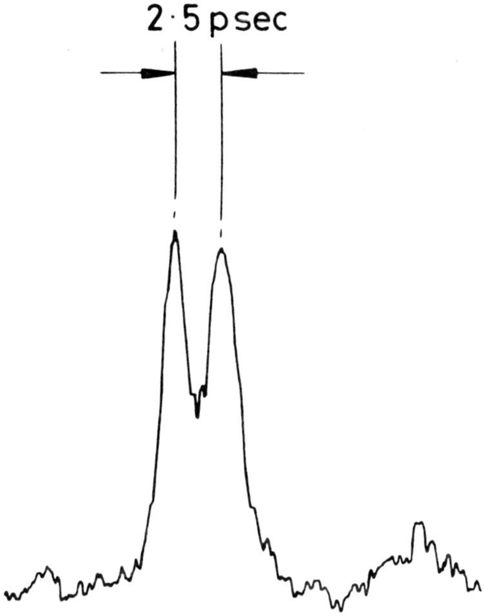

FIG. 6. Microdensitometer trace of an enlarged streak photograph revealing structure with a separation of 2.5 psec within each sub-pulse (from Ref. 11).

diagnostic techniques in particular using the three methods discussed above. The most important pulse parameters are : (1) a reproducible pulse duration, (2) a bandwidth limited pulse (i.e. no chirp), (3) no satellite structure, and (4) a large peak to background ratio. The experimental investigations discussed in a later part of this paper were performed with pulses of parameters according to Table II.

Recently, considerable progress has been made in the design and the performance of cw mode-locked dye lasers[27]. An experimental arrangement

TABLE II. Properties of a single picosecond pulse switched from the leading part of the pulse train

Pulse duration	$t_p = 6$ ps
Contrast ratio	3 ± 0.1
Frequency width	$\Delta\tilde{\nu} = 2.9\ \mathrm{cm}^{-1}$
$t_p \times \Delta\nu =$	0.6
Peak to background ratio	$\sim 10^4$
Pulse intensity	$5 \times 10^8\ \mathrm{W\,cm}^{-2}$
Pulse energy after amplification	4×10^{-3} J

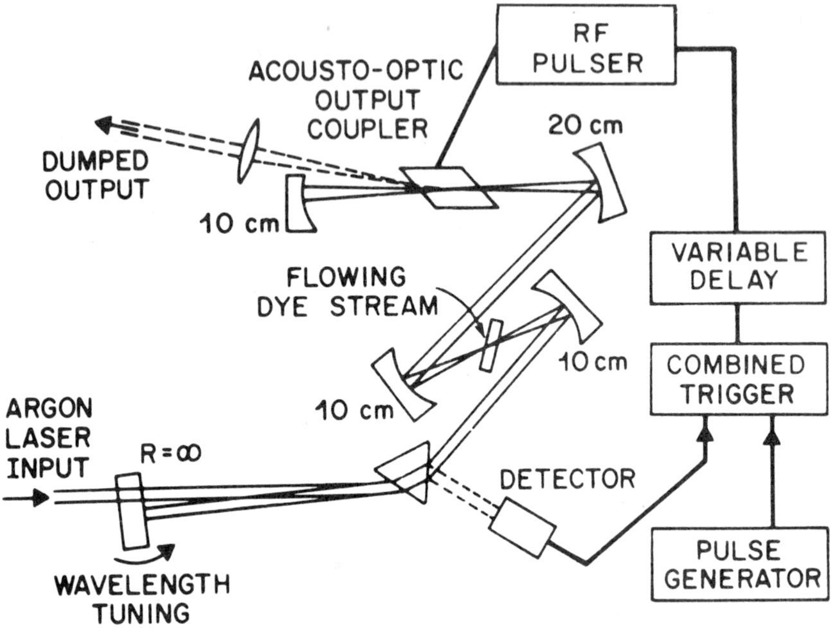

FIG. 7. Mode-locked cw dye laser with arrangement for dumping intracavity pulses (from Ref. 27).

is shown schematically in Fig. 7. An acousto-optic output coupler allows to dump one pulse every 10^{-3} sec. Pulses with a duration of one psec and with peak power values of 100 Watt were generated in a TEM_{00} mode. Further pulse compression to 0.3 psec has been reported[28]. The experiments performed so far used Rhodamine 6G as laser dye over a frequency range of 640 to 600 nm. While the high repetition rate and the tunability of the cw mode-locked dye laser are very valuable for certain applications, the limited peak power precludes many nonlinear optical investigations. On the other hand the fixed frequencies of solid state mode-locked systems are a serious handicap. In particular for spectroscopic investigations tunable light sources are required. Work proceeding in this direction will be discussed in the next section.

Some final remarks should be made on the shape of our pico-second light pulses. It will be apparent later that the pulse shape is of importance for the quality of certain applications. At the beginning of this section it was shown that our single mode-locked laser pulses are well approximated by a Gaussian function. Additional pulse shaping is possible with the help of a saturable absorber, e.g. a dye solution. In Fig. 8 calculated curves are presented of the

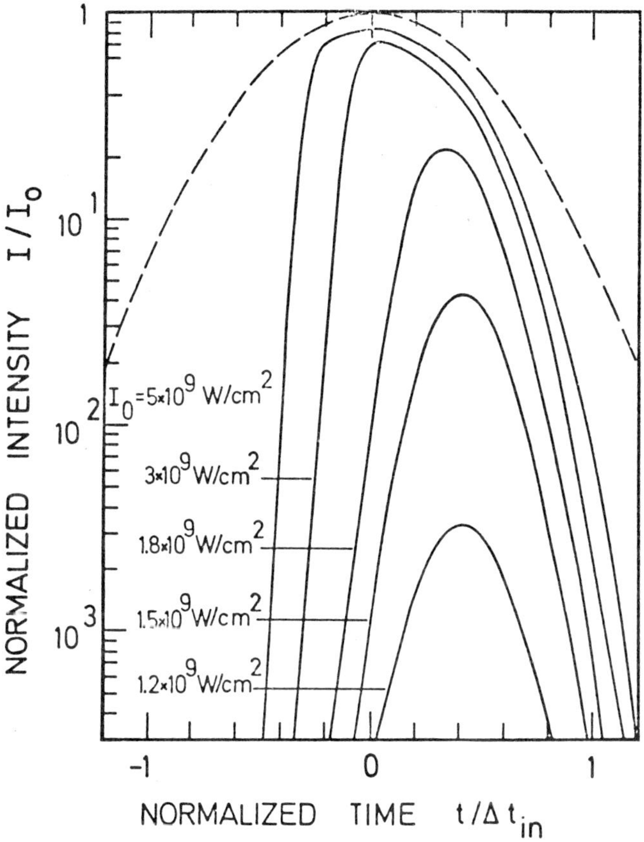

FIG. 8. Changes of the pulse shape after a single pass through a saturable absorber. Broken curve: normalized intensity of the input pulse (Δt_{in} = 8 psec Gaussian pulse shape). Solid curves: normalized intensity of the transmitted pulse for several values of the input peak intensity I_0. Dye parameters: $T_0 = 10^{-7}$, $\sigma = 1.84 \times 10^{-16}$ cm^2, $\tau = 9.1$ psec (from Ref. 29).

single path of a Gaussian input pulse (broken curve) through a saturable dye of well-defined practical parameters[29]. It is readily seen from the figure that the shape of the transmitted pulse is substantially altered compared to the input pulse depending upon the input peak intensity I_0. Of particular interest is an input peak intensity of $I_0 = 5 \times 10^9$ W/cm^2. The leading part of the transmitter pulse is strongly absorbed by the bleachable dye while the peak and trailing part of the pulse are less effected. A pulse shortening by a factor of approximately two is readily obtained by a single passage[29]. We note the asymmetry of the pulse with a steeply rising leading edge.

Experimental investigations of pulse shortening were in agreement with calculations. Of special interest is a multiple transit system where the pulse is amplified to an intensity level for optimum pulse shortening everytime it has passed an absorber cell. In this way pulse reduction from 6 psec to 0.5 psec was achieved after four transits through an absorber amplifier system.

IV. Generation of Ultrashort Pulses at New Frequencies

As a first example in this section we discuss three-photon parametric amplification involving χ_2, the second-order nonlinear susceptibility[30]. A laser pulse enters a nonlinear crystal and generates signal and idler pulses with $\omega_L = \omega_S + \omega_i$. In this way we obtained a useful infrared light source at a picosecond time scale[31]. The same technique has subsequently found application in the visible part of the spectrum.

In Fig. 9 a schematic of our experimental system is depicted which generates intense tunable pulses in the infrared[31]. A single picosecond pulse is switched out of the pulse train of a mode-locked Nd:glass laser and amplified by double passage through an optically-pumped Nd:glass rod. Signal and idler pulses are generated when the intense laser pulse travels through a nonlinear $LiNbO_3$ crystal. Knowing the dielectric properties of $LiNbO_3$ it is possible

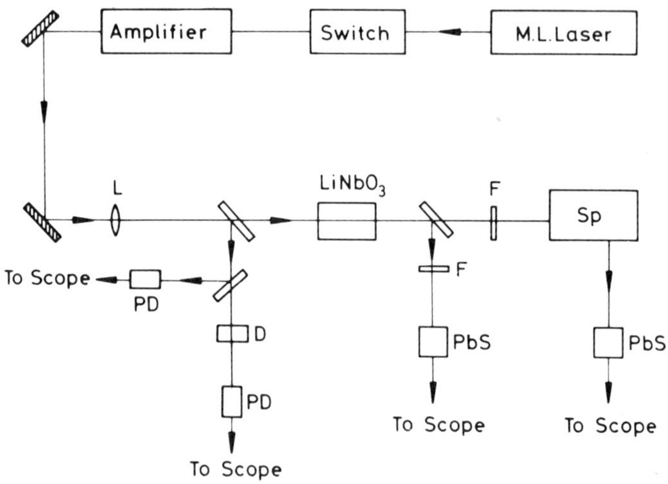

FIG. 9. Schematic of the experimental system: generator crystal $LiNbO_3$, lens, L; filter, F; I.R. spectrometer, Sp; I.R. detector, PbS; nonlinear absorber, D; fast photodiode, PD (from Ref. 31).

to predict the frequencies as a function of crystal orientation. Experimentally determined frequencies were compared with the calculated phasematching curve for $LiNbO_3$ at 293 K. The agreement between experimentally observed frequencies and the calculated curve is excellent[31]. The frequency range extends from 2500 cm^{-1} to 7000 cm^{-1}. The lower frequency limit is determined by beginning infrared absorption of the $LiNbO_3$ crystal. In addition to the tuning range, the conversion efficiency of input power into signal power is of importance for practical applications. The signal power is plotted as a function of input peak intensity in Fig. 10: It is readily seen from the figure that the signal power rises first exponentially over several orders of ten and begins to saturate for very high input peak intensities[31]. Power conversion

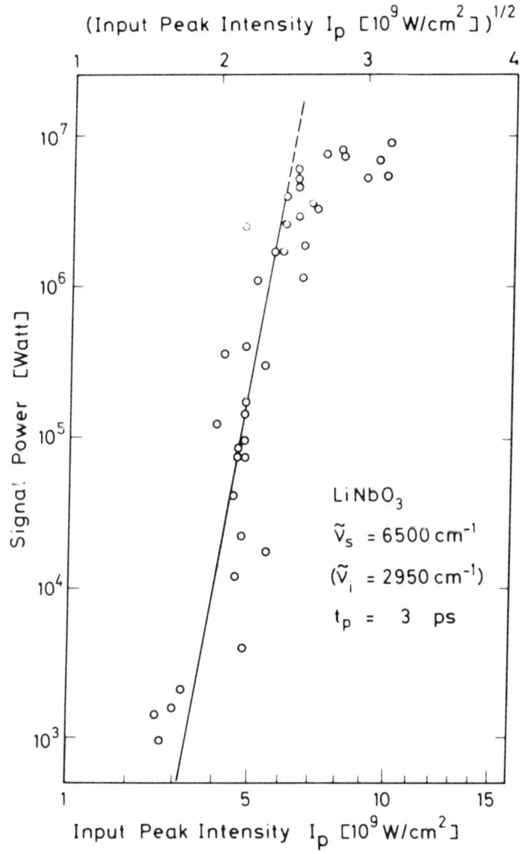

FIG. 10. Measured power of signal pulses at 6500 cm^{-1} as a function of pump intensity I_p. Line calculated using experimental parameters (from Ref. 31).

efficiencies of several per cent have been achieved. This high power conversion makes our travelling-wave parametric generator very attractive for the generation of picosecond pulses of tunable new frequencies.

It should be emphasized that practical conversion efficiencies require very high input peak intensities. Nanosecond pulses or whole trains of mode-locked pulses damage the crystal at these high power levels. Working with single pulses of several picoseconds duration we found no crystal damage even after many hundred measurements.

In Table III various important parameters of our infrared pulses are summarized[32]. The pulse duration of the infrared pulses is shorter than the duration of the pump pulse on account of the highly nonlinear generation process. The slopes of the infrared pulses are expected to rise very steeply with time. This effect is of interest for a high time resolution (see below).

TABLE III. Pulse properties of tunable infrared picosecond pulses

Tunability	2500 to 7000 cm^{-1}
Frequency width	40 cm^{-1}
Divergence	1×10^{-2} rad
Intensity	$\sim 10^9$ W cm^2
Energy conversion	$>5\%$
Pulse duration	2–3 psec
I.R. photons	10^{15}

We now turn to a nonlinear optical process which allows the generation of very broad spectra extending from the infrared to the ultraviolet. In recent years the problem of "superbroadening" has been discussed by several authors[33]. Various explanations have been offered for the phenomenon. For instance, self-phase modulation, stimulated Raman scattering, and optical dielectric breakdown were considered to be responsible for the broad spectra. In most cases the experimental data are difficult to analyse because important pulse parameters such as pulse duration, pulse shape, input bandwidth, and peak intensity were not reported with sufficient accuracy. The question of self-focussing was frequently not investigated.

In the experiments reported here, we worked with well-defined single laser pulses (see above). We are certain that self-focussing and resulting beam distortion were not present in our experiments at $\nu_L = 9\,455$ cm^{-1}. A detailed analysis of our experimental data suggests that the broad frequency spectrum is generated by parametric four-photon processes[34]. Single frequency (infrared) resonances of $\chi^{(3)}$ are important for the signal generation at ω_3

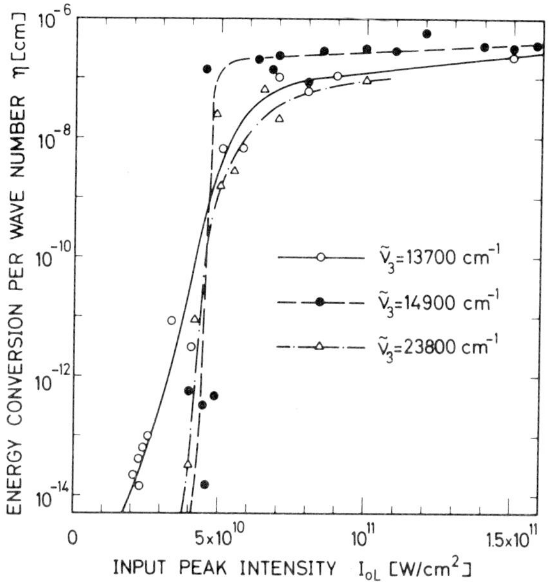

FIG. 11. Energy conversion η of laser emission ($\tilde{\nu}_L = 9455\ \mathrm{cm}^{-1}$) into short wavelength radiation at $\tilde{\nu}_3$. The curves below $\eta = 10^{-9}$ are calculated for $\tilde{\nu}_3 = 13\,700$ and $14\,900\ \mathrm{cm}^{-1}$ (from Ref. 34).

in the primary process $\omega_L + \omega_L \rightarrow \omega_3 + \omega_4$. In Fig. 11 the energy conversion per wavenumber is plotted versus input peak intensity for different frequencies. The rapid rise of energy conversion over seven orders of magnitude within a small interval of input intensity suggests an amplification process of large gain. At higher input intensity a saturation of the output signal sets in. In Fig. 12 the energy conversion[34] is presented for a frequency range of 10 800 to 20 000 cm^{-1}. Four curves of input intensities between 2×10^{10} and 10^{11} W/cm^2 are shown. The striking structure of the spectra should be noted. These spectra and their input dependence are explained by the single-photon resonance structure of $\chi^{(3)}$ of water[34]. In fact, the minima in the energy conversion curves are fully accounted for by the major infrared absorption bands of water.

Of practical interest are the spectra at very high input values[35]. In Fig. 13 the spectral distribution of parametric light is plotted versus frequency for two input intensities of 1×10^{11} and 5×10^{11} W/cm^2. The spectra extend from approximately 4000 cm^{-1} to 40 000 cm^{-1}. The energy conversion is strongest between ω_L and $2\omega_L$ where the primary four-photon process occurs, $\omega_L + \omega_L \rightarrow \omega_3 + \omega_4$. The generated signal wave ω_3 is strong enough to act

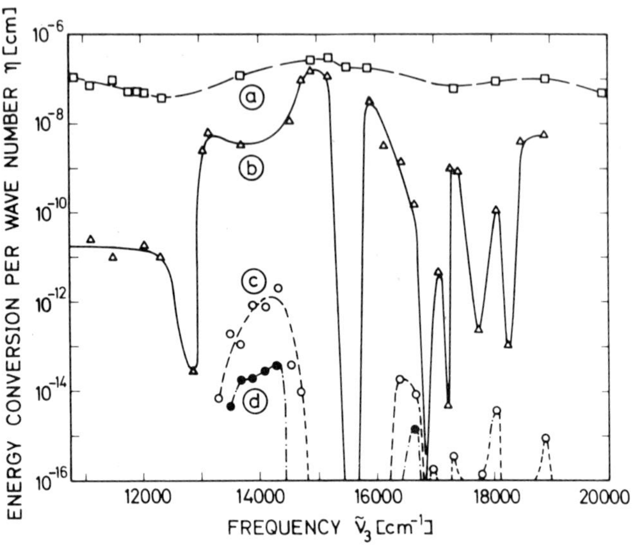

FIG. 12. Experimental energy conversion η versus signal frequency $\tilde{\nu}_3$ for four laser input peak intensities: (a) 1×10^{11}; (b) 5×10^{10}; (c) 3×10^{10}; (d) 2×10^{10} W/cm^2 (from Ref. 34).

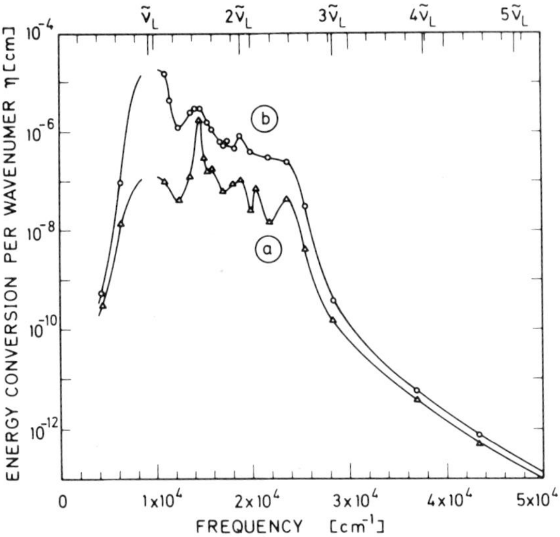

FIG. 13. Energy conversion of laser light. Single picosecond light pulse of $\Delta t_L = 6$ psec, $\lambda_L = 1.06\ \mu$m in water ($l = 2$ cm) (a) $I_{0L} = 1 \times 10^{11}$ W/cm^2 and (b) $I_{0L} = 5 \times 10^{11}$ W/cm^2 (from Ref. 35).

as a broad band pump for secondary processes. Light at frequencies exceeding $2\omega_L$ is produced by frequency conversion $\omega_L + \omega_L + \omega_3 \rightarrow \omega_4'$ and by higher-order parametric four-photon processes $\omega_L + \omega_3 \rightarrow \omega_3'' + \omega_4''$. The frequency conversion process becomes less efficient at shorter wavelengths because of increased phase-mismatch. Furthermore, the efficiency of the parametric four-photon process decreases with increasing order. The fairly smooth spectrum of Fig. 13 is useful for the study of ultrafast dynamical processes, e.g. in liquids.

V. Applications of Picosecond Light Pulses

During the past few years, ultrashort light pulses have stimulated active research in variety of fields. On the low intensity end, mode-locked cw dye lasers offer trains of pulses for high capacity digital communication systems[36]. On the high intensity side, pulses of very high peak power ($\sim 10^{12}$ W) are able to form plasmas of high temperature (~ 1 keV)[37]. Using specially tailored light pulses compression of D-T shells has been suggested for controlled thermonuclear fusion. Work in this field is actively pursued in several laboratories throughout the world.

In this paper we are concerned with applications of ultrashort pulses of peak intensities below 10^{11} W/cm^2. Material destruction does not occur but strong interaction with matter is observed at this high intensity. In many cases the light-matter interaction is strong enough to effect drastically the pulse propagation.

Severe beam distortion results from the self-focussing action of the light pulse[33]. At high light intensities the refractive index of many substances becomes noticeably intensity dependent: $n = n_0 + n_2 I$. The normal variation of the intensity over the beam cross-section leads to an enhanced refractive index in the centre of the beam. As a result the light beam converges with decreasing beam diameter. After the pulse has travelled a certain critical distance through the medium a minimum beam diameter of the order of several 10^{-4} cm is reached[38]. Extremely high light intensities are generated in this self-focussing process which lead to additional nonlinear effects and frequently bring destruction to the material. There exists extensive literature on the self-focussing problem. The reader is referred to various review papers[33,39].

A short remark should be made concerning the pulse length. In mode-locked laser oscillators, it is readily possible to generate pulses of a duration of several picoseconds (10^{-12} sec). Light pulses as short as 0.3 psec have been achieved by subsequent nonlinear pulse reduction. A pulse duration of 3 psec corresponds to a wave packet of 100 μm length. This short pulse length requires, in a number of experiments, a consideration of group velocity of the pulse[40]. The interaction length between pulses of different frequencies is frequently limited by their group velocities.

In this chapter we wish to concentrate on a variety of dynamical processes which have improved our knowledge of rapid molecular phenomena in liquids. Several of these investigations have provided important material parameters which were not accessible by other experimental techniques. It is this latter fact which prompted the general interest in investigations with picosecond light pulses.

The following dynamic processes will be discussed here.

1. Orientational relaxation of polyatomic molecules.
2. Rate of recombination of atoms to molecules.
3. Fluorescence lifetime.
4. Ground state recovery time of dye molecules.
5. Rate of population of the lowest triplet state after singlet excitation.
6. Dephasing time of collectively excited molecular vibrations.
7. Energy relaxation of the first vibrational state.
8. Energy transfer between vibrational levels.
9. Energy relaxation after infrared excitation of defined vibrational modes.

The last four processes refer to interactions in the electronic ground state.

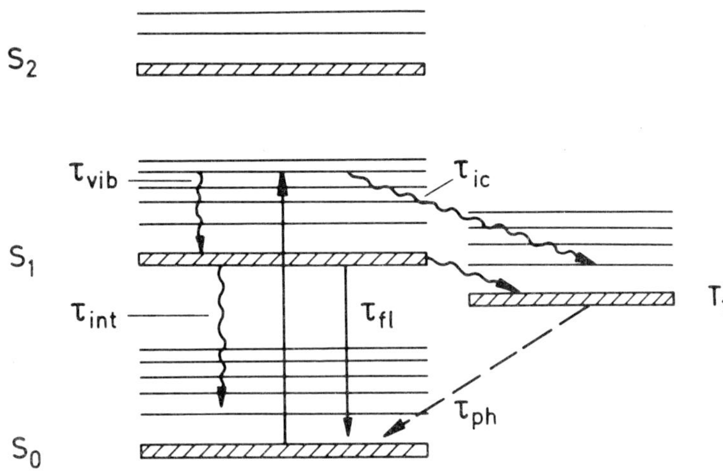

FIG. 14. Schematic of the energy diagram of dye molecules; horizontal bars represent the vibronic manifold of singlet states S_0, S_1, S_2 and of first triplet state T_1; radiating (straight lines) and radiationless transitions (wavy lines) are indicated.

Before discussing different dynamic processes listed above we present a schematic of the energy levels of a large molecule in Fig. 14. Radiating (straight lines) and radiationless transitions (wavy lines) are indicated in the figure. Absorption of light promotes the molecule to an excited electronic level of the same spin multiplicity, e.g. to the first excited singlet state. The electronic levels are split into a manifold of vibrational states represented in Fig. 14 by several horizontal lines. Several decay processes are marked: fluorescence and phosphorescence with time constants τ_{fl} and τ_{ph}, respectively; vibrational lifetime τ_{vib}, intersystem crossing τ_{ic} and internal conversion τ_{int}. Vibrational transitions in the electronic ground state are not indicated in Fig. 14.

A. ORIENTATIONAL RELAXATION OF POLYATOMIC MOLECULES: AN ULTRAFAST LIGHT GATE

Two groups of investigations should be discussed here. First, orientational relaxation measurements of small (non-absorbing) molecules and second, experiments of the relaxation of large dye molecules in various solvents.

It was found in 1964 that powerful optical pulses induce birefringence in a number of liquids[41]. The light pulse gives rise to a small preferential orientation of the anisotropic molecules parallel to the electric field. After the pulse has left the medium the molecules return to an isotropic random distribution. Typical values for the orientational relaxation times are $\tau = 2$ psec for carbondisulphide and $\tau = 32$ psec for nitrobenzene.

Optically-induced birefringence was used for the design of an ultrafast light gate with CS_2[43]. The set-up is very simple: it consists of a liquid cell between crossed polarizers (see Fig. 15). At first, any light will be blocked by the device. When an intense operating light pulse passes through the cell birefringence is

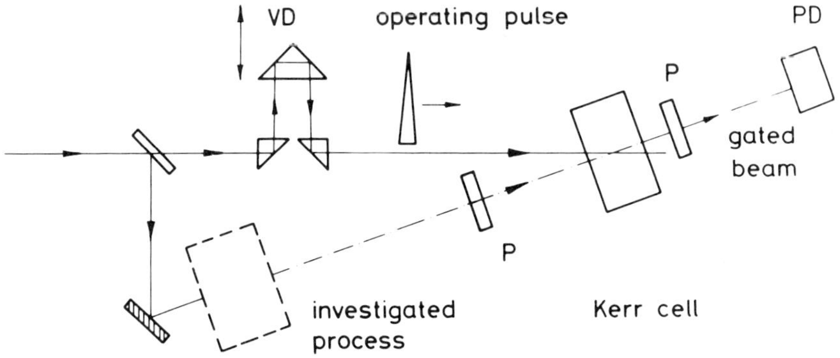

FIG. 15. Schematic of the optical Kerr shutter; the laser pulse operates the Kerr cell and excites the process under investigation, variable delay, VD; polarizer, P; photo-detector, PD.

introduced and a certain fraction (several 10^{-2}) of the light to be investigated passes the second polarizer. The shutter opening time is determined by the duration of the operating pulse and the orientational relaxation time of the medium. In Fig. 16 the transmission of a light gate is shown as a function of time for CS_2 and nitrobenzene[43]. A first short pulse opens the light gate and a second properly delayed probe pulse measures directly the closing time. For nitrobenzene the orientational relaxation time is directly deduced from the figure. The ultrafast light gate has the advantage of being useful over a wide

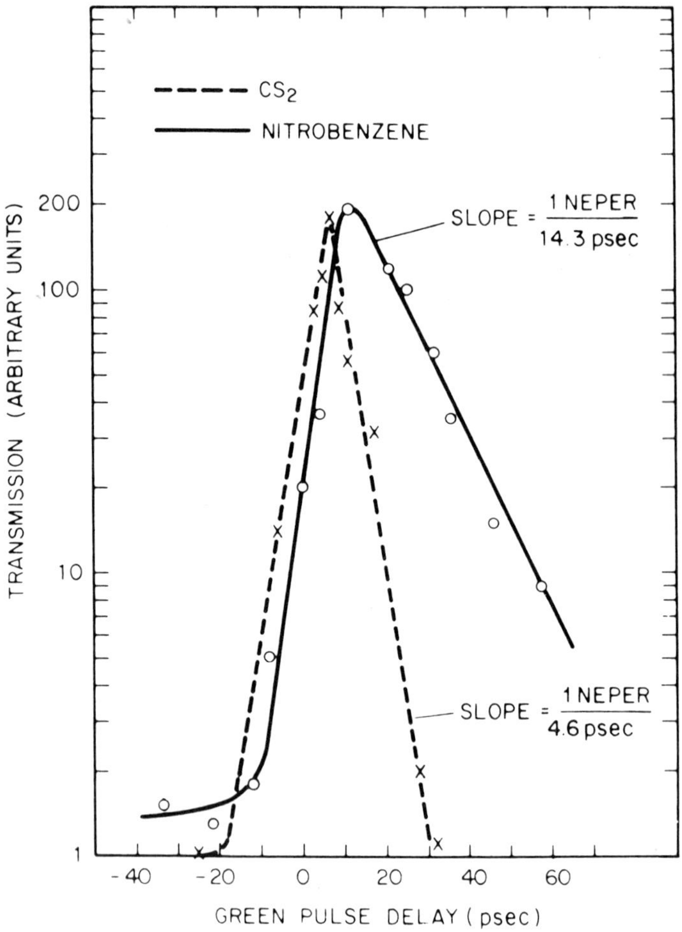

FIG. 16. Transmission curves obtained for CS_2 and nitrobenzene. Each point is the average of ~6 measurements (from Ref. 43).

frequency range where these liquids are transparent. When working with different frequencies for the operating pulse and the measured light, group velocity dispersion has to be considered[44].

The fast light shutter has proven to be a practical device for the study of rapid physical processes. For instance, fluorescence measurements[45] and studies of the self-focussing action of picosecond pulses have applied this fast light gate[42,46].

The orientational relaxation time of dissolved dye molecules was investigated by the following technique[47]. A picosecond pulse excites preferentially those molecules whose transition dipole moment has a large component parallel to the electromagnetic field. Using a second weak probe pulse the return to the isotopic equilibrium distribution was monitored by measuring the polarization dependence of the absorption. The probe transmission versus probe delay time yields the orientational relaxation time τ_{or}. Time constants were found to be proportional to the viscosity of the solution. As an example we quote $\tau_{or} \simeq 1 \times 10^{-10}$ sec for Rhodamine 6G in methanol[48]. Interaction of the dissolved molecules was studied by the same method when large dye concentrations and highly viscous solvents were used[49]. Under these conditions the oriented distribution of excited molecules relaxes via transfer of energy from excited to neighbouring molecules. This interaction was studied for Rhodamine 6G dissolved in glycerol.

B. RECOMBINATION OF ATOMS TO MOLECULES IN A LIQUID CAGE

A single pulse at 5300 Å (and 5 psec duration) excites readily iodine molecules in solution (e.g. in carbon tetrachloride)[50]. The photo-excited I_2 molecules undergo either direct dissociation or a collision-induced predissociation leading to a pair of iodine atoms. The recombination of original fragments is strongly enhanced in the liquid phase since the solvent molecules interfere with the escape of the fragments. The time dependence of the formation of I_2 from the iodine atoms (the return to the original transmission) is monitored by a weaker delayed pulse of the same frequency (see Fig. 17). Chuang *et al.*[50] found the recombination of the original partners to occur with a time constant of 140 psec in carbon tetrachloride. An estimate of the recombination of particles from different "cages" give much longer times of the order of 10^{-8} sec.

Of special interest is the direct observation of the collision-induced predissociation of the excited molecules[50]. It was found that the probe transmission continues to increase for a number of picoseconds after the intense exciting pulse has passed the sample. This result is interpreted as decay of

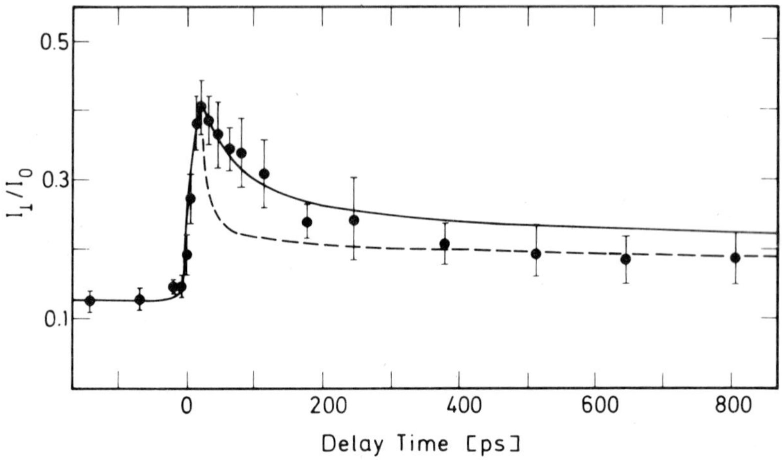

FIG. 17. Probe transmission $I_\perp/I_0$ as a function of time after excitation for iodine in CCl_4. Dashed and solid curves are calculated (from Ref. 50).

excited iodine molecules into iodine atoms. The rate constant for this collision-induced predissociation is found to be approximately 10^{11} sec^{-1}.

C. FLUORESCENCE LIFETIME

Measurements of the fluorescence lifetimes τ_{fl} and of fluorescence spectra are now possible on a picosecond time scale. Two methods should be mentioned here. In a two pulse technique, a first monochromatic pulse excites the dye molecules in a definite electronic state and a second properly delayed pulse opens the Kerr shutter (see above) for observation of the fluorescence signal[45,51,13]. More recently, fluorescence lifetimes have also been determined with an ultra-fast streak camera with a resolution limit of approximately one picosecond[52].

A word of caution should be given here. Since exciting pulses of high power are readily available, it is easily possible to excite the dye to such an extent that fast stimulated emission occurs instead of spontaneous fluorescence. The value of τ_{fl} must be independent of the intensity of excitation.

Flourescence lifetimes have been reported in a number of papers. Noteworthy are the results of Malley and Mourou[45] where the spontaneous fluorescence spectrum of Rhodamine 6G in glycerol was measured with picosecond time resolution. In Fig. 18 five fluorescence spectra are presented which demonstrate clearly the change of the emission peak and the increasing half-width as a function of time after the excitation.

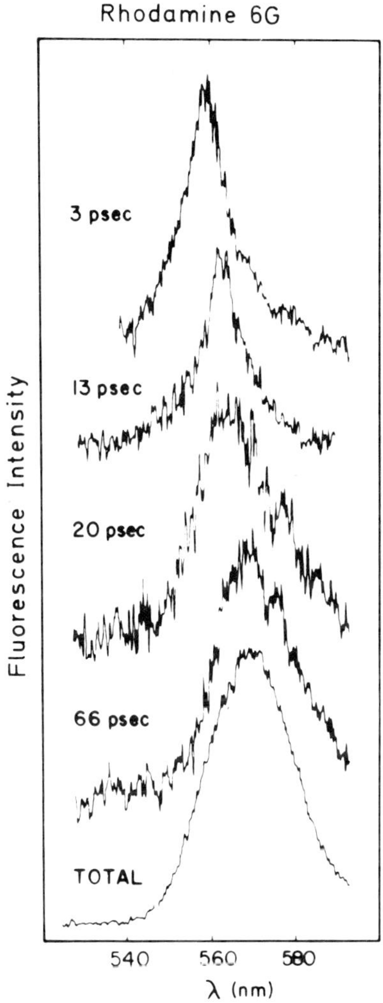

FIG. 18. Spontaneous fluorescence spectra of Rhodamine 6G in glycerol observed with several psec time resolution (from Ref. 45).

D. GROUND STATE RECOVERY TIME OF DYE MOLECULES

As pointed out above, the recovery time of the ground state population τ_{gr} of saturable absorbers (sometimes called the aperture time) is important for the performance of a passive mode-locked system. The repopulation of the ground state following optical excitation depends upon several relaxation processes.

Vibronic relaxation, internal conversion, radiative decay, and intersystem crossing contribute—in the general case—to the ground state recovery time. The value of τ_{gr} can be estimated from various material parameters, e.g. absorption cross-section and saturation intensity or fluorescence lifetime, rate of internal conversion and rate of intersystem crossing. More directly, the ground state recovery time has been measured with a two-pulse technique. A first intense pump pulse bleaches the dye and a second weak probe pulse—properly delayed with respect to the pump pulse—determines the instantaneous transmission of the medium. The probe pulse is kept at a low intensity level to avoid additional bleaching of the dye molecules. Measuring the dye transmission as a function of the probe delay, the dye recovery time τ_{gr} is directly obtained (see Fig. 19). Measurements of this sort have been made by a number of authors[53,54]. As an example, we quote the relaxation time $\tau_{gr} =$ 7 psec for a fast commercial switching dye generally used in Nd:glass mode-locked systems[54]. It should be noted that there are experimental errors as large as a factor of two between different investigations. The use of picosecond pulse trains in earlier measurements instead of single pulses might be the source for experimental uncertainties.

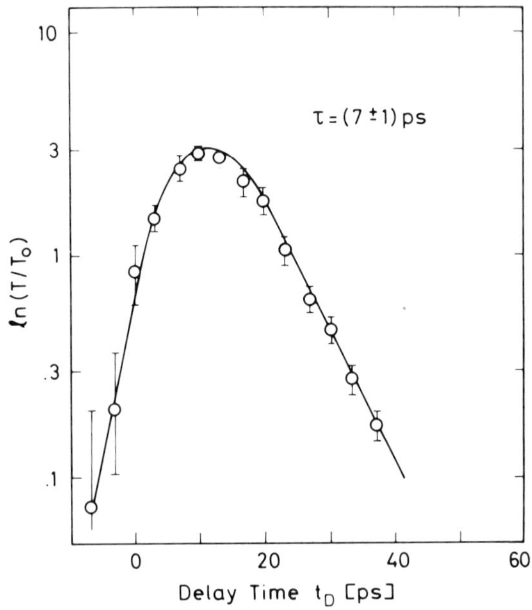

FIG. 19. Logarithm of the measured transmission as a function of delay t_D of the probe pulse (from Ref. 54).

E. RATE OF POPULATION OF THE LOWEST TRIPLET STATE FOLLOWING SINGLET EXCITATION

In molecules with a small energy gap between the lowest excited singlet and the triplet states, effective energy transfer from the singlet to the triplet system has been observed[55]. In order to get more quantitative information of this energy transfer process the build-up of triplet-triplet absorption at 0.53 μm was measured during and subsequent to the excitation by 5 psec pulses of 0.35 μm. Using this technique, Hochstrasser *et al.* investigated benzophenone, where a considerable amount of spectroscopic data is available[55]. In Fig. 20 the growth of absorption due to triplet formation is shown for benzophenone in benzene. The observed change in absorption was interpreted assuming that the initially excited singlet state decays in a single rate-determining step, with rate k, to the lowest triplet state T_1. Rate constants were found of $k^{-1} =$ 16.5 psec and $k^{-1} = 30$ psec for benzophenone in ethanol and benzene, respectively. The authors point out that the rate of build-up of triplet states is not necessarily equal to the rate of intersystem crossing since vibrational relaxation times might or might not be important. Further work is needed to elucidate this point.

In a next step Anderson *et al.*[56] investigated the decay of the triplet state via energy acceptors such as piperylenes and naphthalenes. Triplet decay was

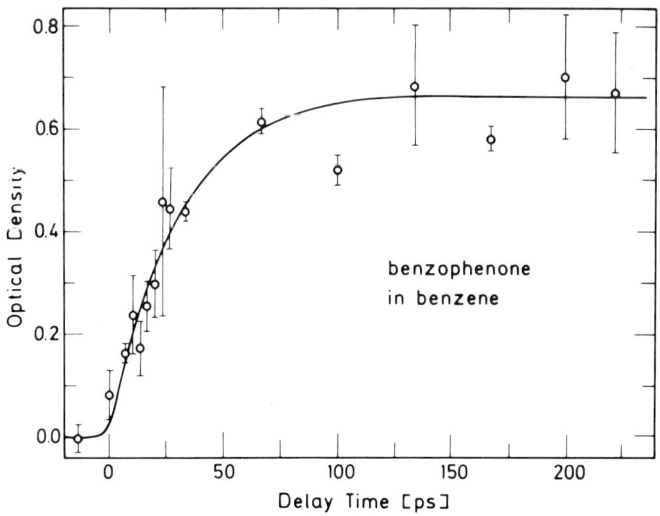

FIG. 20. Optical density at 5300 Å due to triplet formation of benzophenone in benzene as a function of delay time (from Ref. 55).

characterized by a single constant k_0 where $k_0^{-1} = 9$ psec for benzophenone (donor, solute) and cispierylene (acceptor, solvent).

F. DYNAMICAL INVESTIGATIONS OF VIBRATIONAL PROCESSES IN THE ELECTRONIC GROUND STATE

We now turn to the discussion of different dynamical investigations of vibrational molecular processes in liquids. Our various techniques and results are summarized in Fig. 21. Two excitation processes for well-defined normal modes were used successfully in our investigations. First, stimulated Raman scattering which is discussed in more detail in the following part of this chapter and second, direct infrared excitation with ultrashort infrared pulses. Three probing techniques lead to different physical information. The coherent light scattering requires a well-defined scattering geometry between the coherently excited material excitation, the probe pulse and the measured coherent scattering signal. The phasematching angles are calculated from the colour dispersion of the medium to be investigated. Different investigations were made with coherently scattered probe pulses. The dephasing times of normal modes of vibration were measured in a variety of polyatomic molecules and a beating phenomenon of collectively excited molecular vibrations was observed. These investigations will be discussed below.

After the excitation of molecular vibrations a certain number of molecules are promoted to the first excited vibrational state $v = 1$. Spontaneous anti-Stokes scattering, which represents an incoherent scattering process, is directly proportional to the momentary degree of excitation of the system. Measuring the spontaneous anti-Stokes signal as a function of time after the

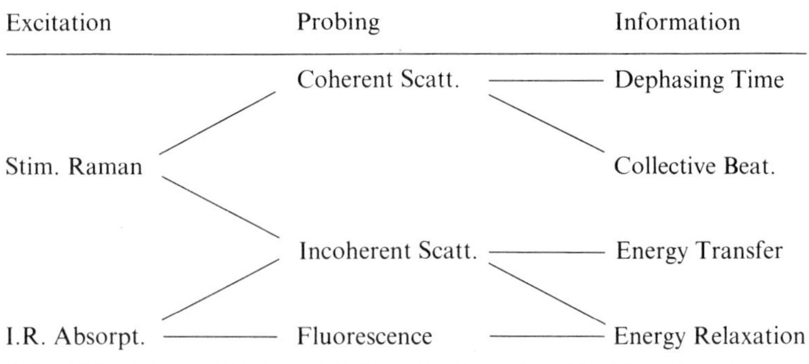

FIG. 21. Physical processes for the excitation and the probing of molecular vibrations in liquids. The dynamical information obtained is listed on the right.

initial excitation allows a direct determination of the population (or energy) relaxation of the vibrational mode. The same incoherent probing technique gives information of the decay routes of an excited mode and of energy transfer processes between different molecules and within one polyatomic molecule.

Finally, we have developed an experimental technique where the energy relaxation time of a well-defined molecular vibration can be determined using a transition to an excited electronic state with observation of a fluorescence signal. The latter system is very sensitive and allows investigations of highly diluted systems. In the following part of this chapter characteristic examples of the different investigations are presented.

In Fig. 22 the excitation of molecular vibrations via stimulated Raman scattering is indicated schematically. When an intense light pulse with peak intensity I_L traverses the medium, a positive gain at the Stokes frequency builds up an intense Stokes wave (I_S) at the end of the sample. In the past, the generated Stokes pulse was the subject of extensive investigations[57–59]. In this chapter we are interested in the material excitation. As indicated in the lower part of Fig. 22, there are two types of material excitation. First, a coherent material excitation of amplitude Q is built up in the stimulated scattering process. As a result, there exists a well-defined phase relationship

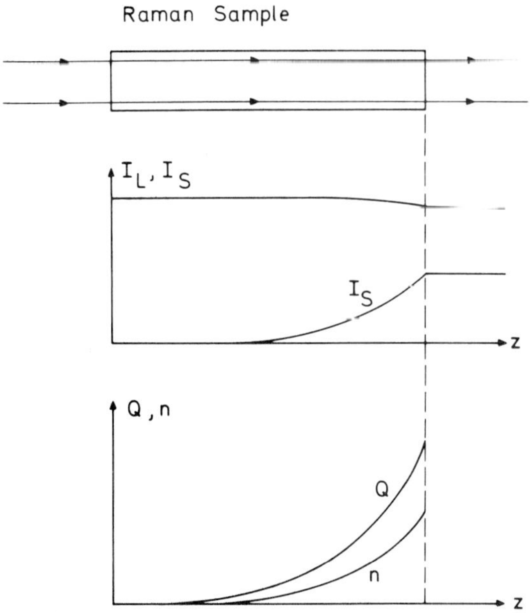

FIG. 22. Light pulses and material excitation in the stimulated Raman scattering as a function of interaction length z.

between molecules in different parts of the excited volume. Second, an excess population n is generated in the first excited vibrational state $v = 1$, which is highest at the end of the interaction path. The differential equations which govern the stimulated Raman process are presented in Fig. 23. From the top to the bottom of the figure we find (1) the wave equation for the total electromagnetic field, (2) a polarization equation which contains the change in

$$\frac{\partial^2 E}{\partial z^2} - \frac{1}{c^2}\frac{\partial^2(\varepsilon E)}{\partial t^2} = \frac{4\pi}{c^2}\frac{\partial^2 P}{\partial t^2}$$

$$P = N\left(\frac{\partial\alpha}{\partial Q}\right)QE$$

$$\frac{\partial^2 Q}{\partial t^2} + \frac{1}{\tau}\frac{\partial Q}{\partial t} + \omega_0^2 Q = \frac{1}{2m}\left(\frac{\partial\alpha}{\partial Q}\right)E^2(1 - 2n)$$

$$\frac{\partial n}{\partial t} + \frac{1}{\tau'}n = \frac{1}{2\hbar\omega_0}\left(\frac{\partial\alpha}{\partial Q}\right)E^2\frac{\partial Q}{\partial t}$$

FIG. 23. Differential equations of transient stimulated Raman scattering.

polarizibility with normal coordinate, $\partial\alpha/\partial Q$, (3) a damped-oscillator equation for the vibrational mode with coherent amplitude Q, and (4) a relaxation equation for the excess population n. It is important to emphasize the two different time constants τ and τ' which are called dephasing time and population relaxation time, respectively. The experimental determination and the understanding of these time constants is the essential part of the following section.

G. DEPHASING TIME OF COLLECTIVELY EXCITED MOLECULAR VIBRATIONS

We have made a detailed theoretical study of the differential equations of Fig. 23[60]. Of special interest for our investigations is the transient excitation where the pulse duration t_p is of the order of the dephasing time τ. In this case the maximum of the material excitation (Q and n) occurs after the maximum of the incident pulse has passed a certain position within the sample. Some numerical results of Q as a function of time are depicted in Fig. 24 for the parameter t_p/τ. It is readily seen from the figure that the coherent excitation rises rapidly during the pumping process and decays during the subsequent free relaxation with the characteristic time constant τ. It should be noted that the measured probe scattering signal is a convolution of the probe pulse with the coherent excitation Q at time t_D.

The experimental system for the determination of the dephasing time τ is schematically depicted in Fig. 25[22]. The incident powerful laser pulse excites

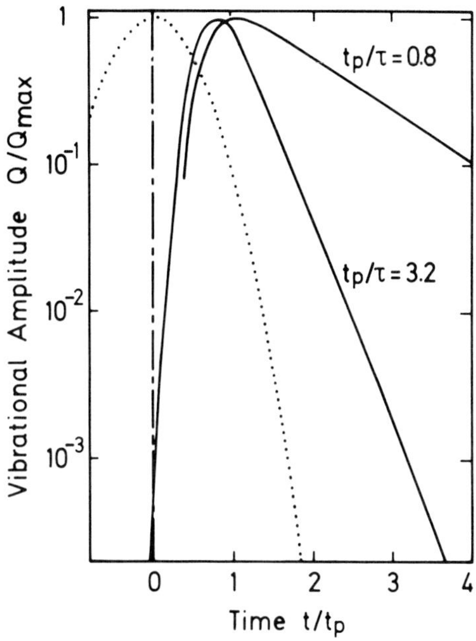

FIG. 24. Coherent vibrational excitation Q in stimulated Raman scattering as a function of time for two values of the ratio pulse duration t_p to dephasing time τ.

one normal mode of the material in the liquid cell, LC, while a weak probing pulse of variable delay, VD, monitors the excited volume. A second probe pulse with constant delay serves as a reference beam to improve the accuracy of the system. The insert on the lower left-hand corner of the figure indicates the phasematching geometry. The pump laser and Stokes pulse generate a coherent material excitation of wave vector $\mathbf{k}_p$. The probe pulse with wave vector $\mathbf{k}_{L2}$ scatters off the material excitation producing a coherent anti-Stokes pulse of wave vector $\mathbf{k}_{AS}$. In practical cases the angles between the various wave vectors are small, i.e. the coherent scattering signal is observed close to the forward direction.

We have measured the coherent scattering signal of a variety of molecules[61–63]. One normal mode of vibration was excited for each molecular system. The mode with the largest scattering cross-section and, as a result, with the largest gain dominates in the stimulated excitation process. A typical experimental result is presented in Fig. 26. The coherent scattering signal is plotted as a function of delay time between pump and probe pulse. Time zero marks the maximum of the pump pulse. The scattered probe signal rises to a delayed maximum and decays with a time constant characteristic of the

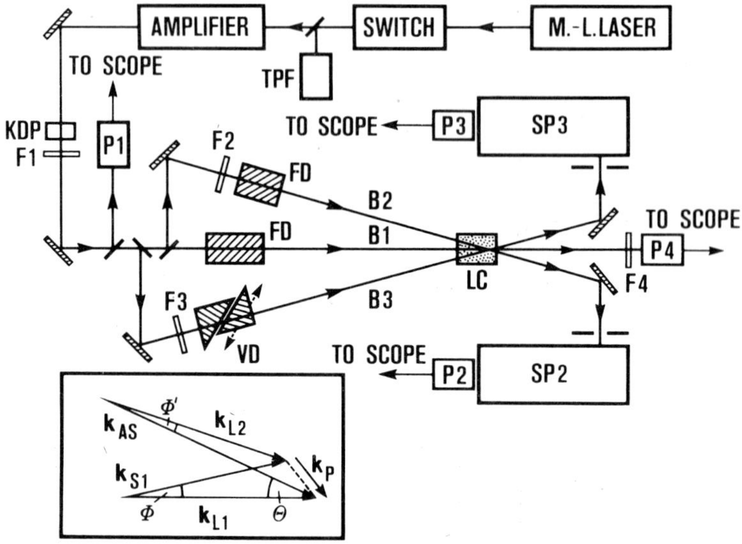

FIG. 25. Schematic of experimental system measuring the vibrational dephasing time. Three light beams B1, B2, B3 interact in the sample. LC, fixed delays, FD; variable delay, D; filters, F; spectrometer, SP; photodetector, P; two-photon fluorescence system, TPF. *Inset*: Schematic of non-collinear phase matched probe anti-Stokes scattering, $\mathbf{k}_{L1}$ and $\mathbf{k}_{L2}$ are the directions of the exciting and probing pulse, respectively (from Ref. 22).

TABLE IV. Experimental data on the dephasing time of molecular vibrations in liquids (from Ref. 63)

		$\tau[10^{-12}$ sec]	
	$\tilde{\nu}_0[\mathrm{cm}^{-1}]$	From line width	Measured
N_2 (77 K)	2326	79 ± 8	75 ± 8
$SnCl_4$	368	2.5 ± 0.5	2.8 ± 0.3
$SiCl_4$	425	2.8 ± 0.5	3.0 ± 0.5
CCl_4	459	3.8 ± 0.5	3.6 ± 0.5
CH_3CCl_3	2939	1.1 ± 0.1	1.0 ± 0.2

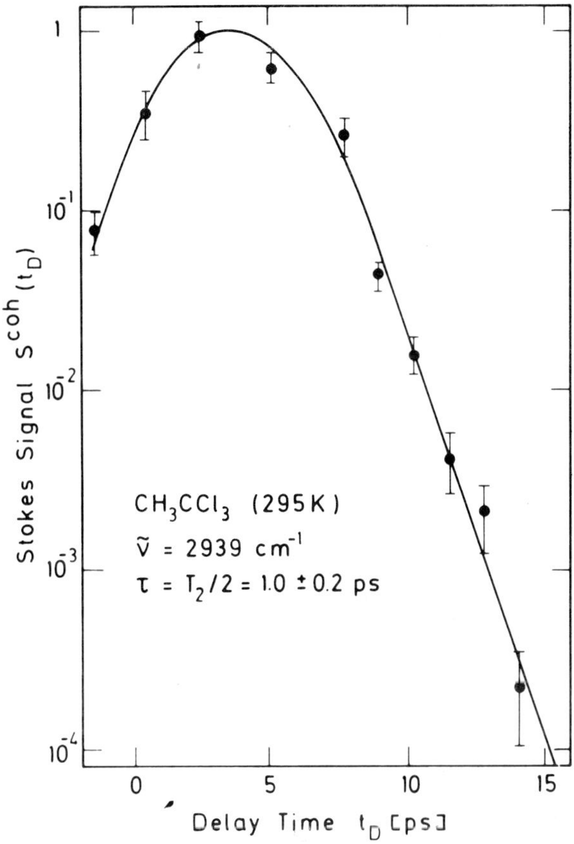

FIG. 26. Coherent probe scattering signal versus delay time between pump and probe pulse. The symmetric CH_3-stretching vibration of CH_3CCl_3 (from Ref. 63).

specific molecular vibration. A relaxation time of 1.0 psec is deduced from the experiment for the symmetric CH_3-vibration of CH_3CCl_3 at 2939 cm^{-1}. In Table IV dephasing times τ obtained from the coherent probe scattering are summarized for five different molecules. The vibrational frequencies cover a range from 370 cm^{-1} to ~ 3000 cm^{-1}. For polyatomic molecules we find time constants of several picoseconds at room temperature. It is interesting to compare our dephasing times τ with time constants calculated from the linewidth of spontaneous Raman scattering according to the equation $\delta\tilde{\nu} = (2\pi c\tau)^{-1}$. The good agreement between the two time values indicates that the Raman linewidth of the vibrations listed here is determined by dephasing processes. Very recently, a mechanism for the phase relaxation based on semi-classical collision theory was studied by Fischer and Laubereau[64]. Due to molecular interaction in elastic collisions the vibrating molecules are shown to experience a change of the vibrational frequency during the collision. The quasi-elastic collision model was treated quantum mechanically. Comparison between numerical phase relaxation times and experimental data for τ give good agreement within a factor of approximately two for a number of liquids.

It was pointed out above that for relatively long pump pulses of $t_p > \tau$ the Raman excitation process is highly selective. Only one molecular vibration with the largest scattering cross-section is excited. The situation is different for the highly transient stimulated Raman process where $t_p \lesssim \tau$. In this case it is possible to excite simultaneously vibrational modes of small frequency differences $\delta\omega$[63]. It can be shown that neighbouring vibrational modes are excited with equal amplitude and phase when the duration of the incident light pulse is sufficiently short, i.e. $t_p \ll 2\pi/\delta\omega$. After the short pump pulse has passed the medium, the collectively vibrating (isotope) components relax freely oscillating with their individual resonance frequencies. A beating between the vibrating systems is predicted where maxima and minima occur after a time interval of $T = 2\pi/\delta\omega$.

Recently we have observed this beating phenomenon of isotope components by measuring the coherent probe scattering as a function of delay time t_D[63]. The probe pulse generates scattered signals at slightly different directions.

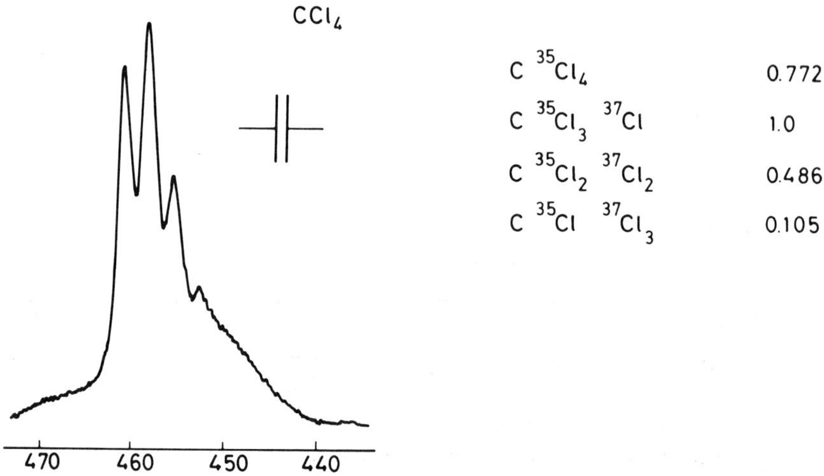

FIG. 27. Isotope splitting of the totally symmetric tetrahedron vibration of CCl_4 in the spontaneous Raman spectrum.

The scattering directions result from the different vibrational frequencies of the isotope components. There are two possibilities to investigate the collectively vibrating molecular systems. (1) A beam direction is selected where one vibrational component has a predominant scattering signal. (2) A direction is chosen where two (or more) beams have scattering intensities of comparable magnitude. In this case we observe as a function of delay time t_D strong beats of the coherent scattering signal. This result was expected since each coherent probe beam reflects the phase information of the respective vibrational component.

In Fig. 27 the spontaneous Stokes Raman spectrum of the totally symmetric tetrahedral vibration of common CCl_4 is depicted. The three major peaks correspond to the three most abundant molecular components which are listed in the figure. A line splitting is observed of $\delta\omega/2\pi c \simeq 3\ cm^{-1}$. We have excited common CCl_4 by transient stimulated Raman scattering using a pump pulse of approximately 3.5 psec duration. In Fig. 28 the coherent probe

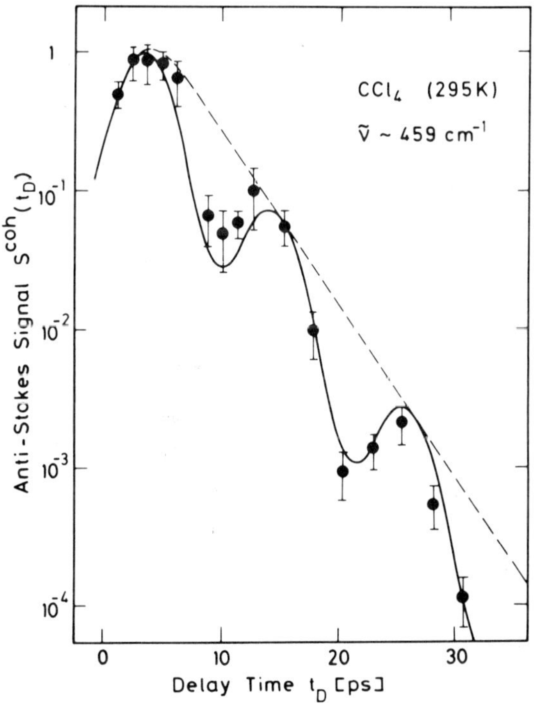

FIG. 28. Coherent scattering signal of common CCl_4 representing the beating of the isotope components separated by $2.9\ cm^{-1}$. The broken line indicates the dephasing of one isotope component (from Ref. 63).

scattering signal of CCl_4 is presented which shows clearly the collective beating of the isotope species with $T = 11.5$ psec[63]. The beat frequency is in good agreement with the spontaneous line splitting. The curve in Fig. 28 is calculated using known parameters of CCl_4 and a dephasing time $\tau = 3.6 \pm 0.5$ psec which was determined in an independent experiment where one isotope component was investigated separately. The good agreement between the calculated curve and our experimental points should be noted.

H. ENERGY RELAXATION OF THE FIRST VIBRATIONAL STATE

We now turn to the discussion of a series of experiments to study the population lifetime τ' of the $v = 1$ state of a normal mode of vibration in the electronic ground state. In a first group of investigations, the molecules were excited via stimulated Raman scattering and the degree of excitation was monitored by spontaneous anti-Stokes scattering of a delayed probe pulse. In a second experimental system ultrashort infrared pulses excite directly the vibrational mode of interest and a second properly delayed probe pulse monitors the excitation using a transition to a fluorescent electronic state.

An experimental set-up for measuring the population relaxation time τ' is depicted schematically in Fig. 29[61]. A single laser pulse enters the system from the left and produces vibrational excitation of the first excited $v = 1$ state. A second pulse generated by the beam splitter serves as the probe pulse

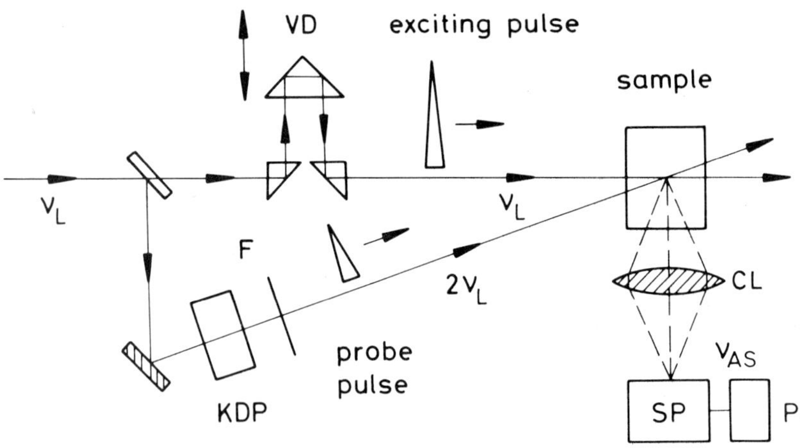

FIG. 29. Experimental system for the measurement of the energy relaxation time τ' of a molecular vibration in liquids. Variable delay. VD; frequency doubling crystal, KDP; collecting lens, CL; for the incoherent probe scattering, spectrometer, SP; photomultiplier, PM (from Ref. 61).

to interrogate the excited volume after a delay time t_D. Spontaneous anti-Stokes scattering is measured at a scattering angle of 90° with the help of a spectrometer and photomultiplier. This signal is directly proportional to the momentary degree of excitation of the $v = 1$ state.

In Fig. 30 the (incoherent) spontaneous anti-Stokes signal is plotted (closed circles) versus t_D for ethanol. An energy relaxation time τ' of 22 psec is deduced from the data[61,65]. It is readily seen from the figure that the coherent (phase-matched) scattering signals (open circles) decrease much faster with time, i.e. the dephasing time is much shorter than the energy relaxation of the same CH_3-stretching vibration of ethanol.

The concentration dependence of the energy (population) relaxation time τ' of the ν_H-stretching mode was studied in several $CH_3CCl_3:CCl_4$ mixtures with mole fractions $x = 1$, 0.8, 0.6, and 0.4 of CH_3CCl_3 (see Fig. 31). It was found that the energy relaxation time of the ν_H-mode of CH_3CCl_3 at $\nu \sim 3000\ \mathrm{cm}^{-1}$ increases strongly from 5.2 psec to 29 psec when x decreases from 1.0 to 0.4[66]. It is interesting to note that the Raman linewidth of the ν_H-mode of CH_3CCl_3 (a measure of the dephasing time in this molecule) does not change by more than 10% in the same mixtures studied, i.e. linewidth measurements do not allow the determination of the strong concentration dependence of τ' found in these investigations.

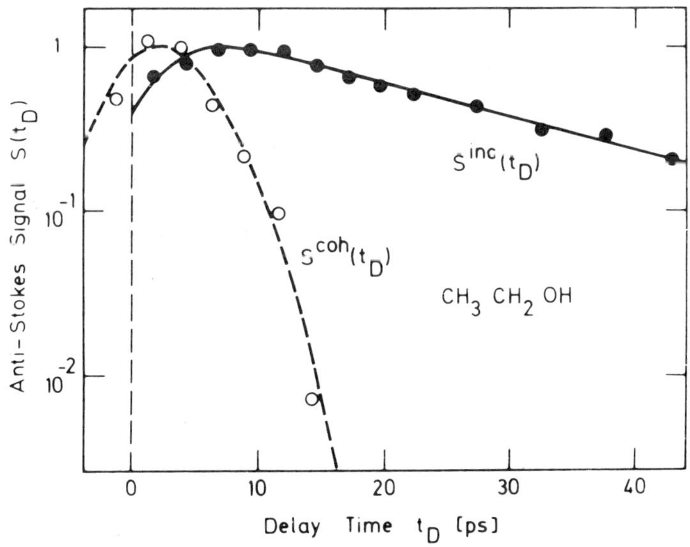

FIG. 30. Measured incoherent scattering $S^{inc}(t_D)/S^{inc}_{max}$ (●) and coherent scattering $S^{coh}(t_D)/S^{coh}_{max}$ (○) versus delay time t_D for ethyl alcohol. The solid and dashed curves are calculated (from Ref. 61).

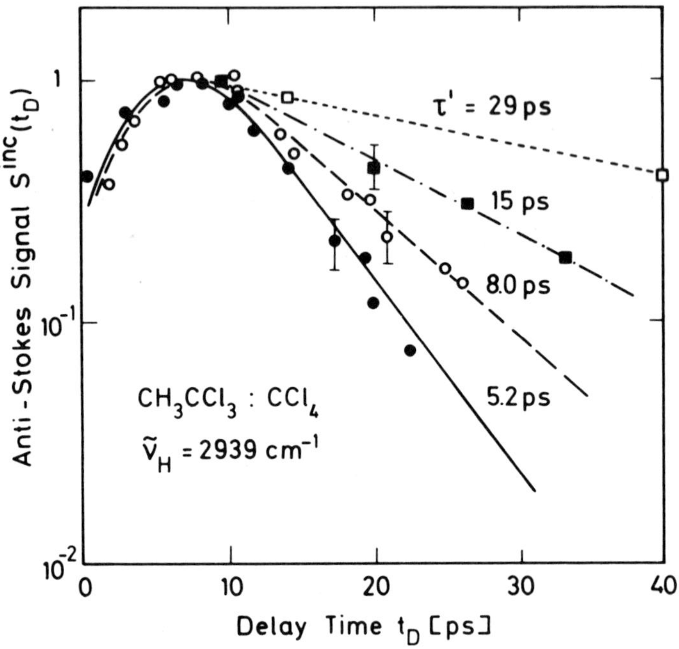

FIG. 31. Incoherent probe signals of the ν_H-vibration versus delay time t_D in various mixtures CH_3CCl_3:CCl_4. Mole fractions of CH_3CCl_3, $x = 0.4$ (□), 0.6 (■), 0.8 (○), 1.0 (●) (from Ref. 66).

I. ENERGY TRANSFER BETWEEN VIBRATIONAL LEVELS

The question now arises what physical processes determine the energy relaxation τ'. Figure 32 serves to illustrate possible decay routes and energy transfer processes. The normal mode frequencies of several molecules are shown. Polyatomic molecules such as CH_3CCl_3 possess a number of vibrational modes. The vibrational level at $\nu_H \sim 2900\ cm^{-1}$ (a CH_3-stretching mode) is primarily excited by the stimulated Raman process. In pure liquid CH_3CCl_3 this ν_H-mode couples effectively to the CH_3-bending mode, δ_H, at approximately half the energy. The decay process $\nu_H \rightarrow 2\delta_H$ is indicated by arrows in the figure. This energy decay route generates a subsequent excess population of the δ_H modes which was experimentally observed[66]. The spectrometer was tuned to the corresponding anti-Stokes frequency position in our incoherent scattering experiment.

Direct information on intermolecular energy transfer was obtained in the mixture CH_3CCl_3:CD_3OD. Figure 32 shows the energy levels of the solvent CD_3OD. The CD_3-stretching vibration of this molecule at $\nu_D = 2227\ cm^{-1}$

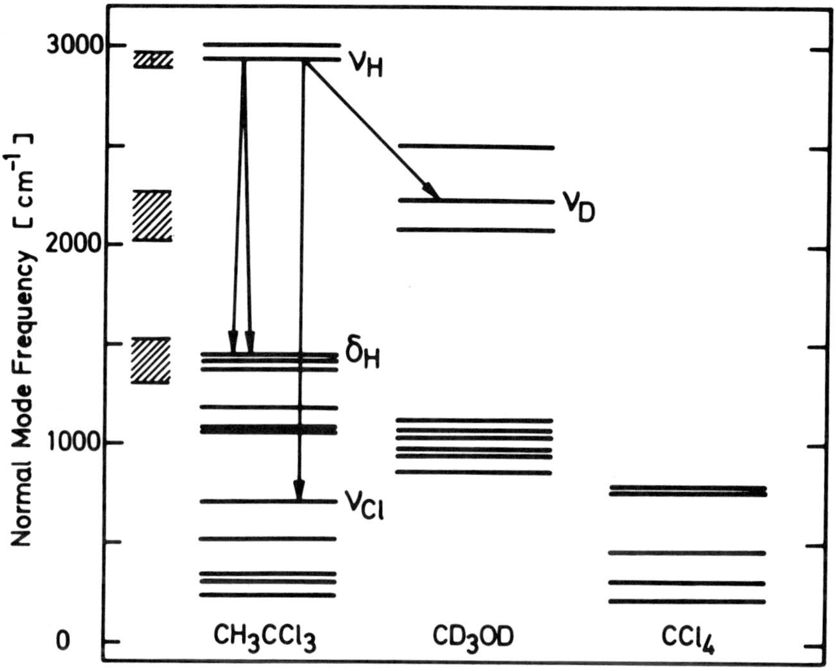

FIG. 32. Frequency values of the normal vibrational modes of the liquids CH_3CCl_3, CD_3OD and CCl_4.

and the frequency $\nu_{Cl} = 713\ cm^{-1}$ of a CCl_3-vibration of CH_3CCl_3 add up with good accuracy to the ν_H-mode of CH_3CCl_3 which is experimentally excited in the mixture. An interaction between CH_3CCl_3 and CD_3OD molecules of the form $\nu_H \rightarrow \nu_D + \nu_{Cl}$ is suggested by energy resonance arguments (see arrows in Fig. 32).

Experimental data on the intermolecular energy transfer from the laser excited ν_H-vibration of CH_3CCl_3 to the ν_D-vibration of the solvent molecule are presented in Fig. 33[66]. The data of the ν_H-vibration (open circles) of CH_3CCl_3 have been discussed above. Of special interest is the incoherent scattering signal observed at a frequency shift of $\sim 2200\ cm^{-1}$ (closed circles) which originates from the subsequent excess population of the ν_D-vibration of CD_3OD. The data display a delayed maximum with a slow decay of 25 ± 10 psec. This experiment gives convincing evidence of energy transfer from the CH_3CCl_3 to the CD_3OD molecules. The results of Fig. 33 represent the first observation of intermolecular transfer of vibrational energy in liquids. The curves in Fig. 33 are calculated from the theory of transient stimulated Raman scattering exciting the ν_H-mode (broken line) and a rate equation-model for

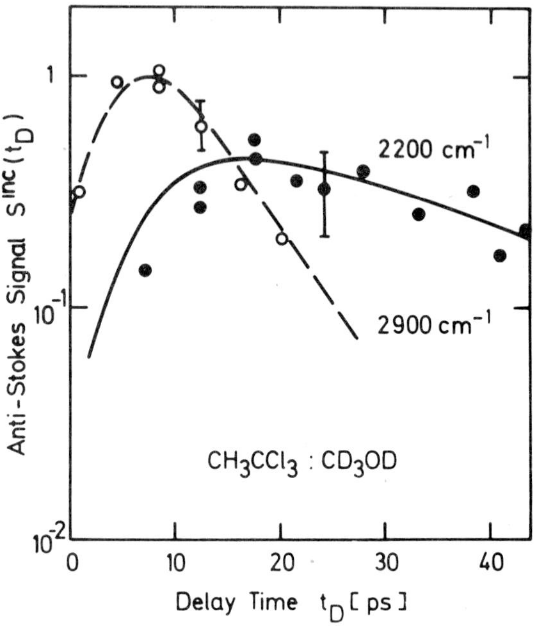

FIG. 33. Measurement of the intermolecular energy transfer in the mixture CH_3CCl_3:CD_3OD. Incoherent anti-Stokes scattering (measuring the population of the first excited vibrational state) is plotted versus delay time t_D between pump and probe pulse. (○) = probe scattering from the CH_3 valence bond vibration (2939 cm^{-1}) of CH_3CCl_3 which is excited by the pump pulse. (●) = scattering observed from the CD_3 valence bond vibration (~2200 cm^{-1}) of CD_3OD, which directly indicates intermolecular transfer of vibrational energy (from Ref. 66).

the energy transfer to the ν_D-vibration (solid line). Knowing the values of the spontaneous cross-sections of the two vibrations, a comparison of the scattering signals yields a quantum efficiency of approximately 60% for the intermolecular transfer process. Our study of the concentration dependence of the energy relaxation of CH_3CCl_3 in mixtures with CCl_4 and CD_3OD show that the $\nu_H \rightarrow 2\delta_H$ interactions have an efficiency of 40%.

Figure 34 shows some of the normal modes of ethanol. In the frequency range around 3000 cm^{-1} we have four CH-stretching vibrations and around 1500 cm^{-1} there are three CH-bending modes of the molecule. The $v = 2$ states of the CH-bending modes are also indicated in the figure. Now, we wish to answer the following question. When one CH-stretching mode is excited above its equilibrium value, is there energy transfer to neighbouring energy states? In particular, is there energy transfer to the $v = 2$ states of the CH-bending modes? In order to answer this question, we have performed

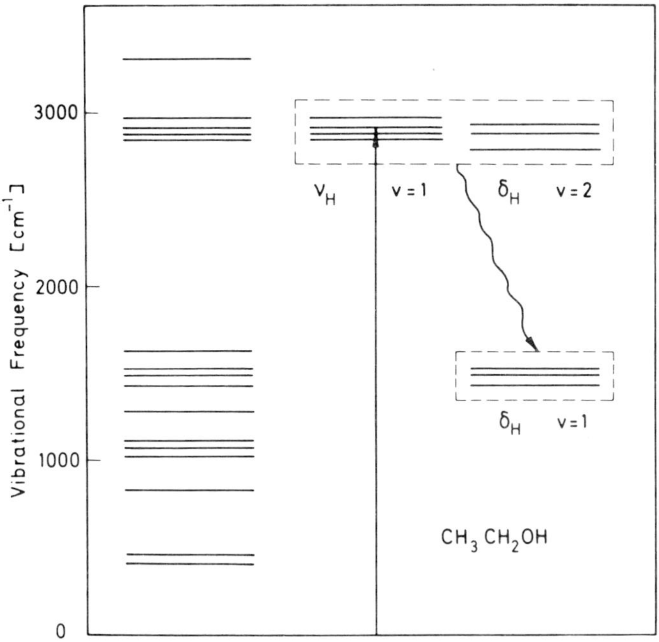

FIG. 34. Schematic of relaxing vibrational states of ethanol CH_3CH_2OH.

three independent experiments which will be discussed below. In Fig. 35(a) the spontaneous Raman Stokes spectrum of ethanol in the range around 3000 cm^{-1} is presented. This spectrum is readily obtained with an argon laser in conjunction with a spectrometer. Three distinct energy states are clearly visible; a fourth one is hidden around the shift of 2900 cm^{-1}, a fifth vibrational state is a degenerate state. We have applied our picosecond probe technique to measure the degree of vibrational excitation of different modes in the frequency range of 3000 cm^{-1}. In Fig. 35(b) the spontaneous anti-Stokes spectrum of probe pulses delayed by $t_D = 11$ psec with respect to the maximum of the input pulse is depicted[65]. Considering the duration of the excitation pulse (here 8 psec) and the transient build-up of the population of $v = 1$, we estimate that the spectrum was taken several picoseconds after the maximum excitation of the pumped mode at 2928 cm^{-1}. It is readily seen from the Fig. 35(b) that the spectrum extends over a frequency range of approximately 150 cm^{-1}, indicating the population of neighbouring vibrational energy states. The curve drawn through our experimental points is calculated under the assumption that the five vibrational modes are populated. The spectral resolution of the specific experiments was taken into account. The

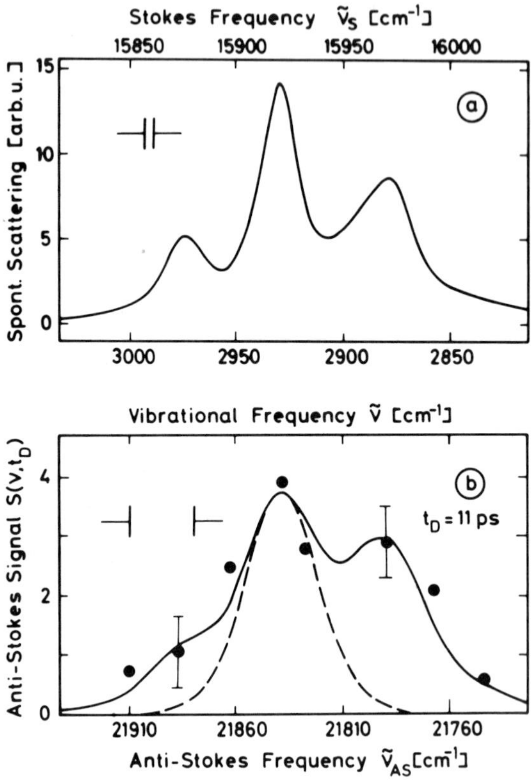

FIG. 35. (a) Spontaneous Stokes spectrum of the CH stretching vibrations; (b) Anti-Stokes probe scattering of CH-stretching vibrations at $t_D = 11$ psec; broken line indicates spectral profile of the pumped mode at 2928 cm^{-1}. Solid line calculated for quasi-equilibrium between levels around 2900 cm^{-1} (from Ref. 65).

good agreement between the experimental points and calculated curve should be noted. In Fig. 36 we discuss the spectral range around 1500 cm^{-1}. At the top, the spontaneous Raman spectrum of the CH-bending modes is presented with the mode 1456 cm^{-1} having the largest scattering cross-section. The experimental anti-Stokes spectrum obtained by our probe pulses with a time delay of 19 psec after the excitation pulse exhibits an interesting spectral distribution[65]. The observed frequency profile is broad with a shift to smaller frequencies. For an analysis we calculated the convolution of the spectrum of Fig. 36(a) with our instrumental function (broken line). Obviously, there is additional scattering around 1400 cm^{-1} which is taken as evidence for anti-Stokes scattering from the δ_H ($v = 2$) states. The shift to smaller frequencies results from the anharmonicity of the δ_H-bending mode.

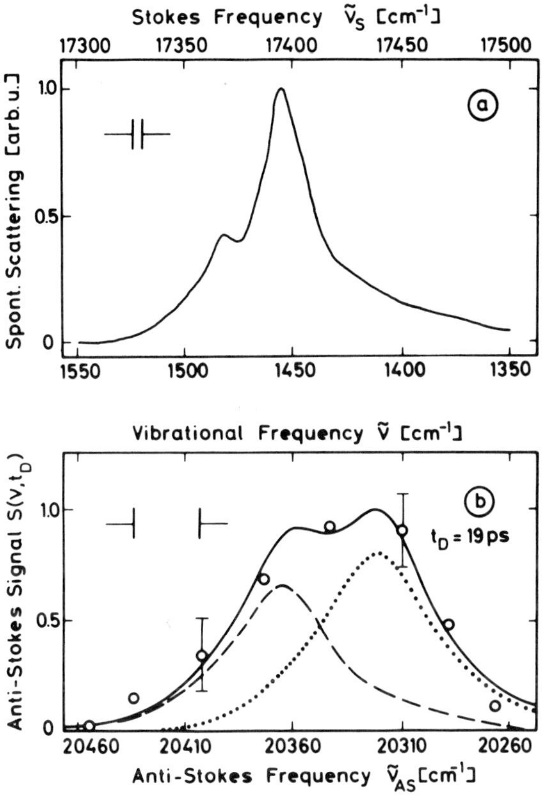

FIG. 36 (a) Spontaneous Stokes spectrum of the CH bending modes δ_H ($v = 1$) (b) Anti-Stokes probe scattering around 1400 cm^{-1}. Calculated curves for δ_H ($v = 1$) scattering (broken line) and δ_H ($v = 2$) scattering (dotted line) (from Ref. 65).

The scattering cross-section of $v - 2 \rightarrow v = 1$ scattering is known to be twice that of $v = 1 \rightarrow v = 0$ scattering. We conclude from this experiment that approximately 20 psec after the excitation the $v = 2$ and $v = 1$ states have a definite excess population due to energy transfer and relaxation processes. A third experiment concerning the energy transfer problem in ethanol is depicted in Fig. 37[67]. Here, the anti-Stokes probe signal around 2900 cm^{-1} is plotted versus delay time for a $CH_3CH_2OH{:}CCl_4$ solution with 0.04 mole concentration of ethanol. The probe signal was measured over a total bandwidth of 90 cm^{-1}. The CH_3-mode at 2928 cm^{-1} was primarily excited by a short (3 psec) infrared pulse of the same centre frequency. The scattered signal shows a very rapid decay with a time constant of one ot two psec and a slower decay with $\tau' = 40$ psec. We interpret this result as follows. The rapid

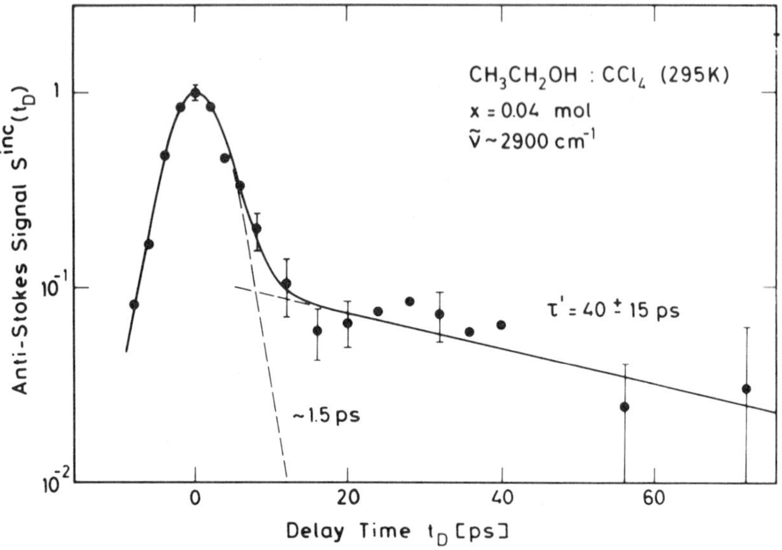

FIG. 37. Incoherent probe scattering signal of ethanol (solvent CCl_4) around 2900 cm^{-1} versus delay time. The CH_3-mode at 2928 cm^{-1} is primarily excited by an infrared pulse of the same frequency (from Ref. 67).

decrease of the scattered signal is due to the fast energy transfer to neighbouring stretching ($v = 1$) and bending ($v = 2$) levels which gives rise to a reduced light scattering around 3000 cm^{-1} of the magnitude observed. The longer time constant of 40 psec corresponds to energy decay to the lower bending modes around 1500 cm^{-1}[68]. This time constant was observed to be 22 psec in pure ethanol[65]. Summing up our observations in ethanol, fast energy transfer between neighbouring energy states appears to occur in this molecule with a slower energy decay to lower energy states.

J. ENERGY RELAXATION AFTER INFRARED EXCITATION OF DEFINED VIBRATIONAL MODES

Very recently we have devised an experimental system for the study of molecular vibrations in highly diluted systems. This technique is particularly suited for investigations of vibrational modes in the electronic ground state of fluorescent molecules. A schematic of our set-up is depicted in Fig. 38[69]. The single picosecond laser pulses enter the system from the left and generate the desired infrared pulse traversing two properly oriented $LiNbO_3$ crystals.

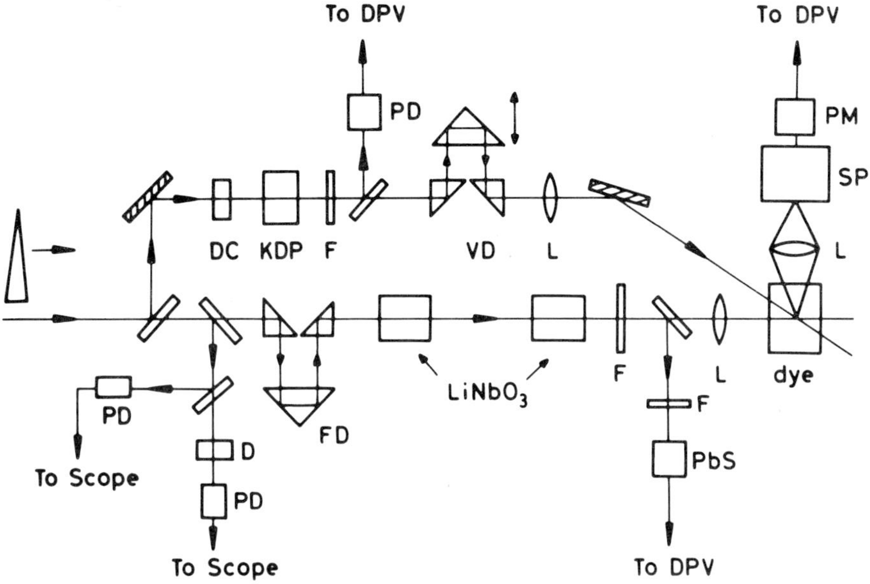

FIG. 38. Schematic of the experimental system measuring the energy relaxation time of a molecular vibration after I.R. excitation by the help of the fluorescence probing technique (from Ref. 69).

A small part of the input pulse produces a green probe pulse at $2\nu_L$ in the KDP crystal. The fluorescence of the sample is measured with spectrometer and photomultiplier. The relevant transitions involved in our experiment are illustrated schematically in Fig. 39. A first short infrared pulse excited the molecular system via infrared absorption. The frequency ν_1 of the infrared pulse is properly selected in order to interact only with one well-defined vibrational mode. Our pump pulse has a duration of approximately 3 psec and contains 10^{14} infrared quanta at a frequency of approximately 3000 cm^{-1}. The molecular system is subsequently interrogated by a second pulse of frequency ν_2. The second probe pulse promotes molecules which are vibrationally excited to a level close to the vibrational ground state of the first excited singlet state S_1. The fluorescence originating from this singlet state is experimentally observed. It should be noted that energy levels smaller than $h\nu_1$ (in the electronic ground state) do not interact with the probing pulse of frequency ν_2. The time-integrated fluorescent radiation serves as a direct measure of the instantaneous vibrational excitation of the energy level $h\nu_1$. The fluorescent signal is measured as a function of delay time t_D between the infrared excitation pulse and the probing light pulse. With this technique it is possible to study well-defined vibrational excitations even in large molecules

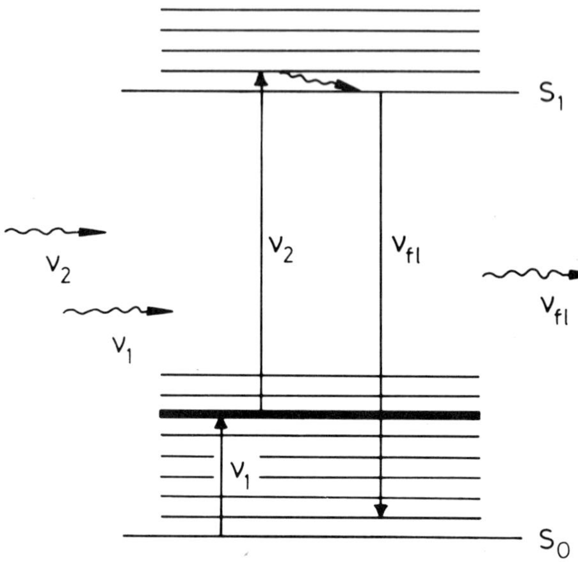

FIG. 39. Fluorescence probing technique. Schematic of the molecular energy levels and transitions during the vibrational excitation and probing process (from Ref. 69).

with complicated vibrational spectra. Of special interest is the high sensitivity of this method for molecules with high fluorescence quantum efficiency.

We have applied this technique to dynamic investigations of dye solutions in a large concentration range of 10^{-6} M to 10^{-3} M[69]. As an example we discuss results on the molecule Coumarin 6 in the solvent CCl_4. The Coumarin 6 molecule is depicted on the top of Fig. 40. Two C_2H_5-groups are bonded to the conjugated ring system. The infrared transmission spectrum of Coumarin 6 around 3000 cm^{-1} is presented in the lower part of Fig. 40. Several absorption peaks are clearly resolved in the spectrum. The bands between 2865 cm^{-1} and 2970 cm^{-1} are readily interpreted as normal vibrational modes of the two ethyl groups of Coumarin 6. The CH-valence bond vibrations of the ring system produce the absorption peak at 3055 cm^{-1}. Most important for our investigations is the absorption maximum at 2970 cm^{-1} which corresponds to the asymmetric CH_3-mode. It is this vibrational mode which is populated by the resonant short infrared pulse. The degree of vibrational excitation is subsequently monitored as a function of time by our fluorescence technique. As frequency of the probing pulse we choose $\nu_2 = 2\nu_L = 18910\ cm^{-1}$ (0.53 μm), i.e. the second harmonic of the Nd:glass laser.

Experimental results of our two-pulse technique are presented in Fig. 41 for Coumarin 6 in CCl_4 for a concentration of 3×10^{-4} M[69]. The observed

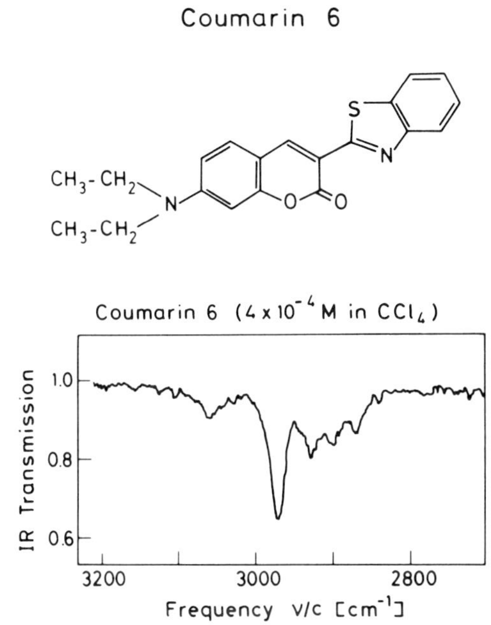

FIG. 40. (Top) Coumarin 6 molecule. (Bottom) Infrared absorption spectrum of Coumarin 6 in CCl_4 around 3000 cm^{-1}. The line at 2970 cm^{-1} is excited in the picosecond experiment.

fluorescence signal, $S(t_D)$, initiated by the probe pulse, is plotted as a function of delay time t_D for two temperatures, 253 K and 295 K. The time scale of a few psec should be noted. The fluorescence signal increases sharply within several psec to a maximum value $t_D \simeq 2$ psec. The build-up of excess population in the vibrational spectrum at $\nu_1 = 2970\ cm^{-1}$ is directly seen from the rise of the signal curve. Of special interest are the decaying parts of the signal curves which extend over two orders of ten. Fast relaxation of the vibrational excitation with slightly different slopes for the two temperatures is clearly indicated by the data. At room temperature a time constant of 1.3 ± 0.4 psec is directly obtained from the exponential slope (open circle). At $-20°C$ the energy relaxation time is measured to be $\tau' = 1.7 \pm 0.3$ psec. We have ascertained that these time constants represent true relaxation times. The liquid cell containing the Coumarin molecules was replaced by a nonlinear crystal. The intensity of the sum frequency at $\nu_1 + \nu_2$ was measured as a function of delay time between the two pulses of frequency ν_1 and ν_2. In this way, the time resolution of our system was tested and found to be 0.4 ± 0.2 psec.

At the present time the energy relaxation process is not yet definitely determined. There are two possible physical mechanisms which lead to a

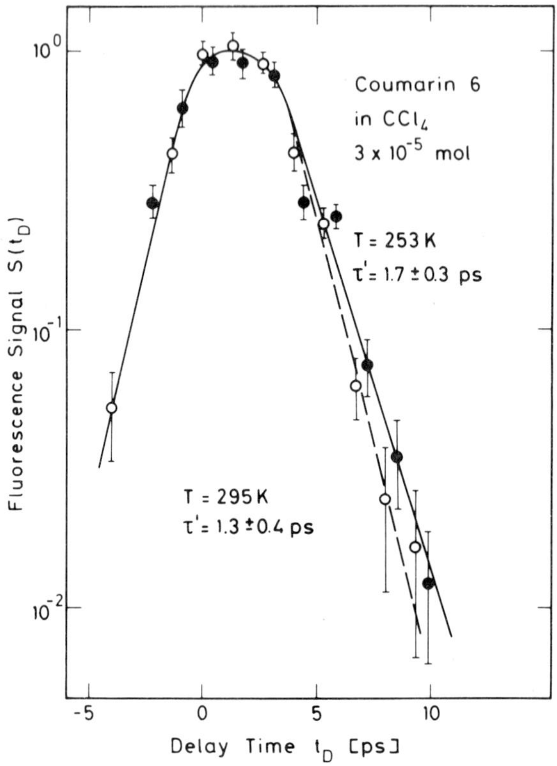

FIG. 41. Ultrafast vibrational relaxation of Coumarin 6 in CCl_4 at 295 K (○) and 253 K (●). The asymmetric CH_3-mode at 2970 cm^{-1} in the electronic ground state is excited and the vibrational excitation is observed as a function of time with a fluorescence probing technique (from Ref. 69).

decay of our fluorescence signal: (1) energy transfer to vibrational states of approximately equal energy with smaller absorption cross-sections to the excited electronic state; (2) energy decay to lower vibrational states which cannot interact with the probe pulse. Work is in progress to elucidate the relaxation process in more detail.

VI. CONCLUSION

In this chapter we focussed our attention on molecular investigations in the liquid state. It was shown that different fundamental molecular time constants are of very short time duration requiring ultrashort light pulses for a quantitative study. While we feel that various investigations are in an early state, a series of new time parameters have been obtained from experimental data. As an example, we pointed to the energy relaxation times which are available for the first time for well-defined normal vibrational modes.

A small number of experiments were concerned with measurements of the relaxation times of optical phonons in solids[70–73]. In diamond, the fundamental TO-phonon was optically excited and the decay time (2.9 psec at 300 K) was found to be as expected from the anharmonicity of the extensively studied lattice[71]. Coherent scattering revealed decay times (4.4 psec at 300 K) of the internal CO_3^{--} symmetrical mode of $CaCO_3$ in agreement with Raman linewidth data[72]. More work is necessary to establish the phonon decay route and phonon decay process in this crystal. A first investigation exists to measure directly the relaxation time of a polariton mode[73]. A complicated experimental system was required to examine a polariton of well-defined frequency and wave vector. The measured polariton in GaP was close to the TO-lattice mode and the observed relaxation time (5.5 psec at 300 K) agreed with a value expected from infrared data[74].

It is the purpose of this chapter to give a brief introduction to a series of new investigations with ultrashort light pulses. The reader is referred to the original papers for details of the experiments. At present, several research groups throughout the world are actively working with ultrashort light pulses. It is a safe prediction that new and interesting results will be published in the years to come.

REFERENCES

1. H. W. Mocker and R. J. Collins. *Appl. Phys. Lett.* **7**, 270 (1965); A. J. DeMaria, D. A. Stetser and H. Heyman. *Appl. Phys. Lett.* **8**, 174 (1966).
2. D. von der Linde, M. Maier and W. Kaiser. *Phys. Rev.* **178**, 11 (1969).
3. A. Laubereau and W. Kaiser. *Ann. Rev. Phys. Chem.* **26**, 83 (1975).
4. V. S. Letokhov. *Sov. Phys. JETP* **27**, 736 (1968); *Ibid.* **28**, 562 (1969); *Ibid.* **28**, 1026 (1969); P. G. Kryukov and V. S. Letokhov. *IEEE J. Quantum Electron.* **QE-8**, 766 (1972).
5. J. A. Fleck. *Jun. Phys. Rev.* **B1**, 84 (1970); E. Mathieu and H. Weber. *Z. Angew. Math. Phys.* **22**, 458 (1971).
6. D. J. Bradley, G. H. C. New and S. J. Caughey. *Phys. Lett.* **30A**, 78 (1969); *Opt. Comm.* **2**, 41 (1970).
7. Kodak Data Release. *See also* Ref. 54.
8. M. A. Duguay and J. W. Hansen. *Opt. Comm.* **1**, 254 (1969).
9. C. R. Giuliano and L. D. Hess. *IEEE J. Quantum Electron.* **QE-3**, 358 (1967). G. Mourou, G. Busca and M. M. Denariez-Roberge. *Opt. Comm.* **4**, 40 (1971).
10. W. Schmidt and F. P. Schäfer. *Phys. Lett.* **26A**, 558 (1968).
11. J. A. Bradley. Review: *Opto-electronics* **6**, 25 (1974).

12. G. H. C. New. *IEEE J. Quantum Electron.* **QE-10**, 115 (1975).
13. D. Madge and W. M. Windsor. *Chem. Phys. Lett.* **27**, 31 (1974); C. V. Shank and E. P. Ippen. *Appl. Phys. Lett.* **26**, 62 (1975).
14. M. A. Duguay, J. W. Hansen and S. L. Shapiro. *IEEE J. Quantum Electron.* **QE-6**, 725 (1970).
15. R. C. Eckhardt and C. H. Lee. *Appl. Phys. Lett.* **15**, 425 (1969).
16. D. von der Linde. *IEEE J. Quantum. Electron.* **QE-8**, 328 (1972).
17. D. J. Bradley and G. H. C. New. *Proc. IEEE* **62**, 313 (1974).
18. D. von der Linde, O. Bernecker and W. Kaiser. *Opt. Comm.* **2**, 149 (1970).
19. D. von der Linde, O. Bernecker and A. Laubereau. *Opt. Comm.* **2**, 215 (1970).
20. J. A. Giordmaine, P. M. Rentzepis, S. L. Shapiro and K. W. Wecht. *Appl. Phys. Lett.* **7**, 270 (1967).
21. J. R. Klauder, M. A. Duguay, J. A. Giordmaine and S. L. Shapiro. *Appl. Phys. Lett.* **13**, 174 (1968). H. P. Weber. *Phys. Lett.* **27A**, 321 (1968).
22. D. von der Linde, A. Laubereau and W. Kaiser. *Phys. Rev. Lett.* **26**, 255 (1971).
23. D. von der Linde and A. Laubereau. *Opt. Comm.* **3**, 279 (1971).
24. D. H. Auston. *Appl. Phys. Lett.* **18**, 249 (1971).
25. D. J. Bradley, B. Liddy and W. E. Sleat. *Opt. Comm.* **2**, 391 (1971).
26. A. A. Malyutin and M. Ya. Shchelev. *JETP Lett.* **9**, 266 (1969).
27. E. P. Ippen, C. V. Shank and A. Dienes. *Appl. Phys. Lett.* **21**, 348 (1972); C. V. Shank and E. P. Ippen. *Appl. Phys. Lett.* **24**, 373 (1974).
28. E. P. Ippen and C. V. Shank. *Appl. Phys. Lett.* **27**, 488 (1975).
29. A. Penzkofer, D. von der Linde, A. Laubereau and W. Kaiser. *Appl. Phys. Lett.* **20**, 351 (1972); A. Penzkofer. *Opto-Electronics* **6**, 87 (1974).
30. S. E. Harris. Review: *Proc. IEEE* **57**, 2096 (1969).
31. A. Laubereau, L. Greiter and W. Kaiser. *Appl. Phys. Lett.* **25**, 87 (1974).
32. A. Seilmeier, K. Spanner and A. Laubereau. (To be published)
33. S. A. Akhmanov, R. V. Khokhlov and A. P. Sukhorukov. *In* "Laser Handbook" (F. T. Arecchi and E. O. Schulz-DuBois, Eds), North-Holland, Amsterdam (1972).
34. A. Penzkofer, A. Laubereau and W. Kaiser. *Phys. Rev. Lett.* **31**, 863 (1973).
35. A. Penzkofer, A. Seilmeier and W. Kaiser. *Opt. Comm.* **14**, 363 (1975).
36. A. Hasegawa and F. Tappert. *Appl. Phys. Lett.* **23**, 142 (1973). **23**, 171 (1973). D. N. Payne and W. A. Gambling. *Opto-electronics* **5**, 297 (1973).
37. N. G. Basov, P. G. Kriukov, S. D. Zakharov, Yu. V. Senatskii, and S. V. Tchekalin. *IEEE J. Quantum Electron.* **QE-4**, 864 (1968); J. Nuckolls, L. Wood, A. Thiessen and P. Zimmerman. *Nature* **239**, 139 (1972).
38. J. H. Marburger. *In* "Progress in Quantum Electronics" (J. H. Sanders and S. Stenholm, Eds), Vol. 4, Part 1, Pergamon, Oxford (1975); O. Rahn and M. Maier. *Phys. Rev.* **A9**, 1427 (1974).
39. O. Svelto. *In* "Progress in Optics" (E. Wolf, Ed.), North-Holland, Amsterdam, Vol. 12, p. 1 (1974).
40. M. Born and E. Wolf. "Principles of Optics", Pergamon Press, Oxford, 2nd ed. (1964).
41. G. Mayer and F. Gires. *Compt. Rend.* **258**, 2039 (1964).
42. F. Shimizu and B. P. Stoicheff. *IEEE J. Quantum Electron.* **QE-5**, 544 (1969).
43. M. A. Duguay and J. W. Hansen. *Appl. Phys. Lett.* **15**, 192 (1969).
44. M. R. Topp and G. C. Orner. *Chem. Phys. Lett.* **32**, 407 (1975); *Opt. Comm.* **13**, 276 (1975).
45. M. M. Malley and G. Mourou. *Opt. Comm.* **10**, 323 (1974).

46. J. Reintjes and R. L. Carman. *Phys. Rev. Lett.* **28**, 1697 (1972).
47. K. B. Eisenthal and K. H. Drexhage. *J. Chem. Phys.* **51**, 5720 (1969).
48. T. J. Chuang and K. B. Eisenthal. *Chem. Phys. Lett.* **11**, 368 (1971).
49. K. B. Eisenthal. *Chem. Phys. Lett.* **6**, 155 (1970).
50. T. J. Chuang, G. W. Hoffman and K. B. Eisenthal. *Chem. Phys. Lett.* **25**, 201 (1974).
51. P. M. Rentzepis, J. Jortner and R. P. Jones. *Chem. Phys. Lett.* **4**, 599 (1970); P. M. Rentzepis, M. R. Topp, R. P. Jones and J. Jortner. *Phys. Rev. Lett.* **25**, 1742 (1970); G. Mourou and M. M. Malley. *Opt. Comm.* **11**, 282 (1974).
52. E. G. Arthurs, D. J. Bradley, A. G. Roddie. *Chem. Phys. Lett.* **22**, 30 (1973).
53. I. W. Shelton and J. A. Armstrong. *IEEE J. Quantum Electron.* **QE-3**, 696 (1967); E. I. Scarlet, J. Figueira and H. Mahr. *Appl. Phys. Lett.* **13**, 71 (1968); M. M. Malley and P. M. Rentzepis. *Chem. Phys. Lett.* **3**, 534 (1969).
54. D. von der Linde and K. F. Rodgers. *IEEE J. Quantum Electron.* **QE-9**, 960 (1973).
55. R. M. Hochstrasser, H. Lutz and G. W. Scott. *Chem. Phys. Lett.* **24**, 162 (1974).
56. R. W. Anderson Jr., R. M. Hochstrasser, H. Lutz and G. W. Scott. *J. Chem. Phys.* **61**, 2500 (1974).
57. R. L. Carman, F. Shimizu, C. S. Wang and N. Bloembergen. *Phys. Rev.* **A2**, 60 (1970).
58. S. A. Akhmanov. *Mater. Res. Bull.* **4**, 455 (1969); S. A. Akhmanov, K. N. Drabovich, A. P. Sukhorukov and A. S. Chirkin. *Sov. Phys. JETP* **32**, 266 (1971).
59. For a review, see W. Kaiser and M. Maier. *In* "Laser Handbook" (F. T. Arecchi and E. O. Schulz-DuBois, Eds.), North-Holland, Amsterdam (1972).
60. A. Laubereau and W. Kaiser. (To be published)
61. A. Laubereau, D. von der Linde and W. Kaiser. *Phys. Rev. Lett.* **28**, 1162 (1972).
62. A. Laubereau. *Chem. Phys. Lett.* **27**, 600 (1974).
63. A. Laubereau, G. Wochner and W. Kaiser. *Phys. Rev.* **A13**, 2212 (1976); G. Wochner, A. Laubereau and W. Kaiser (To be published).
64. S. F. Fischer and A. Laubereau. *Chem. Phys. Lett.* **35**, 6 (1975).
65. A. Laubereau, G. Kehl and W. Kaiser. *Opt. Comm.* **11**, 74 (1974).
66. A. Laubereau, L. Kirschner and W. Kaiser. *Opt. Comm.* **9**, 189 (1973).
67. K. Spanner, A. Laubereau and W. Kaiser. *Chem. Phys. Lett.* **44**, 88 (1976).
68. R. R. Alfano and S. L. Shapiro. *Phys. Rev. Lett.* **29**, 1655 (1972).
69. A. Laubereau, A. Seilmeier and W. Kaiser. *Chem. Phys. Lett.* **36**, 232 (1975).
70. R. R. Alfano and S. L. Shapiro. *Phys. Rev. Lett.* **27**, 802 (1971).
71. A. Laubereau, D. von der Linde and W. Kaiser. *Phys. Rev. Lett.* **27**, 802 (1971).
72. A. Laubereau, G. Wochner and W. Kaiser. *Opt. Comm.* **14**, 75 (1975).
73. A. Laubereau, D. von der Linde and W. Kaiser. *Opt. Comm.* **7**, 173 (1973).
74. D. A. Kleinmann and W. G. Spitzer. *Phys. Rev.* **118**, 110 (1960).

8. Coherent Optical Spectroscopy

R. G. BREWER

IBM Research Laboratory, San Jose, California 95193, U.S.A.

I. INTRODUCTION

This chapter reviews a new branch of optical spectroscopy, what may be called *coherent optical spectroscopy*. Due to the availability of coherent laser light, atomic and molecular quantum states can now be prepared in coherent superposition, in the same way that nuclear spin systems have been prepared for over 25 years in pulsed nuclear magnetic resonance[1]. Coherently prepared samples of this type exhibit a class of transient phenomena that offer new ways to obtain ultra-high resolution optical spectra, and they allow the optical spectroscopist to isolate for the first time individual relaxation processes that have remained hidden within the optical lineshape. Many of these optical transients have now been realized using the recently introduced Stark switching technique[2] and will be reviewed.

We shall treat these coherent transient effects semiclassically using the coupled Schrödinger–Maxwell equations. Our procedure is to first solve the Schrödinger equation, either in density matrix or Bloch form, for a molecular system subject to a pulsed or continuous wave coherent optical field. This preparative stage generates a sample polarization that can radiate coherent light even after the preparation has passed, as prescribed by Maxwell's equations. Each transient effect corresponds to a particular sequence of one or more preparative steps, followed by periods where the sample freely radiates and where the polarization solutions can be traced from one stage to the next. In the case of a gas, the final polarization must be averaged over the Doppler velocity distribution. As long as the radiated field amplitude is small compared to the applied laser field, as in the present measurements, the laser field can be assumed constant; this is the so-called thin sample regime. At high optical densities, other coherence phenomena such as self-induced transparency[3] occur.

A. DENSITY MATRIX EQUATIONS

We consider a molecular two-level quantum configuration where the lower level is labeled 1 and the upper level is 2. In Section III, the corresponding three-level problem is treated. The molecular gas sample encounters a laser field

$$E_x(z, t) = E_0 \cos(\Omega t - kz) \tag{I.1}$$

of frequency Ω and amplitude E_0, polarized along the x-axis and propagating in the z-direction. The time-dependent behaviour of the density matrix ρ is given by the Schrödinger equation of motion

$$i\hbar\dot{\rho} = [H, \rho] + \text{relaxation terms.} \tag{I.2}$$

The Hamiltonian

$$H = H_0 + H_I \tag{I.3}$$

contains the free molecule part H_0 where the eigen-energies are

$$\begin{aligned} \langle 1|H_0|1\rangle &= \hbar\omega_1, \\ \langle 2|H_0|2\rangle &= \hbar\omega_2, \end{aligned} \tag{I.4}$$

and the 1–2 level splitting in angular frequency units is

$$\omega_2 - \omega_1 = \omega_{21}. \tag{I.5}$$

The molecule-optical field interaction term

$$H_I = -\boldsymbol{\mu} \cdot \mathbf{E}_x(z, t) \tag{I.6}$$

has the electric-dipole matrix element

$$\langle 1|\mu_x|2\rangle = \mu_{12} \neq 0. \tag{I.7}$$

For this two-level problem, Equation (I.2) becomes

$$\dot{\rho}_{11} = i\chi(\rho_{21} - \rho_{12})\cos(\Omega t - kz) - \frac{\rho_{11} - \rho_{11}^0}{T_1} \tag{I.8a}$$

$$\dot{\rho}_{12} = i\chi(\rho_{22} - \rho_{11})\cos(\Omega t - kz) - \left(i\omega_{12} + \frac{1}{T_2}\right)\rho_{12} \tag{I.8b}$$

where we have introduced the phenomenological decay times T_1, assumed equal for the diagonal elements ρ_{11} and ρ_{22}, and T_2 for ρ_{12}. The source term

ρ_{11}^0 denotes that molecules enter level 1 via relaxation from other quantum states and we have defined the Rabi flopping frequency as

$$\chi = \frac{\mu_{12} E_0}{\hbar}. \tag{I.9}$$

A similar equation in $\dot{\rho}_{22}$ can be obtained from (I.8a) by the index interchange $1 \leftrightarrow 2$ and an equation in $\dot{\rho}_{21}$ from (I.8b) by taking its complex conjugate.

The rapidly oscillating factors of the off-diagonal elements can be removed with the substitution

$$\rho_{12} = \tilde{\rho}_{12} e^{i(\Omega t - kz)} \tag{I.10}$$

and by neglecting nonresonant high frequency terms that oscillate as $e^{2i(\Omega t - kz)}$. This procedure, known as the rotating-wave approximation, reduces (I.8) to

$$\frac{d\rho_{11}}{dt} = \frac{i\chi(\tilde{\rho}_{21} - \tilde{\rho}_{12})}{2} - \frac{\rho_{11} - \rho_{11}^0}{T_1} \tag{I.11a}$$

$$\left(\frac{d}{dt} - i\Delta + \frac{1}{T_2}\right)\tilde{\rho}_{12} = \frac{i\chi(\rho_{22} - \rho_{11})}{2} \tag{I.11b}$$

where

$$\Delta = -\Omega + kv_z + \omega_{21}. \tag{I.12}$$

We also have recognized that the time derivative

$$\dot{\rho} = \left(\frac{\partial}{\partial t} + v_z \frac{\partial}{\partial z}\right)\rho \tag{I.13}$$

includes a second term because of a molecular velcoity component v along the z-axis that results in the Doppler shift kv_z of (1.12). Note that at resonance $\Delta = 0$.

B. BLOCH EQUATIONS

It is sometimes convenient to work with combinations of (I.11) and the corresponding equations for $\dot{\rho}_{22}$ and $\dot{\tilde{\rho}}_{21}$. For example, we may obtain

$$\dot{\tilde{\rho}}_{12} + \dot{\tilde{\rho}}_{21} = -i\Delta(\tilde{\rho}_{21} - \tilde{\rho}_{12}) - \frac{(\tilde{\rho}_{12} + \tilde{\rho}_{21})}{T_2}, \tag{I.14a}$$

$$i(\dot{\tilde{\rho}}_{21} - \dot{\tilde{\rho}}_{12}) = \Delta(\tilde{\rho}_{12} + \tilde{\rho}_{21}) + \chi(\rho_{22} - \rho_{11}) - \frac{i(\tilde{\rho}_{21} - \tilde{\rho}_{12})}{T_2}, \tag{I.14b}$$

$$\dot{\rho}_{22} - \dot{\rho}_{11} = -i\chi(\tilde{\rho}_{21} - \tilde{\rho}_{12}) - \frac{(\rho_{22} - \rho_{11})}{T_1} + \frac{(\rho_{22}^0 - \rho_{11}^0)}{T_1}. \tag{1.14c}$$

These are the Bloch equations and may be written as

$$\dot{u} + \Delta v + \frac{u}{T_2} = 0, \tag{I.15a}$$

$$\dot{v} - \Delta u - \chi w + \frac{v}{T_2} = 0, \tag{I.15b}$$

$$\dot{w} + \chi v + \frac{(w - w^0)}{T_1} = 0, \tag{I.15c}$$

using the variables

$$u = \tilde{\rho}_{12} + \tilde{\rho}_{21}, \tag{I.16a}$$

$$v = i(\tilde{\rho}_{21} - \tilde{\rho}_{12}), \tag{I.16b}$$

$$w = \rho_{22} - \rho_{11}. \tag{I.16c}$$

The set of Equations I.15 can be written compactly as

$$\frac{d\mathbf{B}}{dt} = \boldsymbol{\beta} \times \mathbf{B} \tag{I.17}$$

where these vectors have components

$$\mathbf{B} = \mathbf{i}u + \mathbf{j}v + \mathbf{k}w = [u, v, w], \tag{I.18}$$

$$\boldsymbol{\beta} = -\mathbf{i}\chi + \mathbf{k}\Delta = [-\chi, 0, \Delta], \tag{I.19}$$

and we have omitted relaxation for the moment[4]. Equation (I.17) has the geometric interpretation of a vector of constant length **B**, the Bloch vector, precessing about an effective field $\boldsymbol{\beta}$ (Fig. 1). In the nuclear magnetic resonance case formulated by Bloch[5], the spins precess in real space about an effective magnetic field, consisting of a static and an oscillating part. In the electric-dipole or optical case considered here, the precession is not in real space, but is only a geometric interpretation of a mathematical result. Relaxation may be included simply in (I.17) if we let $T = T_2 = T_1$, and thus the Bloch vector shrinks with time as $e^{-t/T}$. In general, the Bloch vector will not only be a function of z and t but of other variables such as molecular velocity $\mathbf{v}$, namely, $\mathbf{B}(\mathbf{v}, z, t)$.

The solutions to Equations (I.15), which were discussed initially by H. C. Torrey[6], are of the form

$$M(t) = Ae^{-at} + Be^{-bt}\cos st + Ce^{-bt}\sin st + D \tag{I.20}$$

where $M(t) = u(t)$, $v(t)$ or $w(t)$. The D term expresses steady state behaviour and the other terms are transient in nature. The constants cannot be obtained

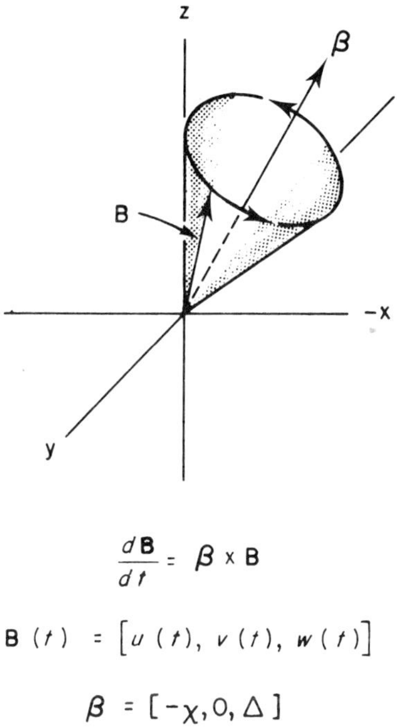

FIG. 1. The Bloch vector **B** in its precessional motion about the effective field **β** provides a simple geometrical representation of the time-dependent behaviour of the molecule-optical field interaction, following directly from Schrödinger's equation.

analytically in general except for certain specialized cases such as

(1) $\Delta = 0$,

(2) $T_1 = T_2$,

(3) $\chi \gg \dfrac{1}{T_1}$ and $\chi \gg \dfrac{1}{T_2}$.

C. POLARIZATION AND FIELD EQUATIONS

The light wave, Equation (I.1), induces in a sample of molecular density N the polarization

$$P(z, t) = N\mu_{12} e^{i(\Omega t - kz)} \langle \tilde{\rho}_{12} \rangle + \text{c.c.} \qquad \text{(I.21)}$$

The bracket $\langle\ \rangle$ denotes an average over the Maxwellian molecular velocity distribution

$$\langle\tilde{\rho}_{12}\rangle = \frac{1}{ku\sqrt{\pi}}\int_{-\infty}^{\infty}\tilde{\rho}_{12}e^{-(\Delta/ku)^2}\,d\Delta \tag{I.22}$$

where Δ is given by (I.12) and u is the rms velocity, not to be confused with the Bloch vector component. Note that $\tilde{\rho}_{12} = (u + iv)/2$. This polarization, in turn, is the origin of a signal field

$$E_s(z, t) = E_{12}(z, t)e^{i(\Omega t - kz)} + \text{c.c.} \tag{I.23}$$

which can be calculated from Maxwell's equations

$$\frac{\partial E_{12}}{\partial z} = -2\pi ikN\mu_{12}\langle\tilde{\rho}_{12}\rangle. \tag{I.24}$$

In Equation (I.24), only the lowest order terms of the wave equation

$$\frac{\partial^2 E(z, t)}{\partial z^2} - \frac{1}{c^2}\frac{\partial^2 E(z, t)}{\partial t^2} = \frac{4\pi}{c^2}\frac{\partial^2 P(z, t)}{\partial t^2} \tag{I.25}$$

have been retained, the so-called slowly-varying-envelope approximation.

D. OPTICAL NUTATION

(a) *Simplified Case*

Consider now the transient effect that results when a molecular sample is suddenly exposed to intense resonant laser light. Molecules begin to execute an alternating absorption and re-emission of laser light as they are driven coherently between upper and lower states (Fig. 2). The effect has been called "optical nutation"[7] by analogy with Torrey's transient "spin nutation"[6].

To illustrate the solution of the Bloch equations for this effect, we first simplify the problem by omitting the decay terms in (I.15) so that

$$\dot{u} + \Delta v = 0, \tag{I.26a}$$

$$\dot{v} - \Delta u - \chi w = 0, \tag{I.26b}$$

$$\dot{w} + \chi v = 0. \tag{I.26c}$$

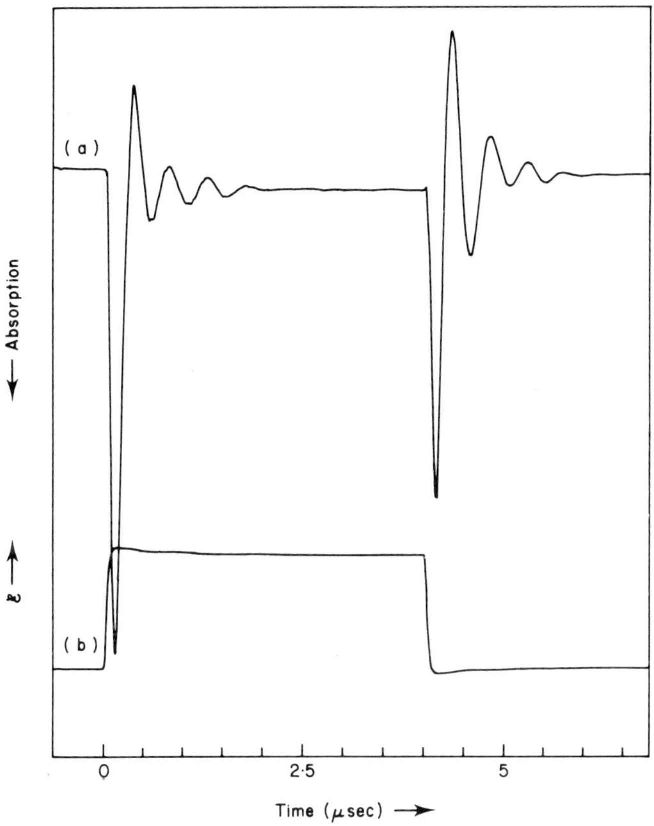

FIG. 2. The optical nutation effect in $^{13}CH_3F$ following a Stark pulse of amplitude $\mathscr{E}$ = 35 volts/cm. (From Brewer and Shoemaker, Ref. 2.)

These are easily solved to give in the régime $t > 0$

$$u(t) = \frac{\Delta\chi w(0)}{\beta^2}(\cos \beta t - 1), \tag{I.27a}$$

$$v(t) = \frac{\chi w(0)}{\beta} \sin \beta t, \tag{I.27b}$$

$$w(t) = w(0)\left[1 + \frac{\chi^2}{\beta^2}(\cos \beta t - 1)\right], \tag{I.27c}$$

where the initial conditions at time $t = 0$

$$\mathbf{B}(0) = [0, 0, w(0)] \tag{I.28}$$

have been utilized. Equation (I.27) states that the amplitude of the Bloch vector

$$B(t) = [u^2(t) + v^2(t) + w^2(t)]^{1/2} = w(0) \tag{I.29}$$

is preserved at all times, and that it precesses about the effective field β of (I.19) with a frequency

$$\beta = \sqrt{\Delta^2 + \chi^2}. \tag{I.30}$$

This motion is especially easy to visualize for the case of exact resonance, when $\Delta = 0$, because $u(t) = 0$ and the Bloch vector executes a right-angle rotation about $\beta = \chi$; see Fig. 1.

We now apply the results of Section IC. According to (I.24), the amplitude of the signal field which exits a sample of length L is

$$E_{12}(L, t) = -2\pi ikNL\mu_{12}\langle\tilde{\rho}_{12}\rangle, \tag{I.31}$$

and the Doppler-averaged density matrix is given by (I.22)

$$\begin{aligned}\langle\tilde{\rho}_{12}\rangle &= \frac{1}{ku\sqrt{\pi}}\int_{-\infty}^{\infty}\frac{1}{2}(u + iv)e^{-(\Delta/ku)^2}\,d\Delta \\ &\approx \frac{i\chi w(0)}{ku\sqrt{\pi}}e^{-(\Delta_1/ku)^2}\int_0^{\infty}\frac{\sin\sqrt{\Delta^2 + \chi^2}\,t}{\sqrt{\Delta^2 + \chi^2}}\,d\Delta \\ &\approx \frac{i\sqrt{\pi}}{2ku}\chi w(0)e^{-(\Delta_1/ku)^2}J_0(\chi t).\end{aligned} \tag{I.32}$$

For simplicity, we have assumed that the excitation is centred near the Doppler peak at $\Delta = \Delta_1$ so that the Bloch component $\langle u\rangle \sim 0$. Furthermore, we assume that the bandwidth excited is narrow compared to the Doppler width ku so that the Gaussian may be taken outside the integral. Combining (I.31) and (I.32), we obtain for the signal field (I.23)

$$E_s(L, t) = \frac{2\pi^{3/2}}{u}(NL)\chi w(0)\mu_{12}e^{-(\Delta_1/ku)^2}J_0(\chi t)\cos(\Omega t - kL). \tag{I.33}$$

The transmitted laser beam thus exhibits a slow oscillation, of frequency $\sim\chi$, that is expressed by the zero-order Bessel function $J_0(\chi t)$ and reflects the rate that molecules are driven between lower and upper states. This is essentially the result that is obtained in a more complete treatment where damping is not neglected, the case we will consider next. Figure 2 shows the "nutation" effect which displays the Bessel function ringing behaviour[2].

(b) *Nutation with Damping*

We now give the results of a similar calculation[6–9] where the damping terms in (I.15) are not neglected. Since the Bloch equations did not yield to an analytical solution in general, we need to assume that $T = T_2 = T_1$. We find in a straightforward manner that for times $t > 0$

$$u(t) = e^{-t/T}\left\{u(0) - \Delta\left[v(0) - \frac{\chi w^0/T}{\chi^2 + \Delta^2 + 1/T^2}\right]\frac{\sin \beta t}{\beta}\right.$$

$$+ \Delta\left[\Delta u(0) + \chi w(0) - \frac{\chi w^0/T^2}{\chi^2 + \Delta^2 + 1/T^2}\right]\frac{\cos \beta t - 1}{\beta^2}$$

$$\left. + \frac{\Delta \chi w^0}{\chi^2 + \Delta^2 + 1/T^2}\right\} - \frac{\Delta \chi w^0}{\chi^2 + \Delta^2 + 1/T^2}, \qquad \text{(I.34a)}$$

$$v(t) = e^{-t/T}\left\{\left[v(0) - \frac{\chi w^0/T}{\chi^2 + \Delta^2 + 1/T^2}\right]\cos \beta t\right.$$

$$\left. + \left[\Delta u(0) + \chi w(0) - \frac{\chi w^0/T^2}{\chi^2 + \Delta^2 + 1/T^2}\right]\frac{\sin \beta t}{\beta}\right\} + \frac{\chi w^0/T}{\chi^2 + \Delta^2 + 1/T^2},$$

$$\text{(I.34b)}$$

$$w(t) = e^{-t/T}\left\{w(0) - w^0 - \chi\left[v(0) - \frac{\chi w^0/T}{\chi^2 + \Delta^2 + 1/T^2}\right]\frac{\sin \beta t}{\beta}\right.$$

$$+ \chi\left[\Delta u(0) + \chi w(0) - \frac{\chi w^0/T^2}{\chi^2 + \Delta^2 + 1/T^2}\right]\frac{\cos \beta t - 1}{\beta^2}$$

$$\left. + \frac{\chi^2 w^0}{\chi^2 + \Delta^2 + 1/T^2}\right\} + w^0\left(1 - \frac{\chi^2}{\chi^2 + \Delta^2 + 1/T^2}\right). \qquad \text{(I.34c)}$$

Here, the initial conditions $u(0)$, $v(0)$ and $w(0)$ at $t = 0$ are not assumed to be zero and we recall that $w^0 = (\rho_{22}^0 - \rho_{11}^0)$ is the occupation probability difference in the absence of external radiation. The last term in Equations (I.34a, b and c) is the steady state value; it will be of interest in the case of steady state preparation. The transient solutions will be of interest for the case of pulse preparation, as in an echo experiment.

Proceeding as before, we adopt the initial conditions (I.28), $\mathbf{B}(0) = [0, 0, w(0)]$, and find that the Doppler-averaged density matrix element is

$$\langle\tilde{\rho}_{12}\rangle = \frac{1}{ku\sqrt{\pi}} \int_{-\infty}^{\infty} \frac{1}{2}(u + iv)e^{-(\Delta/ku)^2}\, d\Delta$$

$$\approx \frac{i\sqrt{\pi}}{2ku} \chi w(0) e^{-(\Delta_1/ku)^2} e^{-t/T} \left\{ J_0(\chi t) + \frac{2w^0/w(0)}{T\sqrt{\chi^2 + 1/T^2}} e^{t/T} \right.$$

$$- \frac{2w^0/w(0)}{\pi T} \int_0^{\infty} \frac{d\Delta}{\chi^2 + \Delta^2 + 1/T^2}$$

$$\left. \times \left(\cos\sqrt{\Delta^2 + \chi^2}\, t + \frac{\sin\sqrt{\Delta^2 + \chi^2}\, t}{T\sqrt{\Delta^2 + \chi^2}} \right) \right\}. \tag{I.35}$$

The leading term of (I.35) is the same result as Equation (I.32) except that it is damped by the factor $e^{-t/T}$. The second term is the steady state value of $\langle\tilde{\rho}_{12}\rangle$ and for the present purpose can be ignored. The remaining two integrals, which cannot be evaluated analytically, can be estimated by replacing the trigonometic functions by unity; they are smaller than the leading term by $\sim 1/\chi T$ and $\sim 1/(\chi T)^2$ respectively; these terms can be neglected usually since $\chi T \sim 10^2$. The nutation signal field then becomes

$$E_s(L, t) = \frac{2\pi^{3/2}}{u} (NL)\chi w(0)\mu_{12} e^{-(\Delta_1/ku)^2} e^{-t/T} J_0(\chi t)\cos(\Omega t - kL) \tag{I.36}$$

E. FREE INDUCTION DECAY

Let us assume that a molecular sample is resonantly excited under steady state conditions by a laser beam and that suddenly the excitation is terminated. Under these conditions, the sample will radiate an intense coherent beam of light which has been called "optical free induction decay"[9] (FID) by analogy with the NMR effect first seen by E. L. Hahn[10]. For the present purpose we shall assume that the preparative stage ends by switching the molecular transition frequency out of resonance with the laser frequency. The switching mechanism is the Stark effect and will be discussed in the next section.

The preparative steady state solutions for the period $t \leq 0$ can be derived from the Bloch equations (I.15) by setting the time derivatives equal to zero, and yield

$$u(0) = -\frac{\Delta\chi w^0}{\chi^2 T_1/T_2 + \Delta^2 + 1/T_2^2}, \tag{I.37}$$

$$v(0) = \frac{\chi w^0/T_2}{\chi^2 T_1/T_2 + \Delta^2 + 1/T_2^2}, \tag{I.38}$$

$$w(0) = w^0\left(1 - \frac{\chi^2 T_1/T_2}{\chi^2 T_1/T_2 + \Delta^2 + 1/T_2^2}\right). \tag{I.39}$$

These are the three time-independent terms of (I.34), but without the restriction that $T_1 = T_2$.

At time $t = 0$, we assume that the molecular transition frequency has been shifted by $\Delta\omega_{21}$ so that

$$\Delta \to \Delta' = \Delta + \Delta\omega_{21}, \tag{I.40}$$

and that the sample is sufficiently far out of resonance with laser light that we may set $\chi = 0$. The Bloch equations are then of the form

$$\dot{u} + \Delta' v + \frac{u}{T_2} = 0, \tag{I.41a}$$

$$\dot{v} - \Delta' u + \frac{v}{T_2} = 0, \tag{I.41b}$$

$$\dot{w} + \frac{w - w^0}{T_1} = 0, \tag{I.41c}$$

and the solutions for $t > 0$ are given by

$$u(t) = [u(0)\cos \Delta' t - v(0)\sin \Delta' t]e^{-t/T_2}, \tag{I.42a}$$

$$v(t) = [u(0)\sin \Delta' t + v(0)\cos \Delta' t]e^{-t/T_2}, \tag{I.42b}$$

$$w(t) = w^0 + [w(0) - w^0]e^{-t/T_1}. \tag{I.42c}$$

To obtain the field radiated, we again take the molecular velocity average expressed by (I.22)

$$\begin{aligned}
\langle \tilde{\rho}_{12} \rangle &= \frac{i}{2\sqrt{\pi}ku} \int_{-\infty}^{\infty} e^{-(\Delta/ku)^2} [u(0)\sin \Delta' t + v(0)\cos \Delta' t] e^{-t/T_2} \, d\Delta, \\
&\approx \frac{i}{2\sqrt{\pi}ku} e^{-(\Delta_1/ku)^2} \chi w^0 e^{-t/T_2} \cos \Delta\omega_{21} t \\
&\quad \times \int_{-\infty}^{\infty} \frac{-\Delta \sin \Delta t + (1/T_2)\cos \Delta t}{\chi^2 T_1/T_2 + \Delta^2 + 1/T_2^2} \, d\Delta, \\
&\approx \frac{i\sqrt{\pi}}{2ku} e^{-(\Delta_1/ku)^2} \chi w^0 e^{-t/T_2(1+\sqrt{\chi^2 T_1 T_2 + 1})} \\
&\quad \times \left(\frac{1}{\sqrt{\chi^2 T_1 T_2 + 1}} - 1 \right) \cos \Delta\omega_{21} t. \qquad \text{(I.43)}
\end{aligned}$$

Here, $\langle u \rangle = 0$ and we have assumed that the Doppler factor can be taken outside the integral. The field amplitude (I.24) associated with the induced polarization is

$$E_{12} = -i\left(\frac{2\pi\Omega}{c}\right) N L \mu_{12} \langle \tilde{\rho}_{12} \rangle, \qquad \text{(I.44)}$$

whereas the total field, including that of the laser, is

$$E_T = \left(E_{12} + \frac{E_0}{2} \right) e^{i(\Omega t - kz)} + \text{c.c.} \qquad \text{(I.45)}$$

The intensity, therefore, contains a cross term or beat

$$\begin{aligned}
(E^2)_{\text{beat}} &= 2E_{12}E_0 \\
&= E_0 Q_{12}(t) \cos \Delta\omega_{12} t \qquad \text{(I.46a)}
\end{aligned}$$

where

$$\begin{aligned}
Q_{12}(t) &= 2\pi^{3/2} N L \mu_{12}^2 E_0 w^0 \left(\frac{1}{\sqrt{\chi^2 T_1 T_2 + 1}} - 1 \right) \Big/ u\hbar \\
&\quad \times e^{-(\Delta_1/ku)^2} e^{-t/T_2(1+\sqrt{\chi^2 T_1 T_2 + 1})}. \qquad \text{(I.46b)}
\end{aligned}$$

The decay behaviour of (I.46) has two contributions[8,9]: (1) a homogeneous part with time constant T_2 and (2) an inhomogeneous part with time constant $T_2/\sqrt{\chi^2 T_1 T_2 + 1}$ that reflects the velocity bandwidth excited during steady state preparation. At moderately high laser intensities (a few watts/cm^2), the inhomogeneous dephasing can be dominant, the free induction signal will then decay rapidly and appear modulated at a frequency given by the Stark

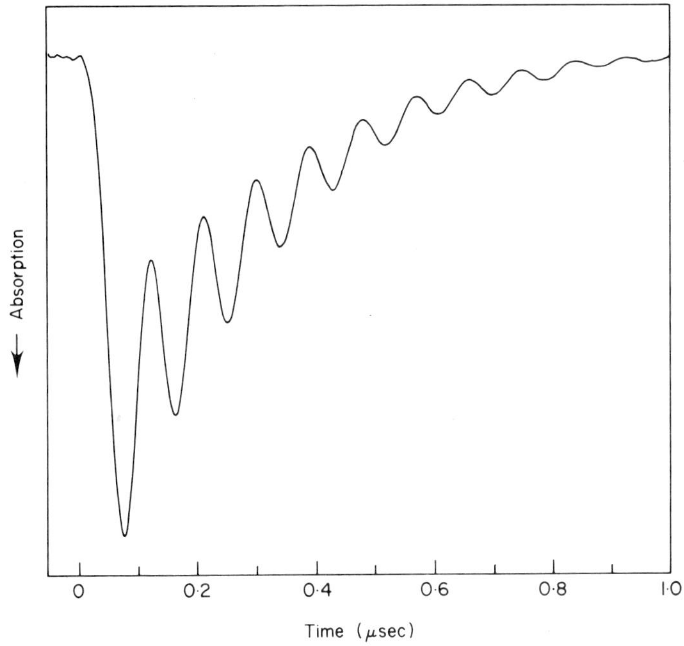

FIG. 3. Optical free induction decay in NH_2D following a step function Stark field. The beat frequency is the Stark shift, and the slowly varying background is a nutation signal of a second velocity group, more clearly shown in Fig. 2. (From Brewer and Shoemaker, Ref. 9.)

shift $\Delta\omega_{21}$. The subtle behaviour near the time origin is discussed elsewhere. An experimental demonstration of the FID effect is shown in Fig. 3.

F. EXPERIMENTS—THE STARK SWITCHING TECHNIQUE†

It may be refreshing at this point to illustrate how these coherent optical transient effects have been observed in our laboratory at IBM. The underlying

† *Note added in proof:* A new and more universal method which compliments the Stark-switching technique of subsection IF and which extends coherent transient measurements into the visible-ultraviolet region was just reported by R. G. Brewer and A. Z. Genack. *Phys. Rev. Lett.* **36**, 959 (1976). In this case, the laser frequency is switched instead of the sample's transition frequency. A stable tunable cw dye laser is frequency-switched by means of an intracavity electro-optic ADP crystal that is driven by a sequence of low voltage pulses. The laser frequency follows the refractive-index variations induced in the ADP crystal, and a sample exposed to this radiation exhibits coherent absorption or emission transients. Hence, the experiment is controlled electronically and in such a way that the advantages inherent in the Stark technique are preserved here as well. In addition, with the broad tuning range available in a dye laser, coherent transient phenomena can now be observed with ease in a large number of optical transitions in various atomic, molecular and solid state systems.

idea rests on the fact that gaseous molecules can exhibit exceedingly sharp optical resonances. Laser light, which is essentially monochromatic, will be strongly absorbed only when its frequency closely matches (to one part in 10^8 or 10^9) the molecular resonance frequency. If the molecular transition frequency is shifted slightly by some mechanism, the absorption can easily be switched on or off. The switching mechanism is the *Stark effect* and allows the molecular resonance to be *tuned* to or away from the laser frequency by subjecting the molecules to a pulsed dc electric field. Instead of applying pulses of laser light, which are difficult to control, the energy level spacing of the molecule is pulsed while the laser's intensity and frequency remain fixed in time. The technique[2] shown in Fig. 4 has been demonstrated in molecules in the infrared spectral region (10 μm) using a stable continuously operating CO_2 laser.

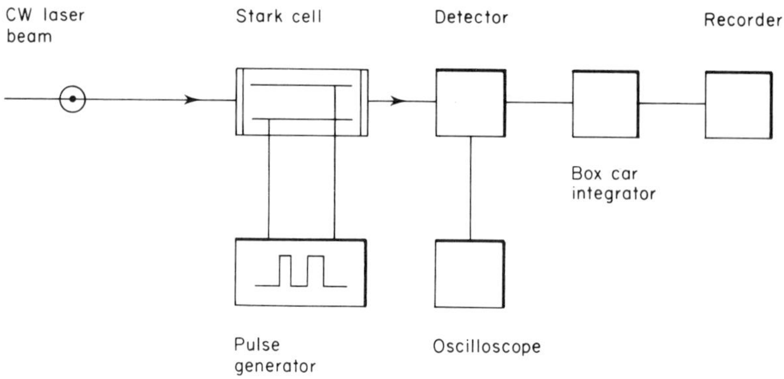

FIG. 4. Method for observing coherent optical transients following a sequence of Stark pulses. (From Brewer and Shoemaker, Ref. 2.)

The first order Stark splitting[11]

$$\Delta w_1 = \frac{-\mu \mathscr{E} MK}{J(J+1)} \tag{I.47}$$

occurs in symmetric top molecules such as methyl fluoride, CH_3F, because of the interaction of the molecule's permanent electric dipole moment μ with a static external electric field $\mathscr{E}$. The molecular rotational quantum states are designated by J and K while M labels the space-quantized levels. In Fig. 5, the laser photon energy matches the splitting of two of these levels, but if an electric pulse is applied, the upper and lower states will shift by different amounts so that laser light can no longer be absorbed.

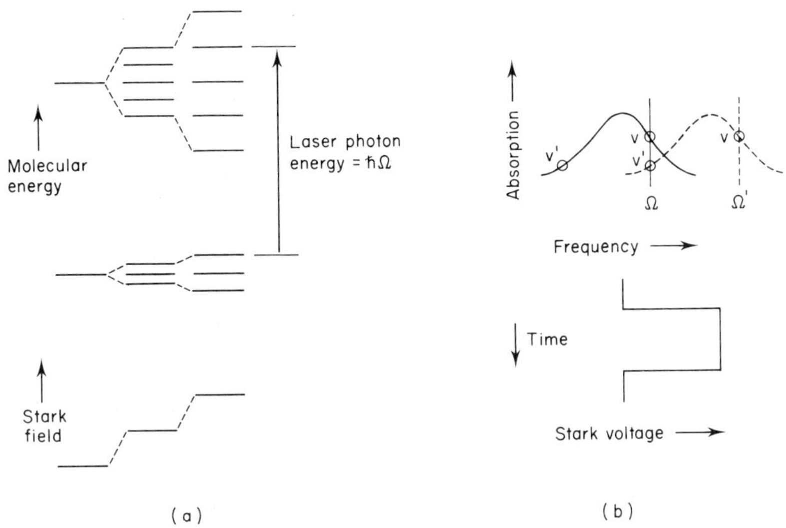

FIG. 5. The Stark effect is illustrated schematically in (a) for a molecule such as CH_3F that possesses a permanent dipole moment and therefore exhibits a first order Stark effect. (The level spacings are not drawn to scale as the Stark splittings are generally much smaller than the quantum levels separated by an optical energy.) In (b), the Stark switching principle devised by Brewer and Shoemaker for observing coherent optical transients is indicated. The sudden application of a dc electric field (a pulse) shifts the molecular level spacing because of the Stark effect and causes the transition frequency to jump from the solid to the dashed curve while the laser frequency Ω remains fixed. The absorption lineshape as shown is Doppler-broadened. Hence, molecules of velocity v (the component along the laser beam) that were initially excited by laser light are switched out of resonance and spontaneously emit a coherent beam of light at Ω' (Fig. 3). The velocity group v' that is suddenly switched into resonance exhibits an optical ringing effect (Fig. 2). When two Stark pulses are applied in succession, the velocity group v' spontaneously emits a photon echo (Fig. 9).

Since the electric pulses (Stark pulses) are generated by conventional electronics, the pulse sequence, amplitude, duration and shape are easily varied—a technique that allows the study of an entire class of coherent optical transient phenomena. This is a much simpler and more versatile method than generating short, very precisely timed laser pulses. Furthermore, since the cw laser itself is highly stable, the transient signals derived from it are highly reproducible. The method also may be applied in other spectral regions, at microwave or visible frequencies for example, and to atoms or solids at low temperature where the quadratic Stark effect would apply.

The Stark-switching technique has now allowed the observation of optical nutation[2], photon echoes[12] involving two[2] or multiple pulses (stimulated[13] and Carr-Purcell echoes[14]), optical free induction decay (FID)[9], coherent Raman beats[15], optical adiabatic fast passage[16,17], the optical analog of spin locking[18], FID interference pulses[19], and quantum beats[20]. With the exception of perhaps Raman beats, all of these effects are the optical analogs of spin transients that have been studied by nuclear magnetic resonance (NMR) over the past 25 years[1,6,10,21,22].

Let us review, by referring to Fig. 5, how a few of these coherent transients may be observed when a two-level system is suddenly subjected to a Stark pulse. Initially, molecules having a longitudinal velocity v are excited in steady state by laser light of frequency Ω, thereby preparing the transition levels in coherent superposition. When a pulse appears, this velocity group is no longer in resonance, but because of its preparation it will freely radiate an intense coherent infrared beam—the free induction decay signal (FID). At the same time, a second velocity group v' may be switched into resonance and will alternately absorb and emit laser radiation. This is the optical nutation effect. When the pulse terminates, the group v is suddenly excited and it too begins to nutate, while the second group v' now emits a FID signal. If two pulses are applied, it is possible to observe a photon echo, the optical analog of a spin echo. The optical nutation, FID and echo effects are illustrated in Figs. 2, 3 and 9, respectively.

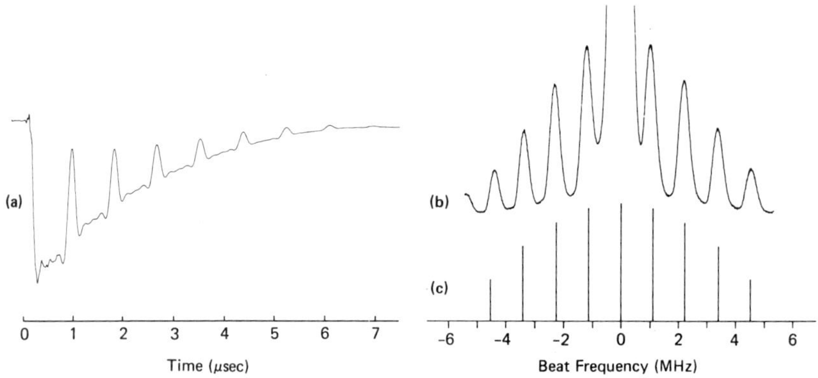

FIG. 6. (a) Free induction decay interference pulses in $^{13}CH_3F$, (b) the corresponding beat spectrum or its Fourier transform, and (c) the predicted spectrum. (From Foster *et al.*, Ref. 19.)

Transient light signals that are emitted by molecules switched out of resonance propagate in the forward direction because of the preparative step (**k** vector conservation), and are monitored together with the transmitted laser beam by a photodetector. Heterodyne detection is possible, as in the FID and photon echo experiments, since the emission signal is Stark-shifted from the laser frequency; heterodyne detection increases the emission signal amplitude 1000-fold and in the infrared, enhances detection sensitivity. In fact, incoherent infrared emission is many orders of magnitude weaker and is rarely observed.

Most of the experiments discussed here involve an infrared vibration-rotation transition of $^{13}CH_3F$. The transition is a fundamental ν_3 band $R(4)$ line, $(J, K) = (4, 3) \rightarrow (5, 3)$ that overlaps the $P(32)$ CO_2 laser line at 1035.474 cm^{-1} (9.66 microns). The CO_2 laser, which is described elsewhere[23], is free running and possesses high amplitude and frequency stability. The output is ~1 watt and the beam is expanded by a Galilean telescope to ~1 cm diameter to lengthen the molecule-optical interaction time. Furthermore, the light is linearly polarized, permitting $\Delta M = 0$ selection rules when its orientation is parallel to the dc Stark field or $\Delta M = \pm 1$ transitions when it is perpendicular to the Stark field.

In the above discussion of FID, we have considered coherent emission in a collection of two-level systems that are not degenerate. A new effect[19] arises, however, when the transition levels are degenerate during the steady state preparative stage. The emission that results after a step-function Stark field is applied is no longer a simple decay but instead appears as a train of sharp pulses regularly spaced in time due to a repetitive interference. The effect is shown in Fig. 6(a) for $^{13}CH_3F$ where the selection rule is $\Delta M = 0$. This situation arises because an entire set of infrared transitions within the Stark-split manifold are initially prepared. This emission, which beats with the laser, produces a heterodyne beat spectrum consisting of a set of regularly spaced frequencies that is the Fourier transform of the slowly decaying pulse progression observed. By means of a spectrum analyzer, the time behaviour of Fig. 6(a) can be Fourier transformed into the spectrum given in Fig. 6(b).

We thus demonstrate what is the optical analog of the well known NMR method of high resolution pulse Fourier spectroscopy[24]. In this simple example, the regular frequency interval between spectral lines is due to the first order Stark shift, but with other molecules, hyperfine structure and other spectral features might be investigated where very high resolution is required. In Fig. 6(b), the linewidth is ~0.5 MHz, considerably narrower than the 66 MHz Doppler width. The usual advantages of pulse Fourier spectroscopy apply, namely: (1) an enhanced signal to noise ratio compared to steady state techniques because of averaging repetitively produced signals, (2) rapid data acquisition, and (3) the possibility of monitoring *spectra* as a function of decay time in a variety of coherent transient experiments.

These interference pulses superficially resemble mode-locking of a laser because several evenly spaced frequency components are involved. However, in this case the regular frequency spacing is inherent in the transitions involved and no interaction between them is needed. The detailed behaviour of the pulse train agrees, in fact, with a FID theory[19] that assumes the transitions to be uncoupled. Using Equation (I.46a), we can show that for $\Delta M = 0$ selection rules, the photodetector will monitor cross terms in the optical intensity of the form

$$(E^2)_{\text{beat}} = \sum_{[1,2]} E_0 Q_{12}(t) \cos \Delta\omega_{12} t = E_0 Q \frac{\sin(9\delta\mathscr{E}t/2)}{\sin(\delta\mathscr{E}t/2)} \tag{I.48}$$

where $\delta\mathscr{E}$ is the frequency interval between two neighbouring lines of Fig. 6(b) due to a dc Stark field $\mathscr{E}$ and the number 9 signifies the number of transitions involved. For simplicity, we have taken all transition matrix elements to be equal. Equation (I.48) shows the characteristics displayed in Fig. 6(a), i.e., a pulse interval given by $T = 2\pi/\delta\mathscr{E}$ and a pulse width $\Delta t = 2\pi/9\delta\mathscr{E} = T/9$ so that their ratio $T/\Delta T = 9$ is just the number of transitions contributing. The more rigorous theory gives $T/\Delta T = 6.25$. The time scale becomes compressed as either the number of transitions, the Stark tuning rate δ or the Stark field $\mathscr{E}$ increases. It is tempting to consider the effect of higher Stark fields. At the present time, pulse widths of ~10 nsec have been realized with a modest Stark field of ~350 V/cm. At a field of ~100 000 V/cm, coherent infrared pulses of ~50 psec might be achieved under suitable conditions.

II. Photon Echoes and Molecular Collisions[25]

A. Conventional Echo Theory

The inhomogeneous dephasing discussed in the FID effect is a reversible phenomenon. Coherent electromagnetic energy stored momentarily in a sample can be partially recovered if the dephasing process is reversed in time. The spin echo[21], which set a precedent for this class of phenomena, provides such a mechanism. It was discovered by Hahn, and the echo concept has been extended to the optical region by Kurnit *et al.*[12].

To understand the dephasing—rephasing echo cycle in its simplest terms, consider Fig. 7. This shows the relative Doppler phase $\mathbf{k}\cdot\mathbf{v}\tau$ of a gaseous molecule moving with velocity $\mathbf{v}$ where $\mathbf{k}$ is the radiation propagation vector and τ is the elapsed time after excitation by an optical pulse.

Assume another pulse applied to the sample at time τ can reverse the sign of the phase from $\mathbf{k}\cdot\mathbf{v}\tau$ to $-\mathbf{k}\cdot\mathbf{v}\tau$. Since the phase continues to increase an amount $\mathbf{k}\cdot\mathbf{v}\Delta t$ in a time Δt, we see that at time $t = 2\tau$ the initial phase at time $t = 0$ will be reproduced. Different velocity groups behave in the same way, although they traverse different phase trajectories, and therefore, all velocity

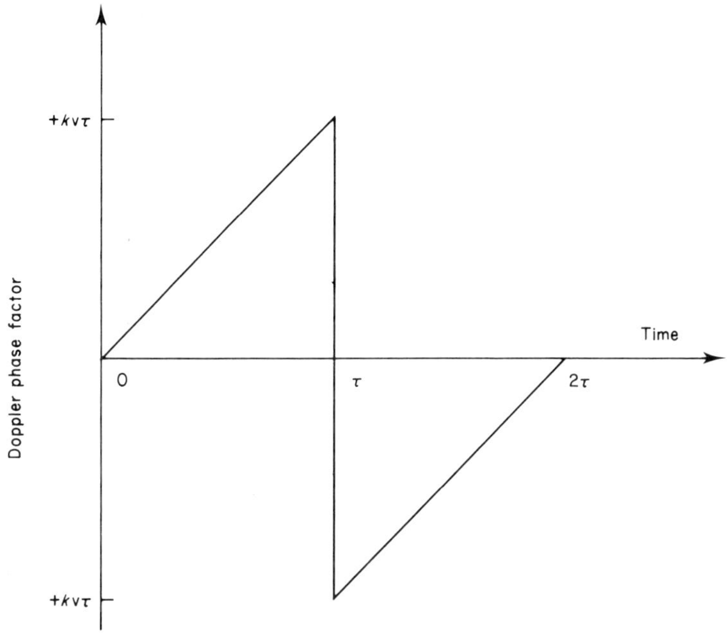

FIG. 7. The relative Doppler phase of molecules with longitudinal velocity v during a two pulse sequence. The molecules are all in phase because of a $\pi/2$-pulse at $t = 0$. The phase then advances but reverses sign at $t = \tau$ by application of a π-pulse. At $t = 2\tau$, the molecular induced dipoles have all rephased again and an echo signal is produced. (From Berman *et al.*, Ref. 25).

groups will come back into phase at $t = 2\tau$. When this happens, the macroscopic sample can coherently radiate a spontaneous burst of light—the photon echo.

We shall first present the conventional echo theory[21] for the Stark pulse sequence shown in Fig. 8. A comparison of these results with the geometrical Bloch vector description[4] will then be made. The role of elastic and inelastic

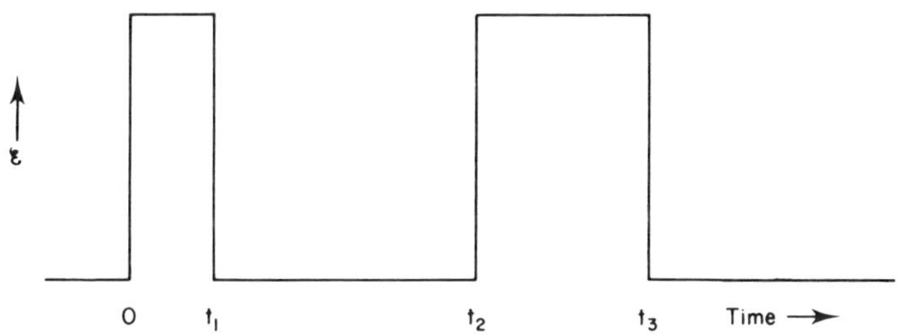

FIG. 8. Echo pulse sequence showing the Stark field amplitude versus time. The two pulses occur over the intervals $t = 0$ to t_1, and $t = t_2$ to t_3.

molecular collisions in photon echo theory will then be treated and compared with recent experiments.

Assume that the molecular gas sample is switched into resonance with the laser beam during the Stark pulse intervals $0 - t_1$ and $t_2 - t_3$ of Fig. 8 and is out of resonance at other times. The tuning parameter (I.12) in the relevant time regimes is

$$0 < t < t_1 \quad \text{and} \quad t_2 < t < t_3 : \Delta \tag{II.1a}$$

$$t_1 < t < t_2 \quad \text{and} \quad t_3 < t : \Delta' = \Delta + \Delta\omega_{21} \tag{II.1b}$$

where $\Delta\omega_{21}$ is the Stark shift.

$0 < t < t_1$—For a sufficiently brief pulse, $t_1 \ll T_2, T_1$, relaxation terms in the Bloch equations (I.15) can be dropped. Furthermore, the w^0 term does not contribute to echo formation and can be removed so that (I.15) reduces to

$$\dot{u} + \Delta v = 0, \tag{II.2a}$$

$$\dot{v} - \Delta u - \chi w = 0, \tag{II.2b}$$

$$\dot{w} + \chi v = 0. \tag{II.2c}$$

The solutions, subject to the initial conditions

$$\mathbf{B}(0) = [0, 0, w(0)] \tag{II.3}$$

are given by the nutation results (I.34) when we set $T = \infty$ and $w^0 = 0$. This procedure yields

$$u(t_1) = \frac{\Delta\chi w(0)}{\beta^2}(\cos\theta_{10} - 1), \tag{II.4a}$$

$$v(t_1) = \frac{\chi w(0)}{\beta}\sin\theta_{10}, \tag{II.4b}$$

$$w(t_1) = w(0) + \frac{\chi^2 w(0)}{\beta^2}(\cos\theta_{10} - 1), \tag{II.4c}$$

where the pulse areas are

$$\begin{aligned} \theta_{10} &= \sqrt{\chi^2 + \Delta^2}\, t_1 \\ \theta_{32} &= \sqrt{\chi^2 + \Delta^2}\,(t_3 - t_2) \end{aligned} \tag{II.5}$$

and we recall that

$$\beta = \sqrt{\chi^2 + \Delta^2}.$$

$t_1 < t < t_2$—Over this interval, it is assumed that the sample is sufficiently far out of resonance that $\chi \sim 0$ and the Bloch equations become

$$\dot{u} + \Delta' v + \frac{u}{T_2} = 0, \qquad \text{(II.6a)}$$

$$\dot{v} - \Delta' u + \frac{v}{T_2} = 0, \qquad \text{(II.6b)}$$

$$\dot{w} + \frac{w}{T_1} = 0. \qquad \text{(II.6c)}$$

The w^0 term has been omitted in (II.6c) again. These are the same equations used to describe FID and the solutions, evaluated at $t = t_2$, are the same as (I.42)

$$u(t_2) = [u(t_1)\cos\Delta'(t_2 - t_1) - v(t_1)\sin\Delta'(t_2 - t_1)]e^{-(t_2-t_1)/T_2}, \qquad \text{(II.7a)}$$

$$v(t_2) = [u(t_1)\sin\Delta'(t_2 - t_1) + v(t_1)\cos\Delta'(t_2 - t_1)]e^{-(t_2-t_1)/T_2}, \qquad \text{(II.7b)}$$

$$w(t_2) = w(t_1)e^{-(t_2-t_1)/T_1}, \qquad \text{(II.7c)}$$

but subject to the initial conditions

$$\mathbf{B}(t_1) = [u(t_1), v(t_1), w(t_1)]. \qquad \text{(II.8)}$$

$t_2 < t < t_3$—For the second pulse, the solutions of (II.2) are again given by the nutation results (I.34) where $T = \infty$, $w^0 = 0$ and the initial condition is

$$B(t_2) = [u(t_2), v(t_2), w(t_2)]. \qquad \text{(II.9)}$$

We find at $t = t_3$ that

$$u(t_3) = u(t_2) - \frac{\Delta}{\beta} v(t_2)\sin\theta_{32} - \frac{2\Delta}{\beta^2}[\Delta u(t_2) + \chi w(t_2)]\sin^2\frac{\theta_{32}}{2}, \qquad \text{(II.10a)}$$

$$v(t_3) = v(t_2)\cos\theta_{32} + \frac{1}{\beta}[\Delta u(t_2) + \chi w(t_2)]\sin\theta_{32}, \qquad \text{(II.10b)}$$

$$w(t_3) = w(t_2) - \frac{\chi}{\beta} v(t_2)\sin\theta_{32} - \frac{2\chi}{\beta^2}[\Delta u(t_2) + \chi w(t_2)]\sin^2\frac{\theta_{32}}{2}. \qquad \text{(II.10c)}$$

$t_3 < t$—For the period following the second pulse, the FID Equations (II.7) yield

$$u(t) = [u(t_3)\cos\Delta'(t - t_3) - v(t_3)\sin\Delta'(t - t_3)]e^{-(t-t_3)/T_2}, \qquad \text{(II.11a)}$$

$$v(t) = [u(t_3)\sin\Delta'(t - t_3) + v(t_3)\cos\Delta'(t - t_3)]e^{-(t-t_3)/T_2}, \qquad \text{(II.11b)}$$

$$w(t) = w(t_3)e^{-(t-t_3)/T_1}. \qquad \text{(II.11c)}$$

The earlier solutions (II.10) and (II.7) when inserted in (II.11) result in terms that contain as a factor $\cos \Delta'(t - 2\tau)$ where

$$2\tau = (t_3 + t_2 - t_1). \tag{II.12}$$

It is clear that these terms are responsible for echo formation at time

$$t = 2\tau \tag{II.13}$$

since the Doppler phase vanishes. The other terms such as $\sin \Delta'(t - 2\tau)$ are zero at $t = 2\tau$ or do not rephase and need not be considered. Another point is that the echo terms of $u(t)$ are an odd function in Δ so that if the optical excitation selects molecules near the Doppler peak, we may consider the Doppler-averaged quantity to be

$$\langle u_c \rangle \sim 0. \tag{II.14}$$

The subscript c designates that only rephasing or echo terms are included. For this case, the polarization (I.21) and the echo field (I.23) that it radiates at time $t = 2\tau$ is due solely to the v_c component of (II.11b). It is of the form

$$\begin{aligned} \langle v_c(t) \rangle = &-\chi^3 w(0) e^{-t/T_2} \cos \Delta\omega_{21}(t - 2\tau) \\ &\times \left\langle \frac{1}{\beta^3} \sin \theta_{10} \sin^2\left(\frac{\theta_{32}}{2}\right) \cos \Delta(t - 2\tau) \right\rangle. \end{aligned} \tag{II.15}$$

Keeping in mind that $\langle \tilde{\rho}_{12} \rangle = \frac{1}{2}\langle u + iv \rangle$, we find that the echo field amplitude

$$E_{12}(L, t) = -2\pi i k N L \mu_{12} \langle \tilde{\rho}_{12}(t) \rangle$$

and the laser field (I.1) add to give (I.45). As in FID, the intensity contains a beat term

$$\begin{aligned} [E_c^2(t)]_{\text{beat}} &= 2E_{12}(t)E_0 \\ &= 2\pi\hbar k N L \chi^4 w(0) e^{-t/T_2} \cos \Delta\omega_{21}(t - 2\tau) \\ &\quad \times \left\langle \frac{1}{\beta^3} \sin \theta_{10} \sin^2\left(\frac{\theta_{32}}{2}\right) \cos \Delta(t - 2\tau) \right\rangle \end{aligned} \tag{II.16}$$

where the beat frequency is the Stark shift $\Delta\omega_{21}$. The echo signal† reaches a maximum value at time $t = 2\tau$ and (II.16) becomes

$$[E_c^2(t = 2\tau)]_{\text{beat}} = 2\pi\hbar k N L \chi^4 w(0) e^{-t/T_2} \left\langle \frac{1}{\beta^3} \sin \theta_{10} \sin^2\left(\frac{\theta_{32}}{2}\right) \right\rangle. \tag{II.17}$$

† *Note added in proof:* The echo Doppler integral of (II.17) has been evaluated analytically for the limiting cases where either a large or a small fraction of the Doppler linewidth is prepared. See A. Schenzle, S. Grossman and R. G. Brewer. *Phys. Rev.* **A 13**, 1891 (1976).

The envelope function of the echo signal therefore decays with a time constant T_2 that is independent of the more rapid inhomogeneous Doppler dephasing encountered in FID. Equation (II.17) also indicates that the signal will be largest when the pulse areas are $\theta_{10} = \pi/2$ and $\theta_{32} = \pi$.

An example[2] of the photon echo effect is shown in Fig. 9 for $^{13}CH_3F$. The echo is shown as the third optical pulse in the upper trace; the first two pulses are manifestations of nutation behaviour (preparation) that accompany the two Stark pulses shown on the lower trace. The various times labeled (a) through (d) of the pulse sequence are described on the right in terms of the Bloch vector's precessional motion. The dephasing-rephasing cycle of the various molecular velocity groups occurs in the following way. For the case of exact resonance, optical excitation during the first Stark pulse causes the Bloch vector to precess about the x-axis. At the end of the pulse (a), the Bloch vector stops, in this case in the x-y plane. Over the time interval (a)–(b), the vector decomposes into a set of components that precess about the z-axis at different speeds, fast (f) and slow (s). Each component vector corresponds to one molecular velocity group emitting at its own unique frequency because of the Doppler shift. The second optical excitation in the interval (b)–(c) causes the "pancake" of vectors to again rotate about the x-axis, thereby placing the slow vectors ahead of the fast ones. Following the second pulse, these

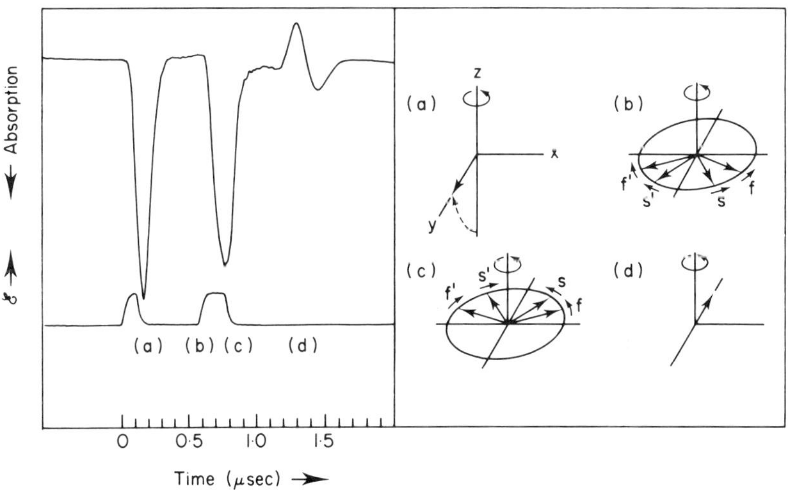

FIG. 9. Left: A photon echo in $^{13}CH_3F$ is shown as the third optical pulse in the upper trace; the first two optical pulses are nutation signals that accompany the two Stark pulses shown on the lower trace (Brewer and Shoemaker, Ref. 2). The various times labeled (a) through (d) of the pulse sequence are described on the right in terms of the Bloch vector's precessional motion.

vectors continue to rotate in the same direction about z, as before. At (d), the fast vectors catch up with the slow ones, and the macroscopic sample now with its induced dipoles all in phase emits a coherent burst of light, the photon echo. Note that the Bloch vector at this stage is reversed in direction from that at (a). Therefore, while the vectors rephase they do not retrace their behaviour in time.

Many of the features predicted by (II.16) and (II.17) appear to be verified by our experiments at IBM. For example, as Fig. 9 illustrates, the echo is observed as a heterodyne beat signal. In addition, the echo signal is observed to be proportional to the number of molecules N, in agreement with (II.17), and hence, the echo intensity varies as N^2. The signal amplitude also goes through a maximum value when the first and second pulse areas are approximately $\pi/2$ and π, respectively.

However, there is one serious disagreement with the observations and that is in the decay behaviour. The echo amplitude for different pulse delay times does not follow the simple exponential decay e^{-t/T_2} of (II.17) but instead behaves as in the lower curve of Fig. 10[14]. The limiting time-dependence is in

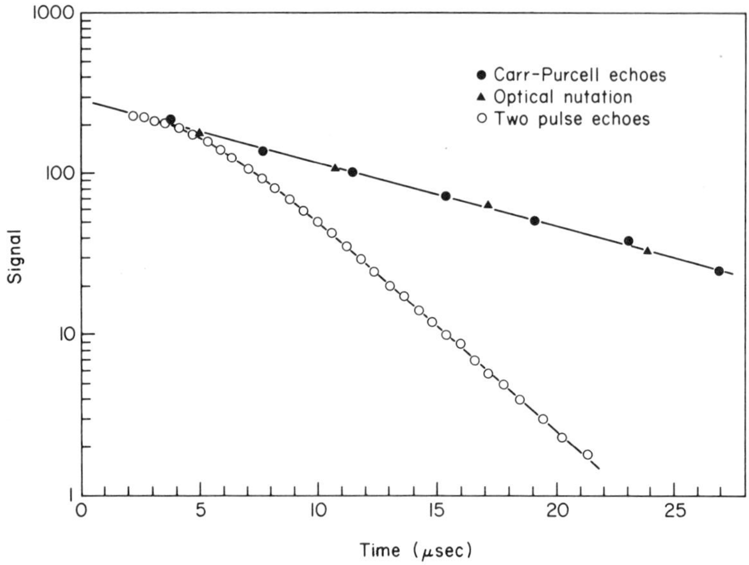

FIG. 10. Decay behaviour of two coherent optical transients in $^{13}CH_3F$. The upper curve is obtained either by multiple pulse photon echoes (Carr-Purcell echoes) or by the delayed nutation technique (Ref. 25) and measures the decay time T_1. The lower curve is due to the two-pulse photon echo; its deviation from the upper curve is due to elastic molecular collisions. (From Schmidt *et al.*, Ref. 14.)

fact of the form

$$\begin{aligned} &e^{-Kt^3} \quad \text{for short times,} \\ &e^{-\Gamma t} \quad \text{for long times.} \end{aligned} \tag{II.18}$$

The reason for the discrepancy with (II.17) is that the traditional echo theory does not allow for elastic collisions between molecular pairs where the linear velocity exhibits a diffusive character through small angle scattering. The transition frequency of the radiating sample is no longer pure therefore, but displays a bandwidth, corresponding to the velocity spread produced by elastic collisions. The velocity packets diverge in their phase relationship causing the echo amplitude to diminish with a characteristic time-dependence that distinguishes the velocity diffusion mechanism from other dephasing processes. An echo theory that includes elastic collisions is presented in the next section.

B. THE EFFECT OF ELASTIC COLLISIONS

To allow for an elastic collision dephasing mechanism in echo formation, we introduce two new terms in the Bloch equations

$$\dot{u}(\mathbf{v}, t) = -\Delta v(\mathbf{v}, t) - (\Gamma_1 + \Gamma)u(\mathbf{v}, t) + \int d^3\mathbf{v}' W(\mathbf{v}' \to \mathbf{v})u(\mathbf{v}', t), \tag{II.19a}$$

$$\dot{v}(\mathbf{v}, t) = \Delta u(\mathbf{v}, t) + \chi w(\mathbf{v}, t) - (\Gamma_1 + \Gamma)v(\mathbf{v}, t) + \int d^3\mathbf{v}' W(\mathbf{v}' \to \mathbf{v})v(\mathbf{v}', t), \tag{II.19b}$$

$$\dot{w}(\mathbf{v}, t) = -\chi v(\mathbf{v}, t) - (\Gamma_1 + \Gamma)w(\mathbf{v}, t) + \int d^3\mathbf{v}' W(\mathbf{v}' \to \mathbf{v})w(\mathbf{v}', t), \tag{II.19c}$$

which follow from a Boltzmann or quantum mechanical transport equation[25–27]. (Note that the Bloch vector component v is not to be confused with molecular velocity $\mathbf{v}$.) In contrast to Section II–A, the Bloch vector $\mathbf{B}(\mathbf{v}, t)$ for an N molecule sample is now weighted by the thermal equilibrium velocity distribution $G(\mathbf{v})$ where

$$\mathbf{B}(\mathbf{v}, t) = \mathbf{B}'(\mathbf{v}, t)e^{-\Gamma_1 t} = \tilde{\mathbf{B}}(\mathbf{v}, t)NG(\mathbf{v})e^{-\Gamma_1 t}, \tag{II.20}$$

$\Gamma_1 = 1/T_1$ and the $e^{-\Gamma_1 t}$ factor will permit us to remove the Γ_1 terms in (II.19). The quantity $-\int d^3\mathbf{v}' W(\mathbf{v} \to \mathbf{v}')\mathbf{B}(\mathbf{v}, t) = -\Gamma\mathbf{B}(\mathbf{v}, t)$ expresses the reduction of $\mathbf{B}(\mathbf{v}, t)$ due to elastic collisions associated with the collision kernel $W(\mathbf{v} \to \mathbf{v}')$ that bring molecules from velocity $\mathbf{v} \to \mathbf{v}'$. The factor Γ is the total rate of elastic collisions. The second term $\int d^3v' W(\mathbf{v}' \to \mathbf{v})\mathbf{B}(v', t)$ represents the inverse process and gives the increase in $\mathbf{B}(\mathbf{v}, t)$ due to elastic collisions that change the velocity from $\mathbf{v}'$ to $\mathbf{v}$.

In (II.19), we have neglected the possibility of phase interrupting collisions[26,27] so that $T_2 = T_1$. That is, for our present purpose, an elastic collision is state-independent. This simplification is suggested by our current experiments in the infrared although it may not be valid in the visible or ultraviolet region. As before, the source term w^0 is dropped in (II.19c) since unprepared molecules that enter the transition states do not contribute to echo formation. We also neglect any spatial variation of $\mathbf{B}$ over the laser beam geometry.

To carry out the calculation, a collision kernel $W(\mathbf{v}' \to \mathbf{v})$ must be selected. *Since the functional time dependence of the results will be independent of the choice of kernel*, we have adopted the Brownian motion kernel of Keilson and Storer[28],

$$W(\mathbf{v}' \to \mathbf{v}) = \Gamma(\pi \Delta u^2)^{-3/2} e^{-(\mathbf{v} - \alpha \mathbf{v}')^2/\Delta u^2} \tag{II.21}$$

where α is a constant close to but less than unity. Equation (II.21) and detailed balancing

$$W(\mathbf{v}' \to \mathbf{v})G(\mathbf{v}') = G(\mathbf{v})W(\mathbf{v} \to \mathbf{v}'), \tag{II.22}$$

impose the constraint that

$$\Delta u^2 = (1 - \alpha^2)u^2 \approx 2(1 - \alpha)u^2 \tag{II.23}$$

where Δu is root-two times the rms change in velocity per collision while u is the most probable speed of the *thermal equilibrium distribution*. This kernel is mathematically simple to treat and the assumption that $\alpha \sim 1$, which corresponds to having very small changes in velocity per collision, will be supported by the experimental data.

(a) *Simplified Case*

To illustrate the physics of the problem, we consider a simplified case where the pulses are sufficiently short that $|\Delta t_1| \ll 1$ and $|\Delta(t_3 - t_2)| \ll 1$ while at the same time the optical field strength is large enough that $\chi t_1 \sim 1$ and $\chi(t_3 - t_2) \sim 1$. Physically, this corresponds to assuming that the pulses uniformly excite the entire thermal distribution of molecules or that they are all on resonance ($\Delta = 0$).

The solutions are now traced for each period as in the previous treatment.

$0 < t < t_1$—Equations (II.19) become

$$\dot{u}'(\mathbf{v}, t) = 0, \quad \text{(II.24a)}$$

$$\dot{v}'(\mathbf{v}, t) = \chi w'(\mathbf{v}, t), \quad \text{(II.24b)}$$

$$\dot{w}'(\mathbf{v}, t) = -\chi v'(\mathbf{v}, t), \quad \text{(II.24c)}$$

where the system is assumed to be on resonance and relaxation for short pulses can be ignored. The solution at $t = t_1$ subject to the initial condition

$$B'(\mathbf{v}, 0) = [0, 0, w'(0)], \quad \text{(II.25)}$$

is the same as Equation (II.4) when we set $\Delta = 0$, namely,

$$u'(\mathbf{v}, t_1) = u'(\mathbf{v}, 0) = 0, \quad \text{(II.26a)}$$

$$v'(\mathbf{v}, t_1) = w'(\mathbf{v}, 0)\sin \chi t_1 = -NG(\mathbf{v})\sin \chi t_1, \quad \text{(II.26b)}$$

$$w'(\mathbf{v}, t_1) = w'(\mathbf{v}, 0)\cos \chi t_1 = -NG(\mathbf{v})\cos \chi t_1. \quad \text{(II.26c)}$$

These results may be simplified further by choosing t_1 such that $\chi t_1 = \pi/2$ to yield

$$u'(\mathbf{v}, t_1) = 0, \quad \text{(II.27a)}$$

$$v'(\mathbf{v}, t_1) = -NG(\mathbf{v}), \quad \text{(II.27b)}$$

$$w'(\mathbf{v}, t_1) = 0. \quad \text{(II.27c)}$$

$t_1 < t < t_2$—In this interval, χ is effectively zero and (II.19) becomes

$$\dot{\tilde{u}}'(\mathbf{v}, t) = -\Delta'\tilde{v}'(\mathbf{v}, t) - \Gamma\tilde{u}'(\mathbf{v}, t) + \int d^3v' W(\mathbf{v} \to \mathbf{v}')\tilde{u}'(\mathbf{v}', t), \quad \text{(II.28a)}$$

$$\dot{\tilde{v}}'(\mathbf{v}, t) = \Delta'\tilde{u}'(\mathbf{v}, t) - \Gamma\tilde{v}'(\mathbf{v}, t) + \int d^3v' W(\mathbf{v} \to \mathbf{v}')\tilde{v}'(\mathbf{v}', t), \quad \text{(II.28b)}$$

$$\dot{\tilde{w}}'(\mathbf{v}, t) = -\Gamma\tilde{w}'(\mathbf{v}, t) + \int d^3v' W(\mathbf{v} \to \mathbf{v}')\tilde{w}'(\mathbf{v}', t). \quad \text{(II.28c)}$$

Due to the Stark shift $\Delta\omega_{21}$

$$\Delta'(\mathbf{v}) = \Delta(\mathbf{v}) + \Delta\omega_{21}$$

Here, we have incorporated (II.20), thereby removing the Γ_1 term in (II.19). In addition, the detailed balancing relation (II.22) has been utilized in the integral of (II.19) so that the factor $NG(v)$ common to every term of this equation can be removed.

From the assumed thermal equilibrium condition, implied by (II.22), the solution of (II.28c) is simply

$$\tilde{w}'(\mathbf{v}, t) = \tilde{w}'(\mathbf{v}, t_1), \tag{II.29a}$$

and from (II.27c)

$$w'(\mathbf{v}, t_2) = 0. \tag{II.29b}$$

That is, the $\pi/2$-pulse equalizes the population in levels 1 and 2, and thereafter it remains constant according to our simplifying assumption.

Equations (II.28a and b) are solved by combining them as in (II.7), i.e.,

$$\begin{aligned}\dot{\tilde{u}}'(\mathbf{v}, t) - i\dot{\tilde{v}}'(\mathbf{v}, t) = {} & -(i\Delta'(\mathbf{v}) + \Gamma)(\tilde{u}'(\mathbf{v}, t) - i\tilde{v}'(\mathbf{v}, t)) \\ & + \int d^3v'\, W(\mathbf{v} \to \mathbf{v}')(\tilde{u}'(\mathbf{v}', t) - i\tilde{v}'(\mathbf{v}', t)).\end{aligned} \tag{II.30}$$

In terms of the density matrix element $\tilde{\rho}_{21} = \frac{1}{2}(u - iv)$, Equation (II.30) reduces to

$$\begin{aligned}\frac{d}{dt}(\tilde{\rho}'_{21}(\mathbf{v}, t)e^{i\Delta'(v)t}) = {} & -\Gamma\tilde{\rho}'_{21}(\mathbf{v}, t)e^{i\Delta'(v)t} \\ & + \int d^3v'\, W(\mathbf{v} \to \mathbf{v}')\tilde{\rho}'_{21}(\mathbf{v}', t)e^{i\Delta'(\mathbf{v})t}.\end{aligned} \tag{II.31}$$

We, therefore, define the new variable

$$\tilde{\rho}'_{21}(\mathbf{v}, t, t_1) = \tilde{\rho}'_{21}(\mathbf{v}, t)e^{i\Delta'(\mathbf{v})(t - t_1)} \tag{II.32}$$

which obeys the equation

$$\dot{\tilde{\rho}}'_{21}(\mathbf{v}, t, t_1) = -\Gamma\tilde{\rho}'_{21}(\mathbf{v}, t, t_1) + \int d^3v'\, W(\mathbf{v} \to \mathbf{v}')\tilde{\rho}'_{21}(\mathbf{v}', t, t_1)\, e^{i\mathbf{k}\cdot(\mathbf{v} - \mathbf{v}')(t - t_1)} \tag{II.33}$$

and is subject to the initial condition

$$\tilde{\rho}'_{21}(\mathbf{v}, t_1, t_1) = \tfrac{1}{2}[\tilde{u}'(\mathbf{v}, t_1) - i\tilde{v}'(\mathbf{v}, t_1)] = \frac{i}{2}. \tag{II.34}$$

Note that the $e^{-i\Delta' t_1}$ factor in (II.32) is arbitrary and is added for convenience, as will be apparent later.

We now try a solution of the form

$$\tilde{\rho}'_{21}(\mathbf{v}, t, t_1) = \frac{i}{2} e^{[i\mathbf{x}(t,t_1)\cdot\mathbf{v} + y(t,t_1)]} \quad \text{(II.35)}$$

where the variables $\mathbf{x}$ and y satisfy initial conditions

$$\mathbf{x}(t_1, t_1) = 0, \; y(t_1, t_1) = 0. \quad \text{(II.36)}$$

Substituting (II.35) into (II.33), there results

$$\begin{aligned} i\dot{\mathbf{x}}(t, t_1)\cdot\mathbf{v} + \dot{y}(t, t_1) &= -\Gamma + \int d^3v' W(\mathbf{v} \to \mathbf{v}') e^{i[\mathbf{k}(t-t_1)-\mathbf{x}]\cdot(\mathbf{v}-\mathbf{v}')} \\ &= -\Gamma + \Gamma e^{i[\mathbf{k}(t-t_1)-\mathbf{x}]\cdot\mathbf{v}(1-\alpha)} e^{-[\mathbf{k}(t-t_1)-\mathbf{x}]^2\Delta u^2/4} \end{aligned} \quad \text{(II.37)}$$

where the collision kernel (II.21) has been used. The second term on the right will contribute only if

$$[\mathbf{k}(t - t_1) - \mathbf{x}]^2 \Delta u^2 \lesssim 1. \quad \text{(II.38)}$$

In that case, using $1 - \alpha = \Delta u^2/2u^2$, the leading exponential of (II.37) is

$$[\mathbf{k}(t - t_1) - \mathbf{x}]\cdot\mathbf{v}(1 - \alpha) = [\mathbf{k}(t - t_1) - \mathbf{x}]\cdot\frac{\mathbf{v}\Delta u^2}{2u^2} \lesssim 1, \quad \text{(II.39)}$$

and this term may be expanded to give

$$i\mathbf{x}\cdot\mathbf{v} + \dot{y} \approx -\Gamma + \Gamma(1 + i(1 - \alpha)[\mathbf{k}(t - t_1) - \mathbf{x}]\cdot\mathbf{v})\, e^{-[\mathbf{k}(t-t_1)-\mathbf{x}]^2\Delta u^2/4}. \quad \text{(II.40)}$$

This equation immediately yields two equations

$$\dot{\mathbf{x}}(t, t_1) = \Gamma(1 - \alpha)[\mathbf{k}(t - t_1) - \mathbf{x}]e^{-[\mathbf{k}(t-t_1)-\mathbf{x}]^2\Delta u^2/4} \quad \text{(II.41a)}$$

$$\dot{y}(t, t_1) = -\Gamma + \Gamma e^{-[\mathbf{k}(t-t_1)-\mathbf{x}]^2\Delta u^2/4} \quad \text{(II.41b)}$$

which may be integrated to find $\mathbf{x}(t, t_1)$ and $y(t, t_1)$. If the exponent is taken to be very small, previous derivations of the spin echo problem result where the starting point is the Fokker-Planck rather than the transport equation[29–31].

It is possible, however, to find an analytic solution to Equations (II.41) for a case of practical interest. Since echo signals are only observable for pulse separations τ such that $\Gamma\tau \lesssim 1$ and since $(1 - \alpha) \ll 1$ for all t of interest, we have

$$\Gamma(1 - \alpha)t = \frac{\Gamma t \Delta u^2}{2u} \ll 1. \quad \text{(II.42)}$$

Applied to (II.41a), this shows that the thermal equilibrium velocity distribution is approached in a time $1/\Gamma(1-\alpha)$ that is orders of magnitude longer than the experimental observation time $1/\Gamma$ given by (II.41b).

To solve (II.41) in this limit, we further make the reasonable assumption (to be justified)

$$|\mathbf{x}| \ll |\mathbf{k}|(t-t_1) \tag{II.43}$$

and (II.41) becomes

$$\dot{\mathbf{x}}(t, t_1) \approx \Gamma(1-\alpha)\mathbf{k}(t-t_1)e^{-[\mathbf{k}\Delta u[t-t_1)]^2/4}, \tag{II.44a}$$

$$\dot{y}(t, t_1) \approx -\Gamma + \Gamma e^{-[\mathbf{k}\Delta u(t-t_1)]^2/4}, \tag{II.44b}$$

leading to the immediate solution at $t = t_2$

$$\mathbf{x}(t_2, t_1) = \frac{2\Gamma(1-\alpha)\mathbf{k}}{(k\Delta u)^2}(1 - e^{-[\mathbf{k}\Delta u(t_2-t_1)]^2/4}), \tag{II.45a}$$

$$y(t_2, t_1) = -\Gamma(t_2 - t_1) + \frac{2\Gamma}{k\Delta u}\int_0^{k\Delta u(t_2-t_1)/2} e^{-\eta^2}\, d\eta. \tag{II.45b}$$

Equation (II.45a) is seen to be consistent with (II.43).

Recalling that

$$\tilde{\rho}'_{21}(\mathbf{v}, t_2, t_1) = \tilde{\rho}'_{21}(\mathbf{v}, t_2)e^{i\Delta'(v)(t_2-t_1)}, \tag{II.46a}$$

$$= \frac{i}{2} e^{i\mathbf{x}(t_2, t_1)\cdot\mathbf{v} + y(t_2, t_1)}, \tag{II.46b}$$

$$\tilde{\rho}'_{21}(\mathbf{v}, t_2) = \frac{i}{2} e^{-i\Delta'(v)(t_2-t_1)} e^{i\mathbf{x}(t_2, t_1)\cdot\mathbf{v} + y(t_2, t_1)}, \tag{II.46c}$$

we obtain

$$u'_{21}(\mathbf{v}, t_2) = \frac{i}{2} NG(\mathbf{v})e^{-i\Delta'(\mathbf{v})(t_2-t_1)} e^{i\mathbf{x}(t_2, t_1)\cdot\mathbf{v} + y(t_2, t_1)} + \text{c.c.}, \tag{II.47a}$$

$$v'_{21}(\mathbf{v}, t_2) = -\frac{N}{2} G(\mathbf{v})e^{-i\Delta'(\mathbf{v})(t_2-t_1)} e^{i\mathbf{x}(t_2, t_1)\cdot\mathbf{v} + y(t_2, t_1)} + \text{c.c.} \tag{II.47b}$$

$t_2 < t < t_3$—Equations (II.24) can be solved again to give

$$u'(\mathbf{v}, t_3) = u'(\mathbf{v}, t_2), \tag{II.48a}$$

$$v'(\mathbf{v}, t_3) = v'(\mathbf{v}, t_2)\cos\chi(t_3 - t_2) + w'(\mathbf{v}, t_2)\sin\chi(t_3 - t_2), \tag{II.48b}$$

$$w'(\mathbf{v}, t_3) = w'(\mathbf{v}, t_2) - v'(\mathbf{v}, t_2)\sin\chi(t_3 - t_2) - 2w'(\mathbf{v}, t_2)\sin^2\frac{\chi(t_3 - t_2)}{2}. \tag{II.48c}$$

For a pulse where $\chi(t_3 - t_2) = \pi$,

$$u'(\mathbf{v}, t_3) = u'(\mathbf{v}, t_2), \tag{II.49a}$$

$$v'(\mathbf{v}, t_3) = -v'(\mathbf{v}, t_2), \tag{II.49b}$$

$$w'(\mathbf{v}, t_3) = w'(\mathbf{v}, t_2) = 0. \tag{II.49c}$$

$t_3 < t$—The procedure is the same as in the $t_1 < t < t_2$ region. Note first that

$$w'(\mathbf{v}, t) = 0.$$

We seek a solution for $\tilde{\rho}'_{21}(\mathbf{v}, t, t_3 + t_2 - t_1)$ which is defined by (II.32), satisfies the differential equation (II.33) and is subject to the initial condition

$$\begin{aligned}
\tilde{\rho}'_{21}(\mathbf{v}, t_3, t_3 + t_2 - t_1) &= \tilde{\rho}'_{21}(\mathbf{v}, t_3)e^{i\Delta'(\mathbf{v})(-t_2+t_1)},\\
&= \tfrac{1}{2}[u(\mathbf{v}, t_3) - iv(\mathbf{v}, t_3)]e^{-i\Delta'(\mathbf{v})(t_2-t_1)},\\
&= \tfrac{1}{2}[u(\mathbf{v}, t_2) + iv(\mathbf{v}, t_2)]e^{-i\Delta'(\mathbf{v})(t_2-t_1)},\\
&= \tilde{\rho}'_{21}(\mathbf{v}, t_2, t_1)^*,\\
&= -\frac{i}{2}e^{-i\mathbf{x}(t_2, t_1)\cdot\mathbf{v} + y(t_2, t_1)}.
\end{aligned} \tag{II.50}$$

The above makes use of Equations (II.49) and (II.46b). We see that the effect of a π-pulse is simply to convert $\tilde{\rho}'_{21}(\mathbf{v}, t_2, t_1)$ to its complex conjugate and that the seemingly arbitrary choice of the argument $t_3 + t_2 - t_1$ makes this relation evident.

We attempt a solution of the form

$$\tilde{\rho}'_{21}(\mathbf{v}, t, 2\tau) = -\frac{i}{2}e^{i\mathbf{x}(t, 2\tau)\cdot\mathbf{v} + y(t, 2\tau)} \tag{II.51}$$

where the mean pulse-delay time

$$\tau = \frac{t_3 + t_2 - t_1}{2}.$$

Equation (II.51) is subject to the initial conditions

$$\mathbf{x}(t_3, 2\tau) = -\mathbf{x}(t_2, t_1), \tag{II.52a}$$

$$y(t_3, 2\tau) = y(t_2, t_1), \tag{II.52b}$$

and satisfies (II.33)

$$\dot{\tilde{\rho}}'_{21}(\mathbf{v}, t, 2\tau) = -\Gamma\tilde{\rho}'_{21}(\mathbf{v}, t, 2\tau) + \int d^3v' W(\mathbf{v} \to \mathbf{v}')\tilde{\rho}_{21}(\mathbf{v}', t, 2\tau)e^{i\mathbf{k}\cdot(\mathbf{v}-\mathbf{v}')(t-2\tau)}. \tag{II.53}$$

Essentially the same solution for $\mathbf{x}$ and y are found as before, although the lower limit of the time integral is t_3 instead of t_1. The result is

$$\mathbf{x}(t, 2\tau) = -2\mathbf{x}(t_2, t_1) + \frac{2\Gamma(1-\alpha)\mathbf{k}}{(k\Delta u)^2}[1 - e^{-[k\Delta u(t-2\tau)]^2/4}], \tag{II.54a}$$

$$y(t, 2\tau) = 2y(t_2, t_1) - \Gamma(t - 2\tau) + \frac{2\Gamma}{k\Delta u}\int_0^{k\Delta u(t-2\tau)/2} e^{-\eta^2}\, d\eta. \tag{II.54b}$$

Using (II.54), (II.51) and (II.32), we find that

$$\tilde{\rho}_{21}(\mathbf{v}, t) = -\frac{i}{2}NG(\mathbf{v})e^{-\Gamma_1 t}e^{i\mathbf{x}(t, 2\tau)\cdot\mathbf{v} + y(t, 2\tau)}e^{-i\Delta'(\mathbf{v})(t-2\tau)}. \tag{II.55}$$

To obtain the signal field

$$E_{21} = i\left(\frac{2\pi\Omega}{c}\right)L\mu_{12}\langle\tilde{\rho}_{21}(\mathbf{v}, t)\rangle, \tag{II.56}$$

we require the Doppler-averaged quantity

$$\begin{aligned}\langle\tilde{\rho}_{21}(\mathbf{v}, t)\rangle &= -\frac{i}{2}Ne^{-\Gamma_1 t}\int_{-\infty}^{\infty}\frac{dv}{\sqrt{\pi}u}e^{-(v/u)^2}e^{i\mathbf{x}(t, 2\tau)\cdot\mathbf{v} + y(t, 2\tau)}e^{-i\Delta'(\mathbf{v})(t-2\tau)},\\ &= -\frac{i}{2}Ne^{-\Gamma_1 t}e^{y(t, 2\tau)}\cos(-\Omega + \omega_{21} + \Delta\omega_{21})(t - 2\tau)\\ &\quad\times e^{-[\mathbf{x}-\mathbf{k}(t-2\tau)]^2u^2/4}.\end{aligned} \tag{II.57}$$

Since the signal and laser fields, (II.56) and (I.1), add to give

$$E_T = \left(E_{21} + \frac{E_0}{2}\right)e^{i(\Omega t - kz)} + \text{c.c.},$$

the beat term in the intensity E_T^2 is

$$\begin{aligned}[E^2(t)]_{\text{beat}} &= 2E_{21}(t)E_0\\ &= (2\pi\hbar k)NL\chi e^{-\Gamma_1 t}e^{y(t, 2\tau)}\cos(-\Omega + \omega_{21} + \Delta\omega_{21})(t - 2\tau)\\ &\quad\times e^{-[\mathbf{x}-\mathbf{k}(t-2\tau)]^2u^2/4}.\end{aligned} \tag{II.58}$$

Equation (II.58) shows that the echo signal exhibits a beat of low frequency $(-\Omega + \omega_{21} + \Delta\omega_{21})$, which is essentially the Stark shift $\Delta\omega_{21}$ since $-\Omega + \omega_{21} \sim 0$. Evaluating the echo beat signal (II.58) at $t = 2\tau$ yields

$$E_c^2(t = 2\tau) = 2\pi\hbar k N L\chi e^{-\Gamma_1 t} e^{2y(t_2, t_1)} e^{-[x(t_2, t_1)u]^2} \tag{II.59a}$$

$$= 2\pi\hbar k N L\chi \exp\left\{ -\Gamma_1 t - \Gamma t + \frac{4\Gamma}{k\Delta u} \int_0^{k\Delta u t/4} e^{-\eta^2}\, d\eta - \left[\frac{2\Gamma(1-\alpha)\mathbf{k}}{(k\Delta u)^2}\right]^2 [1 - e^{-[\mathbf{k}\Delta u(t_2 - t_1)]^2/4}]^2 u^2 \right\}. \tag{II.59b}$$

The last term in the exponent is quite small and may be neglected since

$$\frac{\Gamma}{ku} \ll 1 \tag{II.60}$$

so that the maximum absolute value of echo amplitude normalized to unity and denoted by $\bar{E}_c^2(t = 2\tau)$ is

$$\bar{E}_c^2(t = 2\tau) = \exp\left[-\Gamma_1 t - \Gamma t + \frac{4\Gamma}{k\Delta u} \int_0^{k\Delta u t/4} e^{-\eta^2}\, d\eta \right]. \tag{II.61}$$

The nature of the solution (II.61) depends on the value of the rms collisional change in the Doppler phase $k\Delta u\tau$. The asymptotic limits are

$$k\Delta u\tau \ll 1;\ \bar{E}_c^2(t = 2\tau) \sim \exp\left[-\Gamma_1 t - \Gamma t^3 \frac{(k\Delta u)^2}{48} \right] \tag{II.62a}$$

$$k\Delta u\tau \gg 1;\ \bar{E}_c^2(t = 2\tau) \sim \exp\left\{ -\Gamma_1 t - \Gamma t + \frac{2\pi^{1/2}\Gamma}{k\Delta u} \right\}. \tag{II.62b}$$

The $\exp(-\Gamma_1 t)$ represents the decrease in echo amplitude due to population decay and is easily understood. The other terms in (II.62) represent the effects of velocity-changing collisions. Additional physical insight into these terms may be obtained by looking at a mathematically non-rigorous picture of echo formation.

The maximum echo signal arises when the net Doppler phase factor

$$e^{i\mathbf{k}\cdot\mathbf{v}(t_2 - t_1) - i\mathbf{k}\cdot\mathbf{v}(t - t_3)} \tag{II.63}$$

goes to unity—i.e. the dephasing-rephasing process of echo formation has been accomplished. Collisions will produce a change in the Doppler phase factor which roughly goes as

$$H = \langle e^{i\mathbf{k}\cdot\Delta\mathbf{v}\tau} \rangle \tag{II.64}$$

where $\langle\ \ \rangle$ represents a collision average. Let Δu be $\sqrt{2}$ times the rms change in velocity per collision. If $k\Delta u\tau \gg 1$ any collision produces destructive phase interference so that the only term which survives in Equation (II.64) will be the one in which no collision occurs during the time $t = 2\tau$. Since the associated probability is $e^{-\Gamma t}$, one finds

$$k\Delta u\tau \gg 1;\ H \sim e^{-\Gamma t}, \tag{II.65a}$$

corresponding to (II.62b).

On the other hand, if $k\Delta u\tau \ll 1$, each collision produces only a small phase change such that

$$H \approx 1 - \frac{k^2}{2}\langle(\Delta v)^2\rangle\tau^2$$

where we have assumed $\langle\Delta\mathbf{v}\rangle = 0$ for simplicity. The quantity $\langle(\Delta v^2)\rangle$ will be equal to (number of collisions in time $t = \Gamma t$) $\times$ ($\langle\Delta v^2\rangle$ for one collision $= \Delta u^2/2$) so that

$$k\Delta u\tau \ll 1;\ H \sim 1 - \tfrac{1}{4}k^2\Delta u^2\tau^2\Gamma t \approx \exp\left[-\Gamma t^3\frac{(k\Delta u)^2}{16}\right] \tag{II.65b}$$

Except for numerical factors Equation (II.65) agrees with the asymptotic forms (II.62) of the transport equation solution. Thus, the decrease in echo amplitude due to velocity-changing collisions results simply because the collisions destroy the perfect Doppler phase cancellation which would have occurred had the collisions been absent.

The approximations used to obtain (II.61) require further discussion. First, note by expanding the exponential in (II.37) and carrying through the calculation as above, it is easy to derive the following formula for the echo amplitude valid for any type Brownian motion collision kernel $W(\mathbf{v}' \to \mathbf{v})$ when $k\Delta u\tau \ll 1$

$$k\Delta u\tau \ll 1;\ \bar{E}_c(t = 2\tau) = \exp\left\{-\Gamma_1 t + 2\sum_{m=1}^{\infty}\left[\frac{(-1)^m(\tau)^{2m+1}}{(2m)!}\right]\right.$$
$$\left.\times\int d^3v W(\mathbf{v} \to \mathbf{v}')[\mathbf{k}\cdot(\mathbf{v} - \mathbf{v}')]^{2m}\right\}. \tag{II.66}$$

In addition, Equation (II.65a) will still be valid for $k\Delta u\tau \gg 1$ so that Equations (II.66) and (II.65a) generalize our results to any Brownian motion kernel. This argument substantiates our claim that the t^3 and t dependence associated with velocity-changing collision are independent of the choice of collision kernel.

Previous theories of echo formation employed a Fokker-Planck Equation (FPE) rather than a transport equation to treat collisions[21,29-31]. For times $\Gamma t(1-\alpha) \ll 1$, such theories give the limiting form (II.62a) for the echo amplitude rather than the entire Equation (II.61). It has been shown that the transition from the transport equation to the FPE is valid only if a single collision produces a negligible change in the distribution function or density matrix elements[32]. For the echo problem, the effect of a single collision is negligible only if $k\Delta u\tau \ll 1$. Thus, one can expect the FPE treatment to fail (as it does) when $k\Delta u\tau \gtrsim 1$.

In the actual experiment, it is possible to use field strengths such that the optimal conditions $\chi t_1 \approx \pi/2$ and $\chi(t_3 - t_2) \approx \pi$ are maintained. Due to power broadening, a monochromatic source excites a frequency bandwidth of order χ in a system of two-level molecules. For the echo problem under consideration, the velocity bandwidth that may be excited is of order

$$u_0 = \frac{\chi}{k}.$$

In the simple model above, we assumed that u_0 was much greater than the thermal speed u so that the field interacted with the entire thermal distribution of molecules. Moreover, since $u_0 = \chi/k \gg u$, the power broadening was so great that the excitation spectrum was constant over the entire thermal distribution of velocities.

Experimentally, $u_0 \approx 10^3$ cm/sec while $u \approx 10^4$ cm/sec indicating that only a fraction of the molecules will interact with the laser field. The excitation spectrum for the molecules will *not* be constant but, instead, will have a width of order u_0. This, in turn, implies that if the field represents a π-pulse for one velocity subgroup it will not be a π-pulse for another velocity subgroup. Thus, the simplification of assuming $\pi/2$-and π-pulses for the entire sample can not be maintained.

The fact that only a fraction of the molecules are excited by the laser means that collisions, in addition to producing the changes in the Doppler phase factors discussed above, will also cause the excited velocity distribution of molecules to decay towards equilibrium. However, if velocity diffusion is slow, $\Gamma t\Delta u^2/u_0^2 \ll 1$, as it is in our experiment, changes in the velocity distribution can be neglected and Equation (II.61) is not affected[32]. If collisions did produce large Δu jumps such that $\Delta u > u_0$, such collisions would remove molecules from the excitation bandwidth and would be reflected in an effective increase in the population loss rate Γ_1; however, this situation does not prevail according to our experiments.

The fact that there is an excitation spectrum rather than $\pi/2$- and π-pulses for the entire sample will lead to a modification of the shape of the echo signal, but not any change in its functional time dependence on pulse separation. In other words, the *relative* contribution of each velocity subgroup to the echo amplitude is unchanged as the pulse separation is varied. Thus, Equation (II.61) can still be used to describe the dependence of echo amplitude on pulse separation.

Our observations, as indicated in Fig. 10, are in agreement with (II.61). The observed functionality in time is precisely what is predicted by (II.62), namely,

$$\begin{aligned} &e^{-Kt^3} \quad \text{for short times,} \\ &e^{-\Gamma t} \quad \text{for long times.} \end{aligned} \tag{II.67}$$

The agreement with theory constitutes primary evidence that elastic molecular collisions involving small changes in velocity play a crucial role in photon echo measurements. In Fig. 10, the departure of the echo decay (lower curve) from the T_1 decay (upper curve) is due solely to the velocity-changing collision mechanism. In other measurements[25,33,34], not here described, the T_1 decay channel results from inelastic collisions that do not exhibit velocity diffusion. From the observed decay constants K and Γ of (2.67), the $^{13}CH_3F$—$^{13}CH_3F$ elastic collision parameters are

$$\begin{aligned} \sigma &= 430 \text{ Å}^2 \\ \Delta u &= 85 \text{ cm/sec} \end{aligned} \tag{II.68}$$

where the total elastic collision cross-section is defined by $\sigma = \Gamma/(Nu\sqrt{2})$ and from (II.62a) $\Delta u = 1/k\sqrt{48K/\Gamma}$. We note the smallness of the characteristic velocity jump per collision Δu and the large cross-section which must be associated with the long range permanent dipole-dipole forces. From this value of Δu and (II.23), we see that $(1 - \alpha) = 2.6 \times 10^{-6}$ so that the velocity jumps are closely clustered about the initial velocity as required in a weak collision model. Furthermore, it follows that an arbitrary initial velocity approaches a thermal equilibrium distribution as $e^{-\beta t}$ in a time $1/\beta = 1/[\Gamma(1 - \alpha)] = 5$ sec which contrasts greatly with the time scale of an echo experiment $1/\Gamma \sim 15\ \mu$sec at 1 mTorr pressure.

III. Coherent Two-Photon Processes[35]

In Sections I and II, we treated coherent transient phenomena that can occur in a two-level quantum system. Here, we discuss a class of coherent two-photon[36] processes that can occur in a three-level quantum system under transient or steady state conditions. The problem arises also in nuclear magnetic and quadrupole resonance[37] and has emerged recently at optical frequencies, as in the phenomena of coherent Raman beats[35,38,39], two-photon absorption with oppositely directed beams[40,41] and self-induced transparency[42].

In the case of Raman beats, observed recently by Shoemaker and Brewer[38], three molecular levels are prepared initially in superposition by means of continuous laser radiation. Two of the levels are degenerate and connect optically with a third level during a resonant steady state preparative phase (see Fig. 11a). Upon application of a dc Stark field, the molecular level degeneracy is removed and coherent forward Raman scattering then occurs in the presence of the same laser field during the nonresonant condition. The two beams, laser and Raman light, strike a photodetector where they produce a coherent beat at a frequency that corresponds to the level splitting between initial and final states in this two-photon process (see Fig. 12). The Raman beat effect has been analyzed in terms of a three-level density matrix perturbation treatment, both during the *steady state preparation* and the subsequent transient Raman scattering which follows.

In this section, we treat the Raman beat effect and also the related problem of two-photon absorption of oppositely directed beams for the case of *pulse*

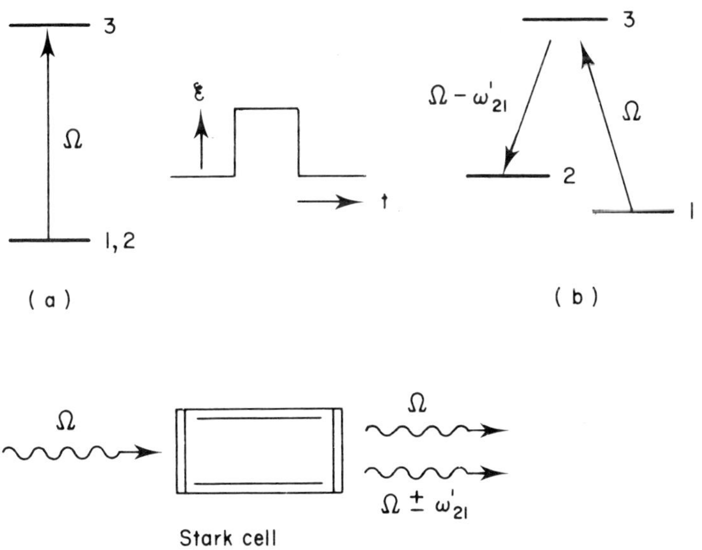

FIG. 11. Molecular energy level diagrams illustrate (a) coherent preparation of a three-level quantum system during a Stark pulse when the levels 1 and 2 are degenerate ($\omega_{21} = 0$), and (b) transient forward scattering following the pulse because of the coherent Raman beat effect. A second frequency component $\Omega + \omega_{21}$, not shown in (b), is also emitted. (From Brewer and Hahn, Ref. 35.)

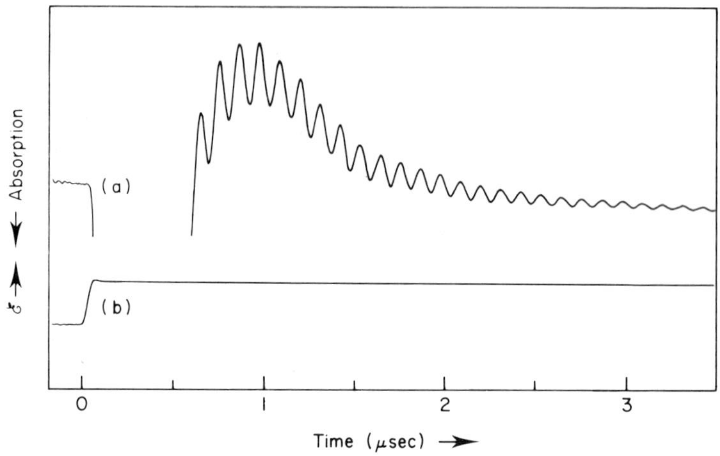

FIG. 12. (a) Coherent Raman beat signal in $^{13}CH_3F$ following (b) a step function Stark pulse (0–46 volts/cm). The modulation or beat frequency corresponds to the level splitting between initial and final states in the two-photon process. The slow decay is independent of Doppler dephasing or elastic collisions. (From Shoemaker and Brewer, Ref. 38.)

preparation. In contrast to previous attempts, an exact pulse solution is attainable for certain experimentally interesting situations. An exact solution is necessary because the relevant density matrix elements swing through large excursions in time and perturbation techniques become invalid. We will show that Bloch-like equations of motion for the three-level problem can be derived to yield a solution quite analogous to the standard two-level case. The transient regime which follows the pulse preparation is still well approximated by a perturbation calculation. Exact steady state solutions for the Raman effect and two-photon absorption are included also, applicable to continuous wave spectroscopy or for transient experiments requiring an initial preparation of quantum states. In general, the form of the solutions is the same for the Raman and two-photon absorption problems, either for transient or steady state conditions.

A. RAMAN BEATS

(a) *Pulse Preparation: Basic Equations*

We consider the Raman beat case first using the molecular level configuration of Fig. 11. The two-photon absorption problem is fundamentally the same and will be handled later. The molecular gas sample is subject to a laser field

$$E_x(z, t) = \tfrac{1}{2}E_0(e^{i(\Omega t - kz)} + \text{c.c.}), \qquad \text{(III.1)}$$

of frequency Ω and amplitude E_0, polarized along the x-axis and propagating in the z-direction. Normally levels 1 and 2 are split by a frequency ω_{12} due to a dc Stark bias field $\mathscr{E}$ applied along the y-axis. However, when a Stark pulse of amplitude $-\mathscr{E}$ appears, levels 1 and 2 become degenerate and the laser field excites the transitions 1–3 and 2–3, placing all three levels in superposition (Fig. 11(a)). Insofar as the Raman beat effect is concerned, the preparation of states is contained in the off-diagonal density matrix element ρ_{12} which persists after the pulse is removed and gives rise to the coherent transient Raman scattering that follows (Fig. 11(b)).

Our objective, therefore, is to obtain a closed form solution for $\rho_{12}(t_w)$ as a result of a Stark pulse of arbitrary duration t_w. This exact solution constitutes the initial condition for transient Raman scattering for times $t > t_w$. For this latter period, a perturbation treatment developed in Ref. 39 suffices and will be referred to again.

The time-dependent behaviour of the density matrix ρ is given by

$$i\hbar\dot{\rho} = [H, \rho] \tag{III.2}$$

where we neglect damping terms, assuming the pulse to be sufficiently short compared to actual molecular decay times. Molecules interact with the laser field, Equation (1), through the dipole interaction part

$$H_I = -\boldsymbol{\mu}\cdot\mathbf{E}(z, t) \tag{III.3}$$

of the total Hamiltonian H.

The electric-dipole matrix elements are given by

$$\begin{aligned} \langle 1|\mu|3\rangle &\equiv \mu_{13} \neq 0 \\ \langle 2|\mu|3\rangle &\equiv \mu_{23} \neq 0 \\ \langle 1|\mu|2\rangle &= 0. \end{aligned} \tag{III.4}$$

The time dependence arising from molecular motion along the z-direction is introduced by defining (I.13)

$$\dot{\rho} = \left(\frac{\partial}{\partial t} + v_z\frac{\partial}{\partial z}\right)\rho, \tag{III.5}$$

We remove the rapidly oscillating factors of the off-diagonal elements by using

$$\rho_{13} = \tilde{\rho}_{13}e^{i(\Omega t - kz)}, \tag{III.6a}$$

$$\rho_{23} = \tilde{\rho}_{23}e^{i(\Omega t - kz)}, \tag{III.6b}$$

and adopt the definitions

$$\omega_{ij} = \frac{E_i - E_j}{\hbar},$$

$$\alpha = \frac{\mu_{13} E_0}{2\hbar},$$

$$\beta = \frac{\mu_{23} E_0}{2\hbar},$$

$$\Delta = \Omega - kv - \omega_{31} = \Omega - kv - \omega_{32}.$$

The equations of motion in the rotating-wave approximation then become

$$\dot{\tilde{\rho}}_{13} + i\Delta\tilde{\rho}_{13} = i\alpha(\rho_{33} - \rho_{11}) - i\beta\rho_{12}, \quad \text{(III.7a)}$$

$$\dot{\tilde{\rho}}_{23} + i\Delta\tilde{\rho}_{23} = i\beta(\rho_{33} - \rho_{22}) - i\alpha\rho_{21}, \quad \text{(III.7b)}$$

$$\dot{\rho}_{12} + i\omega_{12}\rho_{12} = i\alpha\tilde{\rho}_{32} - i\beta\tilde{\rho}_{13}, \quad \text{(III.7c)}$$

$$\dot{\rho}_{11} = i\alpha(\tilde{\rho}_{31} - \tilde{\rho}_{13}), \quad \text{(III.7d)}$$

$$\dot{\rho}_{22} = i\beta(\tilde{\rho}_{32} - \tilde{\rho}_{23}), \quad \text{(III.7e)}$$

$$\dot{\rho}_{33} = i\alpha(\tilde{\rho}_{13} - \tilde{\rho}_{31}) + i\beta(\tilde{\rho}_{23} - \tilde{\rho}_{32}). \quad \text{(III.7f)}$$

An exact solution of Equations (III.7) is easily obtained for the Raman pulse preparation case when $\omega_{12} = 0$, a condition that has been easily satisfied in previous experiments[38].

(b) *Three-Level Bloch Equations*

Pairs of equations in (III.7) may be combined to give Bloch-like functions u, v and w where the matrix elements are

$$u_{ij} = \tilde{\rho}_{ij} + \tilde{\rho}_{ji}, \quad \text{(III.8a)}$$

$$iv_{ij} = \tilde{\rho}_{ij} - \tilde{\rho}_{ji}, \quad \text{(III.8b)}$$

$$w_{ij} = \tilde{\rho}_{ii} - \tilde{\rho}_{jj}. \quad \text{(III.8c)}$$

The condition $\omega_{12} = 0$ is imposed for the Raman problem and we obtain the following set of equations

$$\dot{u}_{13} - v_{13}\Delta - \beta v_{12} = 0 \quad \text{(III.9a)}$$

$$\dot{u}_{23} - v_{23}\Delta + \alpha v_{12} = 0 \quad \text{(III.9b)}$$

$$\dot{v}_{13} + u_{13}\Delta + 2\alpha w_{13} + \beta u_{12} = 0 \quad \text{(III.9c)}$$

$$\dot{v}_{23} + u_{23}\Delta + 2\beta w_{23} + \alpha u_{12} = 0 \tag{III.9d}$$

$$\dot{w}_{13} - 2\alpha v_{13} - \beta v_{23} = 0 \tag{III.9e}$$

$$\dot{w}_{23} - 2\beta v_{23} - \alpha v_{13} = 0 \tag{III.9f}$$

$$\dot{u}_{12} - \alpha v_{23} - \beta v_{13} = 0 \tag{III.9g}$$

$$\dot{v}_{12} - \alpha u_{23} + \beta u_{13} = 0 \tag{III.9h}$$

$$\dot{w}_{12} - \alpha v_{13} + \beta v_{23} = 0. \tag{III.9i}$$

Equations (III.9) can now be symmetrized in the following way. First, (III.9a) and (III.9b) are multiplied respectively by the dimensionless quantities $\alpha_1 = \alpha/\varepsilon$ and $\beta_1 = \beta/\varepsilon$, and then added together. The parameter

$$\varepsilon = 2\sqrt{\alpha^2 + \beta^2}$$

evolves from the treatment below. Similarly, (III.9c) is multiplied by α_1 and (III.9d) by β_1 and the two equations added. In the resulting equations we identify terms of the form

$$U = \alpha_1 u_{13} + \beta_1 u_{23}, \tag{III.10a}$$

$$V = \alpha_1 v_{13} + \beta_1 v_{23}, \tag{III.10b}$$

$$W = 2(\alpha_1^2 w_{13} + \beta_1^2 w_{23} + \alpha_1\beta_1 u_{12}). \tag{III.10c}$$

After substitution of expressions for $\dot{w}_{13}$, $\dot{w}_{23}$ and $\dot{u}_{12}$ from Equations (III.9) into $\dot{W}$, the particular identity

$$\dot{W} = \varepsilon V$$

results. (This ε is not to be confused with the dc Stark field.) A set of reduced Bloch equations may now be written as

$$\begin{aligned} \dot{U} - \Delta V &= 0, \\ \dot{V} + \Delta U + \varepsilon W &= 0, \\ \dot{W} - \varepsilon V &= 0. \end{aligned} \tag{III.11}$$

For constant ε, the well-known solutions[6] to Equation (III.11) are given by

$$U(t) = K\Delta[\cos(\gamma t) - 1], \tag{III.12a}$$

$$V(t) = -K\gamma \sin \gamma t, \tag{III.12b}$$

$$W(t) = W(0) + K\varepsilon[\cos(\gamma t) - 1], \tag{III.12c}$$

with

$$\gamma = \sqrt{\Delta^2 + \varepsilon^2},$$

$$K = \frac{\varepsilon W(0)}{\Delta^2 + \varepsilon^2}.$$

When the Stark pulse is applied at $t = 0$, the initial conditions are $W(t) = W(0)$ and $U(0) = V(0) = 0$.

Note that the quantity γ is an effective precession frequency in the frame of reference rotating at frequency Ω, and γt is the angle through which the three-level Bloch vector turns in a time t.

(c) *Pulse Solution of* ρ_{12}

The next step is to derive solutions for $u_{12}(t)$ and $v_{12}(t)$ using (III.9)–(III.12) since the desired quantity over the pulse interval $0 \leq t \leq t_w$ is

$$\rho_{12}(t) = \tfrac{1}{2}[u_{12}(t) + iv_{12}(t)]. \tag{III.13}$$

These equations ultimately reduce to

$$\dddot{v}_{12} + [\Delta^2 + 2(\alpha^2 + \beta^2)]\ddot{v}_{12} + (\alpha^2 + \beta^2)^2 v_{12} = 0. \tag{III.14}$$

With the initial conditions

$$v_{12}(0) = \dot{v}_{12}(0) = \ddot{v}_{12}(0) = 0,$$
$$\dddot{v}_{12}(0) = 2\Delta\alpha\beta w_{12}(0),$$

we find from (III.14) that

$$v_{12}(t) = \frac{\alpha\beta w_{12}(0)}{\delta}\left[\frac{\sin(\delta - \Delta/2)t}{(\delta - \Delta/2)} - \frac{\sin(\delta + \Delta/2)t}{(\delta + \Delta/2)}\right]$$

$$\delta = \left(\frac{\alpha^2 + \beta^2 + \Delta^2}{4}\right)^{1/2}. \tag{III.15}$$

Note that $w_{12}(0) = \rho_{11}(0) - \rho_{22}(0)$ is the occupation probability difference at time $t = 0$ preceding the pulse. The quantity

$$u_{12}(t) = \frac{8\alpha\beta K}{\varepsilon}(\cos\gamma t - 1)$$
$$- \left(\frac{\alpha\beta}{\delta}\right)\frac{(\alpha^2 - \beta^2)}{(\alpha^2 + \beta^2)} w_{12}(0)\left[\frac{\cos(\delta - \Delta/2)t - 1}{(\delta - \Delta/2)} + \frac{\cos(\delta + \Delta/2)t - 1}{(\delta + \Delta/2)}\right] \tag{III.16}$$

derives from (III.9g) where $v_{23} = \varepsilon(V - \alpha_1 v_{13})/\beta$, V is given by (III.12b), and

$$v_{13} = \frac{-\alpha\beta^2 w_{12}(0)}{\delta(\alpha^2 + \beta^2)}\left[\sin\left(\delta - \frac{\Delta}{2}\right)t + \sin\left(\delta + \frac{\Delta}{2}\right)t\right] - \frac{\alpha K\gamma\varepsilon}{(\alpha^2 + \beta^2)}\sin\gamma t. \tag{III.17}$$

Equation (III.17) follows by taking the time derivative of (III.9h), and then inserting into it expressions for $\dot{u}_{23}$ and $\dot{u}_{13}$ from (III.9b) and (III.9a) and for v_{23} from (III.10b).

Equations (III.15) and (III.16) combine finally to give $\rho_{12}(t_w)$ of Equation (III.13), evaluated at $t = t_w$ at the end of a pulse of duration t_w.†

(d) *Steady State Preparation*

An exact expression for ρ_{12}, under steady state conditions can be derived also and is given by Equation (III.34c). It reduces to the perturbation result[39]

$$\rho_{12}(\text{s.s}) = \frac{2(\tau_2/T_2)\alpha\beta(\rho_{33}^0 - \rho_{22}^0)}{(\omega - \omega_0)^2 + 1/T_2^2} \tag{III.18}$$

when the levels 1 and 2 are degenerate during preparation. For this case, $\omega_0 = \omega_{31} = \omega_{32}$ and the thermal difference in state occupation probability is $\rho_{33}^0 - \rho_{22}^0 = \rho_{33}^0 - \rho_{11}^0$ in the absence of an external field. Equation (III.18) is included here to allow an easy comparison of the steady state case with that of pulse preparation, Equations (III.13), (III.15) and (III.16).

(e) *Transient Decay*

Previously, a perturbation solution was derived for the heterodyne beat signal following preparation under steady state conditions. The result is given in Equation (35) of Ref. 39. In the steady state situation described by Equation (III.18), ρ_{12} is real, but in the present pulse solution for $\rho_{12}(t_w)$, real and imaginary parts of (III.13) contribute to the beat signal so that a slight generalization of our earlier treatment is required (see Section (b) for additional details).

† *Note added in proof:* Essentially the same result can be obtained more simply using wave functions instead of the density matrix. The author thanks M. Sargent, III for bringing this point to his attention. See M. Sargent, III and P. Horwitz. *Phys. Rev.* **A13**, 1962 (1976) for a discussion of the three-level wave function solution and its limitations. In addition D. Gold and E. L. Hahn have verified the oscillations predicted by III.16; *Phys. Rev.* **B** (to be published).

We find for the period $t \geq t_w$ following the pulse that the velocity-averaged beat amplitude of the optical flux in terms of $\rho_{12}(t_w)$ is

$$(E_T E_T^*)^b = -4\pi NL\left(\frac{\hbar\Omega}{c}\right)\alpha\beta e^{-t/\tau_2} \times \left\langle [v_{12}(t_w)\cos\omega'_{12}t + u_{12}(t_w)\sin\omega'_{12}t] \times \left(\frac{1}{\Delta'_{13}} - \frac{1}{\Delta'_{23}}\right)\right\rangle_{\text{vel}}. \quad \text{(III.19)}$$

E_T is the total optical field and levels 1 and 2 are no longer degenerate due to the Stark bias field, so that $\omega'_{31} = \omega_{31} + \Delta\omega_{31}$ and $\omega'_{32} = \omega_{32} + \Delta\omega_{32}$, and

$$\omega'_{12} \neq 0,$$
$$\Delta'_{13} = \Omega - kv_z - \omega'_{31},$$
$$\Delta'_{23} = \Omega - kv_z - \omega'_{32}.$$

The quantity N is the molecular density, L is the sample length and τ_2 is the (Raman) dephasing time associated with ρ_{12}.

For steady state preparation, we see that Equation (III.19) reduces to Equation (35) of Ref. 39

$$(E_T E_T^*)^b_{\text{s.s.}} = -8\pi NL\left(\frac{\hbar\Omega}{c}\right)\alpha\beta e^{-t/\tau_2}\left\langle \rho_{12}(\text{s.s.})\sin\omega'_{12}t\left(\frac{1}{\Delta'_{13}} - \frac{1}{\Delta'_{23}}\right)\right\rangle_{\text{vel}} \quad \text{(III.20)}$$

because $\rho_{12}(\text{s.s}) = u_{12}(\text{s.s})/2$ is real and $v_{12}(\text{s.s}) = 0$ according to (III.18). (Equation (III.20) corrects the error in sign of the Δ'_{23} term in Ref. 39.)

For pulse preparation, considerable simplification of (III.19) results when there is a zero difference in occupation probability $w_{12}(0) \sim 0$. This condition corresponds to the experiments of Shoemaker and Brewer[38] so that (III.15) and (III.16) become

$$v_{12}(t_w) \sim 0$$
$$u_{12}(t_w) \sim \frac{8\alpha\beta W(0)}{\Delta^2 + 4(\alpha^2 + \beta^2)}(\cos\gamma t_w - 1). \quad \text{(III.21)}$$

Note that $W(0) = [\alpha_1^2 w_{13}(0) + \beta_1^2 w_{23}(0)]/\sqrt{\alpha_1^2 + \beta_1^2}$ is large compared to $w_{12}(0)$. The effect of a Stark pulse is now easily visualized as causing a precession of an effective Bloch vector through the angle γt_w during the pulse interval t_w. The maximum beat amplitude in (III.19) will occur for a $\pi/2$-pulse for the resonance case, i.e.,

$$\gamma t_w = 2\sqrt{\alpha^2 + \beta^2}\, t_w = \frac{\pi}{2}. \quad \text{(III.22)}$$

If the level structure has more than three levels so that different three-level combinations are possible, each corresponding to a different $(\alpha^2 + \beta^2)$ term in (III.21), then varying the pulse width t_w will give a series of maxima as the various $\pi/2$ conditions are achieved. Since (III.19) is velocity-dependent through the terms Δ_{13}, Δ'_{13}, Δ_{23} and Δ'_{23}, it must be Doppler-averaged.

The Doppler integral of (III.19) is not analytic in general. However, the case of steady state preparation can be treated and is given here since the integration was not given explicitly by Brewer and Hahn[39]. We begin at an earlier stage than (III.19) using the Raman beat terms of Equation (18) of Ref. 39.

$$\text{(A1)} \qquad \rho_{13}^b(t) \sim \frac{e^{-ikz_0}\left(e^{i(\omega + |\omega_{12}|)t - t/\tau_2} - e^{i|\omega_{13}|t - t/T_2}\right)}{[i(\omega - |\omega_{23}|) + (1/T_2 - 1/\tau_2)][(\omega - \omega_0)^2 + 1/T_2^2]}.$$

The Raman beat signal $(E_T E_T^*)^b$ is proportional to $\rho_{13}^b(t)$ and to $\rho_{23}^b(t)$ which can be obtained from $\rho_{13}^b(t)$ by the index interchange $2 \leftrightarrow 3$. The quantity $z_0 = z - v_z t$ and the laser frequency $\omega' = \omega + kv_z$. The Doppler integral

$$\langle \rho_{13}^b(t) \rangle = \frac{1}{ku\sqrt{\pi}} \int_{-\infty}^{\infty} e^{-(\Delta/ku)^2} \rho_{13}^b(t) d\Delta$$

reduces in the limit where a small fraction of the Doppler width ku is excited, permitting the Gaussian to be factored from the integral, to the form

$$\text{(A2)} \qquad \langle \rho_{13}^b(t) \rangle \sim \frac{T_2 e^{i(\omega' + |\omega_{12}|)t - t/T_2}}{2/T_2 - 1/\tau_2 - i(|\omega_{23}| - \omega_0)}.$$

Here, we have dropped the uninteresting term $e^{i(|\omega_{13}| + \omega' - \omega_0)t - 2t/T_2}$ and set the Gaussian equal to unity. This result, which is obtained by contour integration, is independent of the sign of $2/T_2 - 1/\tau_2$ and does not diverge when $|\omega_{23}| - \omega_0 = 0$. Had we integrated Equation (19) of Ref. 39 or the final result Equation (35) of Ref. 39 instead, we would have omitted the $e^{i|\omega_{13}|t - t/T_2}$ of $\rho_{13}^b(t)$ in (A1) and obtained the result

$$\text{(A3)} \qquad \langle \rho_{13}^b(t) \rangle \sim \frac{T_2 e^{i(\omega' + |\omega_{12}|)t - t/T_2}}{2/T_2 - 1/\tau_2 - i(|\omega_{23}| - \omega_0)} \times \left(1 + \begin{matrix} 0 & \text{for } \dfrac{1}{T_2} - \dfrac{1}{\tau_2} > 0 \\ \dfrac{-2/T_2}{i(\omega_0 - |\omega_{23}|) - 1/\tau_2} & \text{for } \dfrac{1}{T_2} - \dfrac{1}{\tau_2} < 0 \end{matrix} \right).$$

We see, however, that (A2) and (A3) do not differ significantly since $-(2/T_2)/[i(\omega_0 - |\omega_{23}|) - 1/\tau_2] \ll 1$. This is because $\omega_0 - |\omega_{23}| \sim |\omega_{12}| \gg 1/\tau_2, 1/T_2$, i.e., the Raman beat period must be smaller than the decay time otherwise beats would never be observed. Hence, to the degree of approximation involved in this perturbation calculation, either (A2) or (A3) is acceptable. Note the Equation (III.19) also neglects the $e^{i|\omega_{13}|t - t/T_2}$ term of $\rho_{13}^b(t)$ and the corresponding term for $\rho_{23}^b(t)$. The author appreciates a communication received from M. S. Feld who recently has obtained results similar to (A2) and who essentially has pointed out the difference between (A2) and (A3).

B. TWO-PHOTON ABSORPTION AND EMISSION

(a) *Pulse Preparation*

We will show that the exact solutions obtained for pulse preparation of Raman beats also apply to pulsed two-photon absorption when the two light beams propagate collinearly in opposite directions. The transient two-photon emission which follows the pulse is treated in a manner analogous to the Raman problem. In addition, exact steady state solutions for two-photon absorption (preparation) are presented in Section (c) in connection with recent experiments. The appropriate molecular level structure is shown in Fig. 13 where the sample is excited by two cw collinear optical beams

$$E_x(z, t) = E_1 \cos(\Omega_1 t - kz) + E_2 \cos(\Omega_2 t + kz) \qquad \text{(III.23)}$$

polarized in the x-direction and propagating along the z-axis in opposite directions. The frequency Ω_1 is assumed to be in near resonance with the 3–1 transition and Ω_2 is in near resonance with the 2–3 transition. The two frequencies are sufficiently different that each beam excites only one transition rather than both transitions simultaneously. Yet it can be assumed that $\Omega_1 - \Omega_2$ is sufficiently small, say in the microwave region, that we can neglect in (III.23) the difference in propagation vectors $k_i = \Omega_i/c$ ($i = 1,2$). We also retain the optical selection rules expressed by Equation (III.4). The effect of a Stark pulse, directed along the y-axis, is to shift the molecular levels into resonance for a two-photon transition where the frequency condition $\Omega_1 + \Omega_2 = \omega_{21}$ is satisfied (Fig. 13(a)). This process places the three levels in coherent superposition and constitutes the preparative stage over the pulse duration. As in the Raman beat effect, memory of the preparation is contained in the off-diagonal element ρ_{12} and gives rise to a transient two-photon emission following the pulse (Fig. 13(b)). Two light beams are emitted in

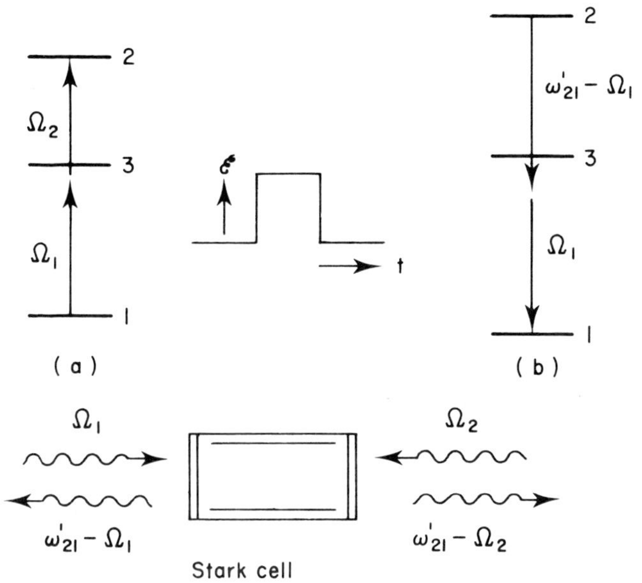

FIG. 13. Molecular energy level diagrams illustrate (a) coherent preparation of a three-level quantum system by two oppositely directed beams (Ω_1 and Ω_2); a Stark pulse shifts the levels so that the condition $\Omega_1 + \Omega_2 = \omega_{21}$ is satisfied; and (b) transient two-photon emission (backward scattering) following the pulse. A second frequency component $\omega_{21} - \Omega_2$, not shown in (b), is also emitted. (From Brewer and Hahn, Ref. 35.)

opposite directions along the z-axis where each beam exhibits a *beat*, due to the Stark shift, that is analogous to the Raman beat.

The appropriate density matrix equations are of the form:

$$\dot{\tilde{\rho}}_{13} + i\tilde{\rho}_{13}\left(\Delta - \frac{i}{T_2}\right) = i\alpha(\rho_{33} - \rho_{11}) - i\beta\tilde{\rho}_{12} \qquad \text{(III.24a)}$$

$$\dot{\tilde{\rho}}_{23} - i\tilde{\rho}_{23}\left(\Delta' + \frac{i}{T_2}\right) = i\beta(\rho_{33} - \rho_{22}) - i\alpha\tilde{\rho}_{21} \qquad \text{(III.24b)}$$

$$\dot{\tilde{\rho}}_{12} + i\tilde{\rho}_{12}\left(\Delta + \Delta' - \frac{i}{\tau_2}\right) = i\alpha\tilde{\rho}_{32} - i\beta\tilde{\rho}_{13} \qquad \text{(III.24c)}$$

$$\dot{\rho}_{11} = i\alpha(\tilde{\rho}_{31} - \tilde{\rho}_{13}) - \frac{\rho_{11} - \rho_{11}^0}{T_1} \qquad \text{(III.24d)}$$

$$\dot{\rho}_{22} = i\beta(\tilde{\rho}_{32} - \tilde{\rho}_{23}) - \frac{\rho_{22} - \rho_{22}^0}{T_1} \qquad \text{(III.24e)}$$

$$\dot{\rho}_{33} = i\alpha(\tilde{\rho}_{13} - \tilde{\rho}_{31}) + i\beta(\tilde{\rho}_{23} - \tilde{\rho}_{32}) - \frac{\rho_{33} - \rho_{33}^0}{T_1} \qquad \text{(III.24f)}$$

where now

$$\Delta = \Omega_1 - kv - \omega_{31}, \tag{III.24g}$$

$$\Delta' = \Omega_2 + kv - \omega_{23}, \tag{III.24h}$$

rapidly oscillating terms are removed with the definitions

$$\rho_{13} = \tilde{\rho}_{13} e^{i(\Omega_1 t - kz)}, \tag{III.25a}$$

$$\rho_{23} = \tilde{\rho}_{23} e^{-i(\Omega_2 t + kz)}, \tag{III.25b}$$

$$\rho_{12} = \tilde{\rho}_{12} e^{i(\Omega_1 + \Omega_2)t}, \tag{III.25c}$$

and anti-resonant terms have been dropped. The quantity ρ_{ii}^0 represents the thermal equilibrium population of level i. The decay time T_1 is assumed to be the same for all diagonal elements, and T_2 is taken to be the same for ρ_{13} and ρ_{23}, but we designate a different decay time τ_2 for ρ_{12}. For Stark pulses shorter than the decay times T_1 and T_2, the relaxation and source terms in (III.24) can be neglected, but they will be of importance in the steady state calculation of Section (c). We notice that if the two-photon resonance condition

$$\Delta + \Delta' = \Omega_1 + \Omega_2 - \omega_{21} = 0, \tag{III.26}$$

is satisfied, Equations (III.24a–f) are identical to Equations (III.7) for the pulse Raman case when

$$\omega_{12} = 0. \tag{III.27}$$

In this situation, the pulse Raman solutions for ρ_{12} (Raman) (Equations (III.13), (III.15) and (III.16)] will be valid also for the $\tilde{\rho}_{12}$ (two-photon) case.

The condition Equation (III.26) is seen to be no more restrictive than that of Equation (III.27). For Equation (III.27) to be satisfied, the dc Stark field must be zero during the pulse and this condition is easily achieved in practice. On the other hand, for Equation (III.26) to be valid it is only necessary that ω_{21} match the sum of the two laser frequencies and this can also be accomplished with a Stark pulse of appropriate magnitude.

(b) *Transient Decay*

For the period following pulse excitation, $t > t_w$, transient two-photon emission will occur. In addition to the two fields of Equation (III.23) which are always present, two new fields will be generated. These are seen in the linearized form of Maxwell's equations.

$$\frac{\partial E_{23}(z,t)}{\partial z} = -2\pi i k_2 N \mu_{23} \langle \tilde{\rho}_{23}(z,t) \rangle_{\text{vel}} \tag{III.28a}$$

$$\frac{\partial E_{13}(z,t)}{\partial z} = -2\pi i k_1 N \mu_{13} \langle \tilde{\rho}_{13}(z,t) \rangle_{\text{vel}} \tag{III.28b}$$

where $\langle \quad \rangle_{\text{vel}}$ denotes a velocity summation over the Doppler distribution. The total field then becomes

$$E_T = \left(E_{13} + \frac{E_1}{2} \right) e^{i(\Omega_1 t - kz)} + \left(E_{23} + \frac{E_2}{2} \right) e^{-i(\Omega_2 t + kz)} + \text{c.c.} \tag{III.29}$$

The matrix elements $\tilde{\rho}_{13}$ and $\tilde{\rho}_{23}$ are obtained from Equations (III.24a, b, and c) in a perturbation treatment following Ref. 39. This assumes that the right hand side of (III.24c) can be neglected, giving

$$\tilde{\rho}_{12}(t) = \tilde{\rho}_{12}(t_w) e^{-i(\Omega_1 + \Omega_2 + \omega'_{12})t - t/\tau_2} \tag{III.30}$$

where ω'_{12} is the level splitting for times $t > t_w$.

Considering only the two-photon (beat) terms of (III.24a and b), it follows that

$$\tilde{\rho}^b_{23}(t) = -\frac{\alpha \tilde{\rho}_{21}(t_w) e^{i(\Omega_1 + \Omega_2 + \omega'_{12})t - t/\tau_2}}{\Omega_1 - kv_z + \omega'_{13}} \tag{III.31a}$$

$$\tilde{\rho}^b_{13}(t) = -\frac{\beta \tilde{\rho}_{12}(t_w) e^{-i(\Omega_1 + \Omega_2 + \omega'_{12})t - t/\tau_2}}{-\Omega_2 - kv_z + \omega'_{23}}. \tag{III.31b}$$

Note that the pulse preparation is contained in $\rho_{12}(t_w)$ as given by Equations (III.13), (III.15) and (III.16).

The absolute square of (III.29) yields beat terms in the intensity

$$\begin{aligned}
(E_T E_T^*)^b &= -4\pi i N L \left\langle \alpha(\hbar\Omega_1/c)(\tilde{\rho}^b_{13} - \tilde{\rho}^b_{31}) + \beta\left(\frac{\hbar\Omega_2}{c}\right)(\tilde{\rho}^b_{23} - \tilde{\rho}^b_{32}) \right\rangle_{\text{vel}} \\
&= -4\pi N L \alpha\beta e^{-t/\tau_2} \Big\langle [-v_{12}(t_w)\cos(\Omega_1 + \Omega_2 + \omega'_{12})t \\
&\quad + u_{12}(t_w)\sin(\Omega_1 + \Omega_2 + \omega'_{12})t] \\
&\quad \times \left(\frac{\hbar\Omega_2/c}{\Omega_1 - kv + \omega'_{13}} + \frac{\hbar\Omega_1/c}{\Omega_2 + kv - \omega'_{23}} \right) \Big\rangle_{\text{vel}}
\end{aligned} \tag{III.32}$$

where the in and out of phase components oscillate at a beat frequency $\Omega_1 + \Omega_2 + \omega'_{12}$. Since $\Omega_1 + \Omega_2 + \omega_{12} = 0$ during the Stark pulse, we see that after the pulse, the beat $\Omega_1 + \Omega_2 + \omega'_{12} = \omega'_{12} - \omega_{12}$ is simply the Stark shift. As in the Raman case, Equation (III.19), the Doppler average must be handled numerically.

Examination of the frequency content of (III.29), using (III.28) and (III.31), shows that there are two possibilities for two-photon emission (as there are in Raman emission). In the first case, a photon of frequency $\omega'_{21} - \Omega_1$ is emitted in the $-z$ direction and one frequency Ω_1 in the $+z$ direction, the direction of emission being contained in the **k** vectors of (III.29). In the second case, a photon of frequency Ω_2 is emitted in the $-z$ direction and one of frequency $\omega'_{21} - \Omega_2$ in the $+z$ direction. Therefore, a photodetector that samples either the $+z$ or the $-z$ propagating beams will detect both two-photon processes simultaneously. The relative contributions of each process are weighted by the off-resonant tuning behaviour expressed in the last two terms of (III.32). Speaking classically, we may say that the coherently prepared sample modulates at a frequency ω'_{21} the Ω_1 beam to produce an oppositely directed sideband at $\omega'_{21} - \Omega_1$; similarly, the sample also modulates the Ω_2 beam to generate an oppositely directed sideband at $\omega'_{21} - \Omega_2$.

It is interesting to speculate on whether or not the beat signal (III.32) would be observed if the two fields E_1 and E_2 of (III.23) prepared the sample but only one of them, say E_1, was allowed to probe the sample afterwards. In such a situation, the total field becomes

$$E_T = \left(E_{13} + \frac{E_1}{2}\right)e^{i(\Omega_1 t - kz)} + E_{23}e^{-i(\Omega_2 t + kz)} + \text{c.c.}, \qquad \text{(III.33a)}$$

instead of (III.29), and the intensity associated with the beat signal is

$$(E_T E_T^*)^b = E_1(E_{13} + E_{13}^*) \qquad \text{(III.33b)}$$

where $E_1 = E_1^*$. The beat term $E_1(E_{13} + E_{13}^*)$, which appears in (III.32) through the term containing $(\Omega_2 + kv - \omega'_{23})$, represents one of the two possible two-photon processes, while the second process contained in the term $(\Omega_1 - kv + \omega'_{13})$ of (III.32) is absent. We see, therefore, that the two fields E_1 and E_2 are not only needed in the preparation stage but in the detection process as well.

Note that $u_{12}(t_w)$ and $v_{12}(t_w)$ in (III.32) do not simplify as in (III.21) for the Raman case because $w_{12}(0) \neq 0$. Nevertheless, most of the features discussed here for two photon-emission apply to the Raman beat problem. A comparison of the Raman beat expression Equation (III.19) with the two-photon expression (III.32) shows indeed that they are of the same form. A unique feature of transient two-photon emission, due to the nature of the preparation, is that the two waves are emitted collinearly in opposite directions. If, on the other hand, the beams of frequency Ω_1 and Ω_2 propagated in the same direction, the two photons would also be emitted in the same direction, but then the Doppler shifts would not cancel and the decay times would be severely shortened because of the Doppler dephasing effect. In the case of coherent Raman beats, only forward scattering is possible where Doppler dephasing does not appear and the incident beam is modulated at frequency ω'_{21} corresponding to the $1 \rightarrow 2$ and $2 \rightarrow 1$ transitions.

We might add that when the E_1 and E_2 fields travel in the same direction, Doppler dephasing could be eliminated in a two-pulse photon echo experiment, in a manner analogous to the two-level photon echo problem.

Another property of Equations (III.19) and (III.32) is that the two-photon emission and Raman beat effects decay as e^{-t/τ_2}, independent of molecular velocity. This result is expected in lowest order perturbation theory because the first order Doppler shift vanishes for these collinear two-photon processes. All of the velocity packets prepared contribute to the beat signal. It follows that elastic collisions should have no effect on the decay time for either two-photon emission of oppositely directed beams or for Raman beats because the superposition of states remains unaffected. Raman beat experiments[18,25] verify this conclusion for molecular vibration-rotation transitions. Whether or not two-photon electronic transitions behave differently because of the influence of phase interrupting elastic collisions on state superposition is not known at present. However, power-dependent frequency shifts can occur in higher order solutions as discussed in the next section, and then these two-photon processes are no longer velocity-independent.

(c) *Steady State Preparation*

Several experiments[41] have been reported recently involving steady state two-photon absorption of two oppositely propagating beams, following the results of an earlier perturbation calculation[40]. We include here an exact treatment of this problem which reveals a significant power-dependent line broadening and frequency shift that is not evident in the perturbation treatment of Vasilenko *et al.*[40]. However, the exact calculation given in this section does verify our previous estimate[39] of the power-dependent shift and broadening for the Raman beat effect. In fact, the steady state two-photon and Raman solutions are again of the same form, as in the transient case.

We consider the two-photon absorption problem first and then the Raman case, and continue to use the optical arrangement and level structure of Fig. 13 where laser radiation of frequency Ω_1 is in near resonance with the

1–3 transition while Ω_2 is in near resonance with the 2–3 transition but $\Omega_1 \neq \Omega_2$. The nine equations, given by the off-diagonal elements Equations (III.24a, b and c) and their complex conjugates and by the diagonal elements Equations (III.24d, e, and f), can then be solved simultaneously in a straightforward manner by setting the time derivatives equal to zero.

The solutions are

$$w_{23} = \frac{Dw_{23}^0 - Pw_{12}^0}{T_1(MP - DQ)} \tag{III.34a}$$

$$w_{12} = \frac{Qw_{12}^0 - Mw_{23}^0}{T_1(MP - DQ)} \tag{III.34b}$$

$$\tilde{\rho}_{12} = a\left(\frac{w_{23}}{\Delta' - i/T_2} + \frac{w_{12} + w_{23}}{\Delta - i/T_2}\right), \tag{III.34c}$$

$$\tilde{\rho}_{23} = \frac{\alpha\tilde{\rho}_{21} + \beta w_{23}}{\Delta' + i/T_2}, \tag{III.34d}$$

$$\tilde{\rho}_{13} = \frac{-\beta\tilde{\rho}_{12} - \alpha(w_{12} + w_{23})}{\Delta - i/T_2}. \tag{III.34e}$$

where $w_{ij}^0 \equiv \rho_{ii}^0 - \rho_{jj}^0$ is the occupation probability difference of levels i and j in the absence of an external field, and we have defined the following quantities

$$a = \alpha\beta\bigg/\left[\left(\Delta + \Delta' - \frac{i}{\tau_2}\right) - \frac{\alpha^2}{\Delta' - i/T_2} - \frac{\beta^2}{\Delta - i/T_2}\right], \tag{III.35a}$$

$$M = a^*b^*e^* + abe^* - \frac{2\alpha^2/T_2}{\Delta^2 + 1/T_2^2} + \frac{2\beta^2/T_2}{\Delta'^2 + 1/T_2^2}, \tag{III.35b}$$

$$Q = a^*b^*j^* + abj^* - \left(\frac{2\alpha^2/T_2}{\Delta^2 + 1/T_2^2} + \frac{4\beta^2/T_2}{\Delta'^2 + 1/T_2^2} + \frac{1}{T_1}\right), \tag{III.35c}$$

$$e = i\alpha\beta\left(\frac{-1}{\Delta + i/T_2} + \frac{1}{\Delta' + i/T_2}\right), \tag{III.35d}$$

$$b = \frac{1}{\Delta - i/T_2} + \frac{1}{\Delta' - i/T_2}, \tag{III.35e}$$

$$D = -\left(\frac{2\alpha^2/T_2}{\Delta^2 + 1/T_2^2} + \frac{1}{T_1}\right) + \frac{a^*e}{\Delta + i/T_2} + \frac{ae^*}{\Delta - i/T_2}, \tag{III.35f}$$

$$j = -i\alpha\beta\left(\frac{2}{\Delta' + i/T_2} + \frac{1}{\Delta + i/T_2}\right), \tag{III.35g}$$

$$P = -\frac{2\alpha^2/T_2}{\Delta^2 + 1/T_2^2} + \frac{a^*j}{\Delta + i/T_2} + \frac{aj^*}{\Delta - i/T_2}. \tag{III.35h}$$

It will be seen that Equations (III.34) simplify considerably for the level configuration of Fig. 13 because the thermal occupation probability $w_{23}^0 \approx 0$ whereas $w_{12}^0 \neq 0$.

The set of solutions (III.34) also apply to the corresponding steady state Raman process of Fig. 11 when the off-resonance tuning terms are defined as

$$\Delta = \Omega - kv - \omega_{31},$$
$$-\Delta' = \Omega - kv - \omega_{32},$$

in place of Equations (III.24g and h). However, now $w_{23}^0 \neq 0$, $w_{12}^0 \approx 0$ and the Raman tuning behaviour is given by $\Delta + \Delta' = \omega_{12}$ in contrast to the two-photon case, Equations (III.26), $\Delta + \Delta' = \Omega_1 + \Omega_2 - \omega_{21}$.

To be specific, we return to the two-photon absorption solution which reduces to the expected two-level solution when we set one of the two external fields equal to zero. For example, when $\beta = 0$ Equation (III.34g) becomes

$$\tilde{\rho}_{13} = -\frac{\alpha w_{13}^0}{\Delta - i/T_2}\left(1 - \frac{4\alpha^2 T_1/T_2}{\Delta^2 + 1/T_2^2 + 4\alpha^2 T_1/T_2}\right), \tag{III.36}$$

which is the result anticipated.

Equation (III.34e) also reduces to the first order two-photon perturbation solution[40,43]

$$\tilde{\rho}_{23}^{(1)} = \frac{\alpha^2 \beta w_{13}^0}{\Delta' + i/T_2} \cdot \frac{1}{\Delta + i/T_2}\left[\frac{1}{\Delta + \Delta' + i/\tau_2} - \frac{2T_1/T_2}{\Delta - i/T_2}\right] \tag{III.37}$$

when one of the external fields is much weaker than the other, corresponding to the condition $\beta \ll \alpha$. The velocity average of (III.37)

$$\begin{aligned}\langle \tilde{\rho}_{23}^{(1)} \rangle_{\text{vel}} &= \frac{1}{\sqrt{\pi} u}\int_{-\infty}^{\infty} e^{-v^2/u^2} \tilde{\rho}_{23}^{(1)}\, dv \\ &\cong \frac{2\sqrt{\pi}}{ku} \alpha^2 \beta w_{13}^0 e^{-v_1^2/u^2} \cdot \frac{\tau_2}{\Delta + \Delta' + i/\tau_2}\end{aligned} \tag{III.38}$$

can be carried out by contour integration assuming that the velocity group v_1 excited has a much narrower frequency band width kv_1 than the Doppler width ku so that the Gaussian may be factored from the integral. For simplicity, we have also assumed that $T_1 = T_2 = \tau_2$. The polarization associated with (III.38) generates a field E_{23} according to (III.28a) that propagates in the same direction as the E_2 laser beam, as in the transient case. A detector viewing this light monitors a cross term in the total intensity E_T of Equation (III.29)

$$|(E_T E_T^*)^b| = 16\pi^{3/2} N L \alpha^2 \beta^2 \frac{e^{-v_1^2/u_2}}{ku} \left(\frac{\hbar\Omega_2}{c}\right) w_{13}^0 \cdot \frac{1}{(\Delta + \Delta')^2 + 1/\tau_2^2} \quad \text{(III.39)}$$

which displays the expected two-photon resonance tuning behaviour having a Doppler-free linewidth $1/\tau_2$ that is independent of the intermediate level and elastic collisions. A second detector viewing light from the opposite direction would show essentially the same behaviour as (III.39) due to the $\langle \tilde{\rho}_{13}^{(1)} \rangle$ source term.

When the exact forms of $\tilde{\rho}_{23}$ and $\tilde{\rho}_{13}$ in Equation (III.34) are to be used in obtaining $(E_T E_T^*)^b$ from (III.28) and (III.29), a numerical evaluation is required as the Doppler average cannot be performed analytically. However, it is still possible to investigate the two-photon resonance denominator or tuning behaviour through (III.35a), namely,

$$a = \alpha\beta \Big/ \left[\left(\Delta + \Delta' - \frac{i}{\tau_2}\right) - \frac{\alpha^2}{\Delta' - i/T_2} - \frac{\beta^2}{\Delta - i/T_2}\right]$$

$$= \alpha\beta \Big/ \left\{\Delta\left(1 - \frac{\beta^2}{\Delta^2 + 1/T_2^2}\right) + \Delta'\left(1 - \frac{\alpha^2}{\Delta'^2 + 1/T_2^2}\right) - i\left[\frac{1}{\tau_2} + \frac{1}{T_2}\left(\frac{\beta^2}{\Delta^2 + 1/T_2^2} + \frac{\alpha^2}{\Delta'^2 + 1/T_2^2}\right)\right]\right\}.$$

This exhibits an effective linewidth

$$\frac{1}{T_2(\text{eff})} = \frac{1}{\tau_2} + \frac{1}{T_2}\left(\frac{\alpha^2}{\Delta'^2 + 1/T_2^2} + \frac{\beta^2}{\Delta^2 + 1/T_2^2}\right) \quad \text{(III.40)}$$

and a frequency shift

$$\nu = -\left(\frac{\beta^2 \Delta}{\Delta^2 + 1/T_2} + \frac{\alpha^2 \Delta'}{\Delta'^2 + 1/T_2^2}\right) \quad \text{(III.41)}$$

that are power-dependent and thus corrects the perturbation result (III.39). Furthermore, we see that the two-photon resonance condition is no longer independent of velocity and elastic collisions. It is interesting that the exact expressions (III.40) and (III.41) agree with an earlier estimate[39] of the power-dependent line broadening and frequency shift in the case of Raman beats.

When the frequency shift (III.41) is not excessively large so that $\Delta + \Delta' \sim 0$, its magnitude

$$\nu \sim \frac{\Delta(\alpha^2 - \beta^2)}{\Delta^2 + 1/T_2^2} \quad \text{(III.42)}$$

can be readily estimated. For instance, in the case $\alpha = \beta$, the frequency shift vanishes. More typically, the condition $\alpha \gg \beta$ (or $\beta \gg \alpha$) prevails, and then the shift

$$\nu \sim \frac{\Delta\alpha^2}{(\Delta^2 + 1/T_2^2)} \tag{III.43}$$

can be quite large where the direction of the shift is determined by the sign of Δ. For example, when $\Delta \gg 1/T_2$, $\nu \sim \alpha^2/\Delta$ which may be orders of magnitude larger the homogeneous linewidth $1/T_2$. Large power-dependent shifts of this kind have been observed in recent two-photon absorption measurements[44].

(d) *Summary*

We have shown that the coherent Raman beat effect with a single beam[38,39] and two-photon absorption of oppositely directed laser beams[40,41] contain identical mechanisms for yielding Doppler-free spectra. The difference between the two cases is only one of level spacing. In the Raman problem, two of the levels are nearly degenerate and connect optically with a third level. In the two-photon problem, the third level is intermediate in eigenenergy with respect to the other two where the selection rules are the same as in the Raman case. Thus far, the Raman beat effect has been seen only in the transient regime with Stark-switching[38] whereas two-photon absorption has been observed only under steady state conditions[41]. However, our solutions apply to both transient and steady state cases for these two problems. We predict that transient two-photon emission should be observed in backward scattering and will exhibit a beat oscillation that is analogous to the Raman beat seen in forward scattering. Exact steady state solutions for the Raman and two-photon problems reveal power-dependent frequency shifts and line broadening that must be considered in precision spectroscopy.

REFERENCES

1. A. Abragam. "The Principles of Nuclear Magnetism", Oxford (1961).
2. R. G. Brewer and R. L. Shoemaker. *Phys. Rev. Lett.* **27**, 631 (1971).
3. S. L. McCall and E. L. Hahn. *Phys. Rev. Lett.* **18**, 908 (1967); *Phys. Rev.* **183**, 457 (1969); *Phys. Rev.* **A2**, 861 (1970).
4. R. P. Feynman, F. L. Vernon and R. W. Hellwarth. *J. Appl. Phys.* **28**, 49 (1957).
5. F. Bloch. *Phys. Rev.* **70**, 460 (1946).
6. H. C. Torrey. *Phys. Rev.* **76**, 1059 (1949).

7. G. L. Tang and B. D. Silverman. "Physics of Quantum Electronics" (P. Kelley, B. Lax and P. E. Tannenwald, Eds), McGraw-Hill, New York, p. 280 (1966).
8. F. A. Hopf, R. F. Shea and M. O. Scully. *Phys. Rev.* **A7**, 2105 (1973).
9. R. G. Brewer and R. L. Shoemaker. *Phys. Rev.* **A6**, 2001 (1972).
10. E. L. Hahn. *Phys. Rev.* **77**, 297 (1950).
11. C. H. Townes and A. L. Schawlow. "Microwave Spectroscopy", McGraw-Hill, New York, p. 248 (1955).
12. N. A. Kurnit, I. D. Abella and S. R. Hartmann, *Phys. Rev. Lett.* **13**, 567 (1964); *ibid. Phys. Rev.* **141**, 391 (1966).
13. R. G. Brewer. *In* "Proceedings of the Rank Prize Fund Symposium, Very High Resolution Spectroscopy" (R. A. Smith, Ed.), Academic Press, London, p. 127 (1976).
14. J. Schmidt, P. R. Berman and R. G. Brewer. *Phys. Rev. Lett.* **31**, 1103 (1973).
15. R. L. Shoemaker and R. G. Brewer. *Phys. Rev. Lett.* **28**, 1430 (1972); R. G. Brewer and E. L. Hahn. *Phys. Rev.* **A8**, 464 (1973); *ibid. Phys. Rev.* **A11**, 1641 (1975).
16. M. M. T. Loy. *Phys. Rev. Lett.* **32**, 814 (1974).
17. J. M. Levy and R. G. Brewer (unpublished).
18. J. Schmidt and R. G. Brewer (unpublished).
19. K. L. Foster, S. Stenholm and R. G. Brewer. *Phys. Rev.* **A10**, 2318 (1974).
20. R. L. Shoemaker and F. A. Hopf. *Phys. Rev. Lett.* **33**, 1527 (1974).
21. E. L. Hahn. *Phys. Rev.* **80**, 580 (1950).
22. H. Y. Carr and E. M. Purcell. *Phys. Rev.* **94**, 630 (1954).
23. C. Freed. *IEEE J. Quantum Electron.* **4**, 404 (1968); **3**, 203 (1967).
24. W. A. Anderson, R. Freeman and H. Hill. "Pure and Applied Chemistry", Butterworths, London, **32**, 27 (1972).
25. This section is abstracted from the article of P. R. Berman, J. M. Levy and R. G. Brewer. *Phys. Rev.* **A11**, 1668 (1975).
26. P. R. Berman. *Phys. Rev.* **A5**, 927 (1972). **A6**, 2157 (1972).
27. P. R. Berman and W. E. Lamb, Jr. *Phys. Rev.* **A2**, 2435 (1972).
28. J. Keilson and J. E. Storer. *Quart. Appl. Math.* **10**, 243 (1952).
29. B. Herzog and E. L. Hahn. *Phys. Rev.* **103**, 148 (1956).
30. M. Scully, M. J. Stephen and D. C. Burnham. *Phys. Rev.* **171**, 213 (1968).
31. H. C. Torrey. *Phys. Rev.* **104**, 563 (1956).
32. P. R. Berman. *Phys. Rev.* **A9**, 2170 (1974).
33. R. L. Shoemaker, S. Stenholm and R. G. Brewer. *Phys. Rev.* **A10**, 2037 (1974).
34. T. Oka. *In* "Advances in Atomic and Molecular Physics" (D. R. Bates, Ed.), Academic Press, New York, Vol. 9, p. 127 (1973).
35. This section is abstracted from the article of R. G. Brewer and E. L. Hahn. *Phys. Rev.* **A11**, 1641 (1975).
36. M. Göppert-Mayer. *Ann. Physik* **9**, 273 (1931).
37. M. Bloom, E. L. Hahn and B. Herzog. *Phys. Rev.* **97**, 1699 (1955); P. A. Fedders and E. Y. C. Lu. *Phys. Rev.* **B8**, 5156 (1973).
38. R. L. Shoemaker and R. G. Brewer. *Phys. Rev. Lett.* **28**, 1430 (1972).
39. R. G. Brewer and E. L. Hahn. *Phys. Rev.* **A8**, 464 (1973); *erratum, ibid. Phys. Rev.* **A9**, 1479 (1974).
40. L. S. Vasilenko, V. P. Chebotaev and A. V. Shishaev, Pis'ma Zh. *Eksp. Teor. Fiz.* **12**, 161 (1970), (*JETP Lett.* **12**, 113 (1970)); I. M. Beterov, Yu. A. Matyugin and V. P. Chebotaev. *Sov. Phys. JETP* **37**, 756 (1973).

41. D. Pritchard, J. Apt and T. W. Ducas. *Phys. Rev. Lett.* **32**, 641 (1974); F. Biraben, B. Cagnac and C. Grynberg. *Phys. Rev. Lett.* **32**, 643 (1974); M. D. Levenson and N. Bloembergen. *Phys. Rev. Lett.* **32**, 645 (1974); T. W. Hänsch, K. C. Harvey, G. Meisel and A. L. Schawlow. *Opt. Comm.* **11**, 50 (1974); N. Bloembergen, M. D. Levenson and M. M. Salour. *Phys. Rev. Lett.* **32**, 867 (1974); J. E. Bjorkholm and P. F. Liao. *Phys. Rev. Lett.* **33**, 128 (1974); T. W. Hänsch, S. A. Lee, R. Wallenstein and C. Wieman. *Phys. Rev. Lett.* **34**, 307 (1975); W. K. Bischel, P. J. Kelly and C. K. Rhodes. *Phys. Rev. Lett.* **34**, 300 (1975).
42. N. Tan-no, K. Yokoto and H. Inaba. *Phys. Rev. Lett.* **29**, 1211 (1972).
43. M. S. Feld and A. Javan. *Phys. Rev.* **177**, 540 (1069).
44. P. F. Liao and J. E. Bjorkholm. *Phys. Rev. Lett.* **34**, 1 (1975).

9. Nonlinear Optical Effects in Plasmas†

T. P. HUGHES

Department of Physics, University of Essex, Wivenhoe Park, Colchester, Essex CO4 3SQ, England

I. INTRODUCTION

Three nonlinear situations may be distinguished in a plasma:

(1) when single electrons and ions interact with two or more photons, (2) in collective processes, when significant coupling takes place between waves in the plasma, (3) when the refractive index of the medium varies with the irradiance.

In the first category we shall discuss Bremsstrahlung and Compton scattering processes. We shall ignore multiphoton excitation and ionization, which play an important part at least initially in plasma generation by lasers (see for example Grey Morgan, 1975). In the second we come nearer

† This chapter is based on material contained in the author's book "Plasmas and Laser Light", with the kind permission of the publishers Messrs. Adam Hilger Limited.

to more conventional nonlinear optics, regarding the plasma as a nonlinear medium. However, the saturation of large-amplitude disturbances induced for example by parametric processes presents additional problems which are unique to plasmas. We shall concentrate on those processes which are energetically important, transferring significant energy from an incident wave. In the third category we shall consider self-focussing effects. Throughout the discussion we shall assume that the plasma is unmagnetized: the introduction of magnetic fields would greatly increase the complexity of the situation, because of the many additional resonances which then occur.

II. Small-Amplitude Waves in Plasma

We begin by summarizing some of the properties of plasmas and considering some of the waves that can propagate in them.

We shall be concerned mostly with fully-ionized gases, in which every atom has lost at least one electron and has a positive charge. Except at the edges of the plasma there is charge neutrality on average. If n_e is the electron density and n_i the density of positive ions of species i, with charge Z_i, $n_e = \Sigma_i n_i Z_i$. Any departure from charge neutrality generates very large electrostatic restoring forces.

The plasmas may be actually formed by heating and ionizing a gas, but they can be formed by heating and vaporizing and ionizing a solid. At present there is a great deal of interest in the use of powerful laser pulses focussed on to small solid targets for producing very dense, hot, plasmas. The irradiance may be up to 10^{15} or more W cm^{-2} with current laser technology, but the pulse duration is only a fraction of a nanosecond and the focal spot is small, of perhaps 100 μm diameter, so that the energy involved is only of the order of 100–1000 J. Under these conditions strongly nonlinear behaviour is to be expected. The practical importance of the nonlinearities is that they influence the fate of radiation energy incident on the plasma. Let us first consider only small-amplitude waves. Maxwell's equations take a simple form in a plasma because we can treat all currents as part of the current density **J**, and all charges as part of the charge density ρ_q. In MKS units we have

$$\nabla \times \mathbf{E} = -\frac{\partial \mathbf{B}}{\partial t}, \tag{1}$$

$$\nabla \times \mathbf{H} = \varepsilon_0 \frac{\partial \mathbf{E}}{\partial t} + \mathbf{J}, \tag{2}$$

$$\nabla \cdot \mathbf{E} = \frac{\rho_q}{\varepsilon_0} \tag{3}$$

$$\nabla \cdot \mathbf{B} = 0 \tag{4}$$

together with the relation

$$\mathbf{B} = \mu_0 \mathbf{H}. \tag{5}$$

From (2) and (3) we obtain the continuity equation:

$$\frac{\partial \rho_q}{\partial t} + \nabla \cdot \mathbf{J} = 0. \tag{6}$$

The current density **J** is related to the electric field by the conductivity $\boldsymbol{\sigma}$, which is a property of the plasma and is in general a tensor. For a small-amplitude wave of frequency ω we may write a generalized Ohm's law:

$$\mathbf{J}(\omega) = \boldsymbol{\sigma}(\omega) \cdot \mathbf{E}(\omega). \tag{7}$$

We introduce the dielectric tensor

$$\boldsymbol{\varepsilon} = \mathbf{1} + \frac{\boldsymbol{\sigma}}{i\omega\varepsilon_0}. \tag{8}$$

Let us suppose **E** and **H** both vary in time and space as exp $i(\omega t - \mathbf{k} \cdot \mathbf{r})$. Then

$$\nabla \times \mathbf{E} = -i\omega\mu_0 \mathbf{H} \tag{9}$$

$$\nabla \times \mathbf{H} = i\omega\varepsilon_0 \boldsymbol{\varepsilon} \mathbf{E}, \tag{10}$$

and we find by taking the curl of (9)

$$\begin{aligned} \nabla \times (\nabla \times \mathbf{E}) &= \mu_0 \varepsilon_0 \omega^2 \boldsymbol{\varepsilon} \mathbf{E} \\ &= \left(\frac{\omega}{c}\right)^2 \boldsymbol{\varepsilon} \mathbf{E}, \end{aligned} \tag{11}$$

where $c = 1/(\mu_0 \varepsilon_0)^{1/2}$ is the velocity of light in free space. This may be written

$$\mathbf{k} \times (\mathbf{k} \times \mathbf{E}) + \left(\frac{\omega}{c}\right)^2 \boldsymbol{\varepsilon} \mathbf{E} = 0. \tag{12}$$

This consists of three homogeneous equations in the components of **E**. The determinant of the coefficients must vanish if there is a solution, that is, if $E \neq 0$. The solution is called the dispersion equation: there may be more than one solution, corresponding to different families of waves. Thus for example we may take transverse waves with $\mathbf{k} \cdot \mathbf{E} = 0$, so that $c^2k^2/\omega^2 = \varepsilon$. Alternatively, we could look for longitudinal waves, with $\mathbf{k} \times \mathbf{E} = 0$, so that $\varepsilon = 0$. We shall assume for simplicity that our plasma is free from any static magnetic fields. If we also assume that the plasma is cold, and ignore collisions between electrons and ions, and suppose that the ions are effectively stationary, we have simply

$$\mathbf{J} = -n_e e \mathbf{v}_e. \tag{13}$$

To lowest order, we may write for the equation of motion of an electron

$$m_e \frac{\partial \mathbf{v}_e}{\partial t} = -e\mathbf{E} \tag{14}$$

so that

$$\mathbf{J} = \frac{n_e e^2}{m_e i\omega} \mathbf{E} \tag{15}$$

and hence

$$\sigma = \frac{n_e e^2}{m_e i\omega} \tag{16}$$

and

$$\varepsilon = 1 - \frac{\omega_{pe}^2}{\omega^2}, \tag{17}$$

where

$$\omega_{pe} = \left(\frac{n_e e^2}{m_e \varepsilon_0}\right)^{1/2} \tag{18}$$

is the electron plasma frequency. For a transverse wave, therefore, we have the dispersion relation

$$k^2 c^2 = \omega^2 \varepsilon = \omega^2 - \omega_{pe}^2, \tag{19}$$

so that transverse waves can propagate provided $\omega^2 > \omega_{pe}^2$ (otherwise k is wholly imaginary). The electron density for which $\omega_{pe} = \omega$ is known as the critical density for that frequency. For a longitudinal wave $\varepsilon = 0$, so $\omega = \omega_{pe}$. This is an oscillation, not a wave, in a cold plasma, because the group velocity $\partial\omega/\partial k = 0$.

If we allow for collisions transferring momentum between electrons and ions, with frequency ν_{ei}, the equation of motion becomes

$$m_e \frac{\partial \mathbf{v}_e}{\partial t} = -e\mathbf{E} - \nu_{ei} m_e \mathbf{v}_e. \tag{20}$$

Assuming that ν_{ei} is independent of velocity,

$$\mathbf{J} = \frac{n_e e^2}{m_e(\nu_{ei} + i\omega)} \mathbf{E}, \tag{21}$$

so

$$\varepsilon = 1 - \frac{\omega_{pe}^2}{(\nu_{ei}^2 + \omega^2)} - \frac{i\omega_{pe}^2 \nu_{ei}}{\omega(\nu_{ei}^2 + \omega^2)}. \tag{22}$$

This gives a modified dispersion relation for transverse waves,

$$k^2c^2 = \omega^2\varepsilon = \omega^2 - \frac{\omega^2\omega_{pe}^2}{(\nu_{ei}^2 + \omega^2)} - \frac{i\omega\omega_{pe}^2\nu_{ei}}{(\nu_{ei}^2 + \omega^2)} \tag{23}$$

but still no propagation for longitudinal waves (when $\varepsilon = 0$). For future reference we may note that for high $\omega \gg \omega_{pe}$ and low $\nu_{ei}^2 \ll \omega^2 - \omega_{pe}^2$ this leads to an *absorption coefficient* (twice the imaginary part of the propagation vector $\mathbf{k}$) of

$$a_\omega \simeq \frac{\nu_{ei}}{\mu c} \frac{\omega_{pe}^2}{\omega^2}, \tag{24}$$

which is proportional to ν_{ei}, and a real refractive index

$$\mu \simeq \left(1 - \frac{\omega_{pe}^2}{\omega^2}\right)^{1/2}, \tag{25}$$

which is independent of ν_{ei}.

In a hot plasma the situation is more complicated: we have to integrate over the electron velocity distribution function. The dielectric tensor $\boldsymbol{\varepsilon}$ now depends on the orientation of the wave electric field relative to the propagation vector $\mathbf{k}$, so we have different values of $\boldsymbol{\varepsilon}$, ε_T for transverse waves and ε_L for longitudinal waves. The expression for ε_L takes the form

$$\varepsilon_L = 1 - \sum_j \frac{\omega_{pj}^2}{\omega^2} f(k, \omega, T_j, m_j) \tag{26}$$

where j describes the various species of particle, with mass m_j and temperature T_j, and $\omega_{pj}^2 = n_j q_j^2/(m_j\varepsilon_0)$, q_j being the charge. ε_L is now k-dependent and propagation is possible. If we ignore the ion motion we find that the real part of the frequency is now given approximately by

$$\omega_{Rk}^2 \simeq \omega_{pe}^2[1 + 3k^2\lambda_D^2] = \omega_{BG}^2 \tag{27}$$

where λ_D is the Debye shielding length:

$$\lambda_D = \left(\frac{\kappa_B T_e \varepsilon_0}{n_e e^2}\right)^{1/2}. \tag{28}$$

ω_{BG} is the Bohm-Gross frequency for electron plasma waves. The phase velocity of the wave is ω/k:

$$V_{\text{phase}}^2 \simeq \frac{\omega_{pe}^2}{k^2} + 3\omega_{pe}^2 \lambda_D^2$$

$$= \frac{\omega_{pe}^2}{k^2} + \frac{3\kappa_B T_e}{m_e}. \tag{29}$$

Provided k is small, $V_{\text{phase}} \gg v_T = (\kappa_B T_e/m_e)^{1/2}$. However, if k becomes large, the phase velocity is reduced and eventually becomes comparable to the electron thermal velocity. Many electrons are then moving at speeds close to that of the wave. Electrons travelling slightly slower than the wave tend to be accelerated up to the phase velocity, while those travelling slightly faster tend to be slowed down. The slope of the distribution is now very important. If there are more slow than fast electrons, as there are with a thermal distribution, the wave loses energy to the electrons—an effect known as Landau damping. Landau damping is only important if $k\lambda_D \gtrsim 0.3$. On the other hand if the slope is in the other direction near the phase velocity the wave gains energy from the electrons and is actually amplified. This effect is known as a two-stream instability, because it can happen if the electron velocity distribution has a second maximum corresponding to the superposition of two groups of particles with a relative drift velocity. Both these effects are collisionless. At high frequencies electron plasma waves are the only important longitudinal waves, because the ions cannot respond to the driving field. At low frequencies, however, they can. In this case

$$\omega_{Rk}^2 \simeq \omega_{pi}^2 \left[\frac{k^2\lambda_D^2}{1 + k^2\lambda_D^2}\right] = \frac{Z\kappa_B T_e}{m_i}\left(\frac{k^2}{1 + k^2\lambda_D^2}\right) = \omega_{ia}^2. \tag{30}$$

These are known as ion acoustic waves. Their phase velocity is given by

$$V_{ia}^2 = \frac{\omega_{ia}^2}{k^2} = \frac{Z\kappa_B T_e}{m_i}\left(\frac{1}{1 + k^2\lambda_D^2}\right) \tag{31}$$

Provided $k\lambda_D \ll 1$,

$$V_{ia} \simeq \left(\frac{Z\kappa_B T_e}{m_i}\right)^{1/2} \tag{32}$$

analogous with the velocity of an ordinary acoustic wave, $(\gamma p/\rho)^{1/2}$, since $p \sim n\kappa_B T_e$ and $\rho \sim nm_i$. These waves are therefore approximately dispersionless. The damping is again by Landau damping, which will be small provided $V_{ia}^2 \gg \kappa_B T_i/m_i$, so that only a few ions have about the same velocity as the wave. From (32) the condition for small damping becomes $T_e \gg T_i$. Damping of ion waves by the electrons is small, because although there are many electrons near the centre of the distribution function the slope of the function is almost zero.

III. Bremsstrahlung Processes in Strong Radiation Fields

The dominant mechanism for the absorption of light of low intensity in a fully-ionized plasma is inverse electron-ion Bremsstrahlung. A free electron colliding with a positive ion absorbs a photon and moves into a more energetic free state. When radiation of frequency ω traverses a plasma with a Maxwellian electron velocity distribution at temperature T_e, the linear inverse Bremsstrahlung absorption coefficient is given by (MKS units)

$$a_\omega = \frac{n_e n_i Z^2 e^6[1 - \exp(-\hbar\omega/\kappa_B T_e)]}{6\pi\varepsilon_0^3 c\hbar\omega^3 m_e^2}\left(\frac{m_e}{2\pi\kappa_B T_e}\right)^{1/2}\frac{\pi}{\sqrt{3}}\bar{g}. \tag{33}$$

In a high-temperature plasma ($\kappa_B T_e \gg \hbar\omega$) the Gaunt factor $\bar{g}$ is given by

$$\bar{g} = \frac{\sqrt{3}}{\pi}\ln\left(\frac{4\kappa_B T_e}{1.781\hbar\omega}\right), \tag{34}$$

and the absorption coefficient becomes

$$a_{\omega T} = \frac{n_e n_i Z^2 e^6}{6\pi\varepsilon_0^3 c\omega^2}\left(\frac{1}{2\pi}\right)^{1/2}\left(\frac{1}{m_e\kappa_B T_e}\right)^{3/2}\ln\left(\frac{4\kappa_B T_e}{1.781\hbar\omega}\right). \tag{35}$$

This expression is only valid provided the radiation field does not perturb the Maxwellian electron distribution significantly.

Light of irradiance I has an electric field of amplitude E_0, given by

$$E_0^2 = \frac{2I}{\varepsilon_0 c},$$

so

$$E_{0(\mathrm{V\,cm^{-1}})} \simeq 27\sqrt{I_{(\mathrm{W\,cm^{-2}})}}. \tag{36}$$

An electron "quivers" in the oscillating field, with a maximum velocity

$$v_E = \left| \frac{eE_0}{m_e \omega} \right| = 25\lambda\sqrt{I} \text{ cm s}^{-1} \tag{37}$$

where λ is in μm and I in W cm^{-2}. The electron thermal velocity is of the order of

$$v_T = \left(\frac{\kappa_B T_e}{m_e} \right)^{1/2}. \tag{38}$$

The ratio of these velocities

$$\eta = \frac{v_E}{v_T} \tag{39}$$

is known as the "field strength parameter". If $\eta \gtrsim 1$, the electron velocity distribution is strongly perturbed by the radiation field and expression (33) for the absorption coefficient is no longer valid.

In the strong field limit, when $\eta \gg 1$, the electron velocity distribution is dominated by the radiation-induced quivering, and classically it might be expected that $\kappa_B T_e = m_e v_T^2$ in expression (35) would be replaced by $m_e v_E^2 = e^2 E_0^2 / m_e \omega^2$, leading to a field-dependent absorption coefficient of form

$$a_{\omega E} = \frac{n_e n_i Z^2 e^3 \omega}{6\pi\varepsilon_0^3 c E_0^3} \left(\frac{1}{2\pi} \right)^{1/2} \frac{\pi}{\sqrt{3}} \bar{g}_E, \tag{40}$$

where $\bar{g}_E$ is a new Gaunt factor.

Quantum-mechanically, in a strong field multiphoton processes must be taken into account. The problem has been discussed by many authors: Rand (1964), Silin (1964), Bunkin and Fedorov (1965), Nicholson-Florence (1971), Brehme (1971), Osborn (1972a), Pert (1972), Seely and Harris (1973), who agree regarding the general form of the absorption coefficient given by (40). Pert has shown that in most cases of practical interest the classical and quantum mechanical treatments yield the same result.

The transition from the low-field absorption coefficient to the high-field coefficient occurs near $\eta = 1$, that is when

$$\frac{e^2 E_0^2}{m_e \omega^2} \simeq \kappa_B T_e. \tag{41}$$

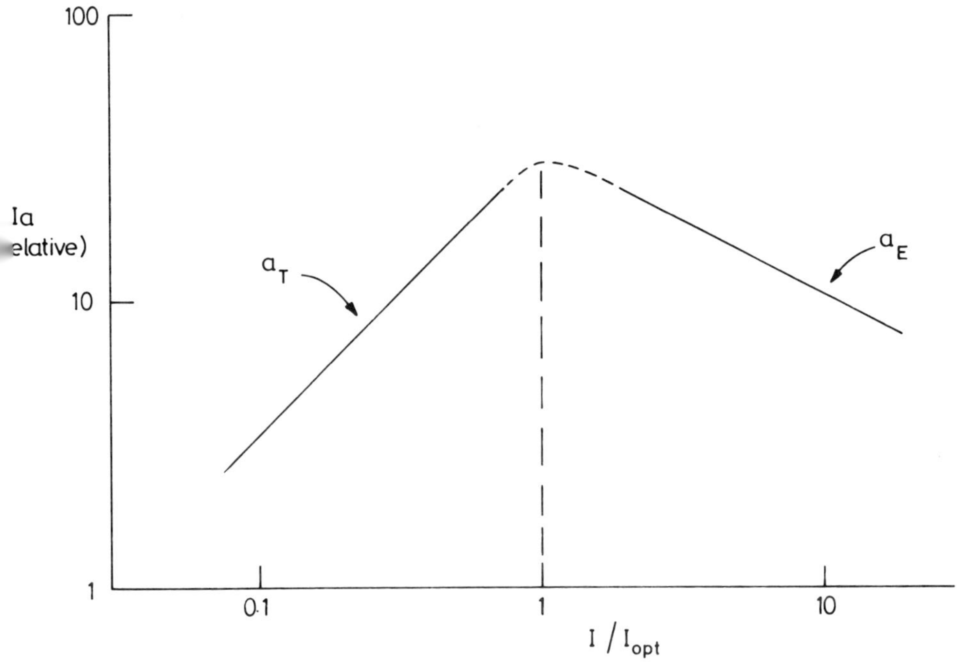

FIG. 1. Power absorbed from a laser beam by Bremsstrahlung per unit volume of a plasma, as a function of the irradiance.

At higher fields, the power absorbed by an element of plasma will actually fall as the irradiance is increased (Fig. 1). There is therefore an optimum irradiance I_{opt} corresponding to $\eta \approx 1$ at which the power absorbed by the plasma due to inverse Bremsstrahlung will be a maximum: if T_e is in eV and λ in microns,

$$I_{\text{opt}} \simeq 3 \times 10^{12} \frac{T_e}{\lambda^2} \text{ W cm}^{-2}. \tag{42}$$

This expression is plotted in Fig. 2. It should be noted that the frequency dependence of the absorption coefficient also changes as the irradiance increases. For low fields $a_{\omega T} \propto (1/\omega^2)$, while for high fields $a_{\omega E} \propto \omega$. This does not necessarily mean that high frequency light is best for plasma heating: assuming that $a_{\omega T}$ is valid up to I_{opt}, the product $I_{\text{opt}} a_{\omega T}$ is approximately independent of ω, neglecting the logarithmic Gaunt factor.

The Bremsstrahlung emission must also be modified by the presence of a strong radiation field (Pert (1973)).

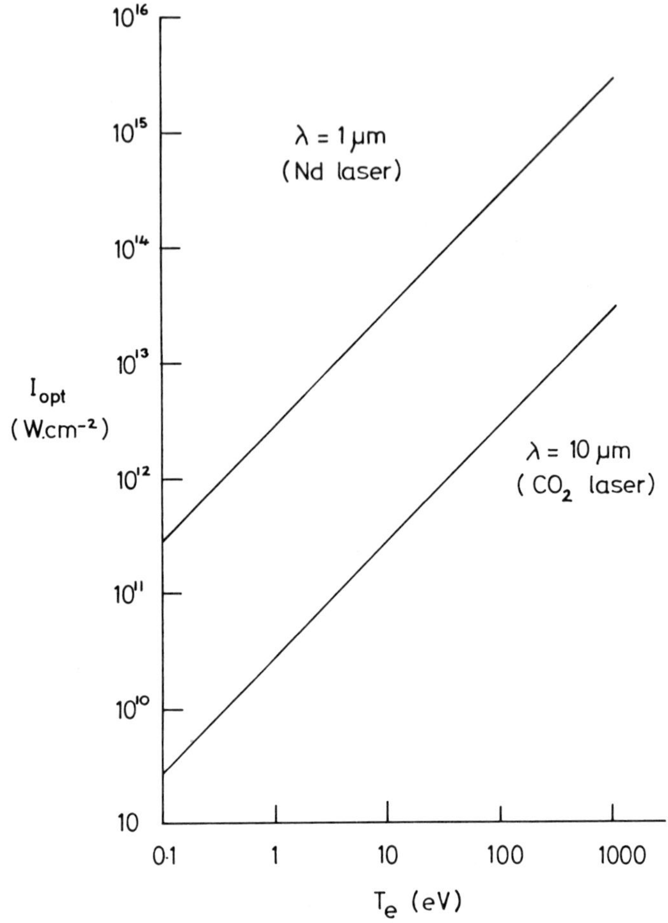

FIG. 2. Optimum irradiance for Bremsstrahlung absorption as a function of the electron temperature, at wavelengths of 1 and 10 μm.

In a plasma at a temperature of 300 eV, for laser light at a wavelength of 1 μm, we find that $I_{opt} \sim 10^{16}$ W cm^{-2}. We shall see later that many other nonlinear effects are expected to occur with thresholds at considerably lower irradinaces: it may prove difficult to isolate the nonlinear Bremsstrahlung processes.

Electron-neutral atom collisions make a significant contribution to the absorption coefficient in partially-ionized gases. Nonlinear effects are unlikely to be apparent for these processes, because at irradiances in the vicinity of I_{opt} the radiation will produce very rapid ionization.

IV. Stimulated Compton Scattering

We may think of the scattering of a photon from a light beam by an electron as absorption followed by emission at a new frequency in a new direction. If a second intense beam coincides in frequency and direction with the scattered light the emission will be stimulated (Dreicer (1964) (Fig. 3(a)). The stimulated scattering rate is proportional to the product of the two beam intensities. If the two beams have frequencies ω_1 and $\omega_2 < \omega_1$ the energy difference $\hbar(\omega_1 - \omega_2)$ is delivered to the scattering electron. The heating rate per particle is independent of density and is proportional to the product of the two beam intensities. Stimulated Compton scattering was discussed in

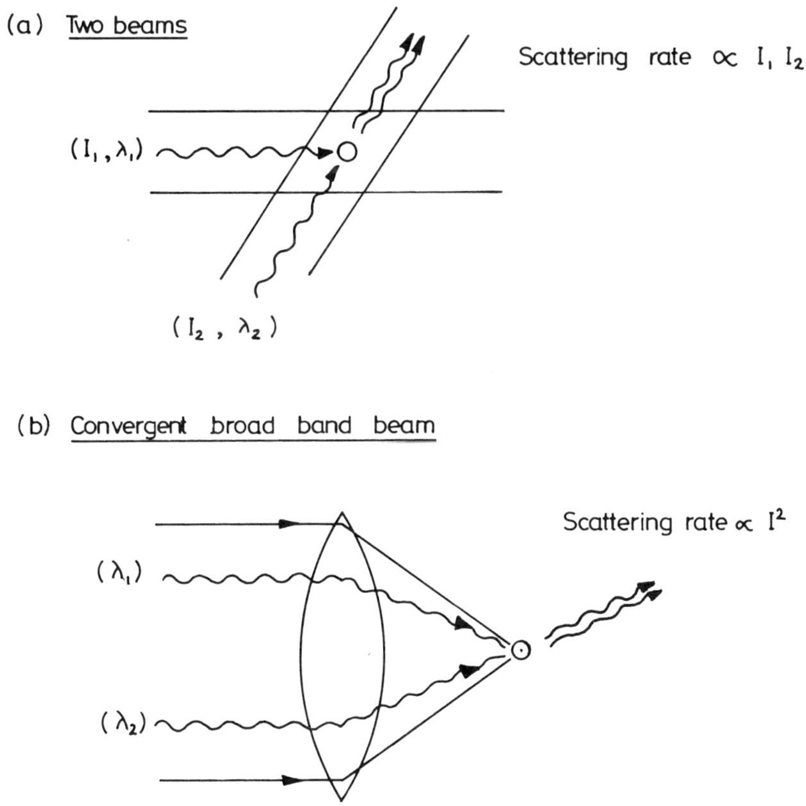

FIG. 3. Stimulated Compton scattering: (a) in two intersecting beams, (b) in a convergent beam of broad bandwidth.

detail by Peyraud (1969). Bunkin and Kazakhov (1970) suggested that the effect could take place in a single converging laser beam of finite bandwidth $\Delta\omega_0$ around ω_0: the heating rate now varies as the square of the converging beam intensity, Fig. 3(b). This process might be helpful for heating very hot, relatively low-density plasmas when inverse Bremsstrahlung is ineffective (Martineau and Pépin (1972), Beaudry *et al.* (1971)). However, the proportion of the incident energy given to the electrons is limited by the bandwidth of the radiation to $\Delta\omega_0/\omega_0$, and is unlikely to exceed 1 %. Further calculations were made by Zel'dovich *et al.* (1972) and by Vinogradov and Pustovalov (1972a and b).

Stimulated Compton scattering may have another more important effect on the heating of dense, isolated plasmas, as was pointed out by Krasyuk *et al.* (1973). When a laser beam is incident on an isolated high-density target some of the light will be reflected from the region of critical plasma density. The incident and reflected waves can then interact through stimulated Compton scattering in the less dense outer layers of the plasma. Although little energy is given to the electrons, the reflected wave is amplified, and if the irradiances are sufficiently great much of the incident wave energy may be converted into reflected wave energy. The spatial variations of the incident and reflected irradiances $I_1(z)$ and $I_2(z)$ are related by the expression

$$\frac{dI_1(z)}{dz} = \frac{dI_2(z)}{dz} = G_c I_1(z) I_2(z), \tag{43}$$

the positive direction of z being towards the laser. Provided the incident bandwidth $\Delta\omega_0$ satisfies $v_T/c \gg \Delta\omega_0/\omega_0$, the parameter G_c is given by

$$G_c = \frac{8\pi^2 r_e^2 c^2}{\omega_0^3} \frac{v_p n_e}{\kappa_B T_e} \left(\frac{m_e}{2\pi\kappa_B T_e}\right)^{1/2} \exp\left(-\frac{m_e v_p^2}{2\kappa_B T_e}\right). \tag{44}$$

Here $r_e = e^2/(4\pi\varepsilon_0 m_e c^2)$ is the classical electron radius and v_p is the plasma velocity along z. The overall intensity reflection coefficient, $\mathscr{R}$, for the incident beam is given by

$$\mathscr{R} = \mathscr{R}_0 \exp G_c I_{10} l(1 - \mathscr{R}), \tag{45}$$

where $\mathscr{R}_0$ is the initial reflection coefficient at $z = 0$, l is the length of the plasma and I_{10} is the original laser irradiance.

Krasyuk *et al.* (1973) solved Equation (45) for conditions approximating those envisaged in an isentropic compression proposal using a neodymium laser: $\kappa_B T_e = 10$ keV, $n_e \sim (0.5 \text{ to } 1) \times 10^{21}$ cm^{-3}, $v_p = 10^8$ cm sec^{-1}, so that $G_c = 3.14 \times 10^{-17}$ cm W^{-1}. The overall reflectivity $\mathscr{R}$ is shown as a function of the initial target reflectivity $\mathscr{R}_0$ for four values of l in Fig. 4(a)

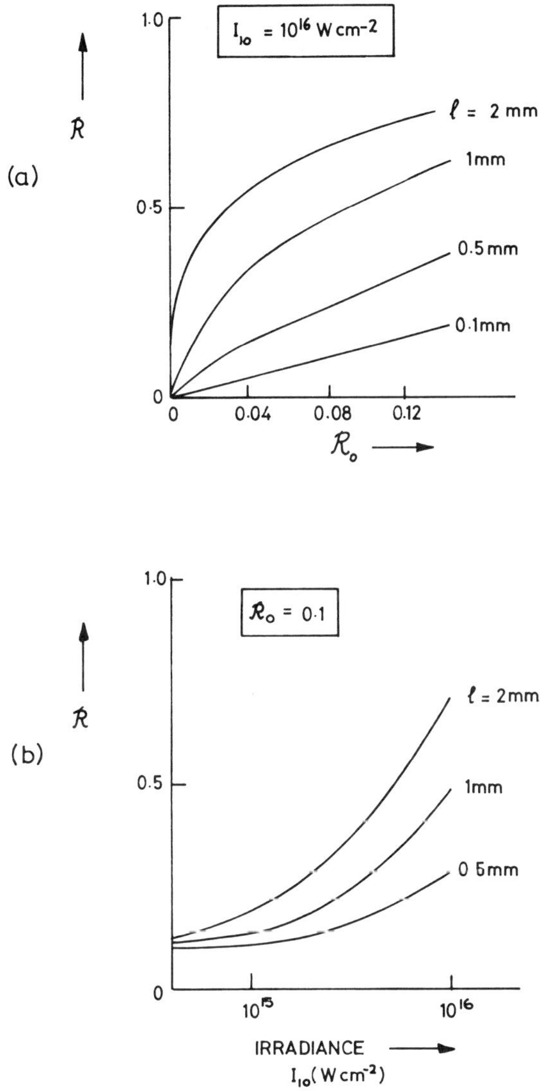

FIG. 4. Enhanced reflectivity of plasma boundary due to stimulated Compton scattering: (a) overall reflection coefficient $\mathscr{R}$ as function of reflection coefficient of critical density layer, $\mathscr{R}_0$, for four values of the plasma thickness l, with incident irradiance $I_{10} = 10^{16}$ Wcm^{-2}, (b) $\mathscr{R}$ as function of I_{10} for $\mathscr{R}_0 = 0.1$ and three values of l. (After Krasyuk *et al.*, 1973.)

while in Fig. 4(b) the overall reflectivity is shown as a function of the incident irradiance for three values of l at $\mathscr{R}_0 = 0.1$.

V. Parametric Interactions

As we saw in Section II both transverse (electromagnetic) and longitudinal (electrostatic) waves can propagate in a warm plasma. We shall be especially interested in parametric processes where an intense electromagnetic pump wave (the incident laser light) transfers its energy to other modes of oscillation such as electron plasma (Langmuir) waves, ion acoustic waves and electromagnetic waves of other frequencies. The strongest interactions will occur at resonant frequencies where damping is least. From a quantum-mechanical point of view we may regard these processes as a conversion of incident laser photons into combinations of "plasmons", or quanta of electron plasma waves; and ion acoustic phonons; and other photons. Energy and momentum balance require that at least three quanta are involved: for a three-wave interaction the selection rules are

$$\begin{aligned} \omega_0 &\simeq \omega_1 + \omega_2 \\ \mathbf{k}_0 &\simeq \mathbf{k}_1 + \mathbf{k}_2 \end{aligned} \tag{46}$$

where ω_0, $\mathbf{k}_0$ are the frequency and wave vector of the pump wave and ω_1, $\mathbf{k}_1$; ω_2, $\mathbf{k}_2$ refer to the driven modes. The quantum-mechanical picture also shows us how the pump wave energy is split between the various driven frequencies (the corresponding wave result is the Manley-Rowe relation): if frequency shifts in the coupled modes are ignored a fraction ω_j/ω_0 of the incident power is transferred to the wave at frequency ω_j.

If T denotes a pump wave photon, T' a photon at a different frequency, E a plasmon and I an ion-acoustic phonon, possible three-wave parametric processes include

$T \rightarrow E + I$,	the "parametric instability"
$T \rightarrow \mathrm{E} + \mathrm{E}$,	the "two-plasmon instability"
$T \rightarrow T' + E$,	stimulated Raman scattering
$T \rightarrow T' + I$,	stimulated Brillouin scattering.

The first two processes convert transverse radiation energy into electron and ion wave energy, and ultimately most probably into particle kinetic energy when these longitudinal waves are damped. The third and fourth processes result in scattering of a pump wave incident on a plasma boundary, which reduces the effective irradiance available for plasma heating. It is necessary to determine the threshold irradiance and the growth rate for each process to find out which is the most important in a given situation. This may be done by considering only the initial stages of development. However, it is also necessary to follow the growth and saturation of the driven modes if we are to estimate their effectiveness for plasma heating or for scattering.

Further complexities must also be expected, because if large-amplitude electron and ion waves are present in a plasma they too may couple non-linearly, generating new waves at other frequencies. Many generations of similar interactions can occur rapidly, producing a broad spectrum of plasma turbulence, plasma turbulence being defined by Kadomtsev as "motion of a plasma in which many collective degrees of freedom are excited". Although analytically intractable, these processes have been simulated by computer experiments which indicate how the pump wave energy is ultimately transformed into particle kinetic energy.

A further possibility is that a transverse wave with a frequency higher than the plasma frequency may be generated by the coupling of large-amplitude plasma waves: this new transverse wave can then escape from the plasma (Lashmore-Davies (1974)).

A. GENERAL THEORY OF PARAMETRIC INTERACTIONS IN A HOMOGENEOUS PLASMA

The general theory of parametric interactions in a homogeneous plasma has been given by Nishikawa (1968a) whose treatment we shall follow here, and others including Lashmore-Davies (1973), Drake *et al.* (1974) and Jorna (1974). Let us consider a pump wave of very long wavelength at frequency ω_0, externally maintained at a large constant amplitude A_0. We suppose that in the absence of the pump wave two independent "normal modes" A_1 and A_2 can exist in the plasma, whose wave equations are

$$\frac{d^2A_1}{dt^2} + 2\Gamma_1 \frac{dA_1}{dt} + \omega_1^2 A_1 \equiv W_1[A_1(t)] = 0 \tag{47}$$

and

$$\frac{d^2A_2}{dt^2} + 2\Gamma_2 \frac{dA_2}{dt} + \omega_2^2 A_2 \equiv W_2[A_2(t)] = 0. \tag{48}$$

Here Γ_1, Γ_2 are the damping constants for the uncoupled waves. We assume that $\omega_1 < \omega_2$. When the pump wave is switched on and coupling is present we may write

$$W_1[A_1(t)] = c_{120} A_2(t) A_0(t) \tag{49}$$

and

$$W_2[A_2(t)] = c_{201} A_0(t) A_1(t). \tag{50}$$

The coupling coefficients c_{120}, c_{201}, which will be discussed later, are assumed to be real and positive. Taking Fourier transforms of (49) and (50) with

$$A_j(t) = \int_{-\infty}^{\infty} e^{i\omega t} A_j(\omega) d\omega \tag{51}$$

we find

$$(\omega^2 - \omega_1^2 + 2i\Gamma_1 \omega) A_1(\omega) + c_{120} A_0 [A_2(\omega - \omega_0) + A_2(\omega + \omega_0)] = 0 \tag{52}$$

and

$$(\omega^2 - \omega_2^2 + 2i\Gamma_2 \omega) A_2(\omega) + c_{201} A_0 [A_1(\omega - \omega_0) + A_1(\omega + \omega_0)] = 0. \tag{53}$$

Thus A_1 at ω couples with A_2 at $(\omega \pm \omega_0)$ and A_2 at $(\omega \pm \omega_0)$ couples with A_1 at ω and also at $(\omega \pm 2\omega_0)$. The sequence is restricted by considering only resonant frequencies. Γ_1 and Γ_2 are assumed to be independent of frequency, which is only true near a resonance. $A_1(\omega \pm 2\omega_0)$ is therefore neglected, and we have the following equations in $A_1(\omega)$, $A_2(\omega - \omega_0)$, $A_2(\omega + \omega_0)$:

$$\begin{aligned}
(\omega^2 - \omega_1^2 + 2i\Gamma_1 \omega) A_1(\omega) + c_{120} A_0 [A_2(\omega - \omega_0) + A_2(\omega + \omega_0)] &= 0 \\
[(\omega - \omega_0)^2 - \omega_2^2 + 2i\Gamma_2(\omega - \omega_0)] A_2(\omega - \omega_0) + c_{201} A_0 A_1(\omega) &= 0 \\
[(\omega - \omega_0)^2 - \omega_2^2 + 2i\Gamma_2(\omega + \omega_0)] A_2(\omega + \omega_0) + c_{201} A_0 A_1(\omega) &= 0.
\end{aligned} \tag{54}$$

In matrix form we may write

$$\begin{bmatrix}
\omega^2 - \omega_1^2 + 2i\Gamma_1 \omega & c_{120} A_0 & c_{120} A_0 \\
c_{201} A_0 & (\omega - \omega_0)^2 - \omega_2^2 + 2i\Gamma_2(\omega - \omega_0) & 0 \\
c_{201} A_0 & 0 & (\omega + \omega_0)^2 - \omega_2^2 + 2i\Gamma_2(\omega + \omega_0)
\end{bmatrix}
\times \begin{bmatrix} A_1(\omega) \\ A_2(\omega - \omega_0) \\ A_2(\omega + \omega_0) \end{bmatrix} = 0. \tag{55}$$

Setting the determinant of the matrix of coefficients equal to zero we find an expression for the complex frequency

$$\omega = \omega_R + i\gamma. \tag{56}$$

The coupled system consists of a wave of amplitude A_1 at real frequency ω_R, and waves of amplitude A_2 at $(\omega_R \pm \omega_0)$, which all grow with the same growth rate γ, drawing energy from the pump wave. Nishikawa solves the determinantal equation for several different conditions. We shall simplify this four-wave system further to a three-wave system by assuming that one of the waves is off resonance. We are interested in energy transfer to A_2 and A_1, so let us neglect the wave at $(\omega_R + \omega_0)$, and suppose that ω_R is near ω_1. Then the determinantal equation for small Γ_1, Γ_2 is approximately

$$\begin{vmatrix} 2\omega_1(\omega - \omega_1 + i\Gamma_1) & c_{120}A_0 \\ c_{201}A_0 & -2\omega_2(\omega - \omega_0 + \omega_2 + i\Gamma_2) \end{vmatrix} = 0. \tag{57}$$

The deviation from exact frequency matching is given by

$$\Delta = \omega_0 - \omega_1 - \omega_2. \tag{58}$$

It is also useful to introduce the parameter

$$\Psi = \frac{c_{120}c_{201}}{\omega_1\omega_2} A_0^2, \tag{59}$$

which is a measure of the pump intensity. The real part of (57) gives

$$(\omega_R - \omega_1)(\omega_R - \omega_1 - \Delta) + \frac{\Psi}{4} = (\gamma + \Gamma_1)(\gamma + \Gamma_2) \tag{60}$$

while the imaginary part gives

$$\omega_R = \omega_1 + \left(\frac{\gamma + \Gamma_1}{2\gamma + \Gamma_1 + \Gamma_2}\right)\Delta. \tag{61}$$

Replacing ω_R in (60) by the expression in (61) yields

$$\Psi = 4(\gamma + \Gamma_1)(\gamma + \Gamma_2)\left[1 + \frac{\Delta^2}{(2\gamma + \Gamma_1 + \Gamma_2)^2}\right]. \tag{62}$$

If the instability is to grow, γ must be positive. The threshold value of Ψ is found when $\gamma = 0$, and is a function of Δ:

$$\Psi_{\text{th}}(\Delta) = \Psi_{\text{th}}(0)\left[1 + \frac{\Delta^2}{(\Gamma_1 + \Gamma_2)^2}\right]. \tag{63}$$

The lowest value of the threshold occurs with perfect frequency matching at

$$\Psi_{\text{th}}(0) = 4\Gamma_1\Gamma_2. \tag{64}$$

If either of the original normal modes is only very slightly damped, the instability grows at very low threshold.

The maximum growth rate depends on Ψ and on Δ: from expression (62) we have for $\Psi \gg \Gamma_1, \Gamma_2$

$$\gamma_{\max} = \frac{\Psi^{1/2}}{2}. \tag{65}$$

If we return to the full determinantal Equation (55) and solve it, we find that provided $\omega_0 < \omega_2$ there is an additional solution of the form $\omega = i\gamma$ implying that A_1 grows at zero frequency while A_2 grows at the pump wave frequency. If $\delta = \omega_0 - \omega_2$, the threshold value of Ψ for this purely growing instability is

$$\Psi_{\text{th}}(\delta) = -\omega_1 \frac{(\Gamma_2^2 + \delta^2)}{\delta}, \tag{66}$$

which is independent of Γ_1 and has a minimum value when $\delta = -\Gamma_2$,

$$\Psi_{\text{th}}(\delta = -\Gamma_2) = 2\omega_1\Gamma_2. \tag{67}$$

For large values of Ψ the maximum growth rate is

$$\gamma_{\max} = \left(\frac{\Psi\omega_1}{2}\right)^{1/3}. \tag{68}$$

B. PARAMETRIC INSTABILITIES

As an example, we shall now apply the results of the previous section to the interaction

$$T \to E + I$$

The uncoupled modes of the plasma are

1. low-frequency ion density fluctuations at frequencies near the ion acoustic frequency, and
2. high-frequency electron density fluctuations at frequencies near the Bohm-Gross frequency.

The wavelength of the pump radiation is assumed to be infinite so that $\mathbf{k}_0 = 0$. Thus $\mathbf{k}_1 = -\mathbf{k}_2$, and the wavenumbers of both the coupled modes are equal and will be denoted by k.

Because the ion acoustic frequency is low and $\omega_0 = \omega_1 + \omega_2$, the pump frequency will be close to the electron plasma frequency, that is, the interaction occurs in a plasma with an electron density close to the critical value for the incident pump light.

A rigorous derivation of the coupling of the electron and ion density fluctuations is given by Nishikawa (1968b) from the hydrodynamic equations. Physically, the interaction may be understood from a simple model (Kruer and Dawson (1972)). A low-frequency ion density fluctuation causes low-frequency variations in the electron density with the same number, because the electrons move to cancel departures from charge neutrality. An electron in the high-frequency electric field of the pump radiation, of amplitude E_0, quivers with a spatial amplitude $x_0 = eE_0/m\omega_0^2$. The pump field thus moves the low-frequency fluctuations rapidly past a given point, resulting in an additional high-frequency contribution to the electron density fluctuations, of amplitude

$$\delta n'_{ek} = n'_{ik}(x + x_0) - n'_{ik}(x). \tag{69}$$

The x-dependence of n'_{ik} is of the form

$$n'_{ik}(x) = n'_{ik} \exp(ikx), \tag{70}$$

so

$$\delta n'_{ek} = x_0 \frac{dn'_{ik}}{dx} = \frac{eE_0}{m_e\omega_0^2} ikn'_{ik}. \tag{71}$$

The second of the coupled equations (50) then becomes (Nishikawa (1968b))

$$\frac{\partial^2 n'_{ek}}{\partial t^2} + \omega_{BG}^2 n'_{ek} + \nu_{ek} \frac{\partial n'_{ek}}{\partial t} = \frac{ikeE_0}{m_e} n'_{ik}, \tag{72}$$

where ν_{ek} is the effective electron collision frequency, allowing for Landau damping. A similar equation is obtained for the ions:

$$\frac{\partial^2 n'_{ik}}{\partial t^2} + \omega_{ia}^2 n'_{ik} + \nu_{ik} \frac{\partial n'_{ik}}{\partial t} = -\frac{ikeE_0}{m_i} n'_{ek}. \tag{73}$$

All the results of Section (VA) for thresholds and growth rates may now be applied by substituting n'_{ik} for A_1, n'_{ek} for A_2, ω_{ia} for ω_1, ω_{BG} for ω_2, kE_0 for A_0, $-ie/m_i$ for C_{120}, ie/m_e for C_{201}, ν_{ik} for $2\Gamma_1$ and ν_{ek} for $2\Gamma_2$.

The solution of form $\omega = i\gamma$ which occurs when the ion oscillations have zero frequency is known as the "oscillating two-stream" instability. The other solution, of form $\omega = \omega_R + i\gamma$, in which a pump photon decays to a plasmon and an ion acoustic phonon, is known as the "parametric decay" instability.

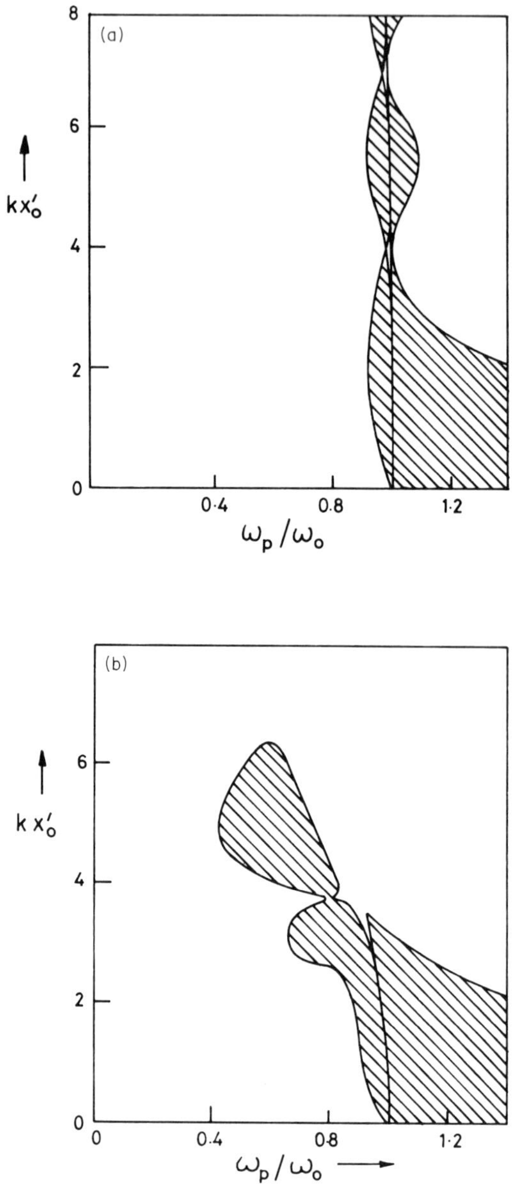

FIG. 5. Parametric instability of plasmas: (a)-(c): regions of instability in (ω_p/ω_0, kx'_0) space for hydrogen plasmas (m/m = 1836); (a) cold plasma, (b) $T_e = 10T_i$, $\eta = 10$, (c) $T_e = 10T_i$, $\eta = 2$: (d) maximum growth rates as functions of ω_p/ω_0 for $m_i/m_e = 1836$ and 100, (e) wave-numbers $k_{(\gamma\,\mathrm{max})}$ corresponding to maximum growth rates (after Freidberg and Marder, 1971).

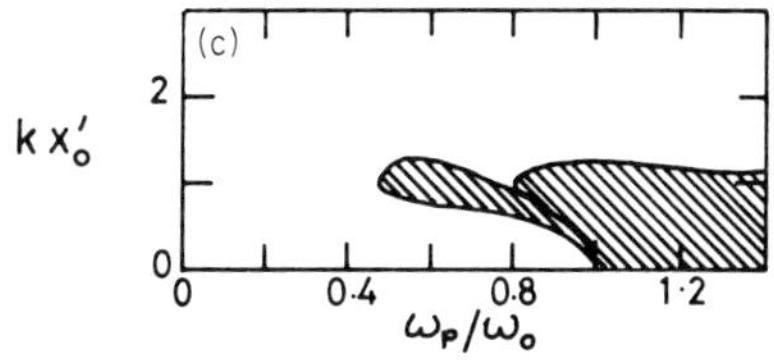
(c)
kx'_0
2
0
0
0·4
0·8
1·2
ω_p/ω_0

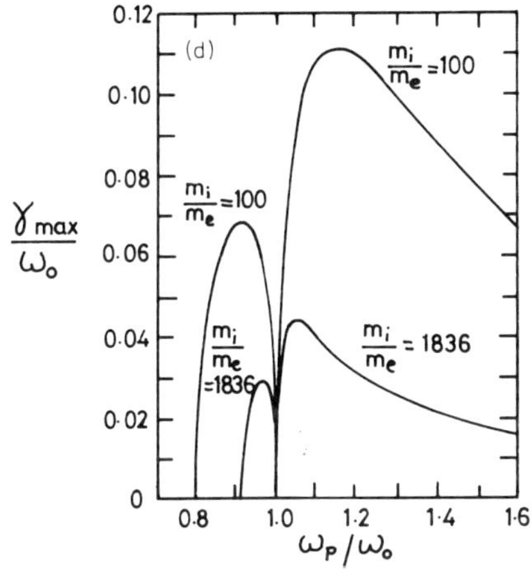
(d)
γ_{max}/ω_0
0·12
0·10
0·08
0·06
0·04
0·02
0
$m_i/m_e = 100$
$m_i/m_e = 100$
$m_i/m_e = 1836$
$m_i/m_e = 1836$
0·8
1·0
1·2
1·4
1·6
ω_p/ω_0

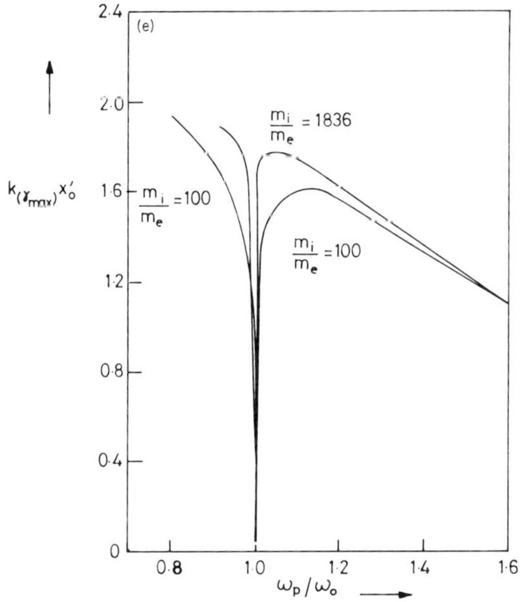
(e)
$k_{(\gamma_{max})}x'_0$
2·4
2·0
1·6
1·2
0·8
0·4
0
$m_i/m_e = 1836$
$m_i/m_e = 100$
$m_i/m_e = 100$
0·8
1·0
1·2
1·4
1·6
ω_p/ω_0

(a) *Growth rates and Thresholds in a Homogeneous Plasma*

Freidberg and Marder (1971) investigated the stability properties of a plasma against both these instabilities for $k_0 = 0$. Fig. 5(a) shows the region of instability in $(\omega_p/\omega_0, kx'_0)$—space for a cold hydrogen plasma. Here $\omega_p = (\omega_{pe}^2 + \omega_{pi}^2)^{1/2}$, k is the wavenumber of the electron oscillation, ω_0 is the incident light wave frequency and

$$x'_0 = \frac{eE_0}{\omega_0^2}\left(\frac{m_i + m_e}{m_i m_e}\right) \simeq \frac{eE_0}{\omega_0^2 m_e} = \frac{v_E}{\omega_{pe}} \tag{74}$$

is a measure of the pump field, being equal to the amplitude of electron excursions in the field. As the plasma temperature rises the unstable regions shrink. The field strength parameters for Figs. 5(b) and (c) are $\eta = 10$ and 2 respectively, with $T_e = 10\ T_i$ in each case to avoid ion wave damping. Instability occurs only at long wavelengths, cutting off at about $k\lambda_D = 0.7$, since

$$\eta = \frac{v_E}{v_T} \simeq \frac{kx'_0}{k\lambda_D}.$$

As kx'_0 increases, instability occurs at lower densities. The Bohm-Gross dispersion relation, expression (27), separates the decay instability (on the left, in the lower density region) from the two-stream instability (which can only grow when $\omega_0 < \omega_{BG}$). At high temperatures or in weak fields the latter extends not only into the overdense region but also into regions of lower density. The maximum growth rates are greatest in the strong field limit and are then independent of temperature: they are shown in Fig. 5(d). The maximum possible growth rate depends on the mass ratio of ions and electrons: for the two-stream instability it is given approximately by

$$\frac{\gamma_{\max}}{\omega_{pe}} \simeq \frac{1}{2}\left(\frac{m_e}{m_i}\right)^{1/3}, \tag{75}$$

while for the decay instability it is given by

$$\frac{\gamma_{\max}}{\omega_{pe}} \simeq \frac{1}{3}\left(\frac{m_e}{m_i}\right)^{1/3} \tag{76}$$

The wave numbers corresponding to maximum growth rates are shown in Fig. 5(e).

Provided the pump wave is not too strong, i.e. if $\eta \lesssim 7$, the wave vectors of the modes with maximum growth rates are parallel to $\mathbf{E}_0$. (De Groot and Katz (1973) found that for $\eta > 7$ the modes with maximum growth rates propagate at an angle φ_E to $\mathbf{E}_0$, such that $\cos\varphi_E \simeq 7/\eta$.)

Provided $\eta \ll 1$, the threshold pump irradiance for the onset of the oscillating two-stream instability in homogeneous plasmas is given by

$$\eta^2 = 4\left(\frac{\nu_{ek}}{\omega_{pe}}\right)\left(1 + \frac{T_i}{T_e}\right), \tag{77}$$

while for the parametric decay mode it is given at high temperatures (such that $\nu_{ek} < \omega_i$) by

$$\eta^2 = 4\left(\frac{\nu_{ek}}{\omega_{pe}}\right)\left(\frac{\nu_{ik}}{\omega_i}\right) \tag{78}$$

and at low temperatures ($\nu_{ek} > \omega_i$) by

$$\eta^2 = \frac{8\sqrt{3}}{9}\frac{\nu_{ek}^2}{\omega_{pe}}\frac{\nu_{ik}}{\omega_i^2}. \tag{79}$$

Here ω_i is the ion wave frequency and ν_{ek}, ν_{ik} are the effective electron-ion and ion-ion collision frequencies respectively, taking Landau damping into account. Yamanaka *et al.* (1972c) have calculated the threshold irradiances for neodymium laser light incident on a deuterium plasma (Fig. 6). At low temperatures ($T_e \lesssim 200$ eV) the two-stream instability has the lower threshold, but at higher temperatures the decay instability is more easily excited, especially when $T_i \ll T_e$. Note that the values of the irradiance are lower than those at which the inverse Bremsstrahlung absorption becomes nonlinear, particularly at high temperatures.

(b) *Effects of Plasma Inhomogeneity*

In an inhomogeneous plasma these threshold conditions must be modified. The unstable region has a finite spatial extent, so that energy is carried away by electron plasma waves and is not replaced by incoming waves, and the threshold pump irradiance rises. Perkins and Flick (1971) considered a plasma with a density gradient scale length L, defined by $n/L = dn/dx$, and found that when $\nu_{ek} \ll \omega_i$ the two-stream instability threshold is given by

$$\begin{aligned}\eta^2 &= \left(1 + \frac{T_i}{T_e}\right)\left(\frac{4\nu_{ek}}{\omega_{pe}} + \frac{2}{k_{\parallel} L}\right)\\ &\geq \left(1 + \frac{T_i}{T_e}\right)\left(\frac{4\nu_{ek}}{\omega_{pe}} + \frac{8\lambda_D}{L}\right),\end{aligned} \tag{80}$$

the second expression being due to $k_{\parallel}$, the component of the electron plasma wave vector parallel to the electric field, having an upper limit of $1/4\lambda_D$ to avoid strong Landau damping.

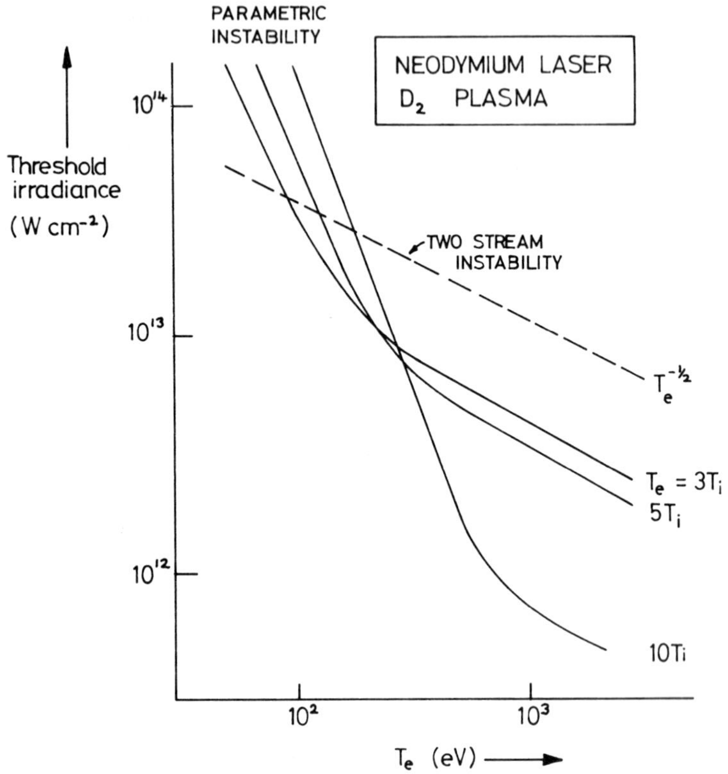

FIG. 6. Threshold irradiances for excitation of parametric decay and two-stream instabilities by neodymium laser light in a deuterium plasma (after Yamanaka *et al.*, 1972).

The decay mode becomes a convective instability in the presence of a density gradient. Although the amplitude of the instability at a fixed point does not grow indefinitely, the limited unstable region acts as an amplifying medium which increases the intensity of waves propagating through it. The threshold may then be defined as the electric field which causes an intensity amplification of e^5, and was found by Perkins and Flick to be given by

$$\eta^2 = \frac{8}{k_{\parallel} L}\left(1 + \frac{3T_i}{T_e}\right)\left(\frac{v_{ik}}{\omega_i}\right)^{1/2} + 3.2\left(\frac{v_{ik}}{\omega_i}\right)\left(\frac{v_{ek}}{\omega_{pe}}\right). \tag{81}$$

When the density gradient is important, unless $T_e \gg T_i$ this threshold is not very different from that for the two-stream instability.

It should be noted that the threshold fields are values within the plasma.

As a laser beam propagates into a plasma its group velocity falls as it approaches the critical density layer. In order to maintain the flow of energy the electric field increases. This "swelling" of the field can be as large as a factor of 10 (Thomson *et al.*, 1974).

(c) *Effects of Incoherent Pump Wave*

The theoretical calculations described above assumed a perfectly monochromatic, coherent pump wave, whereas the output from a real laser has finite bandwidth. The effect of limited pump wave coherence on the growth rate for the parametric decay instability was considered by Tsytovich and Shvartsburg (1967) and by Valeo and Oberman (1973), who showed how the growth rate falls as the coherence deteriorates. The effect of incoherence on the threshold irradiance was discussed by Tamor (1973). If the number of phase jumps in the pump wave per unit time is ν_φ, and all phase jumps are equally probable, the effect on the parametric equations is simply to replace ν_{ek} by $\nu_{ek} + 2\nu_\varphi$. In terms of pump wave bandwidth $\Delta\omega_0$, $\nu_\varphi \sim \Delta\omega_0/2$. For a hydrogen plasma at $T_e = 1$ keV, pumped by neodymium laser light at 1.06 μm, the threshold for the parametric instability is greater by a factor of 2.7 for light with a bandwidth of $10^{-3}\omega_0$ than for perfectly coherent light.

(d) *Development of the Instability*

So far we have calculated the threshold for excitation of the parametric instabilities, and the maximum growth rates for small-amplitude driven waves. We know now the order of magnitude of the irradiance needed to excite them, and we know how quickly they begin to develop But we have no idea yet how they grow, or how they stop growing or "saturate". There clearly must be a limit to growth when the damping of the plasma waves takes energy away as fast as the pump can put it in, or when the electric field of the plasma wave approaches the electric field of the pump. We need to follow the development of the instabilities so that we can find out how important these are as a means of absorbing incident laser light, and also to find out how the absorbed energy is distributed among the plasma particles.

We begin by making a rough estimate of the nonlinear absorption coefficient for laser light passing through a plasma, due to the two-stream instability, for laser pump fields not too far above threshold ($\eta \ll 1$), by means of the following argument due to Kaw and Dawson (1969). The threshold pump field, given by expression (77), varies as the square root of the electron-ion collision frequency ν_{ek}. As the instability develops, strong ion fluctuations occur which interact with the pump wave and as a result (Dawson and Oberman (1962)) the effective collision frequency increases. Eventually the collision frequency rises to a value $\nu_{ek_{\max}}$ at which the pump field is no longer above the threshold value. The instability then ceases to grow. Here $\nu_{ek_{\max}} \simeq \eta^2\omega_{pe}/4$. The absorption coefficient for the pump wave is related to the effective collision frequency:

$$a_\omega \simeq \frac{\nu_{ei}}{c}\frac{\omega_{pe}^2}{\omega^2}, \tag{82}$$

and thus the maximum absorption coefficient due to this instability is given by

$$a_{\max} \simeq \frac{30I}{n_e^{1/2}T_e}\ \mathrm{cm}^{-1}, \tag{83}$$

where I is in watts cm^{-2}, n_e in cm^{-3} and T_e in eV. The normal Bremsstrahlung absorption coefficient, expression (35) may be written approximately, for ω not too far from ω_{pe} and $\bar{g} \sim 5$, as

$$a_{\omega_{pe}} \simeq 2 \times 10^{-15}\frac{n_e}{T_e^{3/2}}\ \mathrm{cm}^{-1} \tag{84}$$

and is therefore exceeded by the parametric absorption when $I > I_x$, given by

$$I_x = 10^{-16}\frac{n_e^{3/2}}{T_e^{1/2}}\ \mathrm{W\ cm}^{-2}. \tag{85}$$

Since $I_x < I_{\mathrm{opt}}$, parametric absorption is stronger than inverse Bremsstrahlung absorption when inverse Bremsstrahlung begins to become nonlinear.

The nonlinear evolution of a parametrically unstable plasma has been treated analytically for weak pump fields using weak turbulence theory and also for moderate pump fields using a theory involving the broadening of the wave-particle resonance by plasma wave turbulence (Bezzerides and Weinstock, 1972). For strong pump fields computer simulations of plasmas have provided some interesting results. The computer is used to follow the self-consistent response of a large number of charged particles (in one-dimension usually) to a driving electric field. Figure 7 shows the energy of the electron plasma wave and the total energy of the plasma computed by Kruer and Dawson (1972) as a function of the duration of a constant oscillatory pump field of amplitude E_0. In this one-dimensional example $\eta = 0.5$, $\omega_0 = 1.04\ \omega_{pe}$, $T_e/T_i = 30$ and (to facilitate computing) $m_i/m_e = 100$. Initially, the plasma heats slowly as the result of binary collisions, while the wave energy rises exponentially. After a time of about $t = 250\ \omega_{pe}^{-1}$ the wave energy begins to saturate and the plasma energy immediately begins to rise rapidly as anomalous heating sets in. The pump wave energy is efficiently converted into plasma oscillations which at large amplitude are rapidly

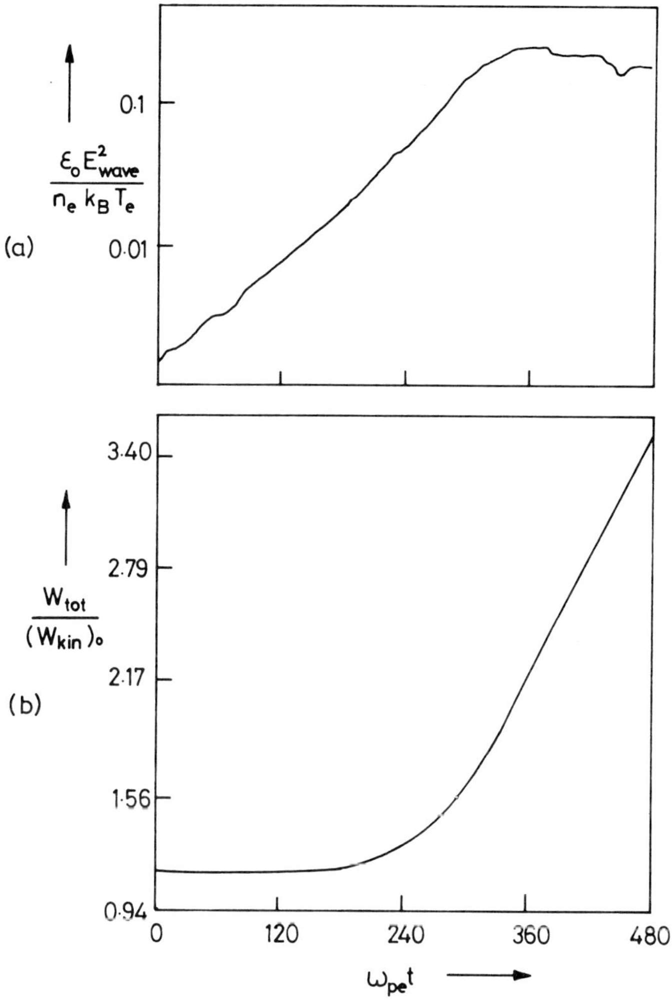

FIG. 7. Computer simulations of plasma heating showing: (a) self-consistent wave energy and (b) total plasma kinetic energy averaged over a plasma oscillation (see text). (After Kruer and Dawson, 1972.)

Landau damped, creating a high energy tail on the electron distribution function. They can trap a significant number of electrons (Dawson and Shanny (1968)). As plasma heating continues it is found that energy is being fed into plasma oscillations of progressively longer wavelengths which, because of their higher phase velocity, experience less Landau damping and become the modes of maximum growth rate.

For large values of η the wave field at saturation becomes strong enough to trap a large proportion of the electrons. One-dimensional computations of the time variation of the wave energy and the electron and ion "temperatures" by De Groot and Katz (1973) are shown in Fig. 8 for a plasma with $m_i/m_e = 100$, $\omega_0 = \omega_{pe}$, (so only the two-stream instability is excited) and $\eta = 7$. The "temperatures" are defined by

$$T_{e,i} \equiv m_{e,i}[\langle v^2 \rangle_{e,i} - \langle v \rangle_{e,i}^2]. \tag{86}$$

In this case the development of the plasma is more rapid, and saturation occurs at a time of about $17 \times 2\pi/\omega_{pe}$. The wavenumber spectrum of the self-consistent electric field is shown in Fig. 9 for two times, the first during the steady growth phase and the second at $t = 40 \times 2\pi/\omega_{pe}$ when turbulence is well developed. The maximum in the early spectrum occurs at a wavenumber very close to $k\lambda_{D_0} = 0.23$ where the most unstable electron plasma mode is predicted by linear theory (λ_{D_0} is the initial value of λ_D). At the later time the maximum has moved to the mode with the lowest available wavenumber. The most unstable modes have trapped electrons and decayed. Strongly nonlinear coupling occurs between modes, and a turbulent state has developed. The wave energy has ceased to grow. The rapid coupling of energy between modes is probably connected with the presence of strong ion density fluctuations which will also enhance the rate of energy transfer from the wave field to particles.

While the electron temperature is low the entire electron distribution function gains energy in Landau damping the large amplitude waves. As a result the electron "temperature" rises rapidly, the electron velocity distribution remaining quasi-Maxwellian. This, however, causes η to fall at constant E_0, so that when $t = 48 \times 2\pi/\omega_{pe}$ the value of η is only 0.52. The situation is then very similar to that of the weak field case of Fig. 7, though at a higher temperature, and a high-energy tail again appears on the electron velocity distribution.

The ion temperature rises much more slowly than the electron temperature. Although the external pump frequency is set equal to the electron plasma frequency, so that only the two-stream instability is excited, a frequency analysis of the turbulent plasma shows a maximum in the spectrum close to the ion plasma frequency. The phase velocity of these waves is much greater than the average ion velocity so no Landau damping is expected unless a very large amplitude is attained, when ion trapping may occur (Sleeper *et al.* (1972)).

Effective collision frequencies ν_{ei}^* for the rate of energy transfer from the laser field to the plasma have been derived from the simulations. A composite

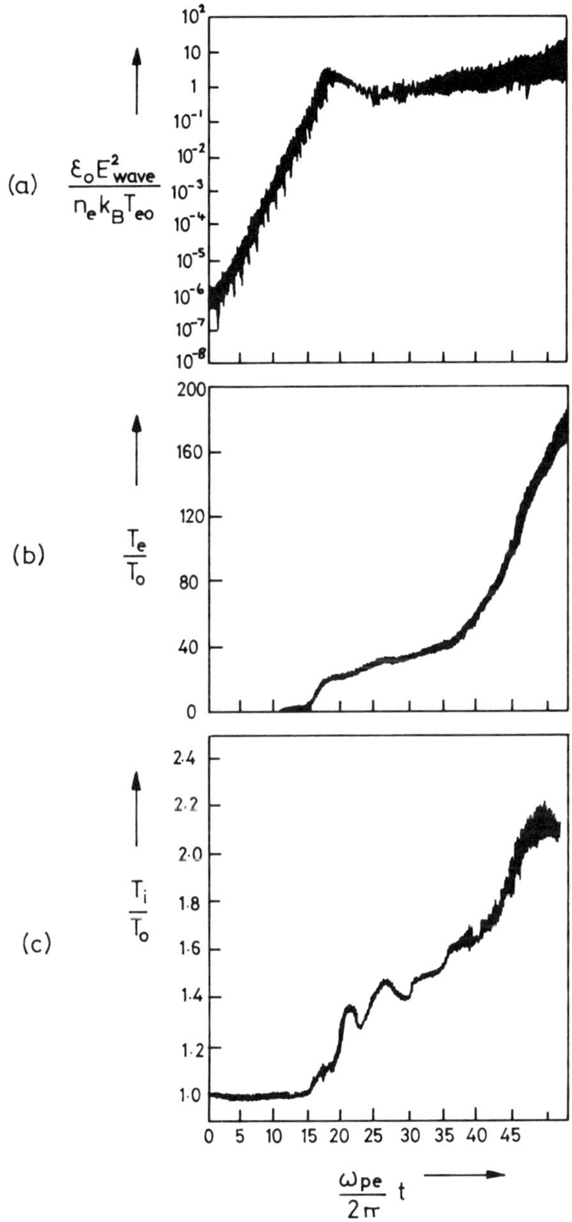

FIG. 8. Simulation of plasma heating showing: (a) self-consistent field energy, (b) electron temperature and (c) ion temperature as functions of time. $T_{i0} = T_{e0} = T_0$ (see text). (After De Groot and Katz, 1973.)

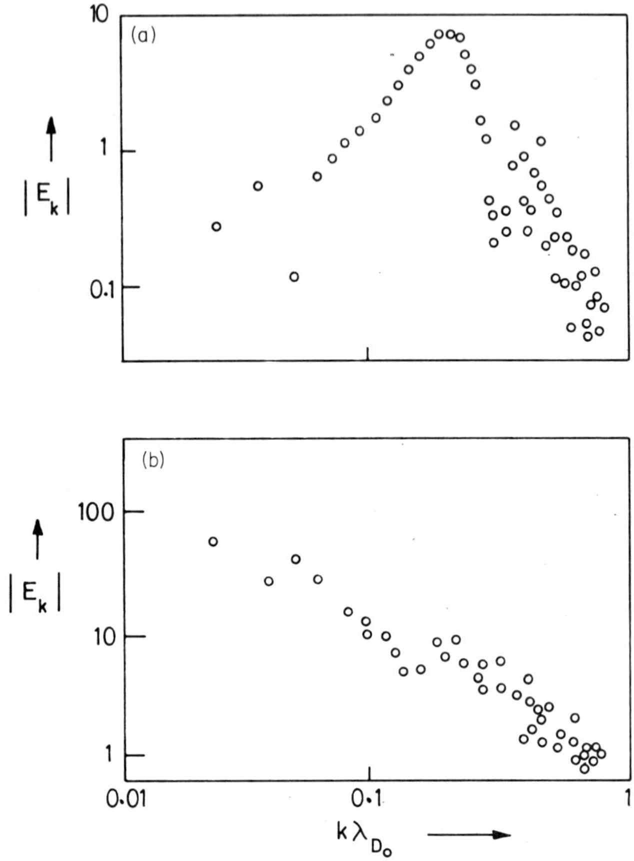

FIG. 9. Spectrum of self-consistent field as function of $k\lambda_D$ from simulation (a) during linear growth, (b) when turbulence is well developed (after De Groot and Katz, 1973).

estimate for a deuterium plasma suggests that ν_{ei}^*/ω_{pe} rises (initially approximately quadratically) with increasing η to a maximum of about 0.04 when η is about 0.5, and then falls linearly to about 4×10^{-3} as η rises to 10. For $\eta > 1$, then,

$$\frac{\nu_{ei}^*}{\omega_{pe}} \simeq \frac{4 \times 10^{-2}}{\eta} \tag{87}$$

The absorption coefficient thus varies as $I^{-1/2}$ for $\eta > 1$, whereas the strong-field inverse Bremsstrahlung absorption coefficient varies as $I^{-3/2}$, so that parametric absorption always exceeds the nonlinear inverse Bremsstrahlung absorption.

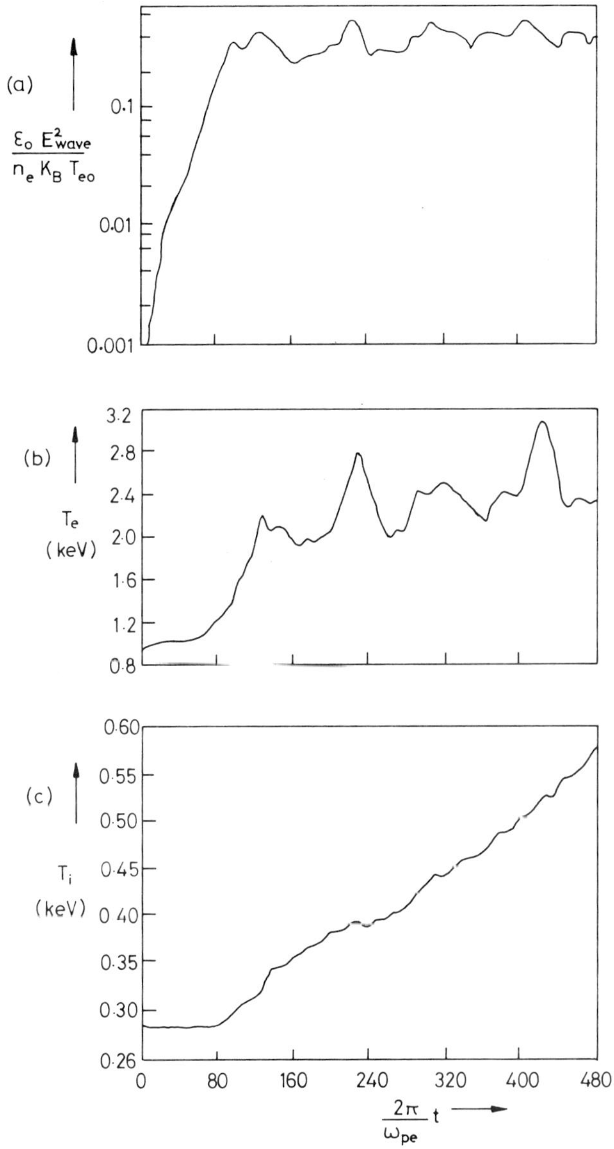

FIG. 10. Simulation of plasma heating allowing for flow of particles through the unstable layer (see text). (After Bodner *et al.*, 1973.)

Shearer and Duderstadt (1973) calculated the efficiency with which a semi-infinite deuterium plasma absorbs normally incident light, allowing for both collisional and parametric processes, and concluded that the overall absorption efficiency increases as the frequency is increased. The inverse Bremsstrahlung absorption coefficient varies approximately as $n_e n_i/\omega^2$ in weak fields and is thus a maximum at the critical density layer, where $n_e = n_{ec} \propto \omega^2$. The maximum absorption coefficient therefore varies as ω^2. To this must be added the parametric absorption.

When a solid target is being heated by a laser pulse, the interaction region is strongly inhomogeneous. There is only a thin unstable region near the critical density layer. Hot electrons flow out of the unstable layer and are replaced by cold ones. As a result the electron temperature in the layer reaches a maximum value. The energy absorbed from the laser ultimately produces an extended region of hot electrons, together with hydrodynamic expansion. Ions move through the unstable layer much more slowly, with approximately the acoustic velocity, and are heated *en route* by turbulence. Bodner *et al.* (1973) showed by a computer simulation that under these conditions the ion temperature can approach the electron temperature. The flow of electrons through the unstable layer was simulated by resetting the velocities of a proportion of the electrons, selected randomly, to their initial values. It was found that the ion heating rate was not very sensitive to the rate at which electrons are reset. The results shown in Fig. 10 are for a plasma of density 10^{21} cm^{-3}, with $m_i/m_e = 100$, $\eta = 0.36$, ($I = 3 \times 10^{14}$ W cm^{-2}), $\omega_0 = \omega_{pe}$ and $T_{e0} = 1$ keV. As may be seen, the electron temperature saturates when $t \simeq 130 \times 2\pi/\omega_{pe}$ and then rises only slowly (the resetting procedure does not properly reduce the high energy tail of the electron distribution). The ion temperature rises almost linearly with the time long after the electron temperature has saturated: the ion distribution remains roughly Maxwellian. The ion heating rate varies with the ion mass as $m_i^{-1/2}$ and is directly proportional to the incident laser irradiance: for a deuterium plasma, with I in watts cm^{-2},

$$\frac{dT_i}{dt} \simeq 10^{-16}\, I \qquad \text{keV per picosecond.} \tag{88}$$

However, the time spent by an ion in the heating region varies roughly as $I^{-1/3}$, so the final ion temperature should vary as $I^{2/3}$.

C. "TWO PLASMON" INSTABILITY

Another parametric process which may be important in plasma heating may occur when $\omega_{pe} \simeq \omega_0/2$, i.e. when $n_e \simeq n_{ec}/4$. It involves the decay of a photon into two plasmons and has been discussed by Goldman (1966), Jackson (1967), Galeev *et al.* (1972), Amano and Okamoto (1969) and Rosenbluth (1972). Only the electrons are involved, so in a homogeneous plasma there can be no instability in the dipole approximation ($k_0 = 0$). Jackson (1967) found that the threshold irradiance for this instability with $k_0 = \omega_0/c$ is

$$I \simeq 4 \times 10^{-2}\left(\frac{v_{ek}}{\omega_{pe}}\right)^2 n_e \text{ W cm}^{-2} \tag{89}$$

and that the most unstable modes are at 45° to $\mathbf{E}_0$ and $\mathbf{k}_0$.

Amano and Okamoto (1969) suggested that in a plasma with a density gradient of scale length L, k_0 could be replaced by $1/L$. Another treatment by Rosenbluth (1972) suggests that the threshold irradiance for significant spatial amplification for this instability is

$$I \gg 1.7 \times 10^{15} \frac{T_e}{L\lambda_0} \text{ W cm}^{-2}, \tag{90}$$

where T_e is in keV and L and λ_0 are in microns.

D. COUPLING OF TWO LASER BEAMS TO A PLASMA WAVE

Two laser beams whose frequencies ω_1 and ω_2, both much greater than the plasma frequency ω_{pe}, differ by ω_{pe} can couple parametrically to a plasma wave (Kroll *et al.* (1964)). The heating due to this process has been discussed by a number of authors. According to Rosenbluth and Liu (1972) the heating rate is greater for opposing (antiparallel) beams than for parallel beams by a factor $4(\omega_{12}/\omega_{pe})^2$. The Manley-Rowe relations show that the maximum proportion of the incident energy that can be converted to plasma wave energy is $\sim\omega_{pe}/\omega_1$. The proportion of incident energy transferred to an ion wave when $\omega_1 - \omega_2 = \omega_{ia}$ will be even smaller. However, a cascade of similar processes is possible which might lead to more effective heating.

E. STIMULATED RAMAN AND BRILLOUIN SCATTERING

As we have seen, a plasma wave may be excited by the interaction of two light waves whose frequencies are separated by ω_{pe}. A single incident light wave scattering from plasma density fluctuations at ω_{pe} is shifted in frequency by ω_{pe}. Scattered light can then interact with incident light to enhance both the plasma fluctuations and the scattered wave. The incident beam thus generates a plasma wave from which it is itself scattered. This process is known as stimulated Raman scattering (or stimulated Thomson scattering) when an electron plasma wave is involved. A similar process can occur involving an ion acoustic wave, and is known as stimulated Brillouin scattering (or stimulated Mandelshtam–Brillouin scattering).

Stimulated Raman scattering occurs only when the plasma electron density is sufficiently low for the scattered light to propagate. If the incident light has frequency ω_0, for which the critical density is n_{ec}, the scattered light will have a frequency not less than $\omega_0/2$, and the maximum electron density in the scattering region will be $n_{ec}/4$. The Manley–Rowe relations show that up to a half of the incident photon energy will be given to the plasma wave. In the case of stimulated Brillouin scattering, the ion acoustic wave frequency is much smaller than the electron plasma frequency, so scattering can take place at densities almost up to n_{ec}: nearly all the incident photon energy is scattered out of the plasma. Drake *et al.* (1974) have shown that in a homogeneous plasma the instability growth rates are greatest and thresholds lowest for back scattering with the two electromagnetic waves polarized in the same direction. For stimulated Raman scattering, the threshold irradiance is found from the relation

$$\frac{v_E^2}{c^2} = \left(\frac{\nu_{ek}}{\omega_{pe}}\right)\left(\frac{a_\omega c}{\omega_0}\right), \tag{91}$$

where a_ω is the linear absorption coefficient for the scattered wave (expression (35)). For stimulated Brillouin scattering the threshold is found from

$$\frac{v_E^2}{c^2} = \frac{2\nu_{ik} a_\omega c}{\omega_{pi}^2} \frac{\omega_0}{k_0 c} \frac{v_{ia}}{c}. \tag{92}$$

The incident wave obeys the usual dispersion relation $\omega_0^2 = \omega_{pe}^2 + k_0^2 c^2$. The ion acoustic wave velocity is given by $v_{ia}^2 = \kappa_B T_e/m_i$, and $\omega_{pi}^2 = n_i e^2 m_i \varepsilon_0$.

In an inhomogeneous plasma, an electron plasma wave eventually travels outwards from the unstable region into regions of lower density where the phase velocity is lower. Waves propagating initially perpendicularly to the density gradient are refracted outwards: waves initially travelling inwards are reflected outwards. The waves are more readily Landau damped when their velocities fall. The wave energy eventually appears as particle kinetic energy. Klein and Manheimer (1974) pointed out that as a consequence the number of very energetic particles accelerated by the waves will be smaller for a plasma with a density gradient than for a homogeneous plasma.

When light is incident on a plasma boundary with a steep density gradient the matching conditions hold only over a limited distance. The propagation of the parametrically-generated waves out of this region of the plasma restricts the growth of the instability (Rosenbluth, (1972); Galeev *et al.* 1973)). This raises the threshold for backscattering of light incident in a direction parallel to the density gradient (Pesme *et al.* (1973)). It was suggested by Forslund *et al.* (1973) that in experiments where light is incident on a target surface and produces a plasma which blows off into vacuum the fastest-growing mode of the backward-scattered Brillouin wave might move at the speed of the expanding plasma, thus giving an absolute instability in the moving plasma frame. However, Brueckner and Jorna (1973) showed that the group velocity of the unstable Brillouin mode is much greater than the plasma expansion velocity, which is near the acoustic velocity, so the unstable mode rapidly moves out of the narrow amplifying region (for a full discussion see Jorna (1974)).

Liu *et al.* (1973) considered sidescattering at 90° to the direction of the incident light. In this case the scattered light stays much longer in the unstable region before refracting out (or reaching the edge of the laser beam) and may again represent a significant reduction in the fraction of incident light energy that reaches the strongly-absorbing critical density layer. The threshold power for Raman sidescattering in an inhomogeneous plasma is lower by a factor $18\kappa_B T_e/m_e c^2$ than that for backscattering, and is comparable to that for Brillouin backscattering. Computer simulations by Klein *et al.* (1973) confirmed these results.

Detailed one-dimensional analyses of backscattering in plasmas with two different density gradients were given by Tsytovich *et al.* (1973). Saturation must be considered (Litvak and Trakhtengerts (1971); Vinogradov *et al.* (1973)). A computer simulation study has been made by Forslund *et al.* (1973) of backscattering of a light wave of fixed amplitude incident from the right on a plasma with the density profile shown in Fig. 11(a). The density is in units of n_{ec}, where $\omega_{pe} = \omega_0$ at n_{ec}. The transmitted wave amplitude is shown in Fig. 11(b) as a function of time, t being in units of $1/\omega_0$. Electron and ion phase space plots show that the first transmission minimum, which occurs at $t \sim 500/\omega_0$, is due to reflection caused by Raman scattering. This saturates as a result of electron trapping and heating. The second transmission minimum, which occurs at $t \sim 1000/\omega_0$ is due to reflection caused by Brillouin scattering. By $t \sim 2000/\omega_0$ the intensity transmission coefficient has settled down to about 0.4.

Biskamp and Welter (1975) made another simulation study of Raman scattering of light entering a plasma with a linear density profile similar to that used by Forslund *et al.* (1973) reaching a maximum of 0.5 n_{ec}. In this simulation electrons hitting the boundary on the dense side were taken out and reintroduced cold to simulate a net heat flow out of the system. This allows a

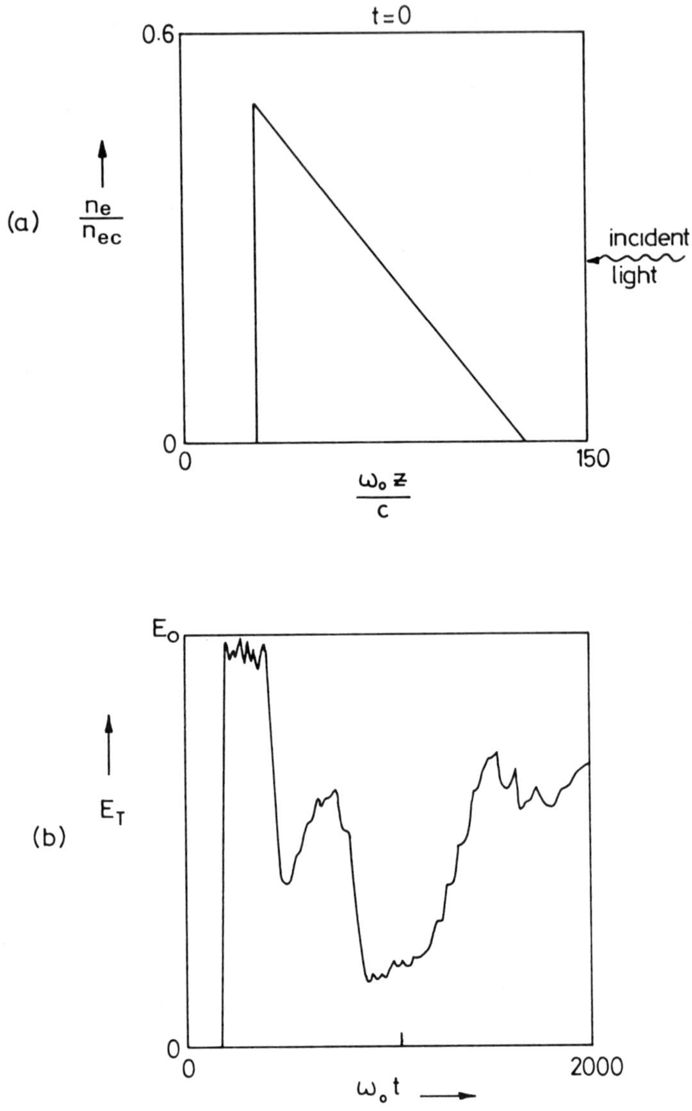

FIG. 11. Simulation of effects of back-scattering from plasma with density profile shown in (a): (b) shows amplitude of transmitted wave as function of time (after Forslund *et al.*, 1973).

quasi-stationary state to develop. The results in one dimension, for back-scattered light, show that a plasma wave is excited, localized near $n_e = n_{ec}/4$, with frequency $\omega_0/2$, ω_0 being the frequency of the incident light. For $v_E/c = 0.1$ ($I = 10^{16}$ W cm^{-2} for neodymium, 10^{14} W cm^{-2} for CO_2), the light absorption was found to be weak. A wide spectrum of anti-Stokes lines appeared in both the reflected and transmitted light, the spacing being $\omega_0/2$, which is the plasma frequency at $n_{ec}/4$. These anti-Stokes lines in the reflected light must be due to plasma waves moving down the density gradient. The basic initial Raman process involves coupling of laser energy into a plasma wave moving up the density gradient. Such a plasma wave will, however, be reflected at n_e just above $n_{ec}/4$ and may then couple with the incoming laser light to produce anti-Stokes backward scattering.

At higher irradiances large amounts of energy are absorbed in the plasma, generating a turbulent spectrum of electron plasma oscillations. The characteristic of such a turbulent plasma is that it tends to produce diffusion in velocity space, with the result that the high-energy tail of the electron distribution function is enhanced. This alters the dispersion relations for the plasma waves, leading to a reduction in the frequency of the plasma wave which couples to the incident laser light, a shift of the peak in the reflected light spectrum towards the blue, and a general smearing of the subsidiary peaks in the spectrum.

Though much more information is needed about these processes it seems clear that they may deflect away a significant fraction of intense light energy incident on a dense plasma surface.

VI. Optical Resonance Absorption

If the electron density rises monotonically in a plasma boundary to more than the critical density, an incident transverse wave will be reflected. The depth of penetration in the geometrical optics approximation depends on the angle of incidence φ_i of the wave at the zero-density surface: at oblique incidence the wave will be refracted, reaching the turning point at a depth in the plasma where $\omega_{pe}^2 = \omega_0^2 \cos \varphi_i$. If the obliquely incident wave has an electric field component in the plane of incidence, then provided it approaches reasonably close to the critical density layer when it reaches the turning point it can tunnel through to drive electron plasma oscillations along the direction of the density gradient and thus loses energy. The effect does not occur at normal incidence or near grazing incidence. The theory has been worked out in considerable detail (see for example Ginsburg (1970)) and has been discussed in the context of laser-heated plasmas by several authors, (see for example, Godwin (1972), Mueller (1973)).

Mueller included collisional effects into the calculation of the absorption at oblique incidence in a plasma with a linearly increasing electron density, making use of the usual low-field Bremsstrahlung expression for $\eta < 1$ and the strong-field expression when $\eta > 1$. The results confirm that strong absorption can be expected over a narrow range of angles. This may influence the choice of focussing optics for heating up a plasma. However, the calculations must first be modified to take parametric processes into account.

Optical resonant absorption is essentially a linear process, in the sense that the absorption coefficient is normally independent of the incident irradiance, and there is no threshold. However, at high irradiances nonlinear parametric coupling to plasma waves must be expected. These waves will propagate parallel to the density gradient into regions of lower density, so that k/k_D increases and Landau damping occurs. Parametric decay at oblique incidence has been discussed by White *et al.* (1973) who concluded that for a density scale length of 1 mm and $T_e = 10$ keV the growth rate would be substantial over a range of incidence of about 10° for 1.06 μm and 15° for 10.6 μ. The maximum ratio of power absorbed to incident power was calculated to be about 1:2.

VII. Experimental Studies of Nonlinear Optical Effects in Plasmas

So far we have been concerned with theoretical ideas, although we have related them to situations which are of practical interest. Let us consider how we may investigate optical nonlinearities in plasmas experimentally. To begin with we shall need high power lasers. We shall need enough energy in each laser pulse to maintain the power for a reasonable time, long enough to allow the plasma—especially the ions—to respond to the driving field, and long enough to allow instabilities to grow to significant amplitudes. Also, it is desirable to have a reasonably monochromatic laser beam, well collimated, with a good mode structure. It would be helpful to have two or more laser beams at different frequencies. We could use different lasers, or perhaps exploit some parametric process to frequency-convert one wavelength into various others. In this way we could perhaps observe the frequency-dependence of the nonlinear effects we are studying. However, there are so many co-existent effects that if we change the wavelength we may also change the entire nature of the process we observe.

Next we shall need plasmas of known characteristics, with a wide range of high densities and temperatures, preferably containing a variety of different ions on different occasions. We shall want to arrange for the plasma to have a predetermined density distribution, at a fixed position, at the time when the main laser pulse arrives. We may be able to approximate these requirements using plasmas formed by laser pulses—probably additional to those used for the interaction studies—on solid targets.

It is desirable to determine:

1. The frequency spectra of density fluctuations in the interaction region, for both the electrons and the ions.
2. The particle velocity distributions in the interaction region, distinguishing thermal from turbulent velocities.
3. The spectrum of light emerging from the interaction region.

One of the great difficulties is that for a given instability the interaction region is very small, and we have to be quite sure that we do not confuse it with other regions where other effects may occur.

A variety of diagnostic methods are available. These include:

1. *Optical spectroscopy*. The input laser pulse spectrum, and the spectrum of light emitted in various directions are relatively easy to determine. Time resolved spectroscopy is desirable, because we want to follow the development of the nonlinear interaction. We shall probably need a resolving time of at most 10 picoseconds. This is possible with image-convertor techniques.

2. *Thomson scattering*. In orthodox plasma physics Thomson scattering has proved to be an extremely powerful diagnostic technique. In a plasma, electrons are the most important scatterers, and electron density fluctuations determine the observed spectrum of scattered light. The scattering process may be thought of as a Raman process, between an incident photon (ω_0, k_0) and a plasmon (ω, k), producing a scattered photon (ω_s, k_s), with the usual selection rules. The intensity of the scattered light depends on the product of the intensity of the incident light, and that of the fluctuations.

Provided the wavelength of the density fluctuation is short compared to the Debye length, i.e. $k\lambda_D \gg 1$, screening effects are slight, and if the electron velocities have a Maxwellian distribution only thermal fluctuations are represented in the scattered spectrum. But if $k\lambda_D \ll 1$, collective effects dominate the spectrum. Ion density fluctuations are also represented in the spectrum because the electrons follow the ions closely. If one uses light at wavelength λ_0 to study a plasma of electron density n_e at temperature T_e, the value of $k\lambda_D$ represented in the light scattered through an angle θ is given by

$$\frac{1}{k\lambda_D} \simeq 1.6 \times 10^{-12}\lambda_0\left[\frac{n_e}{T_e(1 - \cos\theta)}\right]^{1/2} \tag{93}$$

where T_e is in eV, λ_0 in Å.

Thus by looking at light scattered through different angles we can determine the spectrum of fluctuations with different values of k. One of the advantages of Thomson scattering is that the volume element from which the scattering takes place can be defined accurately.

The choice of a suitable light source depends on the density of the plasma we wish to probe. The usual source for low-density plasmas is the ruby laser, and this may be satisfactory when we study the interaction region for CO_2 laser light at 10.6 μm, where $(n_{ec})_{CO_2} \simeq (n_{ec})_{ruby}/200$. But if we wish to study the interaction of neodymium laser light with plasmas it will probably be necessary to use harmonics of the ruby laser, and refraction effects may still be troublesome. The scattering light source need not be very monochromatic for dense plasma experiments, but it must be powerful, because the Thomson cross-section is small.

3. *X-ray spectroscopy.* At the temperatures which are of most interest at present, in the region 100 to 1000 eV, the bulk of the radiation emitted by a plasma is Bremsstrahlung in the soft X-ray region. Some information can therefore be obtained about the electron velocity distribution function by observing the X-ray emission spectrum. It will be apparent, for example, if there is an anomalously large high-energy component in the emission from a homogeneous plasma that there are some non-thermal high energy electrons around. However, we must be very careful about interpreting these results, because inevitably we will be observing the integrated emission along the line of sight of the spectrometer. Thus for example if there are two more-or-less distinct emitting regions, both with Maxwellian electron velocity distributions but at different temperatures, the sum of their emission spectra might be interpreted, wrongly, as being due to a single plasma with a non-Maxwellian distribution having a strong high-energy tail.

A. MICROWAVE OBSERVATIONS

Up to the present time the most detailed and convincing tests of the theory of parametric processes in plasmas have been done with microwave frequencies. Under these conditions, with no shortage of space and time to do the experiments, in cool, low-density plasmas, it has been possible to measure growth rates, frequencies and wavenumbers and check the theory with considerable precision. The agreement between theory and experiment is excellent. Thus for example Phelps *et al.* (1971, 1973) demonstrated nonlinear coupling between electromagnetic, electron plasma and ion acoustic waves in a potassium plasma and verified the frequency and wavenumber selection rules and the magnitudes of the coupling coefficients. Detailed studies of the parametric decay instability were made by Stenzel and Wong (1972) in an argon plasma ($T_e = 2$ eV $= 10\ T_i$, $n_e = 10^9$ cm^{-3}).

B. EXPERIMENTS WITH LASER LIGHT

Several authors have measured the reflectivities of plasmas formed by laser light from solid targets in vacuum. Caruso *et al.* (1970a) measured the specular reflectivity of a nickel target for 100 psec neodymium laser pulses at 1 μm and found that it fell rapidly as the irradiance was increased above 10^{13} W cm^{-2}. Similar measurements by Andrews *et al.* (1972) with 10 psec neodymium laser pulses showed an almost constant reflectivity up to an irradiance of about 10^{13} W cm^{-2}, above which the reflectivity fell, as may be seen in Fig. 12. The reflectivity was independent of the polarization of the incident beam, and the threshold irradiance for the reflectivity to fall was the same for copper, bismuth and boron targets .The experimental arrangement is shown in Fig. 13. The diffusely reflected light energy was found to be approximately proportional to the incident pulse energy. Again, Yamanaka *et al.* (1973) using a solid deuterium target and 1 nsec neodymium laser pulses, found a definite reflectivity maximum at about 10^{14} W cm^{-2} (Fig. 14). However, the K.M.S. group (Charatis *et al.* (1974)) found that with 300 psec, approximately

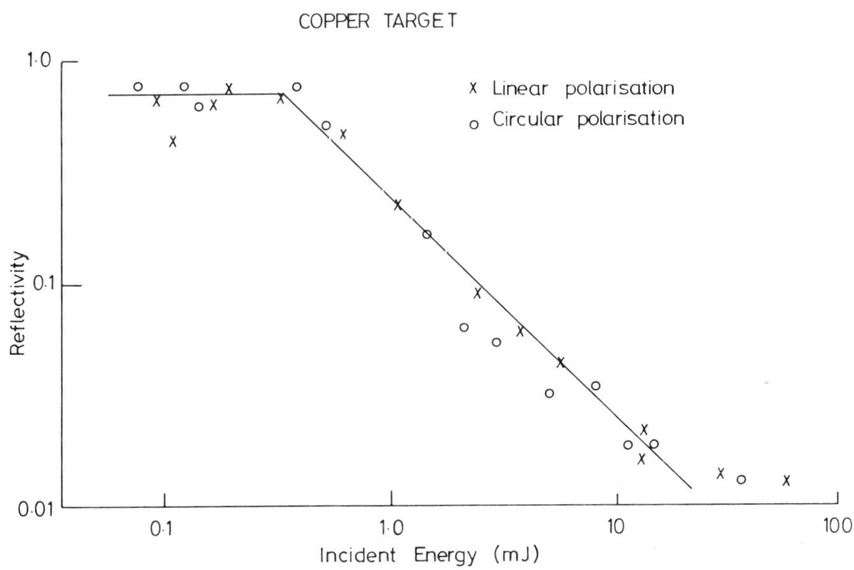

FIG. 12. Reflectivity of a copper target for 10 psec pulses at 1.06 μm as a function of pulse energy (after Andrews *et al.*, 1972).

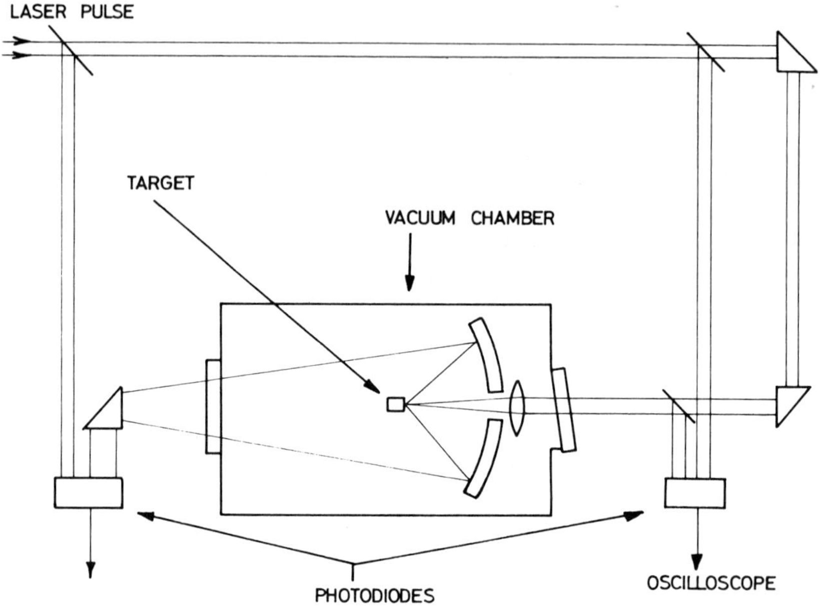

FIG. 13. Arrangement for measuring reflectivity (after Andrews *et al.*, 1972).

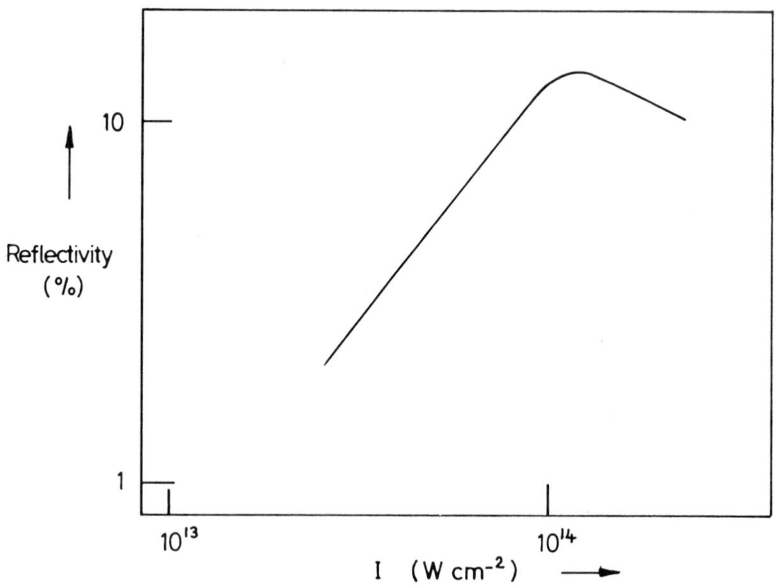

FIG. 14. Reflectivity of a solid deuterium target for 1 nsec pulse at 1.06 μm as a function of irradiance (after Yamanaka *et al.*, 1974).

spherically-convergent neodymium laser light pulses incident on small gas-filled glass balloons the reflectivity continued to rise slowly between 4×10^{13} and 4×10^{15} W cm^{-2}. This may have been due to the optical imperfections which caused light to fall on a point on the target surface not purely normally but from a wide range of angles of incidence.

Similar measurements have also been made at 10.6 μm with carbon dioxide lasers. Dyer *et al.* (1974) at Hull found that there was virtually no change in the reflectivity of a carbon target over the range of irradiances from 5×10^8 to 4×10^{11} W cm^{-2} (Fig. 15). However, Yamanaka *et al.* (1974b) found a distinct maximum in the reflectivity under similar conditions at 4×10^{10} W cm^{-2}, after which the reflectivity fell steadily up to an irradiance of 8×10^{10} W cm^{-2} (Fig. 16). Hall and Negm (1975) obtained results very similar to those of Yamanaka *et al.* (1974b). The discrepancy between these results and those of Dyer *et al.* (1974) has yet to be explained. The occurrence of a maximum in the reflectivity has been interpreted as being due to the onset of a parametric instability which enhances the absorption of light by the plasma.

Krasyuk *et al.* (1970) focussed a 50 psec mode-locked ruby laser pulse in helium at a pressure of 3 atmospheres and examined the incident and transmitted spectra. The transmitted line profile was found to be slightly distorted towards longer wavelengths. When a thin aluminized polyester film was placed at the focus of the laser the transmitted line profile was again slightly displaced to the red, and a secondary maximum appeared on the long wavelength side. These changes in the line profile were ascribed to stimulated Compton scattering: from the redistribution of the energy spectrum the

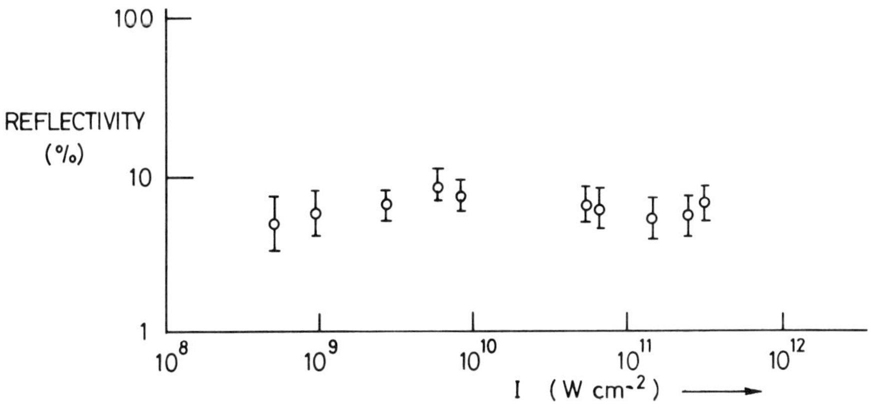

FIG. 15. Reflectivity of a carbon target for 10.6 μm radiation as a function of irradiance (after Dyer *et al.*, 1974).

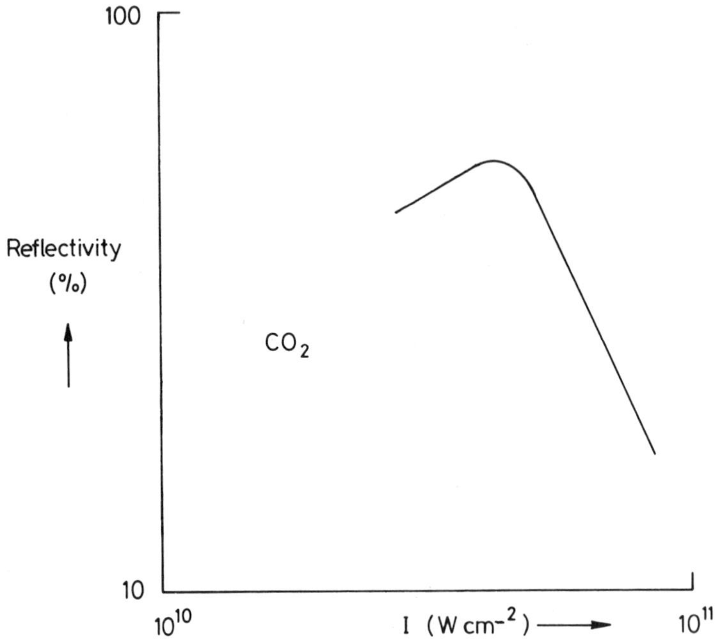

FIG. 16. Reflectivity of a carbon target for 10.6 μm radiation as a function of irradiance (after Yamanaka *et al.*, 1974b).

fraction of the incident light energy absorbed by the electrons was found to be (1 to 2) × 10^{-5}.

Decroisette *et al.* (1970, 1972) focussed a 1 GW, 15 to 20 nsec neodymium laser pulse on to a 2mm thick solid D_2 target with an $f/1$ lens to give an irradiance of about 10^{12} W cm^{-2}. The incident and transmitted light spectra were recorded with time resolution. The transmitted line profiles showed a number of maxima superimposed on the smooth incident light profile on the long wavelength side only. Both the incident and the transmitted light beams were almost completely polarized. Doppler shifts as the result of plasma motion are not seen in forward scattering, and Decroisette *et al.* (1970, 1972) suggest that if the effect was due to a parametric interaction with ion acoustic waves the modified spectrum would show some depolarization. The results were also discussed by Babuel–Peyrissac (1972): it was concluded that stimulated Compton scattering was involved.

Kazakov *et al.* (1971) divided a 20 to 200 picosecond ruby laser pulse into two beams and arranged for them to collide, at the coincident foci of two coaxial lenses, in argon at 0.4 atm. The stronger beam produced an irradiance I_1 of 3 × 10^{14} W cm^{-2} at the focus. After allowing for Bremsstrahlung absorption it was found that the weaker beam was amplified by a factor of 1.32 when $I_2 = 0.2\, I_1$, and by a smaller factor as I_2 was increased. The laser light was strongly broadened on passing through the interaction region. Again, the effect was ascribed to stimulated Compton scattering. The gain calculated for the conditions of the experiment was in good agreement with observation.

Stimulated Compton scattering has also been reported for carbon dioxide laser light in a hydrogen plasma (Offenberger and Burnett, 1973). However, in this and many of the other experiments mentioned above the evidence for Compton scattering is not conclusive and parametric processes involving ion waves may well prove to be responsible for the effects observed.

The spectrum of reflected light was also examined by Caruso *et al.* (1970b) who found that the second harmonic of the incident neodymium light wavelength was generated from perspex and from solid deuterium targets. More complex harmonics were recorded by Bobin *et al.* (1973) in light emitted from the vicinity of a solid hydrogen target at the focus of a 15 nsec, 1 GW neodymium laser. The peak irradiance was about 10^{12} W cm^{-2}. The emitted light was collected at an angle of 135° to the incident beam axis. Three broad lines were observed, at $3\omega_0/2$, $2\omega_0$ and $\omega_0/2$ in order of decreasing intensity: their profiles are shown in Fig. 17. Bobin *et al.* (*see also* Martineau and Bobin, 1974) suggested that the harmonics and subharmonics are generated in the region of the strong density gradient as the result of parametric interactions with electron and ion waves, the line profile being characteristic of the ion wave spectrum. The occurrence of harmonics of $\omega_0/2$ is almost certainly due to stimulated Raman scattering or the two-plasmon instability at $n_{ec}/4$, followed by further mixing processes in the plasma.

Lee *et al.* (1974) measured the relative intensities of light scattered at the neodymium laser frequency ω_0 and at $\omega_0/2$, $3\omega_0/2$, $2\omega_0$, $5\omega_0/2$ and $3\omega_0$ from plasmas produced from solid polyethylene targets. Table I shows the results.

In the 90° non-specular measurements the target normal was at 13° to the beam axis: for the other measurements this angle was 45°. For each geometry the intensities are normalized to that of the fundamental frequency ω_0. No measurable intensity was observed at $\omega_0/2$, $5\omega_0/2$ or at $3\omega_0$. The spectra at ω_0, $3\omega_0/2$ and $2\omega_0$ showed blue shifts from the expected wavelengths for back-scattered light at 180°. The shifts decreased with decreasing scattering angle until at 90° a red shift was observed. Strong structure with peaks separated by 3 to 10 Å was observed in the spectra at all three frequencies.

Ripin *et al.* (1974) investigated the angular distribution of neodymium laser light scattered from a plane solid target, and found that when the target was inclined to the beam axis the near-specularly reflected component had an angular extent of ±40° and contained a total of 5 to 10% of the incident laser energy. The light energy scattered backwards into the $f/14$ focussing lens was

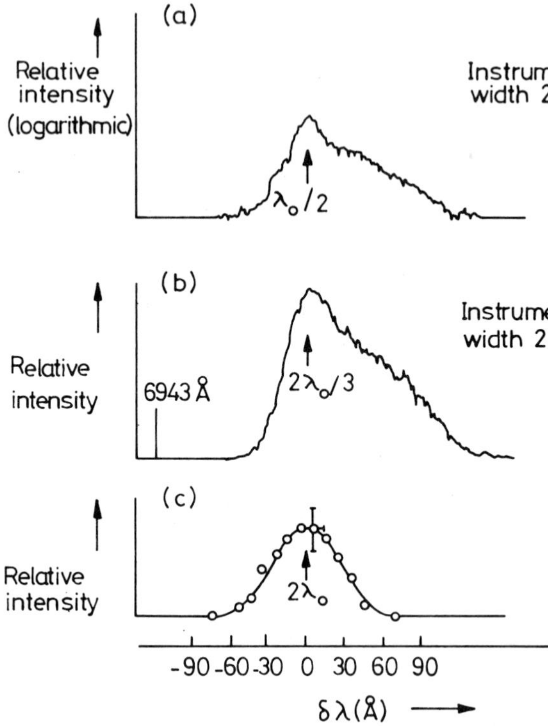

FIG. 17. Profiles of harmonics of the incident neodymium laser wavelength emitted from vicinity of a solid hydrogen target (after Bobin *et al.*, 1973).

TABLE I. (After Lee *et al.*, 1974)

Direction of scattering	Frequency	Relative intensities
Backscatter (180°)	ω_0	1
	$2\omega_0$	5×10^{-4}
	$\frac{3}{2}\omega_0$	2×10^{-3}
135° (non-specular)	ω_0	1
	$2\omega_0$	7×10^{-4}
	$\frac{3}{2}\omega_0$	3×10^{-3}
90° (non-specular)	ω_0	1
	$2\omega_0$	1
	$\frac{3}{2}\omega_0$	—
90° (specular)	ω_0	1
	$2\omega_0$	2×10^{-3}
	$\frac{3}{2}\omega_0$	7×10^{-3}

comparable, but its intensity was much greater because it was contained in a cone of only $\pm 2^{\circ}$. The backscattered light was almost completely polarized in the same plane as the incident light. The backscattered energy with an $f/1.9$ focussing lens increased exponentially as the laser pulse energy was increased up to 50J in a 900 psec pulse. With an $f/14$ lens the backscattered energy increased exponentially at first and saturated when the laser pulse energy exceeded 10J. The observations were very similar for copper, deuterated polyethylene and aluminium targets. Time-integrated spectra of the backscattered light showed a broadening of the spectrum by 10 to 50 Å as compared to the incident light ($\Delta\lambda \sim 1$ Å), and a displacement of several Ångstroms, to the blue at normal incidence but to the red at an angle between the beam axis and the target normal of 45°, in agreement with the observations of Lee *et al.* (1974). Time-resolved spectra showed that the initial displacement was to the red. These results were interpreted as indicating the onset of Brillouin scattering, with ion waves propagating inwards away from the laser with $\omega_{ia} \sim 4 \times 10^{12}$ rad s^{-1}, from an interaction region which is accelerating towards the laser and in which the electron density is typically $n_{ec}/4$. A line structure which occasionally appeared in the backscattered spectrum with a line spacing of about 2Å (similar to that observed by Lee *et al.* (1974)) has not been explained.

Eidmann and Sigel (1975) made a detailed study of second harmonic generation in plasmas produced by neodymium laser pulses of 5 nsec duration with energies up to 20J, focussed on plane copper, glass and tungsten targets at normal incidence. The results are shown schematically in Fig. 18. The second harmonic energy increased quadratically with the laser pulse energy. Only obliquely incident light of p polarization generates the second harmonic, which is emitted in the same direction as the specularly reflected wave. Similar studies at $3\omega_0/2$ would be very interesting.

Two high-power monochromatic transverse waves were mixed in a plasma by Stansfield *et al.* (1971) to produce plasma waves at the difference frequency, as suggested by Kroll *et al.* (1964). A third wave was scattered from the interaction volume and showed enhanced scattering in the electron plasma line when the difference frequency equalled the plasma frequency. The two interacting beams were obtained from a single dye laser pumped by a ruby laser.

Yamanaka *et al.* (1974a) made studies of plasmas formed from solid deuterium and hydrogen by Nd-glass and by Nd-YAG lasers. The backscattered light contained second harmonic components when the irradiance exceeded a certain threshold. When a YAG laser, with a relatively narrow (6 Å) bandwidth was used, the scattered light near the second harmonic showed a structure which depended on the isotope used as the target (Fig. 19). The side peaks at 204 and 319 cm^{-1} are interpreted as being due to

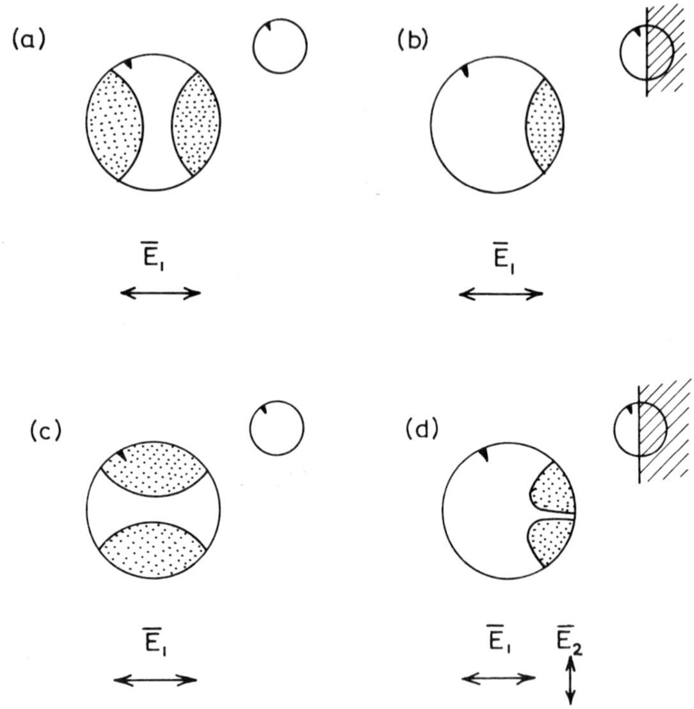

FIG. 18. Distribution of intensity of specularly generated second harmonic light in the plane of a lens focussing linearly polarized neodymium laser light (electric vector $\bar{E}_1$) on a target. In (a), (b) and (d) $\bar{E}_1$ is horizontal: in (c) it is vertical. In (a) and (c) laser light fills the lens: in (b) and (d) the right-hand half is obscured. In (d) the second harmonic is recorded after passing through a linear polarizer which does not transmit $\bar{E}_2$ when parallel to $\bar{E}_1$. The harmonic is generated only by obliquely incident light with $\bar{E}_1$ in the plane of incidence: it is emitted in the same direction as the reflected laser light, with the same polarization (after Eidmann and Sigel, 1975).

Brillouin scattering, because the ion wave frequency depends on the ion mass as $(m_i)^{-1/2}$, so the shift should be $\sqrt{2}$ times bigger for H than for D, as is observed. Sidebands were not observed near the fundamental frequency. When the glass laser, with a 60 Å bandwidth, was used, the fine structure was smeared out.

Less direct evidence for nonlinear effects comes from studies of the behaviour of laser-heated plasmas. Yamanaka *et al.* (1972a, 1972b) produced plasmas from solid deuterium targets using 2 nsec neodymium laser pulses of up to 40J. Figures 20(a) and (b) show the electron temperature, measured by the X-ray absorption method, and the ion temperature derived from time-of-flight velocity profiles. At irradiances above about 3×10^{13} W cm^{-2} a fast

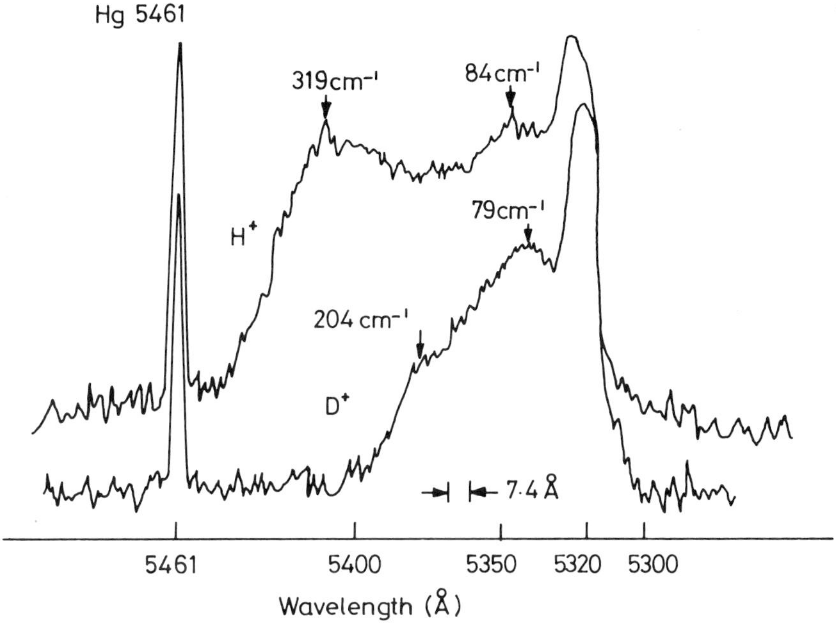

FIG. 19. Spectrum of backscattered second harmonic radiation from a Nd–YAG laser, with solid hydrogen and deuterium targets (see text). (After Yamanaka *et al.*, 1974a.)

group of ions was observed: their appearance coincided with the appearance of neutrons from the plasma, and with a sudden increase in the plasma reflectivity from 4 % to 20 % of the pulse energy. Here the electron temperature was about 300 eV. Fig. 6 shows that these conditions coincide with the threshold for the oscillating two-stream instability. At a slightly lower laser irradiance, about 2×10^{12} W cm^{-2}, when $T_e \simeq 200$ eV, the reflected laser power often oscillated at about 10^9 Hz (Waki *et al.* 1972). Yamanaka *et al.* have suggested that the onset of the instability produces a sharp boundary of very dense plasma which increases the effective reflectivity, but other interpretations are possible.

The X-ray emission spectrum from the plasma produced by a 20 psec neodymium laser from a polyethylene target was observed by Ripin *et al.* (1975) using thirteen detectors behind different filters. The observed spectrum, though not consistent with a single Maxwellian electron velocity distribution, was found to be consistent with the predictions of a hydrodynamic computation which took into account the effects of spontaneously-generated magnetic fields. Such fields may arise if the gradients of temperature and density do not

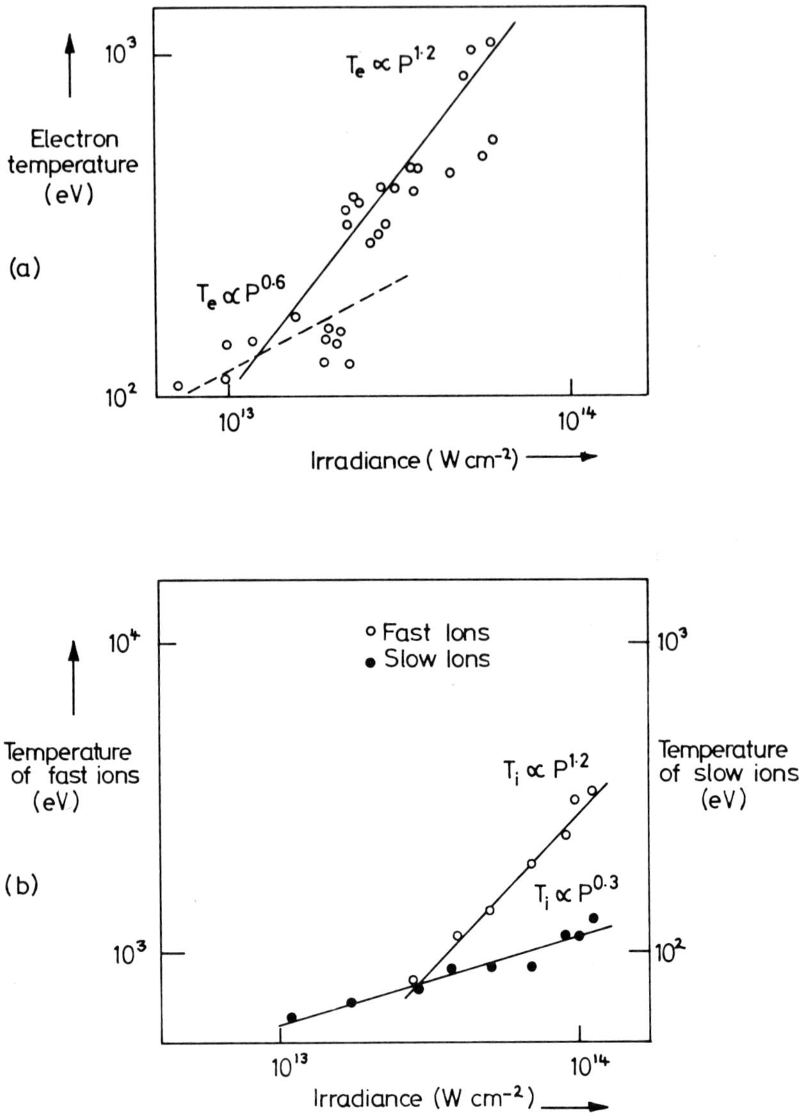

FIG. 20. Electron and ion "temperatures" of plasmas formed from solid deuterium targets by 2 nsec neodymium laser pulses as functions of irradiation (after Yamanaka *et al.*, 1972a, b).

coincide: $\partial \mathbf{B}/\partial t \propto \nabla T_e \times \nabla n_e$. The effect of a magnetic field is to reduce the thermal conductivity of the plasma, and to produce a group of hotter electrons in a region near the critical density surface. Fields of the order of 10^6 Gauss were predicted.

VIII. Self-focussing of Laser Beams in Gases

A. SELF-FOCUSSING PROCESSES

This nonlinear effect is important in the development of the plasma formed by a laser pulse in a cold gas, and may also be important in the development of plasmas formed from solid targets. It is well-known that if the refractive index of a medium increases with intensity a Gaussian beam may be self-focussed, because the beam produces its own positive lens (Askaryan, 1962; Chiao *et al*., 1964; Talanov, 1964). Let us consider a nonlinear medium in which the refractive index $\mu_{\text{total}} = \mu + \mu_2 E^2$, where E is the amplitude of the radiation field. In the absence of diffraction, a plane Gaussian beam of initial radius r and power $P = \pi r^2 \mu^2 \varepsilon_0 c E^2/2$ will be brought to a focus in a distance (Kelley, 1965)

$$z_f \simeq r^2 \left(\frac{\pi c \varepsilon_0 \mu^3}{2\mu_2 P} \right)^{1/2}. \tag{94}$$

In the absence of self-focussing, diffraction would double the beam radius in a distance

$$z_d = \frac{\sqrt{2}\pi r^2}{\lambda}, \tag{95}$$

λ being the wavelength of the light. The threshold for self-focussing, when the focussing effect just balances the diffraction spreading of the beam, is found by equating z_f and z_d. Then

$$P = P_{sf} = \frac{c \varepsilon_0 \mu^3 \lambda^2}{4\pi\mu_2} \tag{96}$$

in MKS units $[(\mu_2)_{\text{esu}} = 9 \times 10^8 \, (\mu_2)_{\text{MKS}}]$. As usual, it is the beam power, not the irradiance, that determines the threshold. However, at threshold the self-focussing length, allowing for diffraction, is infinite. In a real situation the nonlinear medium is of limited extent and geometrical constraints must be considered.

Calculations of the self-focussing process have been made for a Gaussian wavefront (Dyshko *et al.* (1947); Ledenev *et al.* (1972); Abramov *et al.* (1969)) which indicate that at powers well above threshold several focal points will occur on the beam axis. Bespalov and Talanov (1966) have shown that well above threshold a Gaussian beam is unstable against perturbations due to inhomogeneities and aberrations and will decay into several self-focussing channels. As the focussing proceeds the local irradiance in the beam increases and additional non-linearities may be involved: heating, excitation (including multiphoton excitation) and ionization are to be expected and stimulated scattering may become important (Kelley and Gustafson (1973)). Taking into account also the time dependence of the beam power a very complex situation arises in the subsequent development of the channel. The motion of the self-focussed spot as the beam power varies may give rise to the appearance of filaments: this has been discussed by Lugovoi and Prokhorov (1968).

Self-focussing may be caused by several different mechanisms, which can be divided into two groups. In the first group the nonlinear susceptibility is associated with the orientation (optical Kerr effect) or the excitation of the atoms or molecules of the medium. The response is very rapid. In the second group the effect is due to displacement of the medium resulting from electrodynamic or thermal forces, and is therefore relatively slow.

Among the fast processes, the optical Kerr effect has a response time constant of order 10^{-11} sec: coefficients have been given for the inert gases (New and Ward (1967)). The effects of excitation were discussed by Askaryan (1966). An excited atom is more easily polarized than the same atom in the ground state, so excitation leads to an increase in the refractive index. Javan and Kelley (1966) have pointed out that at a frequency just above resonance with an absorption transition, the anomalous dispersion causes the refractive index to increase with intensity as a result of saturation of the transition. This too may lead to self-focussing. Key *et al.* (1970) estimated the cubic nonlinear polarizability of an atom when resonances occur at energies near both the photon energy and twice the photon energy above the ground state, and concluded that this could be more important than the linear polarizability. The nonlinear polarizability of molecules as a result of electronic distortion in strong fields has been discussed by Brewer and Lee (1968).

It should be noted that excitation is often accompanied by ionization: the presence of free electrons, which causes a decrease in refractive index, can more than compensate the self-focussing effect of excited states, leading instead to an expansion of the beam.

The radial gradient of the electric field in a Gaussian beam produces electrodynamic, or "ponderomotive", forces on the medium which may cause self-focussing both in a neutral gas and in a fully-ionized plasma. The time-averaged force per unit volume on a medium of dielectric constant ε and permeability μ' is given by (Landau and Lifshitz (1960))

$$\bar{\mathbf{F}}_m = \frac{(\varepsilon - 1)\varepsilon_0}{2} \nabla(\overline{E^2}) + \frac{(\mu' - 1)\mu_0}{2} \nabla(\overline{H^2}) + \frac{(\varepsilon\mu' - 1)}{c^3} \frac{\partial}{\partial t} \overline{(\mathbf{E} \times \mathbf{H})}, \quad (97)$$

ignoring collisional effects. For high-frequency fields $\mu' = 1$ so the second term is zero. The third term is only important while the field is being established. In a neutral gas $\varepsilon > 1$ and the first term produces electrostriction—a compression into regions of higher electric field. In a fully-ionized gas $\varepsilon \simeq 1 - \omega_{pe}^2/\omega^2 < 1$ and so the first term now represents a force expelling plasma from the higher field regions ("electrodilation"). However, in both cases the refractive index will increase towards the centre of a Gaussian beam, and self-focussing is possible in principle. This effect has been discussed by Hora (1969) and by Palmer (1971, 1972) who has examined the close connection between the onset of self-focussing and the occurrence of stimulated scattering processes in a plasma. The time required for ponderomotive forces to produce self-focussing is of the order of the beam diameter divided by the sound speed.

Heating due to absorption of light in a Gaussian beam may give rise to a number of thermal effects. If $(\partial\varepsilon/\partial T)_\rho > 0$, ρ being the mass density of the medium, a rapid transient self-focussing effect may occur (Petrishchev and Talanov (1971)). Thereafter, pressure and temperature gradients will affect the density distribution. Thermal effects will accompany electro-dynamic ponderomotive effects and will tend to reduce the density near the beam axis: in a neutral gas they may be expected to tend to cancel the self-focussing effect of electrostriction, whereas in a fully-ionized plasma they may reinforce the effect of electrodilation. The balance of all these effects can alter greatly during a single laser shot.

B. EXPERIMENTAL OBSERVATIONS OF SELF-FOCUSSING IN GASES

Basov *et al.* (1967, 1968) found that when a 1 GW, 15 nsec neodymium laser pulse was focussed by a long focal length lens (2.5 m) in air, several hundred separate sparks were produced over a distance of more than 2m. Self-focussing was suggested as a possible explanation for the effect. On a much smaller scale, Korobkin *et al.* (1967) observed that breakdown in air due to a 100 MW 15 nsec ruby laser pulse focussed by a 50mm lens occurred at several discrete points and suggested that this too might be due to self-focussing. Evidence for the occurrence of self-focussing associated with breakdown in air at S.T.P. was found by Korobkin and Alcock (1968) using a 3 MW *Q*-switched ruby laser operated in a single longitudinal and single transverse mode with a pulse duration of 8 nsec. Photographs of the focal volume, taken in ruby laser radiation scattered at 90° to the direction of the axis of the focussed beam, showed bright filaments or points at breakdown. The filaments were less than 5 μm in diameter and ran parallel to the beam axis. Several filaments were observed when the laser pulse was focussed with a 10 cm focal length lens giving a focal spot diameter of 50 μm, but only one was produced by a 3.5 cm focal length lens. Time-resolved streak photography in laser light using an image convertor camera indicated that the filaments recorded without time resolution were formed by the motion of very bright breakdown points towards the laser, as had been suggested previously in connection with breakdown in solids by Lugovoi and Prokhorov (1968). Streak photographs were also taken in the light of the plasma formed at breakdown, which suggested that the small breakdown point was very hot. The angular distribution of forward scattered laser light was studied, after blocking the directly transmitted beam. The scattered light was found to be concentrated into a cone of 30° included angle. It was coherent, it occurred at the moment of breakdown and it contained a third of the laser power, suggesting that it was not in fact scattered, but diffracted from a narrow channel. Spectroscopic studies of this light (Alcock *et al.* (1969b)) showed that it was broadened and displaced towards longer wavelengths by about 0.12 Å: a very similar effect is observed when self-focussing occurs in liquids. These observations are consistent with the occurrence of self-focussing leading to the formation of channels with diameters of about 1.7 μm, in which the power density reached 4.4×10^{13} W cm^{-2}, corresponding to an E field of 1.3×10^8 V cm^{-1}. (The beam diameter might have been even smaller since it was derived on the assumption that the light was diffracted from a plane circular aperture, whereas in fact it would probably be more gradually defocussed (Brewer *et al.*, 1968). This radiation field would be sufficient to cause full ionization in a fraction of a nanosecond, leading locally to strong absorption and rapid heating of the plasma. Direct multiphoton ionization would be the most important ionizing process: diffusion losses would be rapid from so narrow a channel and would hinder the development of an avalanche.

Complex filamentary structures were observed by Tomlinson (1969) in several gases in the region of the focal volume of a ruby laser. The filaments were shown to grow towards the lens during the 30 nsec laser pulse. However, the breakdown thresholds were in agreement with cascade theory and did not appear to be influenced by self-focussing effects, and Tomlinson concluded that if self-focussing was responsible for the filaments it must develop after the ionization processes have begun. A similar conclusion was reached by Key *et al.* (1970) who made schlieren observations of self-focussed filaments produced in argon by a 6 nsec neodymium laser pulse, and also by Alcock *et al.* (1970) and Richardson and Alcock (1971a,b, *see also* Alcock, 1972) from a study of breakdown in several gases by a single mode ruby laser, using photographic, spectroscopic and interferometric methods. Ionization probably occurs by a stepwise process involving excited atoms, so that at some stage during the avalanche process leading to breakdown a partially ionized gas may exist with sufficient excited atoms to overcome the defocussing effect of the free electrons. Key *et al.* (1970) found it necessary to invoke the cubic nonlinear polarizability of excited atoms to account for their results on this basis. Alternatively the ponderomotive effect proposed by Hora (1969) might have been responsible for self-focussing in these experiments, though if this were the case self-focussing should also have been observed in some Thomson scattering experiments where no anomalies were in fact reported (Alcock *et al.*, 1970).

Schlieren observations of self-focussed filaments were also described by Belland *et al.* (1971) with a multimode neodymium laser producing 1.6 nsec pulses of 1.6 GW. A single filament appeared initially, accompanied later by two more forming an arrow. This fork-like effect was noted earlier by Savchenko and Stepanov (1968) and by Alcock *et al.* (1970) at powers well above the breakdown threshold. About 80 % of the incident light was scattered into a cone outside the laser beam itself, and 10 % was found within the solid angle of the laser beam. Only about 10 % was therefore absorbed in the breakdown plasma.

Evidence for self-focussing of very short mode-locked pulses, lasting only a few picoseconds, in air was obtained by Alcock *et al.* (1969a), who produced breakdown with trains of pulses from a ruby laser and also from a neodymium laser. Individual pulses produced very small separate breakdown regions: most of the forward scattered light was contained in a cone of 10° to 20° included angle, suggesting filament diameters of a few microns. The spectrum of the forward-scattered neodymium laser light was broadened to 200 Å compared to the 75 Å breadth of the incident light, and slightly displaced to longer wavelengths.

Bunkin *et al.* (1971) found that self-focussing occurred in air, nitrogen and argon with a single mode-locked ruby laser pulse of 20 to 100 psec duration. With a 15 cm lens many breakdown points were observed over a distance of several centimetres along the beam axis near the focus. In air and nitrogen at STP, when the laser pulse power was about 1.5×10^9 W strong scattering of laser light occurred without being accompanied by breakdown. A two-stage self-focussing process was proposed, the first being due to the nonlinear polarizability of molecules and the second due to the presence of excited molecules.

Self-focussing of CO_2 laser light in BCl_3 and in SF_6 was first reported by

Karlov *et al.* (1973). The effect was observed in BCl_3 at pressures above 0.1 atm with 5 to 10 kW pulses lasting 20 μs. The value of μ_2 derived from the experiments was $(4.6 \pm 0.7) \times 10^{-7} p_{atm}$ cgs esu. In SF_6 the effect was considerably greater: μ_2 was estimated to be about $5 \times 10^{-5}\ p_{atm}$ cgs esu. In each case the self-focussing was ascribed to saturation of the anomalous dispersion in resonant absorption. The laser radiation produced dissociation of the gas, which was followed by recombination accompanied by visible luminescence: this displayed the shape of the laser beam clearly in BCl_3.

IX. Implications for Laser-Compressed Plasma Experiments

If a small solid target, which may be a sphere or a hollow shell, is placed at the centre of a spherically-convergent laser pulse, the light forms a very inhomogeneous plasma at the surface of the target, which blows off into the surrounding vacuum, resulting in a pressure on the remaining target material. If the pulse is shaped correctly, it should be possible to arrange for the target core to be compressed almost isentropically, which will give the highest density for a given pressure. Current laser technology may enable us to create plasmas at densities up to 10^4 gm cm^{-3}, which will be of considerable scientific interest. There is a possibility too that using hydrogen isotope targets useful amounts of fusion energy might be extracted from the resulting nuclear reactions (see for example Brueckner and Jorna (1974)). However, very high irradiances will be needed, which will induce strongly nonlinear behaviour in the plasma.

In general, it appears that stimulated Raman and Brillouin scattering will be a nuisance, because they will reflect out much of the incident light. However, if Raman scattering at quarter critical density predominates, half of the incident light might be converted to plasma wave energy, which would be a reasonable absorption efficiency. The worst situation will occur if Brillouin scattering is important, because hardly any of the incident energy will then be given to the plasma. We need to know much more about the effects of wave-front curvature and wave refraction on these interactions. We might also consider ways of preventing the growth of the instabilities, for example by modulating the laser frequency.

The parametric decay and the parametric two-stream instabilities are probably beneficial, because as we saw they lead to improved absorption. However, the form in which the energy is absorbed may not be very satisfactory. If a significant proportion goes into small numbers of fast particles, these will "decouple" from the rest because they will have a reduced collision cross-section, and their energy may be expended in heating the dense compressed core prematurely, making subsequent compression more difficult.

Self-focussing, or more generally spatial degradation of the laser wavefront, may be a problem. If the spherical geometry of the wave becomes significantly distorted as a result of a self-focussing type instability many new effects must be expected. The compression geometry will be distorted, leading to poor compression. But also (as has already been mentioned) there is a possibility that very large magnetic fields may be generated spontaneously. The effect of such fields is to inhibit thermal conduction, and this may lead to a decoupling between the outer layers of plasma, which will get extremely hot, and the inner regions, which will stay relatively cool, again resulting in poor compression.

There is still much work to be done before we have a satisfactory understanding of all these processes, theoretically and especially experimentally where there are some very stimulating problems to be solved.

References

Abramov, A. A., Lugovoi, V. N. and Prokhorov, A. M. *Zh. Eksp. i. Teor. Fiz. Pis'ma* **9**, 675 (1969). Transl. *JETP Lett.* **9**, 419 (1969).

Alcock, A. J. Experiments on self-focussing in laser-produced plasmas. *In* "Laser interaction and related plasma phenomena", (H. J. Schwarz and H. Hora, Eds), Plenum, New York, Vol. 2, pp. 155–75 (1972).

Alcock, A. J., DeMichelis, C., Korobkin, V. V. and Richardson, M. C. *Appl. Phys. Lett.* **14**, 145 (1969a).

Alcock, A. J., De Michelis, C., Korobkin, V. V. and Richardson, M. C. *Phys. Lett.* **29A**, 475 (1969b).

Alcock, A. J., De Michelis, C. and Richardson, M. C. *IEEE J. Quantum Electron.* **QE-6**, 622 (1970).

Andrews, A. J., Hall, T. A. and Hughes, T. P. "Vth European Conference on Controlled Fusion and Plasma Physics", Grenoble. Grenoble, Ass. Euratom–CEA, Vol. 1. p. 62 (1972).

Askar'yan, G. A. *Zh. Eksp. i. Teor. Fiz.* **42**, 1567 (1962). Transl. *Sov. Phys. JETP* **15**, 1088 (1962).

Askar'yan, G. A. *Zh. Eksp. Teor. Fiz. Pis'ma* **4**, 400 (1966). Transl. *JETP Lett.* **4**, 270 (1966).

Babuel-Peyrissac, J. P. (1972). *Phys. Lett.* **41A**, 143 (1972).

Basov, N. G., Boiko, V. A., Krokhin, O. N. and Sklizkov, G. V. *Dokl. Akad. Nauk. SSSR* **173**, 538 (1967). Transl. *Sov. Phys. Dokl.* **12**, 248 (1967).

Basov, N. G., Boiko, V. A., Krokhin, O. N. and Sklizkov, G. V. "VIII Int. Conf. on Phenomena in Ionized Gases", Vienna. Vienna, IAEA, contributed papers, p. 261 (1968).

Beaudry, G., Martineau, J. and Pepin, H. "Xth Int. Conf. on Phenomena in Ionized Gases", Oxford. Donald Parsons & Co. Ltd., Oxford, contributed papers, p. 226 (1971).

Belland, P., De Michelis, C. and Mattioli, M. *Opt. Commun.* **4**, 50 (1971).
Bespalov, V. I. and Talanov, V. I. *Zh. Eksp. i. Teor. Fiz. Pis'ma* **3**, 471 (1966). Transl. *JETP Lett.* **3**, 307 (1966).
Bezzerides, B. and Weinstock, J. *Phys. Rev. Lett.* **28**, 481 (1972).
Biskamp. D. and Welter, H. *Phys. Rev. Lett.* **34**, 312 (1975).
Bobin, J. L., Decroisette, M., Meyer, B. and Vitel, Y. *Phys. Rev. Lett.* **30**, 594 (1973).
Bodner, S. E., Chapline, G. F. and De Groot, J. *Plasma Phys.* **15**, 21 (1973).
Brehme, H. *Phys. Rev.* **C3**, 837 (1971).
Brewer, R. G. and Lee, C. H. *Phys. Rev. Lett.* **21**, 267 (1968).
Brewer, R. G., Lifsitz, J. R., Garmire, E., Chiao, R. Y. and Townes, C. H. *Phys. Rev.* **166**, 326 (1968).
Brueckner, K. A. and Jorna, S. *Phys. Fluids* **16**, 2350 (1973).
Brueckner, K. A. and Jorna, S. *Rev. Mod. Phys.* **46** (**2**), 325 (1974).
Bunkin, F. V. and Fedorov, M. V. *Zh. Eksp. i. Teor. Fiz.* **49**, 1215 (1965). Transl. *Sov. Phys. JETP* **22**, 844 (1966).
Bunkin, F. V. and Kazakov, A. E. *Zh. Eksp. i. Teor. Fiz.* **59**, 2233 (1970). Transl. *Sov. Phys. JETP* **32**, 1208 (1971).
Bunkin, F. V., Krasyuk, I. K., Marchenko, V. M., Pashinin, P. P. and Prokhorov, A. M. *Zh. Eksp. i. Teor. Fiz.* **60**, 1326 (1971). Transl. *Sov. Phys. JETP* **33**, 717 (1971).
Caruso, A., De Angelis, A., Gatti, G., Gratton, R. and Martellucci, S. *Phys. Lett.* **33A**, 320 (1970a).
Caruso, A., De Angelis, A., Gatti, G., Gratton, R. and Martellucci, S. *Phys. Lett.* **33A**, 29 (1970b).
Charatis, G. *et al.* "5th IAEA Conf. on Plasma Physics and Controlled Nuclear Fusion Research", Tokyo (1974).
Chiao, R. Y., Garmire, E. and Townes, C. H. *Phys. Rev. Lett.* **13**, 479 (1964).
Dawson, J. and Oberman, C. *Phys. Fluids* **5**, 517 (1962).
Dawson, J. M. and Shanny, R. *Phys. Fluids* **11**, 1506 (1968).
Decroisette, M., Piar, G. and Floux, F. *Phys. Lett.* **32A**, 249 (1970).
Decroisette, M., Peyraud, J. and Piar, G. *Phys. Rev.* **A5**, 1391 (1972).
De Groot, J. S. and Katz, J. I. *Phys. Fluids* **16**, 401 (1973).
Drake, J. F., Kaw, P. K., Lee, Y. C., Schmidt, G., Liu, C. S. and Rosenbluth, M. N. *Phys. Fluids* **17**, 778 (1974).
Dreicer, H. *Phys. Fluids* **7**, 735 (1964).
Dyer, P. E., James, D. J., Ramsden, S. A. and Skipper, M. A. *Phys. Lett.* **48A**, 311 (1974).
Dyshko, A. L., Lugovoi, V. N. and Prokhorov, A. M. *Zh. Eksp. i. Teor. Fiz. Pis'ma* **6**, 655 (1967). Transl. *JETP Lett.* **6**, 146 (1967).
Eidmann, K. and Sigel, R. *Phys. Rev. Lett.* **34**, 799 (1975).
Forslund, D. W., Kindel, J. M. and Lindman, E. L. *Phys. Rev. Lett.* **30**, 739 (1973).
Freidberg, J. P. and Marder, B. M. *Phys. Rev.* **A4**, 1549 (1971).
Galeev, A. A., Oraevskii, V. N. and Sagdeev, R. Z. *Zh. Eksp. i. Teor. Fiz. Pis'ma* **16**, 194 (1972). Transl. *JETP Lett.* **16**, 136 (1972).
Galeev, A. A., Laval. G., O'Neil, T. M., Rosenbluth, M. N. and Sagdeev, R. Z. *Zh. Eksp. i. Teor. Fiz. Pis'ma* **17**, 48 (1973). Transl. *JETP Lett.* **17**, 35 (1973).
Ginzburg, V. L. "The propagation of electromagnetic waves in plasmas", (Translation from 2nd Ed. of V. L. Ginzburg's book). Pergamon, Oxford, 2nd Ed. (1970).
Godwin, R. P. *Phys. Rev. Lett.* **28**, 85 (1972).

Goldman, M. V. *Ann. Phys.* **38**, 95 and 117 (1966).
Grey-Morgan, C. *Rep. Prog. Phys.* **38**, 621 (1975).
Hall, T. A. and Negm, Y. Z. (Private communication)
Hora, H. *Z. Phys.* **226**, 156 (1969).
Jackson, E. A. *Phys. Rev.* **153**, 235 (1967).
Javan, A. and Kelley, P. L. *IEEE J. Quantum Electron.* **2**, 470 (1966).
Jorna, S. *Phys. Fluids* **17**, 765 (1974).
Karlov, N. V., Karpov, N. A., Petrov, Yu. N. and Stel'makh, O. M. *Zh. Eksp. i. Teor. Fiz. Pis'ma* **17**, 337 (1973). Transl. *JETP Lett.* **17**, 239 (1973).
Kaw, P. and Dawson, J. M. *Phys. Fluids* **12**, 2586 (1969).
Kazakov, A. E., Krasyuk, I. K., Pashinin, P. P. and Prokhorov, A. M. *Zh. Eksp. i. Teor. Fiz. Pis'ma* **14**, 416 (1971). Transl. *JETP Lett.* **14**, 280–2 (1971).
Kelley, P. L. *Phys. Rev. Lett.* **15**, 1005 (1965).
Kelley, P. L. and Gustafson, T. K. *Phys. Rev.* **8A**, 315 (1973).
Key, M. H., Preston, D. A. and Donaldson, T. P. *J. Phys.* **B3**, L88 (1970).
Klein, H. H. and Manheimer, W. M. *Phys. Rev. Lett.* **33**, 953 (1974).
Klein, H. H., Manheimer, W. M. and Ott, E. *Phys. Rev. Lett.* **31**, 1187 (1973).
Korobkin, U. V. and Alcock, A. J. *Phys. Rev. Lett.* **21**, 1433 (1968).
Korobkin, V. V., Mandel'shtam, S. L., Pashinin, P. P., Prokhindeev, A. V., Prokhorov, A. M., Sukhodrev, N. K. and Shchelev, M. Ya. *Zh. Eksp. i. Teor. Fiz.* **53**, 116–23 (1967). Transl. *Sov. Phys. JETP* **26**, 79 (1968).
Krasyuk, I. K., Pashinin, P. P. and Prokhorov, A. M. *Zh. Eksp. i. Teor. Fiz. Pis'ma* **12**, 439 (1970). Transl. *JETP Lett.* **12**, 305 (1970).
Krasyuk, I. K., Pashinin, P. P. and Prokhorov, A. M. *Zh. Eksp. i. Teor. Fiz. Pis'ma* **17**, 130 (1973). Transl. *JETP Lett.* **17**, 92 (1973).
Kroll, N. M., Ron, A. and Rostoker, N. *Phys. Rev. Lett.* **13**, 83 (1964).
Kruer, W. L. and Dawson, J. M. *Phys. Fluids* **15**, 446 (1972).
Landau, L. D. and Lifshitz, E. M. "Electrodynamics of Continuous Media", Pergamon, Oxford, p. 242 (1960).
Lashmore-Davies, C. N. *Phys. Rev. Lett.* **32**, 289 (1974).
Lee, P., Giovanelli, D. V., Godwin, R. P. and McCall, G. H. *Appl. Phys. Lett.* **24**, 406 (1974).
Litvak, A. G. and Trakhtengerts, V. Yu. *Zh. Eksp. i. Teor. Fiz.* **60**, 1702 (1971). Transl. *Sov. Phys. JETP* **33**, 921 (1971).
Ledenev, V. I., Sukhorokov, A. P. and Khachatryan, A. M. *Kvantovaya Elektron.* **2**, 90–4 (1972). Transl. *Sov. J. Quantum Electron.* **2**, 163 (1972).
Liu, C. S., Rosenbluth, M. N. and White, R. B. *Phys. Rev. Lett.* **31**, 697 (1973).
Lugovoi, V. N. and Prokhorov, A. M. *Zh. Eksp. i. Teor. Fiz. Pis'ma* **7**, 153 (1968). Transl. *JETP Lett.* **7**, 117 (1968).
Martineau, J. and Bobin, J. L. *Phys. Lett.* **47A**, 43 (1974).
Martineau, J. and Pépin, H. *J. Appl. Phys.* **43**, 917 (1972).
Mueller, M. M. *Phys. Rev. Lett.* **30**, 582 (1973).
New, G. H. C. and Ward, J. F. *Phys. Rev. Lett.* **19**, 556 (1967).
Nicholson-Florence, M. B. *J. Phys.* **A4**, 574 (1971).
Nishikawa, K. *J. Phys. Soc. Japan* **24**, 916 (1968a).
Nishikawa, K. *J. Phys. Soc. Japan* **24**, 1152 (1968b).
Offenberger, A. A. and Burnett, N. H. *Phys. Lett.* **42A**, 527–8 (1973).
Osborn, R. K. *Phys. Rev.* **A5**, 1660 (1972).
Palmer, A. J. *Phys. Fluids* **14**, 2714 (1971).

Palmer, A. J. *In* "Laser interaction and related plasma phenomena", (H. J. Schwarz and H. Hora, Eds), Plenum, New York, Vol. 2, p. 367 (1972).
Perkins, F. W. and Flick, J. *Phys. Fluids* **14**, 2012 (1971).
Pert, G. J. *J. Phys.* **A5**, 506 (1972).
Pert, G. J. *Phys. Lett.* **42A**, 511 (1973).
Pesme, D., Laval, G. and Pellat, R. *Phys. Rev. Lett.* **31**, 203 (1973).
Petrischev, V. A. and Talanov, V. I. *Kvantovaya Elektron*, **1**, 35 (1971). Transl. *Sov. J. Quantum Electron.* **1**, 587 (1972).
Peyraud, J. *J. Physique* **30**, 307, 461 (1969).
Phelps, D., Rynn, N. and Van Hoven, G. *Phys. Rev. Lett.* **26**, 688 (1971).
Phelps, D., Van Hoven, G. and Rynn, N. *Phys. Fluids* **16**, 1078 (1973).
Rand, S. *Phys. Rev.* **136B**, 231 (1964).
Richardson, M. C. and Alcock, A. J. *Appl. Phys. Lett.* **18**, 357 (1971a).
Richardson, M. C. and Alcock, A. J. *Kvantovaya Elektron.* **1**, (5) 37 (1971b). Transl. *Sov. J. Quantum Electron.* **1**, 461 (1972).
Ripin, B. H., McMahon, J. M., McLean, E. A., Manheimer, W. M. and Stamper, J. A. *Phys. Rev. Lett.* **33**, 634 (1974).
Ripin, B. H. *Phys. Rev. Lett.* **34**, 1313 (1975).
Rosenbluth, M. N. *Phys. Rev. Lett.* **29**, 565 (1972).
Rosenbluth, M. N. and Liu, C. S. *Phys. Rev. Lett.* **29**, 701 (1972).
Sanmartin, J. R. *Phys. Fluids* **13**, 1533 (1970).
Savchenko, M. M. and Stepanov, V. K. *Zh. Eksp. i. Teor. Fiz. Pis'ma* **8**, 458 (1968). Transl. *JETP Lett.* **8**, 281 (1968).
Seely, J. F. and Harris, E. G. *Phys. Rev.* **A7**, 1064 (1973).
Silin, V. P. *Zh. Eksp. i. Teor. Fiz.* **47**, 2254 (1964). Transl. *Sov. Phys. JETP* **20**, 1510 (1965).
Shearer, J. W. and Duderstadt, J. J. *Nucl. Fusion* **13**, 401 (1973).
Shearer, J. W., Mead, S. W., Petruzzi, J., Rainer, F., Swain, J. E. and Violet, C. E. *Phys. Rev.* **A6**, 764 (1972).
Sleeper, A., Weinstock, J. and Bezzerides, B. *Phys. Rev. Lett.* **29**, 343 (1972).
Stanfield, B. L., Nodwell, R. and Meyer, J. *Phys. Rev. Lett.* **26**, 1219 (1971).
Stenzel, R. and Wong, A. Y. *Phys. Rev. Lett.* **28**, 274 (1972).
Talanov, V. I. *Izv. Vuz. Radiofiz.* **7**, 564 (1964).
Tamor, S. *Phys. Fluids* **16**, 1169 (1973).
Thomson, J. J., Kruer, W. L. and Bodner, S. E. *Phys. Fluids* **17**, 849 (1974).
Tomlinson, R. G. *IEEE J. Quantum Electron.* **QE-5**, 591 (1969).
Tsytovich, V. N. and Shvartsburg, A. B. *Zh. Tekh. Fiz.* **37**, 589 (1967). Transl. *Sov. Phys. Tech. Phys.* **12**, 425 (1967).
Tsytovich, V. N., Stenflo, L., Wilhelmsson, H., Gustavsson, H. G. and Östberg, K. *Physica Scripta* **7**, 241 (1973).
Valeo, E. J. and Oberman, C. R. *Phys. Rev. Lett.* **30**, 1035 (1973).
Vinogradov, A. V. and Pustovalov, V. V. *Zh. Eksp. i. Teor. Fiz.* **62**, 980 (1972a). Transl. *Sov. Phys. JETP* **35**, 517 (1972).
Vinogradov, A. V. and Pustovalov, V. V. *Kvantovaya Elektron.* **2**, 3 (1972b). Transl. *Sov. J. Quantum Electron.* **2**, 91 (1972).
Vinogradov, A. V., Zel'dovich, B. Ya. and Sobelman, I. I. *Zh. Eksp. i. Teor. Fiz. Pis'ma* **17**, 271 (1973). Transl. *JETP Lett.* **17**, 195 (1973).
Waki, M., Yamanaka, T., Kang, H. B., Yoshida, K. and Yamanaka, C. *J. Appl. Phys. Jap.* **11**, 420 (1972).

White, R. B., Liu, C. S. and Rosenbluth, M. N. *Phys. Rev. Lett.* **31**, 520 (1973).
Yamanaka, C., Yamanaka, T., Kang, H., Yoshida, K., Waki, M. and Shimamura, T. *Phys. Lett.* **38A**, 495 (1972a).
Yamanaka, C., Yamanaka, T., Sasaki, T., Yoshida, K., Waki, M. and Kang, H. B. *Phys. Rev.* **A6**, 2335 (1972b).
Yamanaka, C., Yamanaka, T., Sasaki, T., Mizui, J. and Kang, H. B. *Phys. Rev. Lett.* **32**, 1038 (1974a).
Yamanaka, C., Yokoyama, M., Nakai, S., Sasaki, T., Yoshida, K., Matoba, M., Yamabe, C., Yamanaka, T., Mizui, J., Yamaguchi, N. and Nishikawa, K. "5th IAEA Conf. Plasma Physics and Controlled Nuclear Fusion Research", Tokyo (1974b).
Zel'dovich, Ya. B., Levich, E. V. and Syunyaev, R. A. *Zh. Eksp. i. Teor. Fiz.* **62**, 1392 (1972). Transl. *Sov. Phys. JETP* **35**, 733 (1972).

Subject Index

Materials Index